Theory of Remote Image Formation

In many applications, sensor outputs, such as ultrasonic or X-ray signals, are recorded and then analyzed with digital or optical processors in order to extract information to form images. Such processing requires the development of algorithms of great precision and sophistication. This book presents a unified treatment of the mathematical methods that underpin the various algorithms used in remote image formation.

The author begins with a review of transform and filter theory. He then discusses two- and three-dimensional Fourier transform theory, the ambiguity function, image construction and reconstruction, tomography, baseband surveillance systems, and passive systems (where the signal source might be an earthquake or a galaxy). Information-theoretic methods for image formation in the presence of noise are also covered.

Throughout the book, practical applications illustrate theoretical concepts, and there are many homework problems. The book is aimed at graduate students of electrical engineering and computer science, and practitioners in industry.

Professor Richard E. Blahut is Head of the Department of Electrical and Computer Engineering at the University of Illinois, Urbana-Champaign. He is a Fellow of the Institute of Electrical and Electronics Engineers and the recipient of many awards including the IEEE Alexander Graham Bell Medal (1998), the Tau Beta Pi Daniel C. Drucker Eminent Faculty Award, and the IEEE Millennium Medal. He was named a Fellow of the IBM Corporation in 1980 (where he worked for over 30 years) and was elected to the National Academy of Engineering in 1990.

Theory of
Remote Image Formation

Richard E. Blahut

Henry Magnuski Professor in Electrical and Computer Engineering
University of Illinois at Urbana-Champaign

CAMBRIDGE
UNIVERSITY PRESS

CAMBRIDGE UNIVERSITY PRESS
Cambridge, New York, Melbourne, Madrid, Cape Town,
Singapore, São Paulo, Delhi, Tokyo, Mexico City

Cambridge University Press
The Edinburgh Building, Cambridge CB2 8RU, UK

Published in the United States of America by Cambridge University Press, New York

www.cambridge.org
Information on this title: www.cambridge.org/9781107404526

First published 2004
First paperback edition 2011

A catalogue record for this publication is available from the British Library

Library of Congress Cataloguing in Publication Data

Blahut, Richard E.
Theory of remote image formation/Richard E. Blahut.
p. cm.
Includes bibliographical references and index.
ISBN 0 521 55373 3
1. Computer vision. 2. Image processing – Digital techniques. I. Title.
TA1632.B62 2004
621.36′7–dc22 2004045736

ISBN 978-0-521-55373-5 Hardback
ISBN 978-1-107-40452-6 Paperback

"The story is told that young King Solomon was given the choice between wealth and wisdom. When he chose wisdom, God was so pleased that he gave Solomon not only wisdom but wealth also.
. . . So it is with science."

Arthur Holly Compton

"Where the telescope ends, the microscope begins."

Victor Hugo

Contents

Preface

As processing technology continues its rapid growth, it occasionally causes us to take a new point of view toward many long-established technological disciplines. It is now possible to record precisely such signals as ultrasonic or X-ray signals or electromagnetic signals in the radio and radar bands and, using advanced digital or optical processors, to process these records to extract information deeply buried within the signal. Such processing requires the development of algorithms of great precision and sophistication. Until recently such algorithms were often incompatible with most processing technologies, and so there was no real impetus to develop a general, unified theory of these algorithms. Consequently, it was scarcely noticed that a general theory might be developed, although some special problems were well studied. Now the time is ripe for a general theory of these algorithms. These are called algorithms for *remote image formation*, or algorithms for *remote surveillance*. This topic of image formation is a branch of the broad field of *informatics*.

Systems for remote image formation and surveillance have developed independently in diverse fields over the years. They are often very much driven by the kind of hardware that is used to sense the raw data or to do the processing. The principles of operation, then, are described in a way that is completely intertwined with a description of the hardware. Recently, there has been interest in abstracting the common signal processing and information-theoretic methods that are used by these sensors to extract the useful information from the raw data. This unification is one way in which to make new advances because then the underlying principles can be more clearly understood, and ideas that have already been developed in one area may prove to be useful in others. A unified formulation is also an efficient way to teach the many topics as one integrated subject.

Surveillance theory, which we regard as comprising the various information-theoretic and computational methods underlying remote image formation systems, is only now starting to emerge as an integrated field of study. This is in marked contrast to the development of the subject of communication theory in which radio, telephone, television, and telegraphy were always seen as parts of a general theory of communication. For a long time, communication theory has been treated as an integrated field of study while surveillance theory has not.

This book is devoted to a unified treatment of the mathematical methods that underlie the various sensors for remote surveillance. It is not a book that describes the hardware used to build such systems. Rather, it is a book that describes the mathematics used to design the illumination waveforms and to develop the algorithms that will form the images or extract the desired information.

Because my goal is a unified presentation of the mathematical principles underlying the development of remote surveillance, the book is constructed so that the core is the ubiquitous two-dimensional Fourier transform. In addition, the mathematical topics of coherence, correlation functions, the ambiguity function, the Radon transform, and the projection-slice theorem play important roles. The applications, therefore, are introduced as consequences or examples of the mathematical principles, whereas the more traditional treatment of these subjects begins with a description of a sensing apparatus and lets the mathematical principles arise only as needed. While many would prefer that approach – and certainly it is closer to the history of these subjects – I maintain that it is a less transparent approach and does not advance the goal of unification.

Many of the physical aspects of a surveillance system – such as a radar or sonar system or a microwave antenna – are complicated and difficult to model exactly. Some may conclude from this that it is pointless to develop a mathematical theory that is honed to a sharpness beyond the sharpness of the physical situation. I take exactly the opposite view: If a rich mathematical theory can be developed, it should be developed as far as is possible. It is in the application of the theory that care must be exercised. For example, the interaction of electromagnetic waves with reflectors or with antennas can be quite complex and incompletely understood. Nevertheless, we can model the interaction in some more simple way and then carry the study of that model as far as it will go. To object to this would be analogous to objecting to the study of linear differential equations under the argument that every physical system has some degree of nonlinearity. Thus our primary emphasis is on an axiomatic formulation of the mathematical principles of surveillance theory rather than a phenomenological development from the underlying physical laws. The book treats surveillance theory in a formal way as a topic in applied mathematics. The engineering task, then, is to choose judiciously and combine the elements developed here with due regard to the capricious behavior of physical devices.

The two-dimensional Fourier transform (or the three-dimensional Fourier transform) is central to most of the systems developed, and we shall always take pains to bring the role of the two-dimensional Fourier transform to the forefront. Huygens' principle, which is at the core of much of optics and antenna theory, but often is not treated rigorously, will be presented simply as a mathematical consequence of the convolution theorem of two-dimensional Fourier transform theory. In turn, the study of the far-field diffraction pattern of antennas will be largely reformulated as the study of the two-dimensional Fourier transform. Even the near-field diffraction pattern can be studied with the aid of the Fourier transform.

The book also describes the behavior of an imaging or search radar system from the vantage point of the two-dimensional Fourier transform. Specifically, it views the output of a radar's front-end signal processing as the two-dimensional convolution of a radar ambiguity function and the reflectivity density function of the illuminated scene. This is the elegant formulation, and it is very powerful in that it submerges all the myriad details of image formation while leaving exposed the underlying limitations on resolution and estimation accuracy, as well as the nature of ambiguities and clutter.

I have regularly taught a one-semester course from the manuscript of this book, starting with Chapters 3 through 12 and ending with portions of Chapter 13. This program would treat Chapters 1 and 2 as introductory reading material, intended only for review and motivation, and would return to topics in Chapter 2, principally Sections 2.3 and 2.6, as they arise. Presumably, the Fourier transform is already known, and the study of the two-dimensional Fourier transform in Chapter 3 implicitly provides a review of the one-dimensional case.

Chapters 3, 4, and 5 are closely connected and can be thought of as a long discourse on the two-dimensional Fourier transform; the theory is in Chapter 3, its role in optics is in Chapter 4, and its role in antenna systems is in Chapter 5. Chapter 8 picks up this thread again, studying X-ray diffraction imaging as an application of the three-dimensional Fourier transform.

Chapters 6, 7, and 12 form another closely connected sequence centered on the ambiguity function. Chapter 6 develops the theory of the ambiguity function; Chapters 7 and 12 develop the theory of imaging radar and search radar, respectively, from the point of view of the ambiguity function. Chapter 14 deals with data association and tracking as a sequel to Chapter 12; these problems arise in the postprocessing of radar and sonar data. The ambiguity function also plays a role in some parts of Chapter 13, which may be described as a counterpart to Chapter 12 wherein the "radar" is passive.

The important problems of image construction and reconstruction from fragmentary data are studied in Chapters 9, 10, and 11. Chapter 10 is devoted to the important problem of tomography and to the general task of reconstruction of an image from multiple degraded views. Chapter 9 deals with other topics in image construction and reconstruction including the formation of images from partial Fourier transform data and the formation of images from discrete event data. Chapter 11 introduces the important subject of information-theoretic methods in image formation, a subject which itself can grow to fill a book.

Other kinds of surveillance systems are discussed in Chapter 13. Passive systems, in which the illuminating signal is already in the environment and is not under the control of the system designer, are discussed in Chapter 13. The passive location of emitters may arise as a sonar problem in which the emitters are sources of underwater sound; as a seismic problem in which the emitters may be earthquakes or artificial explosions; or as a radar surveillance problem in which the emitters are noncooperative radars or radios. Passive systems can also be imaging systems; modern radio astronomy is a conspicuous example of a passive imaging system, forming images of distant radio galaxies from passively received radio signals. The book concludes with Chapter 15 where an analysis of phase errors and phase noise appears, and where the phase approximations used elsewhere in the book are studied more closely.

Acknowledgments

Drafts of this book have been in existence for more than twenty years, and the material has been presented to many classes of students at Cornell University and at the University of Illinois. Classroom questions and comments provided the feedback and stimulation that led to revision after revision. Clever homework solutions by the students and many excellent student term papers sharpened my own understanding of the material and made the book better. The book also benefitted from many conversations with the outstanding faculty of the University of Illinois and Washington University. I am particularly indebted to many friends and reviewers who provided valuable criticism and comments, in particular Professor Joseph A. O'Sullivan, Professor Donald L. Snyder, Professor Aaron D. Lanterman, Professor P. Scott Carney, Professor Farzad Kamalabadi, Professor David L. Munson, Jr., Professor Timothy J. Schulz, and Negar Kiyavash. The quality of the presentation has much to do with the editing and composition skills of Mrs Helen Metzinger and Mrs Francie Bridges. And, as always, Barbara made it possible. Finally, Lauren showed me the way to accept all things.

1 Introduction

Our immediate environment is a magnificent tapestry of information-bearing signals of many kinds: some are man-made signals and some are not, reaching us from many directions. Some signals, such as optical signals and acoustic signals, are immediately compatible with our senses. Other signals, as in the radio and radar bands, or as in the infrared, ultraviolet, and X-ray bands, are not directly compatible with human senses. To perceive one of these signals, we require a special apparatus to convert it into observable form.

A great variety of man-made sensors now exist that collect signals and process those signals to form some kind of image, normally a visual image, of an object or a scene of objects. We refer to these as sensors for remote surveillance. There are many kinds of sensors collected under this heading, differing in the size of the observed scene, as from microscopes to radio telescopes; in complexity, as from the simple lens to synthetic-aperture radars; and in the current state of development, as from photography to holography and tomography. Each of these devices collects raw sensor data and processes that data into imagery that is useful to a user. This processing might be done by a digital computer, an optical computer, or an analog computer. The development and description of the processing algorithms will often require a sophisticated mathematical formulation.

In this book, we shall bring together a number of signal-processing concepts that will form a background for the study and design of the many kinds of image formation and remote surveillance systems. The signal-processing principles we shall study include or adjoin the theory of classical radar and sonar systems, electromagnetic propagation, tomography, and physical optics, as well as estimation and detection theory.

1.1 Remote image-formation systems

Mankind has designed a variety of devices that are used to observe the environment by processing electromagnetic radiation in the radio and microwave frequency bands, the infrared band, and the optical or X-ray band; or by processing acoustic, pressure, or magnetic variations. The many varieties of radar and sonar systems are examples of

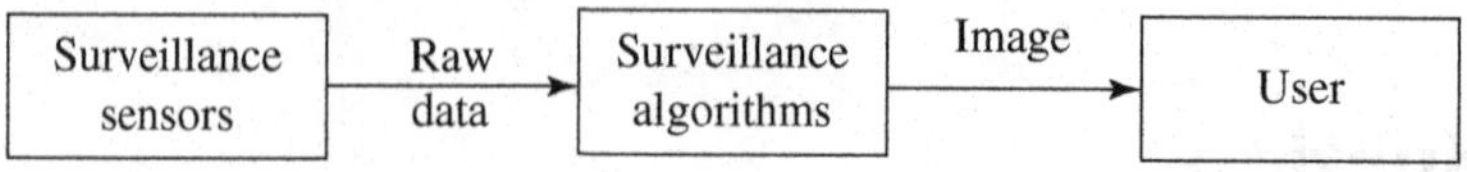

Figure 1.1 Elements of a remote surveillance system

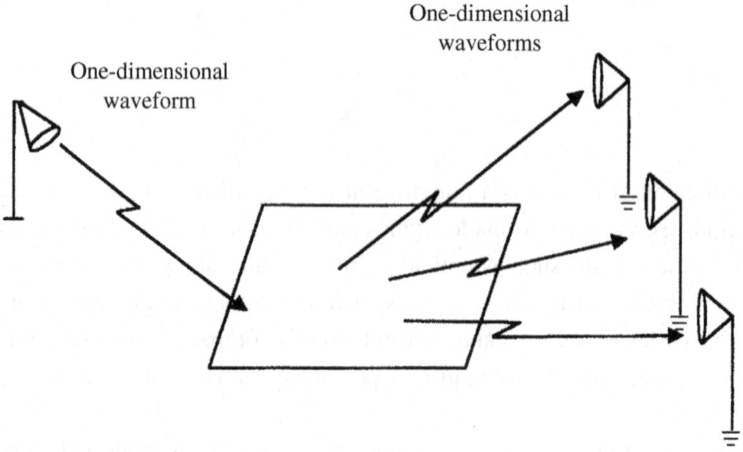

Figure 1.2 Probing a scene with waveforms

such systems. A surveillance system may be active, using its own energy to illuminate the environment, or it may be passive, relying on signals already in the environment.

Surveillance theory studies the design of signals to probe the environment as well as the design of computational procedures for the extraction of information from received signals within which that information may be deeply buried. As such, surveillance theory is that branch of information theory that is explicitly concerned with the design of systems to observe the environment and with the performance of those systems. The theory is concerned specifically with the mathematical structure of the algorithms, such as image-formation algorithms, needed to extract information from received signals.

A precise definition of the term "remote surveillance system" is difficult to formulate. Loosely, we mean any system that collects signals and creates an observable image by processing those signals. Figure 1.1 illustrates how a remote surveillance system can be partitioned into the "sensors" and the "algorithms." We shall be concerned in detail with the image-formation algorithms and with the performance, but not with the detailed physics of the sensors or with the implementation of the algorithms.

The "image," which often is the end product of the surveillance system, is always some kind of depiction of an "actual" scene, usually a two-dimensional or three-dimensional scene, which we denote as $\rho(x, y)$ or $\rho(x, y, z)$. The scene may emit its own signals that the surveillance sensors intercept, or it may be probed by signals generated within the surveillance system. Figure 1.2 shows a representative configuration in which the scene $\rho(x, y)$ is probed by signals generated as one-dimensional waveforms. In this case, the sensors collect one or more reflected one-dimensional waveforms, $s_m(t)$, and from these, the computational algorithms must form a suitable image of the scene. Thus the computational task is to estimate a two-dimensional function, $\rho(x, y)$ (or a three-dimensional function, $\rho(x, y, z)$) when given a set of one-dimensional scattered waveforms, $s_m(t)$, that depend on $\rho(x, y)$ (or $\rho(x, y, z)$). This kind of task is sometimes called an *inverse problem*. Among the most useful mathematical tools that we shall develop for this task are the two-dimensional Fourier transform, the projection-slice theorem, and the ambiguity function.

Another important topic that will be studied is the relationship between the signal at the input aperture of a transducer, such as an antenna, and the wavefront radiated by the antenna. The reflection of waves, however, will be modeled in a simple way. The detailed relationship between the incident wave on a reflecting object and the reflected wave, which is called the *forward problem*, is of interest in this book only insofar as it affects the inverse problem. Simple models for the reflection will be adequate for most of our purposes.

Radar and sonar are among the surveillance systems that will be studied. Originally, radar and sonar systems used simple waveforms and simple processing techniques that could be implemented with simple devices such as filters and threshold detectors. But, over the years, a new level of sophistication began to find its way into many of these systems. By maintaining a precise phase record of long-duration signals, and processing the signals phase-coherently, one can obtain new levels of system performance. Systems that depend on phase coherence over time are called *coherent surveillance* systems. Some early coherent systems in the radar bands were designed to use optical processing. More recently, digital processing of coherent radar waveforms has become practical.

Our goal is to develop the theory of the various kinds of remote surveillance systems in a common mathematical setting. We shall be concerned with both coherent processing techniques, such as are used for forming the images of radar reflectors, for detecting moving objects, and for passively locating radiation sources, and with other processing techniques, such as are used in X-ray tomography, that do not employ phase coherence. The scope of the treatment is illustrated by these examples.

The many forms of remote surveillance imagery, such as radar imagery, may often look very different from visual imagery or conventional photographs of that same scene or object. This means that the user of that sensor may need training and experience in interpreting the output imagery. To the novice, it may seem to be a limitation of the sensor, but a more sophisticated view is that a new sensor opens a new window in our way of perceiving reality. A bat or a dolphin lives in a world that is perceived in large measure by means of acoustic or sonar data. This kind of sensor has nothing like the high angular resolution of our optical world, yet it does have other attributes, such as a strong doppler shift and the ability to resolve objects instantly by their velocities. Because it has a different kind of input data, the dolphin or the bat undoubtedly perceives the world differently from the way in which we do.

One way of defining the kind of surveillance system we shall study is as a system in which raw signals in the environment that the human cannot sense directly are turned into processed signals that are compatible with one of the human senses. Thus a radar receiver converts an electromagnetic wave into a visual image compatible with the human eye, and a tomographic medical scanner turns an X-ray signal into another kind of visual image. Again, the novice may resent the fact that the image is not an exact replica of a photograph. An X-ray image of the human body does not look like a photograph of a human skeleton, but, to the diagnostician, it will be preferred because it contains other kinds of useful information.

1.2 The history of image formation

The subject of remote image formation consists of the common overlap of a number of well-developed subjects such as physical optics and signal processing, and, from a broad point of view, its historical roots go back to the roots of these various subjects. We are interested here in a narrower view of the history of remote surveillance, especially of coherent surveillance systems and tomography. Our brief treatment will serve only to sketch the historical background of the material in this book.

There are many kinds of remote surveillance systems that were developed independently, but which share common fundamentals of signal processing and a common mathematical framework. These include: imaging radars, moving-target detection radars, optical imaging, holography, radio astronomy, sonar beamforming, microscopy, diffraction crystallography, as well as more recent topics such as seismic processing, tomography, and passive source location.

Optical image-formation systems are the earliest, the most developed, and the most readily accepted by the user. Optical images are usually quite sharp, and with high resolution and excellent color contrast. Optical imaging systems may be passive, usually using reflected light and occasionally radiated light, or may be active, using an illumination source. Many such systems are adequately described using geometrical optics. Image-formation systems using radiation in the infrared bands, which are similar to optical systems, form images of temperature variations in a scene because an object emits infrared radiation according to its temperature.

Early radars used simple techniques for processing the received data, while modern radars can use processing that is quite sophisticated. Sophisticated radar signal processing first appeared in the development of those imaging radars known as *synthetic-aperture radars*. Wiley, in 1951, suggested the principle of such radars, although he did not publish his ideas nor did those ideas then result directly in the construction of such a radar. Wiley observed that, whereas the azimuthal resolution of a conventional radar is limited by the width of the antenna beam, each reflecting element within the antenna beam from a moving radar has a doppler frequency shift that depends on the angle between the velocity vector of the radar and the direction to the reflecting element. Thus he concluded that a precise frequency analysis of the radar reflections would provide finer along-track resolution than the azimuthal resolution defined by the antenna beamwidth. The following year, a group at the University of Illinois arrived at the same idea independently, based upon frequency analysis of experimental radar returns. During the summer of 1953, these ideas were reviewed by the members of a summer study, "Project Wolverine," at the University of Michigan and plans were laid for the development of synthetic-aperture radar. It was recognized that the processing requirements placed extreme demands on the technology of the day. Many kinds of analog processors (filter banks, storage tubes, etc.) were tried. Meanwhile, Emmett Leith, at the University of Michigan, turned to the processing ideas of holography and modified the optical processing techniques to satisfy the processing requirements for radar. In 1957, by using optical processing, the first synthetic-aperture radar was successfully demonstrated. Later, Green (1962) proposed the use of range-doppler techniques for remote radar imaging of the surface of rotating planetary objects, a method closely related to synthetic-aperture radar. High-resolution radar images of Venus from the earth gave us our first view of the surface of that planet.

Optical processors[1] are analog processors based on the Fourier transforming property of a lens. These have been the processors of choice for imaging radars because of the sheer volume of data that can be handled. However, optical processors for imaging are slow and may require developing the photographic film twice within the processing. Optical processors are very sensitive to vibration, and they are limited in the form of computation that can be included. Hence attention now has turned to other methods for processing. The advent of high-speed, digital array processors has had a large impact on the processing of synthetic-aperture imagery, and optical processing now plays a diminished or vanishing role.

The development of search radars for the detection of moving targets is spread more broadly, and individual contributions are not as easy to identify. From the first use of radar, it was recognized that the need to detect moving targets could be satisfied by using the doppler shift of the return. A moving reflector causes a doppler-shifted echo. However, the magnitude of the doppler shift is only a very small fraction of the transmitted pulse bandwidth. At that time, the technology did not exist to filter a faint, doppler-shifted signal from a strong background of signals echoed from other stationary emitters. Hence the development of search radars did not depend so much on invention at the conceptual level as it did on the development of technology to support widely understood requirements. By the end of World War II, radars had been developed that used doppler filters to suppress the clutter signal reflected from the stationary

[1] Not to be confused with photonic processors.

background. These early radars used simple delay lines to cancel the stationary return from one pulse with the (nearly identical) return from the previous pulse, thereby rejecting signals with zero doppler shift. In this way, large, rapidly moving objects could be detected from stationary radar platforms.

Later, the requirements for search radars shifted to include moving, airborne radars for observing small, slowly moving target objects at long range. It then became necessary to employ much more delicate techniques for finding a signal return within a large clutter background. These techniques employ coherent processing with the aid of large digital computers.

Meanwhile, astronomers had come to realize that a large amount of astronomical information reaches the earth in the microwave bands. Astronomers are well grounded in optical theory where beamwidths smaller than one arc second are obtained. In the microwave band, a comparable beamwidth requires a reception antenna that is many miles in diameter. Under the impact of wind, ice, and temperature gradients, such an antenna would need to be mechanically rigid to a small fraction of an inch. Clearly, such antennas are impractical. Around 1952, Martin Ryle, at the University of Cambridge, began to study methods for artificially creating such an aperture by combining individual antenna elements, or by allowing the earth's rotation to sweep an array of fixed antenna elements through space. In retrospect, this development of radio astronomy may be viewed as a passive counterpart to the development of active synthetic-aperture radar. The aperture is synthesized by recording the radio signal received at two or more antenna elements and later processing these records coherently within a digital computer. The first such radio telescope was the Cambridge One-Mile Radio Telescope completed in 1964, followed by the Cambridge Five-Kilometer radio telescope in 1971. More recently, other synthetic-aperture radio telescopes have been built and put into operation throughout the world. (The continent-sized Very Large Baseline Array has an angular resolution of 0.0002 arc second.) For the development of synthetic-aperture radio telescopes, Ryle was awarded the 1974 Nobel prize in physics (jointly with Hewish who discovered pulsars with the radio telescope).

The diffraction of X-rays by crystals was demonstrated in 1912 by Max von Laue as a proof of the wave properties of X-rays. Sir William H. Bragg immediately inverted the point of view to turn this diffraction phenomenon into a way of probing crystals, which has since evolved into a sophisticated imaging technique. The 1914 Nobel prize in physics was awarded to von Laue, and the 1915 Nobel prize in physics was awarded to Bragg and his son, Sir W. Lawrence Bragg, who formulated the famous Bragg law of diffraction. This early work was directed toward finding the lattice structure of the crystal as a whole, but was not much concerned with the structure of the individual molecular cell. More recently, attention has turned to the finer question of finding the scattering structure within an individual cell. A difficulty of the task is that, because of the small wavelength of X-rays, the phase of the diffracted X-ray wavefronts cannot be measured. Herbert Hauptman and Jerome Karle (1953) showed how to bypass this problem of missing phase by using prior knowledge about the molecules composing the crystal, for which they received the 1985 Nobel prize in chemistry. Earlier, James Watson and Francis Crick – using the diffraction images produced by Rosalind Franklin – discovered the structure of the DNA molecule, for which they shared the 1962 Nobel prize in medicine.

Closely related to the methods of the Fourier transform and signal processing are many kinds of optical processing, many of them using diffraction phenomena that are describable in terms of the two-dimensional Fourier transform. A method known as the *schlieren method* was proposed by Jean Foucault in 1858 as a way to image air density variations. Frits Zernike in 1935 developed phase-contrast methods to improve microscopy images, for which he was awarded the 1953 Nobel prize in physics. Aaron Klug (1964) developed methods for the imaging of viruses using the diffraction of laser light by electron microscope images, for which he won the 1982 Nobel prize in chemistry.

Dennis Gabor, influenced by the techniques used in microscopy and crystallography, proposed holography in a series of papers in 1948, 1949, and 1951, originally for microscopy but later as a replacement for photography. The work earned Gabor the 1971 Nobel prize in physics. Gabor realized that, whereas conventional photography first processes the optical information to form an image which is then recorded on film, it is also possible to record the raw optical data on the photographic film directly and place the processing in the future with the viewer. He called his method for the photographic recording of the raw optical data a *hologram*. Because the raw optical data contain more information than a final photographic image, in principle the hologram can be used to create images superior to a photograph. Most striking in this regard is the creation of three-dimensional images from a two-dimensional hologram. Holography is technically much more difficult than photography because recording the raw optical data requires precision on the order of optical wavelengths. For this reason, the idea of holography did not immediately draw the attention it deserved. Holography became more attractive after the invention of the laser and also after the more practical reformulation by Leith and Upatnieks (1962), which was strongly influenced by Leith's work on synthetic-aperture radar.

Imaging was introduced into medical diagnostics by Röntgen in 1895 with the invention of X-ray radiography, which exposes a photographic film to transmission X-rays. Edison, by introducing an X-ray-sensitive fluorescent screen or fluoroscope in 1896, eliminated the delay required to process the film. The development of X-ray tomography in the modern sense of computerized image reconstruction for medical applications began in Great Britain. The key feature, based on the projection-slice theorem and the Radon transform, is the algorithmic reconstruction of images from their X-ray projections, first developed by Cormack in 1963 and reduced to practice by Hounsfield in 1971. The 1979 Nobel prize in physiology and medicine was awarded to Hounsfield and Cormack for the development of computerized tomography. The ideas of tomography are closely related to similar methods used in radio astronomy, especially the formulation of reconstruction algorithms by Bracewell (1956). Other kinds of computerized tomography are now in use for medical diagnostic imaging systems. In addition to the method of projection tomography based on X-ray projections, there are the methods of emission tomography and diffraction tomography. Emission tomography based on radioisotope decay was proposed by Kuhl and Edwards (1963).

Magnetic-resonance imaging (MRI) is yet another kind of tomographic imaging system based on magnetic excitation of atomic nuclei and their subsequent radiation. The ground-breaking idea that enables magnetic-resonance imaging, for which Lauterbur and Mansfield (1973) shared the 2003 Nobel prize in medicine, is to use gradient magnetic fields to encode spatial information into the transient response of a spin system after magnetic pulse excitations. Whereas X-ray tomography gives an image of electron density, MRI gives an image of the distribution of hydrogen nuclei (protons), though in principle it may be tuned to observe instead the distribution of other species of nuclei. The physical phenomenon of nuclear magnetic resonance had been observed in 1946 by Bloch and Purcell, independently, for which they received the 1952 Nobel prize in physics. It was later realized that magnetic resonance effects varied with the kind of tissue excited, but it was not known how to use this effect to make images. Lauterbur conceived and demonstrated his method of using the magnetic resonance phenomenon to form images by spatially encoding the magnetic field, for which he received the 2003 Nobel prize in medicine. Since then, MRI has become an important modality in medical diagnosis. By using both static and time-varying magnetic excitation fields, a magnetic-resonance imaging system causes all nuclei of a given kind, selected by resonance frequency, to oscillate, but with an amplitude and frequency modulation that depend on position as determined by the magnetic excitation at that position. The energy that is radiated by the selected species of nuclei, usually hydrogen nuclei, is measured and sorted by frequency analysis. Because of the spatially-varying magnetic excitation, the frequency distribution of the radiated energy corresponds to the spatial distribution of the sources of radiation, which equates to the spatial density distribution of the target nuclei. Sophisticated algorithms based on the mathematics of tomography have been developed to so extract the image from the frequency distribution of multiple projections of measured data.

1.3 Radar and sonar systems

A radar obtains information about an object or a scene by illuminating the object or scene with electromagnetic waves, then processing the echo signal that is reflected from that object and intercepted by the radar receiving antenna. A sonar obtains information

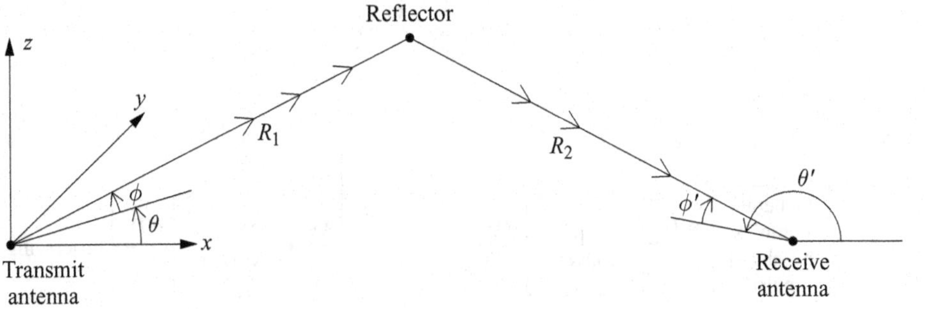

Figure 1.3 Elementary radar/sonar geometry

about an object or a scene by illuminating the object or scene with acoustic waves, then processing the echo signal that is reflected from that object and intercepted by the sonar hydrophones. The geometry of a radar or sonar system is shown in Figure 1.3. The transmitting antenna and the receiving antenna are shown at separate locations although, in most instances, they will be together; they may even be the same antenna, as is the usual case.

By using electromagnetic waves in the microwave bands, a radar is able to penetrate optically opaque media such as clouds, dust, or foliage. In this way, it is possible to form radar images of objects hidden by dust, soil, or snow. Similarly, a sonar or ultrasound system can form images of objects that are optically masked by opaque media. In some cases, the broad beamwidth of a radar antenna or a sonar hydrophone is attractive as a way of viewing a large region of space. For reasons such as these, radar and sonar have long been popular as surveillance systems.

While there may be a great deal of difference between the propagation of electromagnetic waves and the propagation of pressure waves, there is also a great deal of similarity. This similarity carries over to radar and sonar systems. From our point of view, each is a system that forms a complex baseband pulse, $s(t)$, which is transmitted as the amplitude and phase modulation of the passband pulse $\widetilde{s}(t)$, and receives an echo pulse, $\widetilde{v}(t)$, which is a delayed and frequency-shifted copy of the passband pulse $\widetilde{s}(t)$ and is also contaminated by noise. The transmitted pulse $\widetilde{s}(t)$ propagates at a velocity of c over a path of length R_1 from the transmitter to the reflector, and then over a path of length R_2 from the reflector to the receiver. The received pulse $\widetilde{v}(t)$ is an echo of the transmitted pulse but, because it is contaminated by noise and other impairments, it may be difficult to recognize or use. We shall be interested in methods of processing the received pulse to extract useful information from it. The same fundamental ideas apply equally to radar pulses and to sonar pulses.

Figure 1.4 shows an elementary block diagram of a radar system. The baseband pulse $s(t)$ is "up-converted" to form the passband pulse $\widetilde{s}(t)$ that is transmitted by the antenna. The signal transmitted at angle (ϕ, ψ) is $E_t(\phi, \psi)\widetilde{s}(t)$, where $E_t(\phi, \psi)$ is a function of the angles ϕ and ψ and is called the *antenna pattern* of the transmitting antenna. The

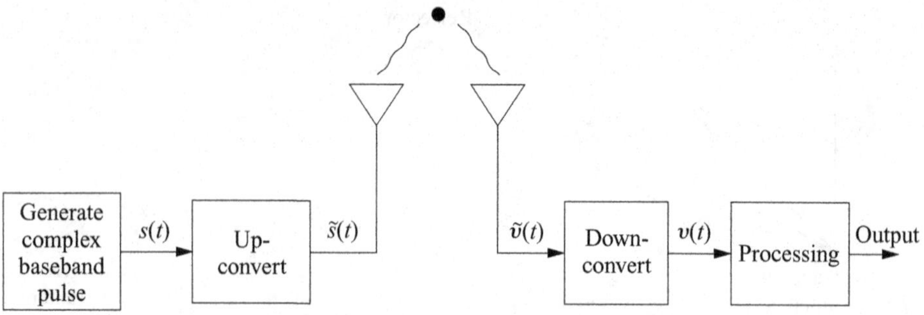

Figure 1.4 Elementary radar block diagram

signal reflected from the reflector is (proportional to[2]) $\rho E_t(\phi, \psi)\widetilde{s}(t - R_1/c)$ where ρ is a parameter called the *reflectivity* of the reflector and is usually dependent on the incident angle and the scattered angle of the signal. The echo signal intercepted by the receiving antenna is

$$\widetilde{v}(t) = E_r(\phi', \psi')\rho E_t(\phi, \psi)\widetilde{s}(t - (R_1 + R_2)/c) + n(t)$$

where $n(t)$ is additive noise, and $E_r(\phi', \psi')$ is the antenna pattern of the receiving antenna. The angular coordinates of the receiving antenna are distinguished by primes. The received echo signal $\widetilde{v}(t)$ is "down-converted" to the complex baseband representation $v(t)$. The signal-processing task is to process the received baseband signal $v(t)$ to obtain information about the reflector such as its position, velocity, size, or even its detailed shape.

The received signal is distributed both in space across the aperture of an antenna, and in time. The distribution in space may be processed, usually by the antenna system, to gather all of the received spatially distributed signal into a single, time-dependent signal. Simple, linear processing of the signal across the aperture is usually summarized by referring to the shape and width of an antenna "beam." The time variations of the received signal are processed so as to determine the time-varying range of the reflecting objects. The space distribution may be processed in other ways to determine the direction of arrival of the signal. We shall study the processing of the space distribution of the signal in Chapter 5 (and Chapter 13) and the processing of the time distribution of the signal in several later chapters.

We define a coherent surveillance system loosely as any system for which coherent processing is fundamental to its operation. We further define coherent processing as the processing of a received passband signal (or signals) by procedures that exploit the carrier phase structure of the waveforms to extract the desired information more effectively. We shall see that the extraction of maximum information from a received

[2] A proportionality constant associated with spherical wave propagation is usually suppressed in a discussion of this kind and is treated separately in a power budget calculation.

passband signal requires coherent processing over long time intervals, and this can lead to the use of sophisticated signal processing.

Whenever the transmitter or receiver is in motion with respect to the reflector of the signal, then coherent processing becomes an especially potent technique because of the detailed structure of the frequency shifts, called *doppler*, in the echo. One system of this kind that is used for imaging is called a *synthetic-aperture radar* because of the heuristic notion of synthesizing a long, fixed antenna by the sequence of positions of a short, moving antenna. A synthetic-aperture imaging system depends on motion. The motion can be described as a time-varying position, which description is often simplified to a straight line and is specified by an initial position and a constant velocity. Whenever this description suffices, the received electromagnetic signal depends on the scene parameters only through the time delay and the doppler frequency shift of that signal.

The typical synthetic-aperture imaging radar transmits a passband microwave signal consisting of a train of uniform pulses. A moving transmitter illuminates a scene with this waveform. This is illustrated in Figure 1.5, which shows the position of the real antenna for each transmitted pulse. The received echo signal corresponding to each pulse is converted to a precision optical or digital replica, maintaining both amplitude and phase modulation. As time progresses, a history of such pulse replicas is accumulated, each at a slightly different position along the trajectory of the receiver. From this history, an image of the illuminated scene is assembled by coherent processing.

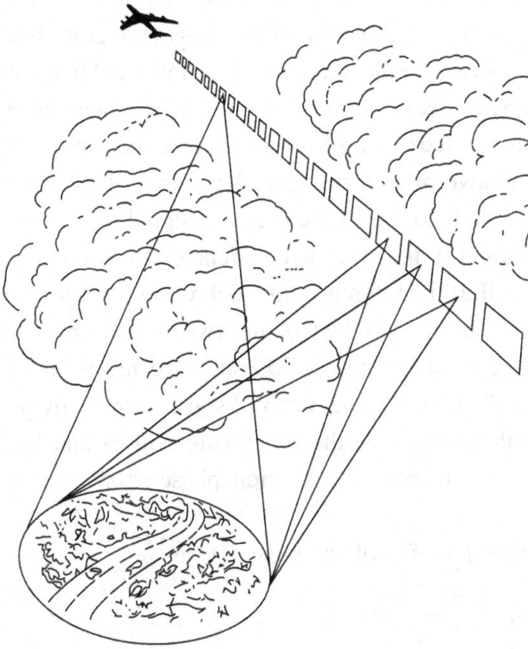

Figure 1.5 Synthetic-aperture radar

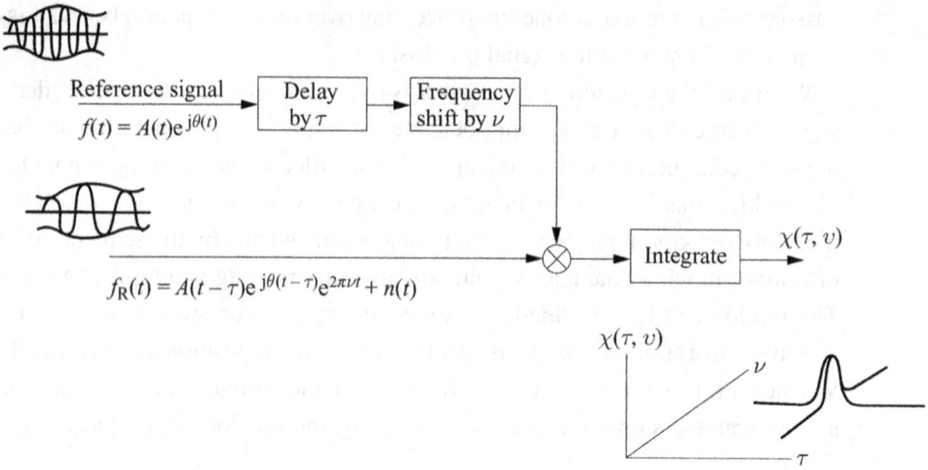

Figure 1.6 Principle of coherent processing

The most important calculation of a traditional coherent radar system is that of the *cross-ambiguity function*, which will be studied in Chapter 6. It has the form

$$\chi_c(\tau, \nu) = \int_{-\infty}^{\infty} v(t)s^*(t - \tau)e^{-j2\pi\nu t}dt$$

where $s(t)$ is the complex baseband representation of the transmitted signal and $v(t)$ is the complex baseband representation of the received echo signal. We shall see that the cross-ambiguity function is useful in many ways and can be motivated in many ways. Figure 1.6 shows the underlying principle of the cross-ambiguity function. The incoming radar signal $v(t)$ from a point reflector is a time-delayed and frequency-shifted replica in noise of the reference waveform $s(t)$. The time delay enters because of the propagation delay between the source of the waveform and the receiver. The frequency shift enters because of the relative motion between the source of the waveform and the receiver. The processing task is to match the received signal $v(t)$ to the reference signal $s(t)$. As shown in Figure 1.6, the reference is delayed, frequency shifted, and beats against the received signal. If the reference time delay and frequency shift match those of the received signal, then the product of the two is real and positive, and integrates to the energy in the received signal. To match the received signal to the reference, it is important that distortion in the received signal, especially phase errors in the receiver, be kept small. Sometimes, the processing will extend literally over billions of cycles of carrier phase, and over this duration, phase errors must be precisely controlled.

We shall define the *ambiguity function* of any pulse $s(t)$ as

$$\chi(\tau, \nu) = \int_{-\infty}^{\infty} s(t + \tau/2)s^*(t - \tau/2)e^{-j2\pi\nu t}dt.$$

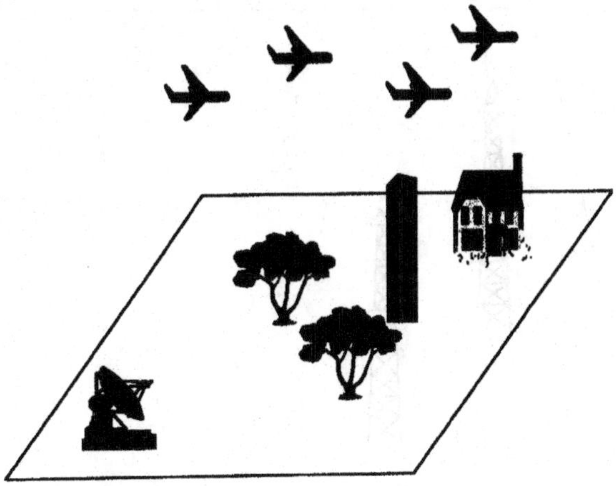

Figure 1.7 Moving-target detection

We shall study the ambiguity function in detail in Chapter 6. There we will see that the ambiguity function of a pulse $s(t)$ is the key to understanding the performance of an imaging or detection radar that uses that pulse.

A *moving-target detection* (MTD) radar is another kind of radar that detects those elements in a scene that are in motion, as shown in Figure 1.7. Moving-target detection originated as a radar technique that separated the signal reflected from a moving reflector from other reflections by a simple bank of doppler filters. Modern formulations of moving-target-detection radars are more sophisticated and much more powerful. In order to detect very slowly moving targets from a moving aircraft, very fine doppler resolution is necessary, and the long system integration time leads to the notion of a synthetic aperture. The task of radar imaging of moving objects can be regarded as a problem in the six-dimensional space of velocity and position.

Another type of radar or sonar surveillance system is one used for the passive imaging of radiation sources. An important example of a passive imaging system is a modern radio telescope that employs multiple separated antennas and correlation techniques to estimate spatial Fourier coefficients of images of remote astronomical objects, such as galaxies. Radio astronomy is described in Chapter 13. A different example of a passive surveillance system is a terrestrial passive-emitter location system, as shown in Figure 1.8. A number of radiation sources of unknown characteristics are distributed in space at unknown locations. These unknown locations are to be estimated. The transmitted signals are incident at various system stations, at which are located passive receivers. From these multiple received signals, jointly processed, selected characteristics of the emitters, such as their locations, are computed.

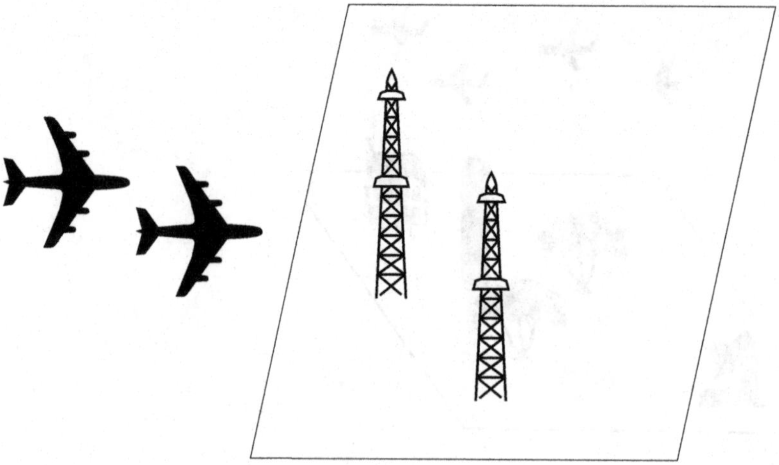

Figure 1.8 Passive location of radiation sources

1.4 Imaging by projections

In many situations, it is convenient to observe certain projections of the two-dimensional object $s(x, y)$ (or the three-dimensional object $s(x, y, z)$), but it is difficult to make more direct observations of that object. A familiar example is an elementary X-ray projection of internal body organs. As a single ray passes along a line, say the y axis, the intensity is attenuated at each y by an amount described by a function $s(y)$. That is, the intensity, denoted I', leaving a very small interval of width Δy, centered at y_1, is related to the intensity, denoted I, entering that interval, by

$$I' = I[1 - s(y_1)\Delta y].$$

This is approximated as

$$I' \approx I\mathrm{e}^{-s(y_1)\Delta y}$$

under the assumption that the attenuation is small. Over two consecutive intervals, each of width Δy, the intensity attenuation is described approximately as

$$I'' = I\mathrm{e}^{-s(y_1)\Delta y}\mathrm{e}^{-s(y_2)\Delta y}.$$

Consequently, over many such intervals

$$I_{out} = I_{in}\mathrm{e}^{-\sum_i s(y_i)\Delta y}$$

and, in the limit as Δy goes to zero,

$$\log \frac{I_{in}}{I_{out}} = \int_{-\infty}^{\infty} s(y)\,\mathrm{d}y$$

where the actual integration limits are defined by the support of $s(y)$. The integral on the right is called the projection of $s(y)$. By introducing a ray in the y direction at each value of x, passing through the two-dimensional function $s(x, y)$, one can define the projection onto the x axis as a function of x:

$$p(x) = \int_{-\infty}^{\infty} s(x, y)\, dy.$$

In the general case, the attenuation of an X-ray at angle θ integrates $s(x, y)$ along each ray in the direction at angle θ, so the projection

$$p_\theta(t) = \int_{-\infty}^{\infty} s(t \cos\theta - r \sin\theta, t \sin\theta + r \cos\theta)\, dr$$

at each t consists of the superposition of $s(x, y)$ at all depths along that ray. By varying the viewing angle θ, such projections can be observed from many directions. One wishes to process such a set of projections to form an estimate of $s(x, y)$ thereby showing the internal organs of the body. The topic of signal processing that reconstructs images of objects based on the measured projections is known as *tomography*. The field of tomography will be studied in Chapter 10. The central theorem of signal processing that underlies tomography is the *projection-slice theorem*, which will be introduced in Chapter 3.

The origins of tomography can be traced back to 1917 when the Austrian mathematician Radon showed that a spatial function, $s(x, y, z)$, can be reconstructed from the complete set of its projections. Because the reconstruction of images from projections arises in many diverse situations, it is not surprising that this mathematical principle, first discovered by Radon, was independently rediscovered many times and in many fields. It has been used in radio astronomy and in the field of electron microscopy. In the context of medical applications, tomography has led to very important advances in the noninvasive imaging techniques available in recent years for medical research and clinical practice. Tomography is used in many other applications, such as geophysical applications, where it can be used for subsurface exploration, or in atmospheric sensing, where it can be used, for example, to form images of pollutant densities in the upper atmosphere.

In Chapter 10, we shall study the mathematical principles underlying tomography, especially the projection-slice theorem, which relates the one-dimensional Fourier transform of the projection to the two-dimensional Fourier transform of the object. We shall also study a number of algorithms for the reconstruction of images. The central idea of these algorithms is the method of back projection. We shall see that reconstruction of an arbitrary object from projections is exact only if the infinitely large set of projections from all angles is known. However, mild prior conditions on the object, such as a spatially bandlimited Fourier transform, can soften this statement. Many good algorithms are known that compute an approximate reconstruction of an image from a finite set of its projections. These images are usually satisfactory for practical applications even when the individual projections are weak and noisy.

In addition to projection tomography, many other important forms of tomography go by names such as *emission tomography*, *diffraction tomography*, *diffusion tomography*, and *coherence tomography*; some of these will be studied in Chapter 10. Emission tomography requires that the scene itself emits some form of radiation that provides the received signal from which the image is computed. This usually means that the scene must receive an excitation that provides the energy for the emission. There are two methods that are in wide use to excite an object to radiate. The very important method of *magnetic-resonance imaging* (MRI) uses a time-varying and spatially-varying magnetic field to provide illumination energy to the scene. This time-varying magnetic field causes isolated protons, or other chosen nuclei, to resonate and thus radiate electromagnetic waves. The radiation is intercepted, measured, and processed, by the methods of tomography, to form an image of the density of isolated protons (hydrogen atoms). Another method of excitation, called *positron-emission tomography*, uses a radioactive isotope that is selectively absorbed by a tissue of interest, usually a diseased tissue. The radioactive isotope then decays, thereby releasing radiation energy in the form of positrons. These positrons immediately combine with electrons to produce photons. The photons are captured by an array of photosensors, and from the positions and times at which these photons are detected, an image is formed. Through this method, a specific tissue can be selectively imaged by its tendency to acquire a particular radioactive isotope.

Diffraction tomography and diffusion tomography deal with other situations in which the geometrical-optics approximation to propagation is not adequate. It may be necessary to treat wave propagation in free space in a more exact way by considering the effect of diffraction. This is particularly important when observing details that are small when measured in wavelengths. Another situation is propagation in a strongly scattering medium, in which case the topic is that of diffusion tomography.

A related form of tomography is *geophysical tomography* in which seismic waves are used to image geophysical features. Then the dispersion of the wave is not true diffraction, but rather is caused by scattering anomalies in the propagation medium.

1.5 Passband and complex baseband waveforms

We shall have frequent occasion to use both baseband signals and passband signals. A real *baseband signal*, $s(t)$, is any real function of time with its spectral energy density concentrated near zero frequency. The baseband signal $s(t)$ may also be called a *baseband waveform*, often when it is regarded as a complicated signal; or a *baseband pulse*, usually when it is regarded as a relatively simple signal of finite energy.

A *passband signal*, which is denoted by $\widetilde{s}(t)$, is a function of the form

$$\widetilde{s}(t) = s_R(t)\cos 2\pi f_0 t + s_I(t)\sin 2\pi f_0 t$$

where f_0 is a constant known as the *carrier frequency*, and $s_R(t)$ and $s_I(t)$ are real functions of time whose Fourier spectra $S_R(f)$ and $S_I(f)$ are zero for $|f| \geq f_0$. The signals $s_R(t)$ and $s_I(t)$ are called the *modulation components* of $\widetilde{s}(t)$. For radar systems, the constant f_0 is usually in the interval from 0.1 to 35 gigahertz, and is most often in the interval from 1 to 10 gigahertz. For sonar systems, f_0 is usually measured in kilohertz. For ultrasound systems, f_0 may be measured in megahertz.

The passband signal $\widetilde{s}(t)$ may also be called a *passband waveform*, usually when it is regarded as a complicated signal; or a *passband pulse*, often when it is regarded as a relatively simple signal of finite energy.

The *complex baseband signal* corresponding to the passband signal $\widetilde{s}(t)$, is

$$s(t) = s_R(t) + \mathrm{j}s_I(t).$$

The passband signal may be expressed in terms of the complex baseband signal as[3]

$$\widetilde{s}(t) = \mathrm{Re}[s(t)\mathrm{e}^{-\mathrm{j}2\pi f_0 t}].$$

We often regard $\widetilde{s}(t)$ and $s(t)$ as essentially the same signal but for the detail of the multiplying complex exponential. To emphasize this, these may be called the *passband representation* and the *complex baseband representation* of the same abstract signal.

There are two reasons for replacing the passband signal $\widetilde{s}(t)$ with the complex baseband signal $s(t)$. From the notational point of view, the complex baseband signal is preferred because the complex baseband signal is notationally more compact than the passband signal, and mathematical manipulations of complex baseband equations exactly mimic mathematical manipulations of the corresponding passband equations and are much easier. Moreover, within a transmitter or receiver, it is often convenient to translate a real passband signal into the complex baseband representation for most of the circuit implementation. Ultimately, the simplest and most rewarding point of view is to think of the complex baseband signal as the more fundamental form, which is temporarily represented as a passband signal for purposes of transmission and reception. While we study it and process it, it is a complex signal; when we transmit it and receive it, it is a passband signal. To convert between the two forms is trivial, and is usually the last operation in a transmitter and the first operation in a receiver.

[3] The sign in the exponent is arbitrary. It is chosen so that Fourier transform relationships in optics and antenna theory have the conventional form. This choice leads to the sign convention used for the passband waveform.

1.6 Temporal and spatial coherence

The words "coherent" and "noncoherent" will continually recur, and it is important to indicate the sense in which these words are used. Every passband signal can be re-expressed as a sinusoid with both amplitude and phase modulation

$$\widetilde{s}(t) = A(t)\cos(2\pi f_0 t + \theta(t))$$

where $A(t)$ is the amplitude modulation, f_0 is the carrier frequency, and $\theta(t)$ is the phase modulation. The complex baseband representation then has the form

$$s(t) = A(t)\mathrm{e}^{-\mathrm{j}\theta(t)}.$$

The phase modulation may be intentional and known, or it may be partially or wholly unintentional and unknown. If the phase angle $\theta(t)$ of the signal $\widetilde{s}(t)$ is known to the extent that knowledge of $\theta(t)$ is critical to the application, then the signal is called *coherent*. Otherwise, $\widetilde{s}(t)$ is called *noncoherent*.

The term "coherent" may also arise in connection with the processing of such waveforms, either in the form of a real passband signal or in the form of a complex baseband signal. The processing may be the kind known as *coherent processing*, which fully uses both $A(t)$ and $\theta(t)$, or the kind known as *noncoherent processing*, which makes only limited – or no – use of $\theta(t)$.

Coherence not only refers to a deterministic relationship between the phase angles of a passband waveform at different time instants, but may also refer to a deterministic relationship between the phase angle of two different waveforms, $\widetilde{s}_1(t)$ and $\widetilde{s}_2(t)$. The former case is then referred to as a *temporally* coherent waveform. The latter case is referred to as a *spatially* coherent waveform. It arises when a common wavefront is incident on two antennas or two lenses at different locations, or at two regions of the same antenna or lens. The deterministic relationship between points in a spatially coherent wavefront is due primarily to the different times at which the wavefront reaches those different points. Two signals, $\widetilde{s}_1(t)$ and $\widetilde{s}_2(t)$, may be spatially coherent even though each is individually temporally noncoherent. For example, in photographic systems, the light from a point source incident on a lens may be temporally noncoherent, but across the lens it is spatially coherent. Otherwise, the lens could not focus the light into an image of that point source. Moreover, when there are multiple point sources, the light emitted by the multiple point sources can be mutually spatially noncoherent because the point sources are mutually noncoherent, yet the light reaching the lens from each individual point source can be spatially coherent.

A common example of a passband radar waveform is a pulse train,

$$\widetilde{p}(t) = \sum_{n=0}^{N-1} s(t - nT_r)\cos(2\pi f_0 t + \theta_0),$$

where $s(t)$ is a pulse, T_r is a constant called the *pulse repetition interval*, and θ_0 is a constant. The pulse train consists of N translates of the pulse $s(t)$ modulated onto the carrier $\cos(2\pi f_0 t + \theta_0)$. In complex baseband notation, the pulse train is denoted

$$p(t) = \sum_{n=0}^{N-1} s(t - nT_r)e^{-j\theta_0}.$$

The pulse train may be called coherent to mean that θ_0 remains constant from pulse to pulse even though it may be unknown. Occasionally, the waveform will be called coherent to mean that the constant θ_0 is known. Hence whether or not a given waveform is called coherent might depend on the circumstances of the application.

A pulse train in which the phase is not the same from pulse to pulse is given in the passband representation by

$$\widetilde{p}(t) = \sum_{n=0}^{N-1} s(t - nT_r)\cos(2\pi f_0 t + \theta_n),$$

and in the complex baseband representation as

$$p(t) = \sum_{n=0}^{N-1} s(t - nT_r)e^{-j\theta_n}.$$

This is called a noncoherent pulse train if the θ_n are random and not related; usually, the θ_n form a sequence of independent random variables, perhaps assuming values uniformly between 0° and 360°. It is important to the usage here that the phase angles are unknown. If the phase angles were known, the waveform would be considered a coherent waveform because the known values of the phase angles could be included in the processing of the waveform. Figure 1.9 compares the coherent pulse train with a noncoherent pulse train.

More generally, $\theta(t)$ may separate into two parts: a phase angle that is known, and a phase angle that is unknown. For an arbitrary waveform in the passband representation, we may write this as

$$\widetilde{s}(t) = A(t)\cos[2\pi f_0 t + \theta_s(t) + \theta_n(t)],$$

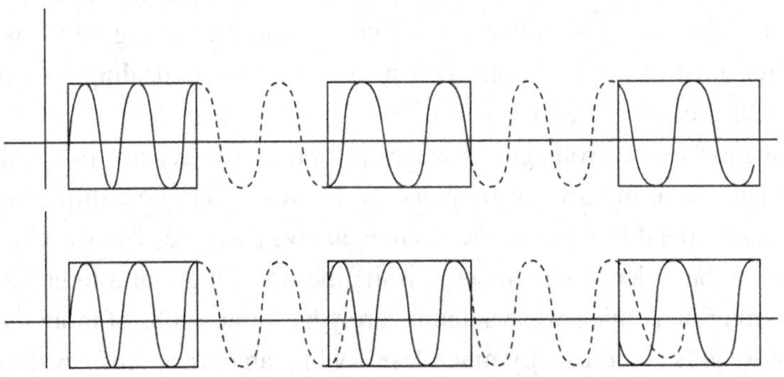

Figure 1.9 A coherent pulse train and a noncoherent pulse train

and in the complex baseband representation it is

$$s(t) = A(t)e^{-j[\theta_s(t)+\theta_n(t)]}$$

where $\theta_s(t)$ is the intentional or known part of the phase modulation associated with the signal, and $\theta_n(t)$ is the unintentional and unknown part of the phase modulation, then called *phase noise*. When described in this way, a coherent waveform usually means a waveform in which $\theta_n(t)$ is negligible, and a noncoherent waveform means a waveform in which $\theta_n(t)$ is not negligible. In some applications, surprisingly large values of unknown phase may be acceptable. Even phase errors as large as one radian will sometimes be tolerated, though with a significant loss of system performance. Chapter 15 is devoted to the quantitative analysis of the effect of phase error on performance, including the notion of coherence in various situations.

1.7 Monodirectional waves

Images are frequently formed by sensing and processing waves reflected from a scene, and many image formation algorithms are based on the structure and behavior of waves. A wavefront propagating in space may have a complicated structure, both temporally and spatially. To gain an understanding of the general case, one can begin with a study of the two simple cases: *plane waves* and *spherical waves*. More complicated situations can be built up from these. The Huygens–Fresnel principle, which will be developed in Chapter 4, is the appropriate statement for describing how an arbitrary planar surface in space can be viewed as the source of any wave that crosses that surface.

Physically, a wave may be a varying electric or magnetic field as is associated with an electromagnetic wave, or it may be the pressure field as is associated with an acoustic wave. A wave may be a vector function, as in the case of the electromagnetic wave, or it may be a scalar function, as in the case of the acoustic wave. Our primary concern is with the mathematical description of the wave. Usually, we will be content to deal with scalar-valued waves because of analytical simplicity. Even when the wave is an electromagnetic wave, the vector-valued character of the wave will not often affect the properties of the propagation that are of interest, and we can regard the wave as a scalar wave for most of our purposes. This means that we are dealing with only one component of the vector-valued wave.

The propagation of electromagnetic waves at optical frequencies obeys the same fundamental laws as at microwave frequencies. However, the great difference in the wavelengths leads to a difference in the phenomena we perceive. The wavelength of a microwave is on the order of centimeters, while the wavelength of a light wave is on the order of a micron. A microwave antenna rarely has dimensions of more than a few hundred wavelengths – and usually much less – while an optical lens has dimensions of more than 10^4 wavelengths. Consequently, an optical beam is usually much sharper than a microwave beam and may often be described adequately by geometrical optics and ray tracing.

Monochromatic monodirectional waves

Mathematically, a spatially uniform, monodirectional, monochromatic, scalar plane wave traveling in the z direction is given by

$$\begin{aligned}\widetilde{s}(t,x,y,z) &= A\cos(2\pi f_0(t-z/c)+\theta)\\ &= A\cos(2\pi f_0 t - kz+\theta)\end{aligned}$$

where the constant $k = 2\pi f_0/c = 2\pi/\lambda$ is called the *wave number* and λ is the *wavelength*. We shall also write this passband wave as

$$\begin{aligned}\widetilde{s}(t,x,y,z) &= \mathrm{Re}[Ae^{-j\theta}e^{-j(2\pi f_0(t-z/c))}]\\ &= \mathrm{Re}[Ae^{-j\theta}e^{-j(2\pi f_0 t-kz)}]\end{aligned}$$

where $Ae^{-j\theta}$ is called the *complex amplitude* of the waveform at $z = 0$. The *complex baseband representation* of this wave at arbitrary z, which is independent of time, is

$$\begin{aligned}s(x,y,z) &= Ae^{-j\theta}e^{j2\pi f_0 z/c}\\ &= Ae^{-j\theta}e^{jkz}.\end{aligned}$$

This is the complex baseband representation of a monodirectional, monochromatic wave moving in the z direction. It has the same value at every point of the x, y plane.

The most general form of a spatially uniform, monodirectional, monochromatic wave that satisfies the wave equation is given by

$$\begin{aligned}\widetilde{s}(t,x,y,z) &= A\cos(2\pi f_0(t-(\alpha x+\beta y+\gamma z)/c)+\theta)\\ &= A\cos(2\pi f_0 t - \alpha kx - \beta ky - \gamma kz + \theta).\end{aligned}$$

The variables α, β, and γ, called *direction cosines*, specify the direction of travel of the plane wave. They are equal, respectively, to the cosines of the angles between the direction of travel of the plane wave and the three coordinate axes:

$$\begin{aligned}\alpha &= \cos\phi_x\\ \beta &= \cos\phi_y\\ \gamma &= \cos\phi_z.\end{aligned}$$

The direction cosines are related to spherical coordinates, as shown in Figure 1.10, by

$$\begin{aligned}\alpha &= \cos\theta\sin\phi = \cos\phi_x\\ \beta &= \sin\theta\sin\phi = \cos\phi_y\\ \gamma &= \cos\phi \qquad\;\;\, = \cos\phi_z,\end{aligned}$$

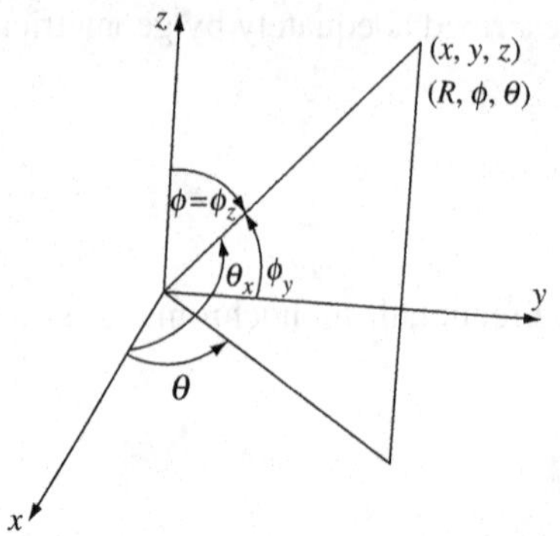

Figure 1.10 Direction cosines and spherical coordinates

and so they are related by

$$\alpha^2 + \beta^2 + \gamma^2 = 1.$$

The complex baseband representation of the monodirectional, monochromatic wave is

$$\begin{aligned} s(x, y, z) &= A\mathrm{e}^{-\mathrm{j}\theta}\mathrm{e}^{\mathrm{j}2\pi f_0(\alpha x+\beta y+\gamma z)/c} \\ &= A\mathrm{e}^{-\mathrm{j}\theta}\mathrm{e}^{\mathrm{j}k(\alpha x+\beta y+\gamma z)} \\ &= A\mathrm{e}^{-\mathrm{j}\theta}\mathrm{e}^{\mathrm{j}(k_1 x+k_2 y+k_3 z)}. \end{aligned}$$

The quantities k_1, k_2, and k_3 are called the *wave numbers* of the plane wave. The wave numbers are related to the direction cosines by

$$\begin{aligned} k_1 &= (2\pi f_0/c)\alpha = (2\pi/\lambda)\alpha \\ k_2 &= (2\pi f_0/c)\beta = (2\pi/\lambda)\beta \\ k_3 &= (2\pi f_0/c)\gamma = (2\pi/\lambda)\gamma. \end{aligned}$$

The vector $\boldsymbol{k} = (k_1, k_2, k_3)$ is called the *vector wave number* of the plane wave.

The complex baseband representation, using complex exponentials, is more convenient to work with than is the passband representation. The passband representation is recovered by

$$\widetilde{s}(t, x, y, z) = \mathrm{Re}[s(x, y, z)\mathrm{e}^{-\mathrm{j}2\pi f_0 t}].$$

Time-varying monodirectional waves

If the complex amplitude $A\mathrm{e}^{-\mathrm{j}\theta}$ is replaced by a time-varying complex amplitude $A(t)\mathrm{e}^{-\mathrm{j}\theta(t)}$, the waveform is no longer monochromatic. A monodirectional waveform with time-varying amplitude and phase has the general form[4]

$$\tilde{s}(t,x,y,z) = A(t-\tau(x,y,z))\cos[2\pi f_0(t-\tau(x,y,z)) + \theta(t-\tau(x,y,z))]$$

where $\tau(x,y,z) = (\alpha x + \beta y + \gamma z)/c$. This is called the *passband representation* of the time-varying monodirectional wavefront. Using the complex amplitude $A(t)e^{-j\theta(t)}$, the passband representation can be written more concisely as

$$\begin{aligned}\tilde{s}(t,x,y,z) &= \mathrm{Re}\left[A(t-\tau(x,y,z))e^{-j\theta(t-\tau(x,y,z))}e^{-j2\pi f_0(t-\tau(x,y,z))}\right]\\ &= \mathrm{Re}\left[s(t,x,y,z)e^{-j2\pi f_0 t}\right]\end{aligned}$$

where $s(t,x,y,z)$ is the complex baseband representation of the wavefront given by

$$s(t,x,y,z) = A(t-(\alpha x+\beta y+\gamma z)/c)e^{-j\theta(t-(\alpha x+\beta y+\gamma z)/c)}e^{j2\pi f_0(\alpha x+\beta y+\gamma z)/c}.$$

At the origin, the passband waveform is

$$\tilde{s}(t,0,0,0) = \mathrm{Re}\left[A(t)e^{-j\theta(t)}e^{-j2\pi f_0 t}\right].$$

Let $S(f)$ be the Fourier transform of $s(t,0,0,0)$. A narrowband wave is one for which $S(f)$ is narrow compared with f_0 insofar as the needs of an application may require. In most of this book, narrowband waves are treated as monochromatic waves, which is a very good approximation in many applications.

1.8 Wavefront diffraction

A wavefront that is not monodirectional is the superposition of waves traveling in multiple directions. We shall see that, if the waveform amplitude is spatially varying in any plane, the waveform is no longer monodirectional; it is a superposition of monodirectional waves traveling in multiple directions. The complex amplitude now depends on direction.

4 A wave of the form

$$\tilde{s}(t,x,y,z) = A(x,y)\cos(2\pi f_0(t-z/c))$$

does not satisfy the wave equation

$$\frac{\partial^2 \tilde{s}}{\partial x^2} + \frac{\partial^2 \tilde{s}}{\partial y^2} + \frac{\partial^2 \tilde{s}}{\partial z^2} = \frac{1}{c^2}\frac{\partial^2 \tilde{s}}{\partial t^2}.$$

Consequently, it is not among the waves we are studying. Sometimes, we may find it convenient to write a wave in this form. In such cases, it should only be regarded as an (geometrical optics) approximation to a wave that satisfies the wave equation.

Space-varying monochromatic waves

If multiple monochromatic waves are simultaneously traveling in a finite number of directions, indexed by ℓ, then the composite wave at complex baseband is

$$s(x, y, z) = \sum_{\ell=1}^{L} A_\ell e^{-j\theta_\ell} e^{j2\pi f_0(\alpha_\ell x+\beta_\ell y+\gamma_\ell z)/c}.$$

If monochromatic waves are simultaneously traveling in all directions with an infinitesimal amplitude in each direction, then the complex baseband representation of the wavefront becomes an integral over the extent of wavefront directions. A wavefront that is monochromatic, but has a continuum of directions, has the complex baseband representation

$$s(x, y, z) = \int_{-\infty}^{\infty}\int_{-\infty}^{\infty} a(\alpha, \beta) e^{j2\pi f_0(\alpha x+\beta y+\gamma z)/c} d\alpha \, d\beta$$

where the pair (α, β) specifies a direction with $\gamma = \sqrt{1-\alpha^2-\beta^2}$, and $a(\alpha, \beta)\, d\alpha\, d\beta$ is the infinitesimal complex amplitude of the wave propagation in direction α, β. Even though the direction cosines range only between -1 and 1, the limits of integration have been written from $-\infty$ to ∞. This allows some important flexibility later. For now, the excess region of integration can be suppressed by requiring that $a(\alpha, \beta) = 0$ for $|\alpha| > 1$ or $|\beta| > 1$.

If the wave $s_0(x, y) = s(x, y, 0)$ in the plane $z = 0$ is specified, then because

$$s_0(x, y) = \int_{-\infty}^{\infty}\int_{-\infty}^{\infty} a(\alpha, \beta) e^{j2\pi(\alpha x+\beta y)/\lambda} \, d\alpha \, d\beta$$

the term $\alpha(\alpha, \beta)$ is implicitly defined by $s_0(x, y)$. This equation can be interpreted as an instance of the inverse two-dimensional Fourier transform. The function $a(\alpha, \beta)$ is called the *angular spectrum* of the input signal $s_0(x, y) = s(x, y, 0)$ in the plane $z = 0$. The angular spectrum completely describes the propagation of a monochromatic wave. The "input" at $z = 0$ is $s_0(x, y)$, which determines the angular spectrum $a(\alpha, \beta)$. In turn, in the plane $z = d$, the complex amplitude $s_d(x, y) = s(x, y, d)$ is given in terms of $a(\alpha, \beta)$ by

$$s_d(x, y) = \int_{-\infty}^{\infty}\int_{-\infty}^{\infty} a(\alpha, \beta) e^{j(2\pi/\lambda)\sqrt{1-\alpha^2-\beta^2}} e^{j2\pi(\alpha x+\beta y)/\lambda} \, d\alpha \, d\beta.$$

We shall follow this thought in Chapter 4 to derive the important Huygens–Fresnel principle.

Evanescent waves

There is also a less-familiar, monochromatic and monodirectional solution of the wave equation called an *evanescent wave*. An evanescent wave is a wave of the form

$$\widetilde{s}(t, x, y, z) = \cos(2\pi f_0(t - (\alpha x + \beta y)/c))e^{-2\pi f_0 \gamma z/c}$$

where now the term involving z is a real exponential. This has the complex baseband representation

$$s(x, y, z) = e^{j2\pi f_0(\alpha x+\beta y)/c}e^{-2\pi f_0 \gamma z/c}$$

with

$$\alpha^2 + \beta^2 - \gamma^2 = 1,$$

satisfying the wave equation. The evanescent wave is an exponentially decreasing wave in the z direction. It may be needed to meet certain boundary conditions that cannot be met with a propagating wave in the z direction.

Clearly, the amplitude of the evanescent wave becomes infinite as z goes to negative infinity. Therefore an evanescent wave can only exist in a half-space and requires special boundary conditions on the boundary of this half-space. If the half-space is taken to be the half-space for which z is nonnegative, then the boundary conditions are on that plane for which $z = 0$. The evanescent wave decays quickly with increasing z, becoming negligible after a few wavelengths. Along the plane at $z = 0$, in the direction specified by α and β, runs an evanescent wave with a velocity of $c/\sqrt{\alpha^2 + \beta^2}$, which is smaller than c.

With the introduction of evanescent waves, the equation

$$s(x, y, z) = \int_{-\infty}^{\infty}\int_{-\infty}^{\infty} a(\alpha, \beta)e^{j2\pi f_0 z\sqrt{1-\alpha^2-\beta^2}/c}e^{j2\pi f_0(\alpha x+\beta y)/c}\, d\alpha\, d\beta$$

introduced earlier, now can be interpreted more generally. The infinite limits of integration allow the direction cosines to be larger than one. Physically, this allows evanescent waves to be included within the angular spectrum $a(\alpha, \beta)$.

Vector-valued waves

Besides scalar-valued waves, there are also vector-valued waves. A vector-valued wave may be regarded as three scalar-valued waves comprising the three components of the vector in a suitable coordinate system. A monodirectional, monochromatic vector wave at complex baseband has the form

$$\boldsymbol{s}(x, y, z) = [a_x\boldsymbol{i}_x + a_y\boldsymbol{i}_y + a_z\boldsymbol{i}_z]e^{j2\pi f_0(\alpha x+\beta y+\gamma z)/c}$$

where $(\boldsymbol{i}_x, \boldsymbol{i}_y, \boldsymbol{i}_z)$ forms a triad of orthogonal unit vectors along the three axes of the coordinate system. If the three scalar components a_x, a_y, and a_z can be independently specified, then such a wave amounts to nothing more than three independent scalar waves. However, vector waves of frequent physical interest, called *transverse waves*, are those that satisfy an additional constraint. Electromagnetic waves are of this kind.

A transverse wave is a vector wave that only takes values perpendicular to its direction of propagation. For the wave to be a transverse wave, the vector direction must be perpendicular to the direction of propagation. This means that the side condition

$$a_x\alpha + a_y\beta + a_z\gamma = 0$$

must be satisfied because

$$\alpha \boldsymbol{i}_x + \beta \boldsymbol{i}_y + \gamma \boldsymbol{i}_z$$

specifies the direction of propagation.

1.9 Deterministic and random models

The purpose of an image-formation system, of course, is to form an image, but it may be difficult to formulate a precise statement to this effect because a general criterion of optimality can be elusive. This is due partly to the fact that the underlying physical reality is much richer than the desired image, and it is difficult to state the real goal as an abstraction of the physical reality, and also because there may be prior knowledge or prior assumptions about the image that must be accommodated. Such considerations require that a model of the problem be developed: such a model may be either deterministic or random. In the early chapters of this book, deterministic models of the image will usually be used. We then assume that a "true" image does exist, and our task is to estimate that image by processing the measured data. Randomness does enter the problem in those chapters because the measurements can be random or noisy, but not because the image is modeled as random.

In later chapters, we shall turn to a more abstract view of imaging, regarding the problem as one of selecting an image from a space of possible images. In that more abstract view, an image is a realization of a random variable characterized by a probability distribution on a predefined space of images. The goal is not to pick the "true" image, but to select that image from the space of images that best explains the observed data.

For these reasons, both to model measurement noise and to model a random image, probability theory inevitably enters the subject of this book. We shall require the use of the notions of a *random variable* and a *random process*. For this reason, we will briefly review some of the fundamentals of probability that will be used.

A random variable, X, consists of a set of values that the random variable can take and a probability distribution on this set of values. A random process, $X(t)$, consists of a set of functions that the random process can take and a probability measure on this set of functions. A random variable, X, may be restricted to a finite number of values, in which case it is called a *discrete random variable*. Then it is characterized by a *probability vector*, $\boldsymbol{p}$, with a finite number of components denoted p_j. Likewise, a pair of discrete random variables, (X, Y), is associated with a joint probability distribution, $\boldsymbol{P}$, with an array of components denoted P_{jk}. A joint probability distribution is associated with *marginals*, defined by $p_j = \sum_k P_{jk}$ and $q_k = \sum_j P_{jk}$, and *conditionals*, defined by $Q_{k|j} = P_{jk}/p_j$ and $P_{j|k} = P_{jk}/q_k$. This leads to the *Bayes formula*

$$Q_{k|j} = \frac{q_k P_{j|k}}{\sum_k q_k P_{j|k}}.$$

A *real random variable* is a random variable that takes values in the set of real numbers. A real random variable may take values in a finite set of real numbers, in which case it is called a *discrete real random variable*, or in a continuous set of real numbers, in which case it is called a *continuous real random variable*. Whereas a discrete random variable is described by a probability vector, a continuous random variable is described by a function, $p(x)$, called the *probability density function*, or $p(x|y)$, called the *conditional probability density function*. We shall consider only discrete random variables and continuous random variables. A probability may depend on a nonrandom parameter γ. Then, the notation is $p_j(\gamma)$, $p(x;\gamma)$, or $p(x|y;\gamma)$ as is appropriate.

A discrete or continuous real random variable has a mean, $\overline{x}$, denoted by

$$\overline{x} = \Sigma_j p_j x_j,$$

or

$$\overline{x} = \int_{-\infty}^{\infty} x p(x)\,\mathrm{d}x,$$

and a variance, σ^2, denoted by

$$\sigma^2 = \Sigma_j (x_j - \overline{x})^2,$$

or

$$\sigma^2 = \int_{-\infty}^{\infty} (x - \overline{x})^2 p(x)\,\mathrm{d}x.$$

Similar definitions can be made for *complex random variables*, which are random variables taking values in the set of complex numbers.

An important random variable is a *gaussian random variable*, which is the only example of a random variable given in this section. We shall meet other random variables later in the book. The real gaussian random variable is defined by its probability density function

$$p(x) = \frac{1}{\sqrt{2\pi}\,\sigma} \mathrm{e}^{-(x-\overline{x})^2/2\sigma^2}.$$

The gaussian random variable has the mean $\overline{x}$ and variance σ^2. Likewise, the complex gaussian random variable $X = X_R + jX_I$ has probability density function

$$p(x_R, x_I) = \frac{1}{2\pi\sigma^2} e^{-|x-\overline{x}|^2/2\sigma^2}$$

where $\sigma^2 = E[X_R^2] = E[X_I^2] = E[XX^*]/2$ for the complex random variable.[5]

A real (or complex) multivariate random variable, $\boldsymbol{X} = (X_1, \ldots, X_n)$, also called a *vector random variable*, has a probability density function, $p(x_1, \ldots, x_n)$, and a covariance matrix, $\boldsymbol{\Sigma}$, whose ij entry is the expectation $E[X_i X_j]$ (or $E[X_i X_j^*]$). A covariance matrix is always nonnegative-definite because[6] $\boldsymbol{a}\boldsymbol{\Sigma}\boldsymbol{a}^\dagger = \boldsymbol{a}E[\boldsymbol{X}\boldsymbol{X}^\dagger]\boldsymbol{a}^\dagger = E[(\boldsymbol{a}\boldsymbol{X})^2]$, which is nonnegative.

A random process is a more elaborate quantity. A random process, $X(t)$, is characterized both by its amplitude structure and its time structure. The random process $X(t)$ is a random variable for each fixed value of t. A real random process, $X(t)$, is a real random variable for each fixed value of t. As such, for each fixed value of t, it has a probability description of its amplitude. The amplitude structure of $X(t)$ is described by a probability density function, $p(x)$. This probability density function will be independent of t if the random process is a *stationary random process*. The time structure of a stationary random process may be described, in part, by the *correlation function*, defined by

$$\phi(\tau) = E[X(t)X(t+\tau)],$$

(or $E[X(t)X^*(t+\tau)]$) and by its *power density spectrum* $N(f)$, which is the Fourier transform of $\phi(\tau)$. Thus we have the Fourier transform pair

$$\phi(\tau) \leftrightarrow N(f).$$

If $N(f)$ is a constant, by convention written as $N(f) = N_0/2$, then the random process is called *white noise*.

The reason for introducing random variables and random processes is to deal with randomness in a specific situation. The theory assumes that quantities such as probability density functions and correlation functions are meaningful and known. However, these are not always easy to specify. Eventually, we shall deal instead with the doubly vague situation in which the realization of the random variable X is unknown and the probability density function $p(x)$ associated with X is unknown as well. Such formulations provide structure to a problem leading to useful procedures for image formation.

[5] An alternative notation, which we do not use, is that $E[XX^*] = \sigma^2$.

[6] The symbol † denotes the transpose or the complex transpose, as is appropriate.

2 Signals in one dimension

We shall study waveforms in one dimension in this chapter and waveforms in two dimensions in the next chapter. We shall call these waveforms *one-dimensional signals* and *two-dimensional signals*, respectively.

A waveform that is used by a surveillance system to probe the environment is usually a function of the single variable – time. The collected sensor data will be a set of one or more of such one-dimensional waveforms. However, the environment usually is a two-dimensional or three-dimensional spatial region on which is defined a two-dimensional or three-dimensional signal of interest. Thus the computational task of image formation frequently consists of estimating an unknown two-dimensional or three-dimensional function when given a set of one-dimensional signals as the measured data.

Every waveform of finite energy is associated with another function, known as its *Fourier transform*, that describes a decomposition of the waveform into an infinite number of sinusoids. The Fourier transform constitutes an alternative representation of the function in the frequency domain. The frequency-domain representation is often much easier to work with than the original time-domain representation of the waveform. The Fourier transform is so pervasive in our studies that, in time, we shall scarcely be able to decide whether the original waveform or its Fourier transform is the more fundamental concept. For this reason, it is important to develop good intuition regarding the Fourier transform.

2.1 The one-dimensional Fourier transform

A real-valued or complex-valued function of time, $s(t)$, whose energy

$$E_p = \int_{-\infty}^{\infty} |s(t)|^2 \, dt$$

is finite, is called a *signal*. We sometimes call the signal $s(t)$ a *pulse* when we regard it as a simple signal, and call it a *waveform* when we regard it as a complicated signal. The *support* of $s(t)$ is the set of t at which $s(t)$ is nonzero. The pulse $s(t)$ has *bounded support* if the support is contained in a finite interval.

The Fourier transform $S(f)$ of the signal $s(t)$ is defined as

$$S(f) = \int_{-\infty}^{\infty} s(t)e^{-j2\pi ft}\,dt.$$

The Fourier transform pair will be indicated by the notation

$$s(t) \leftrightarrow S(f).$$

The two-way arrow implies that $s(t)$ determines $S(f)$, and $S(f)$ determines $s(t)$. The Fourier transform could be denoted by the functional notation

$$S(f) = \mathcal{F}[s(\cdot)](f).$$

Then

$$s(t) = \mathcal{F}^{-1}[S(\cdot)](t)$$

denotes the inverse Fourier transform.

The Fourier transform is a linear operation, which means that if

$$s(t) = as_1(t) + bs_2(t)$$

for any complex constants, a and b, then

$$S(f) = aS_1(f) + bS_2(f).$$

If $S(f)$ is the Fourier transform of the signal $s(t)$, then it is not the Fourier transform of any other signal.[1] Consequently, the Fourier transform can be inverted. This is proved in the following theorem, which gives an explicit formula for the inverse Fourier transform.

Theorem 2.1.1 (Inverse Fourier Transform) *If $S(f)$ is the Fourier transform of $s(t)$, then*

$$s(t) = \int_{-\infty}^{\infty} S(f)e^{j2\pi ft}\,df$$

for all t at which $s(t)$ is continuous.

Proof: Rather than give a rigorous proof, we shall use the formal symbolism of the impulse function $\delta(t)$ to facilitate an informal development. (The impulse function is discussed in Section 2.2.) By definition,

$$S(f) = \int_{-\infty}^{\infty} s(\xi)e^{-j2\pi f\xi}\,d\xi.$$

[1] Technically, there can be violations, known as Gibbs phenomena, on a finite set of points. Therefore, to make the statement precise, the Fourier transform is defined on the function space of equivalence classes of square-integrable functions. If the difference of two functions has zero energy, then those two functions are regarded as the same element of the function space. In this sense, any two functions with the same Fourier transform are equivalent.

Therefore

$$\int_{-\infty}^{\infty} S(f)\mathrm{e}^{\mathrm{j}2\pi ft}\,\mathrm{d}f = \int_{-\infty}^{\infty}\left[\int_{-\infty}^{\infty} s(\xi)\mathrm{e}^{-\mathrm{j}2\pi f\xi}\,\mathrm{d}\xi\right]\mathrm{e}^{\mathrm{j}2\pi ft}\,\mathrm{d}f$$
$$= \int_{-\infty}^{\infty} s(\xi)\left[\int_{-\infty}^{\infty} \mathrm{e}^{-\mathrm{j}2\pi(\xi-t)f}\,\mathrm{d}f\right]\mathrm{d}\xi.$$

The inner integral is infinite if $\xi = t$. Otherwise, it is the integral of a sine wave, and it integrates to zero. Therefore

$$\int_{-\infty}^{\infty} S(f)\mathrm{e}^{\mathrm{j}2\pi ft}\,\mathrm{d}f = \int_{-\infty}^{\infty} s(\xi)\delta(\xi - t)\,\mathrm{d}\xi$$
$$= s(t),$$

and the theorem is "proved." □

The formula for the inverse Fourier transform has the same structure as the formula for the Fourier transform except for a change in sign in the exponent. Consequently, there is a duality in Fourier transform pairs. Specifically, if

$$s(t) \leftrightarrow S(f)$$

is a Fourier transform pair, then

$$S(t) \leftrightarrow s(-f)$$

is also a Fourier transform pair.

The properties

$$S(0) = \int_{-\infty}^{\infty} s(t)\,\mathrm{d}t \qquad s(0) = \int_{-\infty}^{\infty} S(f)\,\mathrm{d}f$$

are trivial. Other useful properties are developed in the following theorems.

Theorem 2.1.2 (Scaling Property) *If*

$$s(t) \leftrightarrow S(f),$$

then

$$s(at) \leftrightarrow |a|^{-1}S(f/a).$$

Proof: The proof consists simply of a change of variables in the defining integral. □

In the next several theorems, we will develop a general *shift property*, which is known as the *translation property* (or *delay property*) when used to shift the time origin, and as the *modulation property* when used to shift the frequency origin.

Theorem 2.1.3 (Translation Property) *If $s(t)$ has the Fourier transform $S(f)$, then $s(t-t_0)$ has the Fourier transform $S(f)\mathrm{e}^{-\mathrm{j}2\pi t_0 f}$.*

Proof: Use the change of variable $\tau = t - t_0$ to write

$$\begin{aligned}\int_{-\infty}^{\infty} s(t-t_0)\mathrm{e}^{-\mathrm{j}2\pi ft}\,\mathrm{d}t &= \int_{-\infty}^{\infty} s(\tau)^{-\mathrm{j}2\pi f(\tau+t_0)}\,\mathrm{d}\tau \\ &= S(f)\mathrm{e}^{-\mathrm{j}2\pi t_0 f}.\end{aligned}$$

□

Theorem 2.1.4 (Complex Modulation Property) *If $s(t)$ has the Fourier transform $S(f)$, then $s(t)\mathrm{e}^{-\mathrm{j}2\pi f_0 t}$ has the Fourier transform $S(f+f_0)$.*

Proof:

$$\begin{aligned}\int_{-\infty}^{\infty} s(t)\,\mathrm{e}^{-\mathrm{j}2\pi f_0 t}\mathrm{e}^{-\mathrm{j}2\pi ft}\,\mathrm{d}t &= \int_{-\infty}^{\infty} s(t)\mathrm{e}^{-\mathrm{j}2\pi(f+f_0)t}\,\mathrm{d}t \\ &= S(f+f_0).\end{aligned}$$

□

Corollary 2.1.5 (Modulation Property) *If $s(t)$ has the Fourier transform $S(f)$, then $s(t)\cos 2\pi f_0 t$ has the Fourier transform $\frac{1}{2}[S(f+f_0)+S(f-f_0)]$, and $s(t)\sin 2\pi f_0 t$ has the Fourier transform $\frac{1}{2}\mathrm{j}[S(f+f_0)-S(f-f_0)]$.*

Proof: The proof follows immediately from the theorem and the relations

$$\begin{aligned}\cos 2\pi f_0 t &= \tfrac{1}{2}\left(\mathrm{e}^{-\mathrm{j}2\pi f_0 t}+\mathrm{e}^{\mathrm{j}2\pi f_0 t}\right) \\ \sin 2\pi f_0 t &= \tfrac{1}{2}\mathrm{j}\left(\mathrm{e}^{-\mathrm{j}2\pi f_0 t}-\mathrm{e}^{\mathrm{j}2\pi f_0 t}\right).\end{aligned}$$

□

Theorem 2.1.6 (Energy Relation) *If $s(t)$ has the Fourier transform $S(f)$, then the energy E_p of the pulse $s(t)$ satisfies*

$$E_p = \int_{-\infty}^{\infty} |s(t)|^2\,\mathrm{d}t = \int_{-\infty}^{\infty} |S(f)|^2\,\mathrm{d}f.$$

Proof:

$$\begin{aligned}\int_{-\infty}^{\infty} |s(t)|^2\,\mathrm{d}t &= \int_{-\infty}^{\infty} s(t)\left[\int_{-\infty}^{\infty} S(f)\mathrm{e}^{\mathrm{j}2\pi ft}\,\mathrm{d}f\right]^*\mathrm{d}t \\ &= \int_{-\infty}^{\infty} S^*(f)\left[\int_{-\infty}^{\infty} s(t)\mathrm{e}^{-\mathrm{j}2\pi ft}\,\mathrm{d}t\right]\mathrm{d}f \\ &= \int_{-\infty}^{\infty} |S(f)|^2\,\mathrm{d}f.\end{aligned}$$

□

The energy theorem is a special case of the following theorem.

Theorem 2.1.7 (Parseval's Formula) *If pulses $s_1(t)$ and $s_2(t)$ have the Fourier transforms $S_1(f)$ and $S_2(f)$, respectively, then*

$$\int_{-\infty}^{\infty} s_1(t)s_2^*(t)\,\mathrm{d}t = \int_{-\infty}^{\infty} S_1(f)S_2^*(f)\,\mathrm{d}f.$$

Proof: The proof is similar to the proof of Theorem 2.1.6. □

A linear filter with the impulse response $g(t)$ and the input signal $p(t)$ has the output $s(t)$ given by the convolution

$$s(t) = \int_{-\infty}^{\infty} p(\xi)g(t-\xi)\,\mathrm{d}\xi.$$

Because multiplication is simpler than convolution, it is often easier to treat a filtering problem in the frequency domain. The next theorem is a fundamental theorem for the study of the effect of a linear filter on a signal.

Theorem 2.1.8 (Convolution Theorem) *If $p(t)$, $g(t)$, and $s(t)$ have the Fourier transforms $P(f)$, $G(f)$, and $S(f)$, respectively, then they are related by the convolution*

$$s(t) = \int_{-\infty}^{\infty} p(\xi)g(t-\xi)\,\mathrm{d}\xi$$

if and only if

$$S(f) = P(f)G(f).$$

Proof:

$$\begin{aligned} s(t) &= \int_{-\infty}^{\infty} g(t-\tau)\left[\int_{-\infty}^{\infty} P(f)e^{j2\pi f\tau}\,\mathrm{d}f\right]\mathrm{d}\tau \\ &= \int_{-\infty}^{\infty} P(f)e^{j2\pi ft}\left[\int_{-\infty}^{\infty} g(t-\tau)e^{-j2\pi f(t-\tau)}\,\mathrm{d}\tau\right]\mathrm{d}f. \end{aligned}$$

Let $\eta = t - \tau$. Then

$$\begin{aligned} s(t) &= \int_{-\infty}^{\infty} P(f)e^{j2\pi ft}\left[\int_{-\infty}^{\infty} g(\eta)e^{-j2\pi f\eta}\mathrm{d}\eta\right]\mathrm{d}f \\ &= \int_{-\infty}^{\infty} P(f)G(f)e^{j2\pi ft}\mathrm{d}f. \end{aligned}$$

Consequently, by the invertibility of the Fourier transform,

$$S(f) = P(f)G(f).$$

The theorem is proved in the reverse direction by tracing this argument in the reverse direction. □

Closely related to the convolution of two signals is their correlation. Given any two signals, $p(t)$ and, $g(t)$ their *correlation function* is defined by

$$s(t) = \int_{-\infty}^{\infty} p(\xi)g^*(\xi - t)\,\mathrm{d}\xi$$
$$= \int_{-\infty}^{\infty} p(\xi + t)g^*(\xi)\,\mathrm{d}\xi,$$

and the *correlation* between $p(t)$ and $g(t)$ is the correlation function at $t = 0$.

Corollary 2.1.9 (Correlation Theorem) *If $p(t)$, $g(t)$, and $s(t)$ have the Fourier transforms $P(f)$, $G(f)$, and $S(f)$, respectively, then $s(t)$ is the correlation of $p(t)$ with $g(t)$ if and only if*

$$S(f) = P(f)G^*(f)$$

in the Fourier transform domain.

Proof: Replace $G(f)$ by $G^*(f)$ in the theorem, noting that $g^*(-t) \leftrightarrow G^*(f)$. □

An immediate consequence of the correlation theorem is that the convolution of $s(t)$ with $s^*(-t)$ has the transform $|S(f)|^2$:

$$s(t) * s^*(-t) \leftrightarrow |S(f)|^2.$$

The dual of this Fourier transform pair is

$$|s(t)|^2 \leftrightarrow S(f) * S^*(-f).$$

Theorem 2.1.10 (Differentiation Property) *If the pulse $s(t)$ has the Fourier transform $S(f)$, then the derivative $s'(t)$ has the Fourier transform $\mathrm{j}2\pi f S(f)$.*

Proof:

$$\frac{\mathrm{d}s(t)}{\mathrm{d}t} = \frac{\mathrm{d}}{\mathrm{d}t}\int_{-\infty}^{\infty} S(f)\mathrm{e}^{\mathrm{j}2\pi f t}\,\mathrm{d}f$$
$$= \int_{-\infty}^{\infty} [\mathrm{j}2\pi f S(f)]\mathrm{e}^{\mathrm{j}2\pi f t}\,\mathrm{d}f.$$

□

2.2 Transforms of some useful functions

A useful list of some one-dimensional Fourier transform pairs is given in Table 2.1. Several of these pairs are developed in this section. Others may arise only later in the book.

Table 2.1 *A table of one-dimensional Fourier transform pairs*

$s(t)$	$S(f)$
$\mathrm{rect}(t)$	$\mathrm{sinc}(f)$
$\mathrm{sinc}(t)$	$\mathrm{rect}(f)$
$\delta(t)$	1
1	$\delta(f)$
$\mathrm{comb}(t)$	$\mathrm{comb}(f)$
$\mathrm{comb}_N(t)$	$\mathrm{dirc}_N(f)$
$e^{-\pi t^2}$	$e^{-\pi f^2}$
$e^{j\pi t^2}$	$\frac{1+j}{\sqrt{2}} e^{-j\pi f^2}$
$\cos 2\pi t$	$\frac{1}{2}\delta(f+1) + \frac{1}{2}\delta(f-1)$
$\sin 2\pi t$	$\frac{j}{2}\delta(f+1) - \frac{j}{2}\delta(f-1)$
$e^{j2\pi t}$	$\delta(f-1)$
$e^{-\lvert t\rvert}$	$\frac{2}{1+(2\pi f)^2}$
$J_0(2\pi t)$	$\frac{\mathrm{rect}(f/2)}{\pi(1-f^2)^{\frac{1}{2}}}$
$\mathrm{jinc}(t)$	$\sqrt{1-4f^2}\,\mathrm{rect}(f)$

The most elementary pulse is the *rectangle function* or *rectangular pulse*, defined as

$$\mathrm{rect}(t) = \begin{cases} 1 & \text{if} \quad |t| \le 1/2 \\ 0 & \text{if} \quad |t| > 1/2. \end{cases}$$

The Fourier transform of a rectangular pulse is readily evaluated. Let $s(t) = \mathrm{rect}(t)$. Then

$$\begin{aligned} S(f) &= \int_{-1/2}^{1/2} e^{-j2\pi f t}\, dt \\ &= \frac{1}{-j2\pi f}\left[e^{-j\pi f} - e^{j\pi f}\right] \\ &= \mathrm{sinc}(f) \end{aligned}$$

where the *sinc function* or *sinc pulse* is defined as

$$\mathrm{sinc}(x) = \frac{\sin \pi x}{\pi x}.$$

The sinc function is shown in Figure 2.1. By l'Hôpital's rule, we find that $\mathrm{sinc}(0) = 1$. (This also follows from the fact that $S(0)$ is the integral of $s(t)$.) The zeros of $\mathrm{sinc}(x)$ are the nonzero integers. The magnitude of $\mathrm{sinc}(x)$ is bounded as $|\mathrm{sinc}(x)| \le 1/\pi x$.

The Fourier transform of a rectangular pulse of width $T > 0$ is obtained easily by using the scaling property. This Fourier transform pair is written concisely as

$$\mathrm{rect}\left(\frac{t}{T}\right) \leftrightarrow T\,\mathrm{sinc}(Tf).$$

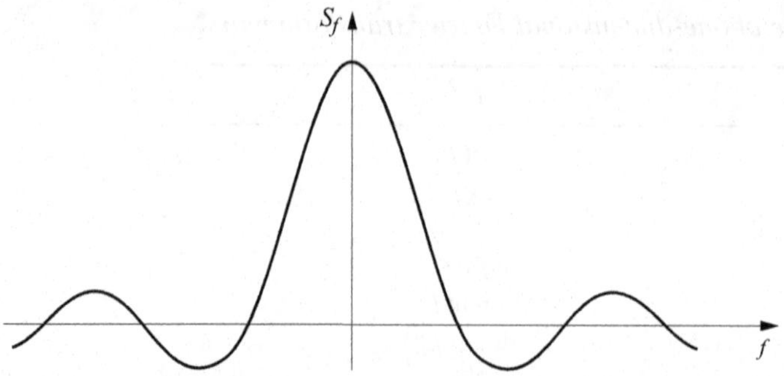

Figure 2.1 Fourier transform of a rectangular pulse

By the duality property of the Fourier transform, for a band of width $W > 0$, we can immediately write

$$W \operatorname{sinc}(Wt) \leftrightarrow \operatorname{rect}\left(\frac{f}{W}\right).$$

The *impulse function* (or *delta function*) $\delta(t)$, though not a true function,[2] is very useful in formal manipulations. The Fourier transform of the impulse function is evaluated as follows. Let

$$s(t) = \begin{cases} \frac{1}{T} & |t| \le T/2 \\ 0 & |t| > T/2. \end{cases}$$

This pulse has a unit area and becomes infinitely high and infinitely thin as T goes to zero. The impulse function $\delta(t)$ is defined formally as the limit of this sequence of rectangular pulses as T goes to zero,

$$\delta(t) = \lim_{T \to 0} \frac{1}{T} \operatorname{rect}\left(\frac{t}{T}\right).$$

The impulse function $\delta(t)$ can be understood in the sense of an operator defined by

$$v(\tau) = \int_{-\infty}^{\infty} v(t)\delta(t-\tau)\, dt.$$

The Fourier transform of $\delta(t)$ does not exist in the sense of the original definition, but it can be defined formally as the limit, as T goes to zero, of the sequence of Fourier transforms of increasingly narrow rectangular pulses. Thus

$$\begin{aligned} S(f) &= \lim_{T \to 0} \frac{1}{T}(T \operatorname{sinc} fT) \\ &= 1. \end{aligned}$$

[2] The impulse function is an example of a *generalized function* and, as such, is defined by the "sifting property" $\int_{-\infty}^{\infty} s(t)\delta(t)\, dt = s(0)$. One of its properties is $\delta(t/T) = T\delta(t)$ for $T > 0$.

This "Fourier transform pair" is written concisely as

$$\delta(t) \leftrightarrow 1.$$

This pair is quite useful and so is appended to the list of Fourier transform pairs.

The rectangular pulse has finite support. In contrast, the *gaussian pulse*

$$s(t) = \mathrm{e}^{-\pi t^2}$$

does not because it has tails that go on forever. The Fourier transform of the gaussian pulse is

$$S(f) = \int_{-\infty}^{\infty} \mathrm{e}^{-\pi t^2} \mathrm{e}^{-\mathrm{j}2\pi f t}\, \mathrm{d}t.$$

Because $s(t)$ is an even function, this becomes

$$\begin{aligned} S(f) &= \int_{-\infty}^{\infty} \mathrm{e}^{-\pi t^2} \cos 2\pi f t\, \mathrm{d}t \\ &= \mathrm{e}^{-\pi f^2} \end{aligned}$$

where the last line follows from consulting a table of definite integrals. Thus we have the Fourier transform pair

$$\mathrm{e}^{-\pi t^2} \leftrightarrow \mathrm{e}^{-\pi f^2}.$$

More generally,

$$\mathrm{e}^{-at^2} \leftrightarrow \sqrt{\frac{\pi}{a}}\, \mathrm{e}^{-\pi^2 f^2/a},$$

which follows from the scaling property.

A pulse can also have complex values. An important complex-valued pulse is the *chirp pulse*. A chirp pulse is also called a *quadratic-phase pulse*. It is

$$\mathrm{chrp}(t) = \mathrm{e}^{\mathrm{j}\pi t^2}.$$

The chirp pulse has infinite energy, so the Fourier transform is not defined. In contrast, the truncated chirp pulse, defined by

$$s_T(t) = \mathrm{e}^{\mathrm{j}\pi t^2} \operatorname{rect}\left(\frac{t}{T}\right),$$

has finite energy. The Fourier transform

$$S_T(f) = \int_{-T/2}^{T/2} \mathrm{e}^{\mathrm{j}\pi t^2} \mathrm{e}^{-\mathrm{j}2\pi f t}\, \mathrm{d}t$$

is defined for every positive value of T. Hence, it is natural to define the Fourier transform of the chirp pulse as

$$S(f) = \int_{-\infty}^{\infty} e^{j\pi t^2} e^{-j2\pi ft} \, dt.$$

To evaluate the integral, complete the square in the exponent

$$S(f) = e^{-j\pi f^2} \int_{-\infty}^{\infty} e^{j\pi(t-f)^2} \, dt.$$

By a change of variables in the integral, and using Euler's formula, this becomes

$$S(f) = e^{-j\pi f^2} \left[\int_{-\infty}^{\infty} \cos \pi t^2 \, dt + j \int_{-\infty}^{\infty} \sin \pi t^2 \, dt \right].$$

The integrals are now standard tabulated integrals, and each is equal to $1/\sqrt{2}$. Therefore the Fourier transform of the chirp pulse is

$$S(f) = \frac{1+j}{\sqrt{2}} e^{-j\pi f^2}$$

which also has infinite energy.

Notice that in both the derivation of the Fourier transform of a chirp pulse and the derivation of the Fourier transform of an impulse, the Fourier transform of a limit of a sequence of functions is defined as the limit of the sequence of Fourier transforms. In this manner, the set of functions that have Fourier transforms can be enlarged in a natural and satisfying way. We will freely enlarge the set of Fourier transform pairs in this way, but we remark that this limiting procedure needs careful justification. This book is not the place to give the formal mathematical setting to make these manipulations precise.

2.3 The dirichlet functions

One way to construct a complicated waveform is by the repetition of a simple waveform. This kind of construction arises in the study of many engineering subjects, such as antenna arrays or pulse trains, and also in digital signal processing. Common mathematical phenomena, underlying all of these constructions, can be understood in terms of the solutions of some simple Fourier transform constructions.

Let $s(t)$ be any pulse, and let $S(f)$ be its Fourier transform. Consider the pulse doublet

$$s'(t) = s(t + T/2) + s(t - T/2).$$

The doublet has the Fourier transform

$$\begin{aligned} S'(f) &= S(f)e^{j\pi fT} + S(f)e^{-j\pi fT} \\ &= 2S(f) \cos \pi fT. \end{aligned}$$

Similarly, the pulse quadruplet can be viewed as a doublet of doublets,

$$s''(t) = s'(t+T) + s'(t-T).$$

This has the Fourier transform

$$\begin{aligned} S''(f) &= 2S'(f)\cos 2\pi fT \\ &= 4S(f)\cos \pi fT \cos 2\pi fT. \end{aligned}$$

This doubling procedure can be repeated multiple times to obtain the Fourier transform whenever the number of pulses in a pulse train is a power of two. For a uniform pulse train of arbitrary length, given by

$$p(t) = \sum_{\ell=0}^{N-1} s(t - \ell T + \tfrac{1}{2}(N-1)T),$$

a more direct derivation is needed.

Theorem 2.3.1 *A uniform pulse train consisting of N equispaced, identical pulses centered at the origin has the Fourier transform*

$$P(f) = S(f)\frac{\sin N\pi fT}{\sin \pi fT}$$

where T is the pulse spacing and S(f) is the Fourier transform of the individual pulse s(t).

Proof: Use the translation property on each pulse to write

$$\begin{aligned} P(f) &= \sum_{\ell=0}^{N-1} S(f)\mathrm{e}^{-\mathrm{j}2\pi f\ell T + \mathrm{j}\pi f(N-1)T} \\ &= S(f)\mathrm{e}^{\mathrm{j}\pi f(N-1)T}\sum_{\ell=0}^{N-1}\mathrm{e}^{-\mathrm{j}2\pi f\ell T}. \end{aligned}$$

Now use the relationship

$$\sum_{\ell=0}^{N-1} x^\ell = \begin{cases} \frac{1-x^N}{1-x} & x \neq 1 \\ N & x = 1 \end{cases}$$

to obtain the identity

$$\begin{aligned} \sum_{\ell=0}^{N-1}\mathrm{e}^{-\mathrm{j}2\pi f\ell T} &= \frac{1-\mathrm{e}^{-\mathrm{j}2\pi fTN}}{1-\mathrm{e}^{-\mathrm{j}2\pi fT}} \\ &= \left[\frac{\mathrm{e}^{\mathrm{j}\pi fTN} - \mathrm{e}^{-\mathrm{j}\pi fTN}}{\mathrm{e}^{\mathrm{j}\pi fT} - \mathrm{e}^{-\mathrm{j}\pi fT}}\right]\mathrm{e}^{-\mathrm{j}\pi f(N-1)T}. \end{aligned}$$

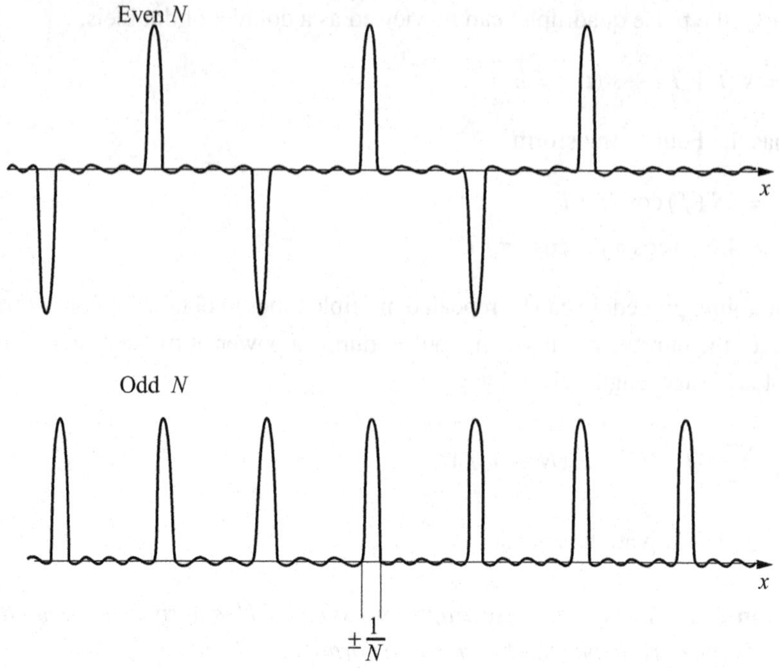

Figure 2.2 Illustrating the dirichlet functions

Then

$$P(f) = S(f)\frac{\sin N\pi fT}{\sin \pi fT},$$

as was to be proved. □

The transform of a pulse train is the product of two parts: one due to a single pulse, and one due to the array. Because this latter term appears so frequently, it deserves its own name. For each integer N, define the *dirichlet function* as

$$\text{dirc}_N(x) = \frac{\sin \pi Nx}{\sin \pi x}.$$

Dirichlet functions are illustrated in Figure 2.2.

It is easily shown by l'Hôpital's rule that $\text{dirc}_N(0) = N$. The first zero of $\text{dirc}_N(x)$ occurs at $x = 1/N$. The portion of the dirichlet function between $x = \pm 1/N$ is called the *main lobe* of the dirichlet function. When x is small, the denominator can be approximated by using the small-angle approximation for the sine. Then

$$\begin{aligned}\text{dirc}_N(x) &\approx \frac{\sin \pi Nx}{\pi x}\\ &\approx N \operatorname{sinc} Nx.\end{aligned}$$

Thus, near the origin, the dirichlet function looks like a sinc function. (For this reason the dirichlet function is sometimes called the *periodic sinc*.) The amplitude at the origin is N, and the first zeros are at $\pm 1/N$.

Next, notice that, for any integer k,

$$\begin{aligned}\text{dirc}_N(x+k) &= \frac{\sin \pi N(x+k)}{\sin \pi (x+k)}, \\ &= \pm \text{dirc}_N(x)\end{aligned}$$

which implies that

$$\text{dirc}_N(k) = \pm N.$$

Moreover, if N is an odd integer, the dirichlet function satisfies

$$\text{dirc}_N(x+k) = \text{dirc}_N(x),$$

and if N is an even integer

$$\text{dirc}_N(x+k) = (-1)^k\, \text{dirc}_N(x).$$

In particular, the main lobe of the dirichlet function appears periodically replicated but, if N is even, there is a sign reversal on alternate grating lobes. The periodic copies of the main lobe are called *grating lobes*.

The statement of Theorem 2.3.1 can now be rewritten as

$$P(f) = S(f)\,\text{dirc}_N\,(fT).$$

Thus $P(f)$ displays the grating lobes of the dirichlet function modulated by the complex function $S(f)$. Because of the modulation theorem, the grating lobes of the dirichlet function can be translated along the frequency axis by phase modulating the pulses. A uniform, linearly phase-modulated pulse train is a pulse train of the form

$$p(t) = \sum_{\ell=0}^{N-1} s(t - \ell T + (N-1)T/2)\mathrm{e}^{\mathrm{j}2\pi \ell f_0 T}.$$

The ℓth pulse has phase $\ell f_0 T$.

Corollary 2.3.2 *A uniform, linearly phase-modulated pulse train has the Fourier transform*

$$P(f) = S(f)\text{dirc}_N(f - f_0)T.$$

Proof: The delay theorem immediately gives

$$P(f) = S(f)\mathrm{e}^{\mathrm{j}\pi f(N-1)T} \sum_{\ell=0}^{N-1} \mathrm{e}^{-\mathrm{j}2\pi(f-f_0)\ell T},$$

which reduces to the statement of the theorem. □

If $s(t)$ is an impulse, then the pulse train $p(t)$ is a finite train of N impulses. It is convenient to name both the finite and the infinite train of impulses as follows.

Definition 2.3.3 *The comb function is given by*

$$\text{comb}(t) = \sum_{\ell=-\infty}^{\infty} \delta(t - \ell).$$

The finite comb function is given by

$$\text{comb}_N(t) = \sum_{\ell=0}^{N-1} \delta(t - \ell + \tfrac{1}{2}(N-1)).$$

The Fourier transform pair

$$\text{comb}_N(t) \leftrightarrow \text{dirc}_N(f)$$

follows from Theorem 2.3.1. Because the grating lobes of $\text{dirc}_N(f)$ become larger and thinner as N goes to infinity, they become more like impulses. If N is odd and very large, then the grating lobes of the dirichlet function will resemble a comb of impulses. This suggests the Fourier transform pair

$$\text{comb}(t) \leftrightarrow \text{comb}(f).$$

To develop this transform pair more directly, define

$$\text{comb}(t) = \lim_{\tau \to 0} \lim_{N \to \infty} \sum_{\ell=-N}^{N} \frac{1}{\tau} \,\text{rect}\left(\frac{t-\ell}{\tau}\right).$$

The Fourier transform of the right side is given by

$$\lim_{\tau \to 0} \lim_{N \to \infty} \text{sinc}(f\tau)\text{dirc}_{2N+1}(f) = \text{comb}(f),$$

which gives the Fourier transform pair mentioned previously. Of course, comb(t) is not an ordinary function. Rather, it is a generalized function and is interpreted by its role under an integral sign. Then

$$\int_{-\infty}^{\infty} s(t)\text{comb}(t)\,\mathrm{d}t = \sum_{\ell=-\infty}^{\infty} s(\ell)$$

for any $s(t)$ of finite energy.

The scaling property applied to the Fourier transform pair

$$\text{comb}(t) \leftrightarrow \text{comb}(f)$$

gives

$$\text{comb}\left(\frac{t}{T}\right) \leftrightarrow T\,\text{comb}(Tf)$$

for $T > 0$. This is written explicitly as

$$\sum_{\ell=-\infty}^{\infty} \delta\left(\frac{t}{T} - \ell\right) = T \sum_{\ell=-\infty}^{\infty} \delta(Tf - \ell).$$

By using the scaling properties of the impulse function, this can be rewritten in the more convenient form

$$\sum_{\ell=-\infty}^{\infty} \delta(t - \ell T) = \frac{1}{T} \sum_{\ell=-\infty}^{\infty} \delta\left(f - \frac{\ell}{T}\right).$$

2.4 Passband signals and passband filters

The Fourier transform $S(f)$ of a typical baseband signal is nonzero in some interval about the frequency origin, as illustrated in Figure 2.3. A signal is called a passband signal if its transform is zero in some interval about the frequency origin and is zero for sufficiently large f, as illustrated in Figure 2.4. From the modulation theorem for

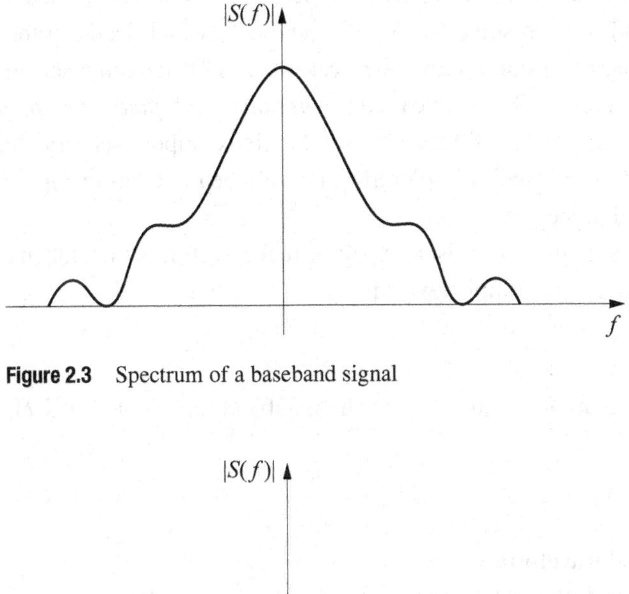

Figure 2.3 Spectrum of a baseband signal

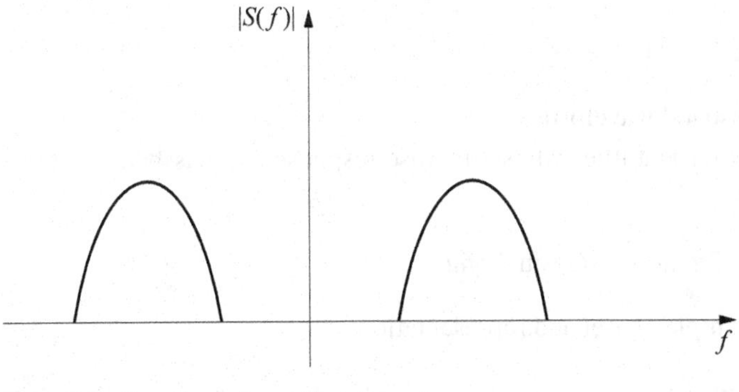

Figure 2.4 Spectrum of a passband signal

the Fourier transform, it is easy to see that one can construct a passband signal $\widetilde{s}(t)$ by multiplying the baseband signal $s(t)$ by $\cos 2\pi f_0 t$ or by $\sin 2\pi f_0 t$ for a sufficiently large f_0, provided $S(f)$ has a bounded support. Moreover, any two baseband signals, $s_R(t)$ and $s_I(t)$, whose transforms $S_R(f)$ and $S_I(f)$ have bounded support, can be used to define the passband signal

$$s(t) = s_R(t)\cos 2\pi f_0 t + s_I(t)\sin 2\pi f_0 t$$

for some sufficiently large f_0. The Fourier transform of this $s(t)$ can be formed by using the modulation theorem to write

$$s_R(t)\cos 2\pi f_0 t \leftrightarrow \tfrac{1}{2}\left[S_R(f+f_0) + S_R(f-f_0)\right]$$

$$s_I(t)\sin 2\pi f_0 t \leftrightarrow \tfrac{1}{2}\mathrm{j}\left[S_I(f+f_0) - S_I(f-f_0)\right],$$

so that

$$\widetilde{S}(f) = \tfrac{1}{2}[S_R(f-f_0) - \mathrm{j}S_I(f-f_0)] + \tfrac{1}{2}[S_R(f+f_0) + \mathrm{j}S_I(f+f_0)].$$

This construction of a passband signal from two baseband signals is the most general construction of this kind in the sense that any passband waveform could have been constructed in this way. That is, the positive and negative frequency segments of $\widetilde{S}(f)$ can be separated and used to solve for $S_R(f)$ and $S_I(f)$ which leads to the desired decomposition. Thus a passband signal $\widetilde{s}(t)$ can always be decomposed into $s_R(t)$ and $s_I(t)$ of the stated form. These are called the *in-phase* and *quadrature modulation components*, respectively, of $\widetilde{s}(t)$. Because the modulation components vary much more slowly than $\widetilde{s}(t)$ itself, it is usually simpler to work with the modulation components in applications of signal processing.

The modulation components consist of a pair of real waveforms that, taken together, can be used to form a single complex waveform,

$$s(t) = s_R(t) + \mathrm{j}s_I(t).$$

The complex representation is equivalent to the passband representation $\widetilde{s}(t)$, and the expression

$$\widetilde{s}(t) = \mathrm{Re}[s(t)\mathrm{e}^{-\mathrm{j}2\pi f_0 t}]$$

recovers the passband waveform.

A passband filter is a filter whose impulse response is a passband pulse. It has the form

$$\widetilde{g}(t) = g_R(t)\cos 2\pi f_0 t + g_I(t)\sin 2\pi f_0 t,$$

which has the complex baseband representation

$$g(t) = g_R(t) + \mathrm{j}g_I(t).$$

Theorem 2.4.1 *A passband filter, $\widetilde{g}(t)$, with a passband input, $\widetilde{s}(t)$, has a passband output, $\widetilde{r}(t)$, whose complex baseband representation $r(t)$ is the output of the complex baseband filter $g(t)$ with the complex baseband input $s(t)$.*

Proof: This is an easy manipulation in the Fourier transform domain. □

Theorem 2.4.2 *If the phase angle of the carrier of a passband signal at the input to a passband filter is changed by θ, then the phase angle of the carrier of the signal at the output of the filter is changed by θ as well, but otherwise the output signal is unchanged.*

Proof: The passband signal with a phase offset in the carrier is

$$\widetilde{s}'(t) = s_R(t)\cos(2\pi f_0 t + \theta) + s_I(t)\sin(2\pi f_0 t + \theta).$$

Theorem 2.4.1 tells us that the passband pulse $\widetilde{s}'(t)$ can be passed through the passband filter $\widetilde{g}(t)$ by representing the passband signal and passband filter at complex baseband as $s'(t)$ and $g(t)$. Because at complex baseband, the phase angle appears in a convenient way as

$$s'(t) = [s_R(t) + \mathrm{j}s_I(t)]\mathrm{e}^{-\mathrm{j}\theta} = s(t)\mathrm{e}^{-\mathrm{j}\theta}$$

and

$$s'(t) * g(t) = [s(t) * g(t)]\mathrm{e}^{-\mathrm{j}\theta},$$

we can conclude that

$$\widetilde{s}'(t) * \widetilde{g}(t) = [s(t) * g(t)]_R \cos(2\pi f_0 t + \theta) + [s(t) * g(t)]_I \sin(2\pi f_0 t + \theta),$$

as was to be proved. □

2.5 Baseband and passband sampling

Digitization is the process of converting an analog waveform on a finite time interval into a representation that uses a finite number of bits. Digitization is commonly partitioned into two parts, called sampling and quantization. *Sampling* represents a continuous function of time by its values, called *samples*, on a discrete set of time points. *Quantization* represents a finite block of samples by a finite number of bits. A *scalar quantizer* represents each sample independently by a finite number of bits. A *vector quantizer* represents a block of samples by a finite number of bits. We shall study sampling, but not quantization.

We shall examine the conditions under which it is sufficient to sample a waveform $s(t)$ at a discrete set of time values. In particular, the set of samples $s_\ell = s(\ell T)$ for

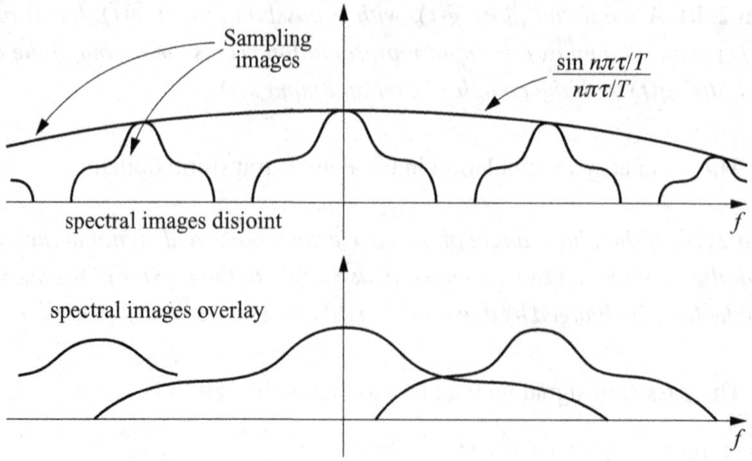

Figure 2.5 Sampling images

$\ell = 0, \pm 1, \pm 2, \ldots$ completely describes a waveform, $s(t)$, whose spectrum is limited to B hertz, provided $2BT < 1$.

Suppose that $s(t)$, a baseband waveform with finite energy, has a spectrum that satisfies $S(f) = 0$ for $f \geq \frac{1}{2}$. The following discussion shows that the set of samples $s_\ell = s(\ell)$ for $\ell = 0, \pm 1, \pm 2, \ldots$ then completely describes $s(t)$. Let

$$s'(t) = s(t)\text{comb}(t) = \sum_{\ell=-\infty}^{\infty} s_\ell \delta(t - \ell).$$

From the formal properties of the Fourier transform, this equation has the transform

$$S'(f) = S(f) * \text{comb}(f) = \sum_{k=-\infty}^{\infty} S(f - k).$$

The spectrum $S'(f)$ consists of an infinite number of translated copies of $S(f)$, the kth copy translated in frequency by k. Each of these translations of the spectrum is called a *sampling image*. Sampling images are shown in Figure 2.5. Because we have assumed that $S(f) = 0$ for $f \geq \frac{1}{2}$, these sampling images do not overlap. This means that

$$S'(f)\text{rect}(f) = S(f)$$

so that $S(f)$ can be completely recovered from $S'(f)$.

A similar conclusion holds for any signal confined to a finite bandwidth, as can be seen by invoking the scaling property of the Fourier transform. Thus suppose that the baseband waveform $s(t)$ has a spectrum that satisfies $S(f) = 0$ for $f \geq B$, for some constant B. The samples, called *Nyquist samples*, for this case are $s_\ell = s(\ell T)$ where T should be chosen so that $2BT < 1$. Again, as shown in the top half of Figure 2.5, the sampling images do not overlap. Hence an ideal lowpass filter of bandwidth B

will reject all sampling images other than $S(f)$ itself. Thus an ideal lowpass filter will recover $s(t)$. Moreover, if the samples s_ℓ are not Nyquist samples, they do not uniquely determine $s(t)$.

The following theorem describes how to recover $s(t)$ in the time domain.

Theorem 2.5.1 (Sampling theorem) *The signal $s(t)$ can be recovered from its Nyquist samples by the Nyquist–Shannon interpolation formula*

$$s(t) = \sum_{\ell=-\infty}^{\infty} s_\ell \operatorname{sinc}\left(\frac{t}{T} - \ell\right).$$

Proof: As we have seen, if $B = \frac{1}{2}$ and $T = 1$, then the sampled waveform

$$s'(t) = s(t)\operatorname{comb}(t) = \sum_{\ell=-\infty}^{\infty} s_\ell \delta(t - \ell)$$

has the Fourier transform

$$\begin{aligned} S'(f) &= S(f) * \operatorname{comb}(f) \\ &= \sum_{k=-\infty}^{\infty} S(f - k). \end{aligned}$$

Because $S(f) = 0$ for $|f| > \frac{1}{2}$, these images do not overlap, and $S(f)$ can be recovered simply by writing

$$S(f) = S'(f)\operatorname{rect}(f).$$

The product in the frequency domain becomes a convolution in the time domain

$$\begin{aligned} s(t) &= s'(t) * \operatorname{sinc}(t) \\ &= \left[\sum_{\ell=-\infty}^{\infty} s_\ell \delta(t - \ell)\right] * \operatorname{sinc}(t) \\ &= \sum_{\ell=-\infty}^{\infty} s_\ell \operatorname{sinc}(t - \ell). \end{aligned}$$

To complete the proof of the theorem, replace t by t/T and use the scaling property.

□

To obtain perfect reconstruction, sinc interpolation requires an infinite number of samples, which implies infinite interpolation delay. In practice, one uses only a finite number of samples for a reconstruction, so one must be satisfied with less than perfect interpolation. Usually, the interpolation error can be kept satisfactorily small by over-sampling – that is, sampling faster than the Nyquist rate – and using a fairly simple alternative to the sinc pulse.

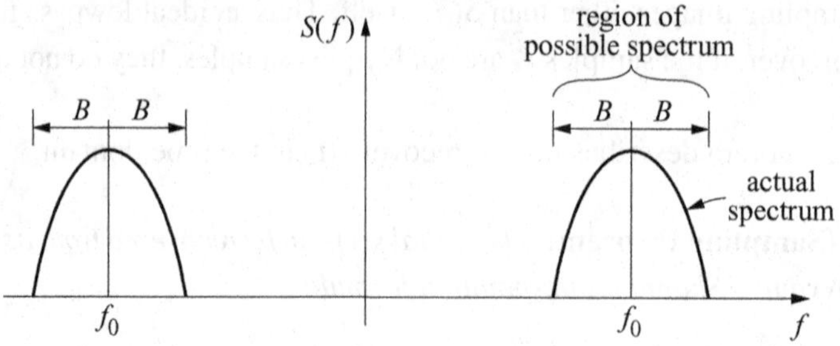

Figure 2.6 A passband spectrum

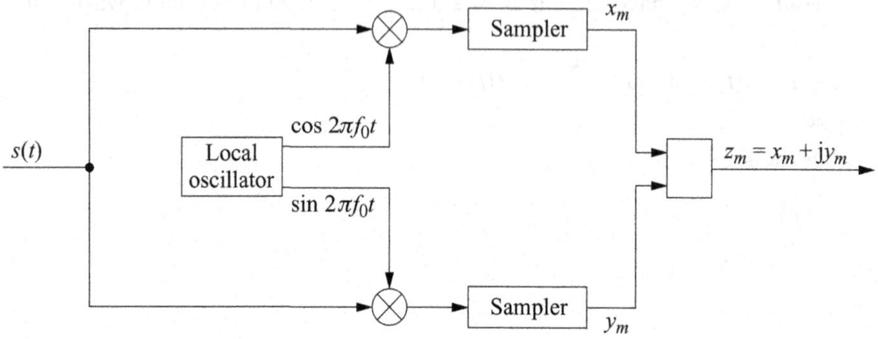

Figure 2.7 Baseband sampling of passband signals

Sometimes prior information about $s(t)$ may be available that may be used to reduce the sampling requirements. An important instance of this is the passband waveform. Figure 2.6 shows that a direct application of the sampling theorem would lead one to sample at $2(f_0 + B)$ samples per second. On the other hand, $s(t)$ can be reduced to its in-phase and quadrature components, each of which can be sampled at $2B$ samples per second. This allows $s(t)$ to be sampled with $4B$ samples per second. The difference is that the first scheme allows one to reconstruct any arbitrary signal whose support is in the region $|f| \leq f_0 + B$. The second scheme applies only to functions with such support that are also zero for $|f| \leq f_0 - B$.

Figure 2.7 gives a sampling scheme at complex baseband for passband signals. It requires two baseband channels that must be precisely matched.

2.6 Signal space

The set of all complex, finite energy functions of the single variable t comprises a space of functions called *signal space*. In defining signal space, it is convenient and conventional to regard the two functions $s(t)$ and $s'(t)$ to be the same element of signal space whenever

$$\int_{-\infty}^{\infty} |s(t) - s'(t)|^2 \, dt = 0.$$

This means that two functions are considered equivalent if their difference has zero energy. With this convention, every element $s(t)$ of signal space has a unique Fourier transform, $S(f)$.

Signal space has the properties[3] that the sum of two elements of the space is again an element of the space, and a scalar multiple of any element of the space is again an element of the space. A set with these two properties is called a *vector space*. Signal space is a vector space, and thus it has all of the general properties that hold in every vector space. Signal space also has the strong property of orthogonality that is defined only in special vector spaces.

Two elements $s(t)$ and $r(t)$ of the signal space are called *orthogonal* if

$$\int_{-\infty}^{\infty} s(t)r^*(t) \, dt = 0.$$

The *inner product* of $r(t)$ and $s(t)$ is

$$\langle r(t), s(t) \rangle = \int_{-\infty}^{\infty} s(t)r^*(t) \, dt.$$

Thus $r(t)$ and $s(t)$ are orthogonal if their inner product is zero. The *norm* of $s(t)$ is defined as $\|s(t)\| = \langle s(t), s(t) \rangle^{\frac{1}{2}}$.

Theorem 2.6.1 *Given any two signals, $r(t)$ and $s(t)$, the signal $r(t)$ has a unique decomposition of the form*

$$r(t) = \alpha s(t) + s^{\perp}(t)$$

where α is a constant and $s^{\perp}(t)$ is orthogonal to $s(t)$.

Proof: The theorem is trivial if $s(t) = 0$. For any other $s(t)$, the specifications

$$\alpha = \frac{\langle r(t), s(t) \rangle}{\langle s(t), s(t) \rangle}$$

and

$$s^{\perp}(t) = r(t) - \alpha s(t)$$

give a representation in the stated form. The decomposition is unique because, if $s(t)$ is written in this way for any α and $s^{\perp}(t)$ orthogonal to $s(t)$, then these definitions can only recover that same α and $s^{\perp}(t)$. □

The following theorem gives an important inequality in signal space.

[3] One must show that the concepts of sum and scalar multiple are uniquely defined in signal space.

Theorem 2.6.2 (Schwarz Inequality) *Let $r(t)$ and $s(t)$ be finite energy pulses, real-valued or complex-valued. Then*

$$\left|\int_{-\infty}^{\infty} r(t)s^*(t)\,\mathrm{d}t\right|^2 \leq \int_{-\infty}^{\infty} |r(t)|^2\,\mathrm{d}t \int_{-\infty}^{\infty} |s(t)|^2\,\mathrm{d}t$$

with equality if and only if $r(t)$ is a constant (real or complex) multiple of $s(t)$.

Proof: If $s(t) = 0$, the statement is immediate. Otherwise, $\langle s(t), s(t)\rangle \neq 0$. By Theorem 2.6.1, let $r(t)$ be decomposed as

$$r(t) = \alpha s(t) + s^{\perp}(t)$$

where $s^{\perp}(t)$ is orthogonal to $s(t)$, and

$$\alpha = \frac{\langle r(t), s(t)\rangle}{\langle s(t), s(t)\rangle}.$$

Then

$$\begin{aligned}\langle r(t), r(t)\rangle &= \langle \alpha s(t) + s^{\perp}(t), \alpha s(t) + s^{\perp}(t)\rangle \\ &= \alpha^2\langle s(t), s(t)\rangle + \langle s^{\perp}(t), s^{\perp}(t)\rangle \\ &\geq \alpha^2\langle s(t), s(t)\rangle \\ &= \frac{\langle r(t), s(t)\rangle^2}{\langle s(t), s(t)\rangle}.\end{aligned}$$

Hence

$$\langle r(t), r(t)\rangle\langle s(t), s(t)\rangle \geq \langle r(t), s(t)\rangle^2,$$

which is the statement of the theorem. □

The correlation between $r(t)$ and $s(t)$ has been defined as $\int_{-\infty}^{\infty} r(t)s^*(t)\,\mathrm{d}t$. The *correlation coefficient* γ between the two signals $r(t)$ and $s(t)$ is defined as

$$\gamma = \frac{\int_{-\infty}^{\infty} r(t)s^*(t)\,\mathrm{d}t}{\left[\int_{-\infty}^{\infty} |r(t)|^2\,\mathrm{d}t \int_{-\infty}^{\infty} |s(t)|^2\,\mathrm{d}t\right]^{1/2}}.$$

An immediate consequence of the Schwarz inequality is that the magnitude $|\gamma|$ of the correlation coefficient is not larger than one.

Theorem 2.6.3 (Triangle Inequality) *Given any two complex signals, $s(t)$ and $r(t)$,*

$$\|s(t) + r(t)\| \leq \|s(t)\| + \|r(t)\|.$$

Proof: Using the Schwarz inequality and $\mathrm{Re}[z] \leq |z|$, we can write

$$\begin{aligned}\|s(t)+r(t)\|^2 &= \langle s(t)+r(t), s(t)+r(t)\rangle \\ &= \langle s(t), s(t)\rangle + \langle r(t), r(t)\rangle + 2\mathrm{Re}\left[\langle s(t), r(t)\rangle\right] \\ &\leq \|s(t)\|^2 + \|r(t)\|^2 + 2\|s(t)\|\,\|r(t)\| \\ &= (\|s(t)\| + \|r(t)\|)^2\end{aligned}$$

which is the statement of the theorem. □

The statement of the theorem can be expressed in terms of the pulse energy as

$$\sqrt{E_{s+r}} \leq \sqrt{E_s} + \sqrt{E_r}.$$

A pulse, $s(t)$, of finite energy must be largely concentrated in some region of the t axis, so every pulse in signal space has the imprecise notion of width. Often the term width is used only in a qualitative way without a precise definition. To make the term precise, one can define the width of the pulse $s(t)$ in various ways. One is the Gabor timewidth, which we now define. Others are discussed in Section 2.8.

Definition 2.6.4 *The Gabor timewidth of the pulse $s(t)$ is given by*

$$T_G = \sqrt{\overline{t^2} - \overline{t}^2}$$

where

$$\overline{t^2} = \int_{-\infty}^{\infty} t^2 \frac{|s(t)|^2}{E_p}\,\mathrm{d}t \qquad \overline{t} = \int_{-\infty}^{\infty} t \frac{|s(t)|^2}{E_p}\,\mathrm{d}t.$$

The Gabor timewidth is also called the *root-mean-squared width* of the pulse $s(t)$. The equation for the Gabor timewidth has the same mathematical form as the equation for the standard deviation of the probability density function $p(t)$ given by $p(t) = |s(t)|^2/E_p$.

Similarly, the spectrum $S(f)$ of the pulse $s(t)$ also has the imprecise notion of width, which we can make precise in a variety of ways.

Definition 2.6.5 *The Gabor bandwidth of the pulse $s(t)$ is defined as*

$$B_G = \sqrt{\overline{f^2} - \overline{f}^2}$$

where

$$\overline{f^2} = \int_{-\infty}^{\infty} f^2 \frac{|S(f)|^2}{E_p}\,\mathrm{d}f \qquad \overline{f} = \int_{-\infty}^{\infty} f \frac{|S(f)|^2}{E_p}\,\mathrm{d}f.$$

Definition 2.6.6 *The Gabor skew parameter of the pulse $s(t)$ is defined as*

$$\rho = \frac{1}{T_G B_G}\mathrm{Re}[\overline{tf} - \overline{t}\,\overline{f}]$$

where

$$\overline{tf} = \frac{\mathrm{j}}{2\pi}\int_{-\infty}^{\infty} t\frac{s(t)\dot{s}^*(t)}{E_p}\,\mathrm{d}t = \frac{\mathrm{j}}{2\pi}\int_{-\infty}^{\infty} f\frac{S(f)S'^*(f)}{E_p}\,\mathrm{d}f,$$

and where $\dot{s}(t) = \mathrm{d}s(t)/\mathrm{d}t$ *and* $S'(f) = \mathrm{d}S(f)/\mathrm{d}f$.

The equality of the two forms in the definition follows from Parseval's formula. If the pulse $s(t)$ is purely real (or has any constant phase), then the skew parameter ρ is zero. This is because, if $s(t)$ is real, $\overline{tf}$ is purely imaginary and $S(f) = S^*(-f)$, so $\overline{f}$ is zero. By duality, if $S(f)$ is purely real (or has any constant phase), then ρ is again zero.

The three parameters T_G, B_G, and ρ are the *Gabor parameters* of the pulse $s(t)$.

Proposition 2.6.7 *For any* t_0 *and* f_0, *the pulses* $s(t)$ *and* $s(t - t_0)\mathrm{e}^{-\mathrm{j}2\pi f_0 t}$ *have the same Gabor parameters.*

Proof: We will prove the theorem only for the Gabor timewidth. The proofs for the other parameters are similar. We start with T_G^2 for the pulse $s(t)$ and manipulate it as follows:

$$\begin{aligned}
T_G^2 &= \overline{t^2} - \overline{t}^2 = \overline{(t+t_0)^2} - (\overline{t}+t_0)^2 \\
&= \int_{-\infty}^{\infty}(t+t_0)^2\frac{|s(t)|^2}{E_p}\,\mathrm{d}t - \left[\int_{-\infty}^{\infty}(t+t_0)\frac{|s(t)|^2}{E_p}\,\mathrm{d}t\right]^2 \\
&= \int_{-\infty}^{\infty} t^2\frac{|s(t-t_0)\mathrm{e}^{-\mathrm{j}2\pi f_0 t}|^2}{E_p}\,\mathrm{d}t - \left[t\frac{|s(t-t_0)\mathrm{e}^{-\mathrm{j}2\pi f_0 t}|^2}{E_p}\,\mathrm{d}t\right]^2,
\end{aligned}$$

which is the Gabor timewidth for the second pulse. □

Theorem 2.6.8 (Uncertainty Principle for Pulses)[4] *Let* $s(t)$ *be a differentiable pulse of finite energy. The Gabor timewidth* T_G *and Gabor bandwidth* B_G *of* $s(t)$ *satisfy*

$$T_G B_G \geq \frac{1}{4\pi}$$

with equality if and only if the pulse is of the form

$$s(t) = A\mathrm{e}^{-a(t-t_0)^2}\mathrm{e}^{-\mathrm{j}2\pi f_0 t}.$$

Proof: The theorem is trivial if either T_G or B_G is infinite, so we may assume that both are finite. Without loss of generality, we may assume that the time center of the pulse and the frequency center of the pulse have been chosen so that $\overline{t}$ and $\overline{f}$ are zero.

[4] Despite the name, there is nothing uncertain about the uncertainty principle for pulses.

We begin with the equation

$$\frac{d}{dt}(t|s(t)|^2) = |s(t)|^2 + ts(t)\frac{ds^*(t)}{dt} + t\frac{ds(t)}{dt}s^*(t).$$

Take the integral of both sides

$$\int_{-\infty}^{\infty} \frac{d}{dt}(t|s(t)|^2)\,dt = \int_{-\infty}^{\infty} |s(t)|^2\,dt + \int_{-\infty}^{\infty} ts(t)\frac{ds^*(t)}{dt}\,dt + \int_{-\infty}^{\infty} ts^*(t)\frac{ds(t)}{dt}\,dt.$$

Because the pulse has finite energy, $|s(t)|^2$ must go to zero faster than $1/t$. Thus the term on the left is evaluated as

$$\int_{-\infty}^{\infty} d(t|s(t)|^2)\,dt = t|s(t)|^2\Big|_{-\infty}^{\infty} = 0.$$

Therefore

$$\begin{aligned} E_p &= 2\text{Re}\left[-\int_{-\infty}^{\infty} ts(t)\frac{ds^*(t)}{dt}\,dt\right] \\ &\leq 2\left|\int_{-\infty}^{\infty} ts(t)\frac{ds^*(t)}{dt}\,dt\right| \end{aligned}$$

where the inequality holds because, for any complex z, $\text{Re}[z] \leq |z|$. Now use the Schwarz inequality to bound this as

$$\begin{aligned} E_p^2 &\leq 4\left|\int_{-\infty}^{\infty} ts(t)\frac{ds^*(t)}{dt}\,dt\right|^2 \\ &\leq 4\int_{-\infty}^{\infty} |ts(t)|^2\,dt \int_{-\infty}^{\infty} \left|\frac{ds(t)}{dt}\right|^2\,dt. \end{aligned}$$

The first integral is equal to $E_pT_G^2$ and, by Parseval's formula, the second integral is equal to $(4\pi)^2E_pB_G^2$. Therefore

$$\left(\frac{1}{4\pi}\right)^2 \leq T_G^2B_G^2,$$

which proves the inequality condition of the theorem. Equality occurs in the Schwarz inequality if and only if the derivative of $s(t)$ is equal to $cts(t)$. This is a differential equation solved by the gaussian pulse, which leads to the equality condition of the theorem provided $t_0 = f_0 = 0$. But the pulse can be translated in time and then in frequency without changing T_G and B_G, so the general equality condition follows.

□

2.7 The matched filter

Let $s(t) = s_R(t) + js_I(t)$ be a known complex signal of finite energy E_p. Suppose that a received signal, $v(t)$, consists of the signal $s(t)$ in additive complex noise,

$$v(t) = s(t) + n(t).$$

We require that the complex noise

$$n(t) = n_R(t) + jn_I(t),$$

has real and imaginary noise components $n_R(t)$ and $n_I(t)$ that are independent, identically distributed, random processes; each has the same correlation function, $\phi(\tau)$, and power density spectrum, $N(f)$. Often the noise $n(t)$ is gaussian noise, but we need not make this assumption now; only the second-order properties of $n(t)$ will be used in this section.

The noise in the received signal $v(t)$ can be partially removed by passing $v(t)$ through a linear filter, $g(t)$, but this will also affect the signal $s(t)$. The output of the linear filter is

$$\begin{aligned} u(t) &= g(t) * v(t) \\ &= g(t) * [s(t) + n(t)]. \end{aligned}$$

We shall find a compromise between maximizing the signal in $u(t)$ and rejecting the noise in $u(t)$ by restricting attention to a single sampling instant, t_0, and finding the filter $g(t)$ that maximizes the signal-to-noise ratio at this time instant. We do not restrict $g(t)$ to be a causal filter, so we are free to choose $t_0 = 0$ as the sampling instant. Any other choice of sampling instant corresponds to a simple translation of $g(t)$.

Because the real and imaginary parts of the noise into the filter are uncorrelated and with the same correlation function, the real and imaginary components of the noise out of the filter will also be uncorrelated and with the same correlation function (see Problem 2.24). The variance of the noise per component at the filter output is called the *noise power* and is denoted N,

$$\begin{aligned} N &= \tfrac{1}{2}\mathrm{var}[u(t)] \\ &= \tfrac{1}{2}\mathrm{E}[|u(t) - \mathrm{E}[u(t)]|]^2. \end{aligned}$$

The noise is stationary, so N does not depend on time. Then

$$\begin{aligned} N &= \tfrac{1}{2}\mathrm{E}\left[\int_{-\infty}^{\infty} g(\xi_1)n(-\xi_1)\,\mathrm{d}\xi_1 \int_{-\infty}^{\infty} g^*(\xi_2)n^*(-\xi_2)\,\mathrm{d}\xi_2\right] \\ &= \int_{-\infty}^{\infty}\int_{-\infty}^{\infty} g(\xi_1)g^*(\xi_2)\phi(\xi_1 - \xi_2)\,\mathrm{d}\xi_1\,\mathrm{d}\xi_2 \end{aligned}$$

because $E[n(-\xi_1)n^*(-\xi_2)] = 2\phi(\xi_1 - \xi_2)$. Let $\tau = \xi_1 - \xi_2$ and $\xi = \xi_1$. Then

$$N = \int_{-\infty}^{\infty} \phi(\tau) \int_{-\infty}^{\infty} g(\xi)g^*(\xi - \tau)\, d\xi\, d\tau.$$

The second integral has the Fourier transform $|G(f)|^2$, so by Parseval's formula,

$$N = \int_{-\infty}^{\infty} N(f)|G(f)|^2\, df.$$

If the noise is white noise, then $N(f) = N_0/2$ and $\phi(\tau) = (N_0/2)\delta(\tau)$. Then

$$N = \frac{N_0}{2} \int_{-\infty}^{\infty} g(\xi)g^*(\xi)\, d\xi.$$

The signal power at the sampling instant is the square of the magnitude of the expected filter output

$$\begin{aligned} S &= |E[u(0)]|^2, \\ &= \left| \int_{-\infty}^{\infty} g(\xi)s(-\xi)\, d\xi \right|^2. \end{aligned}$$

The ratio S/N of signal power to noise power is called the *signal-to-noise ratio*. The filter $g(t)$ is to be chosen to maximize the signal-to-noise ratio at the filter output. Theorem 2.6.1 applied to $g(t)$ and $s^*(-t)$ says that any $g(t)$ can be decomposed as

$$g(t) = \alpha s^*(-t) + s_\perp^*(-t)$$

where

$$\alpha = \frac{\langle g(t), s(t)\rangle}{\langle s(t), s(t)\rangle}.$$

Then the signal power at the output of the filter is

$$\begin{aligned} S &= \left[\alpha \int_{-\infty}^{\infty} |s(t)|^2\, dt + \int_{-\infty}^{\infty} s(t)s_\perp^*(t)\, dt \right]^2 \\ &= \alpha^2 \left[\int_{-\infty}^{\infty} |s(t)|^2\, dt \right]^2 \end{aligned}$$

because the second term is zero. Similarly, in white noise, the noise power is

$$\begin{aligned} N &= \frac{N_0}{2} \int_{-\infty}^{\infty} g(\xi)g^*(\xi)\, d\xi \\ &= \frac{N_0}{2} \left[\alpha^2 \int_{-\infty}^{\infty} |s(t)|^2\, dt + \int_{-\infty}^{\infty} |s_\perp(t)|^2\, dt \right]. \end{aligned}$$

We now have expressions for both signal power S and noise power N at the sampling instant. We are ready to choose the filter to maximize the ratio S/N.

Theorem 2.7.1 *In white noise of power density spectrum N_0 watts per hertz, the maximum signal-to-noise power ratio is achieved by the filter*

$$g(t) = s^*(-t).$$

Proof: The output signal-to-noise ratio is

$$\frac{S}{N} = \frac{\alpha^2 \left[\int_{-\infty}^{\infty} |s(t)|^2 \, dt\right]^2}{\left[\alpha^2 \int_{-\infty}^{\infty} |s(t)|^2 \, dt + \int_{-\infty}^{\infty} |s_\perp(t)|^2 \, dt\right] N_0/2}.$$

To maximize this, the second term in the denominator should be made equal to zero by choice of $g(t)$. This implies that $g(t) = s^*(t)$. □

The filter $g(t) = s^*(-t)$ is called the *matched filter*. In white noise, the signal-to-noise ratio at the output of the matched filter is

$$\begin{aligned}\frac{S}{N} &= \int_{-\infty}^{\infty} \frac{|S(f)|^2}{N_0/2} \, df \\ &= 2E_p/N_0\end{aligned}$$

where E_p is the total energy in the signal pulse. This is a general conclusion and does not depend on the shape of the pulse $s(t)$. The signal pulse contributes to the output signal-to-noise ratio in white noise only through its energy E_p. Because the shape of the signal pulse plays no role in calculating the output signal-to-noise ratio, any pulse shape that is convenient for any reason may be used.

The matched-filter response $g(t) = s^*(-t)$ for an example of a real pulse, $s(t)$, is shown in Figure 2.8. This is an example of a matched filter that is not causal. It can be made causal by delaying the sampling instant to time t_0, as in Figure 2.9.

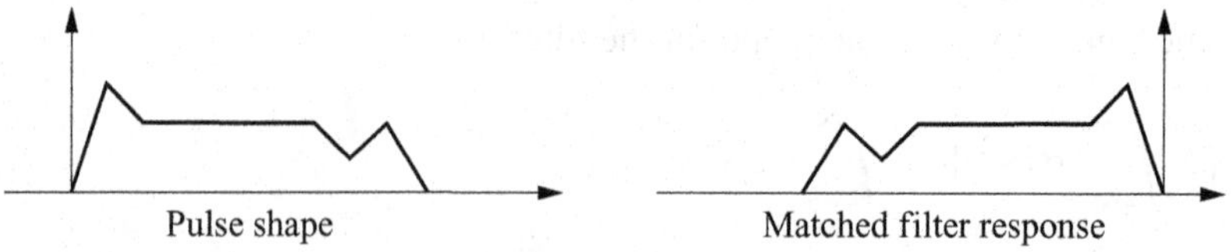

Figure 2.8 Response of a matched filter

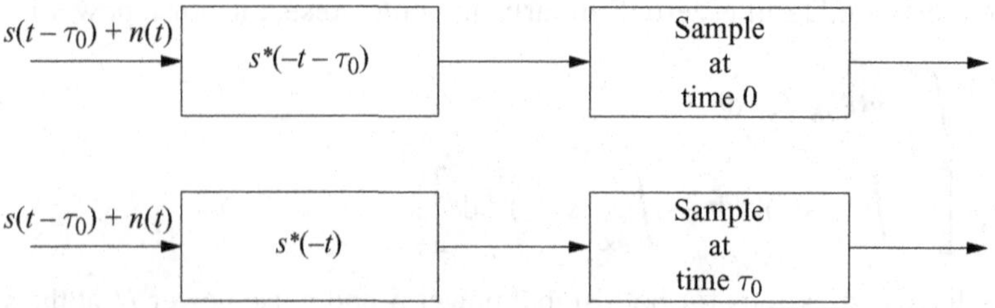

Figure 2.9 Equivalent ways to use a matched filter

If we wish to control the output signal or the output noise individually, then we can write the matched filter as $Cs^*(-t)$ where C is any complex constant. Then, at the sampling instant, the output signal is

$$S = \left| \int_{-\infty}^{\infty} C|s(t)|^2 \, dt \right|^2 = |C|^2 E_p^2,$$

and the output noise power is

$$\begin{aligned} N &= \int_{-\infty}^{\infty} N(f)|G(f)|^2 \, df \\ &= \frac{N_0}{2}|C|^2 E_p \end{aligned}$$

which does not change the ratio S/N. One choice for the normalizing constant C is $1/\sqrt{E_p}$. Then $S = E_p$ and $N = N_0/2$. However, the factor of $1/\sqrt{E_p}$ is inconvenient to carry when only the signal-to-noise ratio is of interest.

To find the maximizing filter for an arbitrary noise power density spectrum $N(f)$ we will decompose the filter $g(t)$ into two filters, $g(t) = g_1(t) * g_2(t)$ where $g_1(t)$ is chosen to whiten the noise and $g_2(t)$ is the subsequent matched filter. This decomposition leads to the following theorem:

Theorem 2.7.2 *The maximum signal-to-noise power ratio at the output of filter* $g(t)$ *is*

$$\frac{S}{N} = \int_{-\infty}^{\infty} \frac{|S(f)|^2}{N(f)} \, df,$$

which is achieved by the filter

$$G(f) = \frac{S^*(f)}{N(f)}.$$

Proof: The noise is whitened by passing it through the filter $g_1(t)$ such that $G_1(t) = 1/\sqrt{N(f)}$. The signal out of this filter $g_1(t) * s(t)$ has Fourier transform $S(f)/\sqrt{N(f)}$. The matched filter for $g_1(t) * s(t)$ is given by

$$G_2(f) = \frac{S^*(f)}{\sqrt{N(f)}}.$$

The cascade of $G_1(f)$ and $G_2(f)$ is then

$$G(f) = \frac{S^*(f)}{N(f)}$$

as was to be proved. □

The filter of the theorem is known as a *whitened matched filter*. It maximizes signal-to-noise ratio for any covariance-stationary noise of power density spectrum $N(f)$. No assumption has been made that the noise is gaussian. For white noise, $N(f) = N_0/2$, so this can be reduced to

$$G(f) = S^*(f)$$

as we have already discussed.

The derivation of the matched filter holds for any pulse with finite energy. The passband pulse,

$$\widetilde{s}(t) = s_R(t) \cos 2\pi f_0 t + s_I(t) \sin 2\pi f_0 t,$$

has finite energy. Therefore,

$$\begin{aligned}\widetilde{g}(t) &= \widetilde{s}(-t)\\ &= s_R(-t) \cos 2\pi f_0 t - s_I(-t) \sin 2\pi f_0 t\end{aligned}$$

is the matched filter for $\widetilde{s}(t)$. The corresponding complex baseband pulse has a matched filter that is the complex representation of the matched filter for the passband pulse. Thus, the matched filter for the complex baseband pulse is

$$\begin{aligned}g(t) &= s_R(-t) - \mathrm{j}s_I(-t)\\ &= s^*(-t).\end{aligned}$$

Theorem 2.7.3 *Two filters matched to orthogonal pulses have uncorrelated noise outputs at time zero if the input is white noise, and independent noise outputs if the noise is also gaussian.*

Proof: The correlation is

$$\begin{aligned}\mathrm{E}[n_1 n_2^*] &= \mathrm{E}\left[\frac{1}{E_p}\int_{-\infty}^{\infty} n(\xi)s_1(-\xi)\,\mathrm{d}\xi \int_{-\infty}^{\infty} n^*(\xi')s_2^*(-\xi')\,\mathrm{d}\xi'\right]\\ &= \frac{1}{E_p}\int_{-\infty}^{\infty}\int_{-\infty}^{\infty} s_1(-\xi)s_2^*(-\xi')\frac{N_0}{2}\delta(\xi - \xi')\,\mathrm{d}\xi\,\mathrm{d}\xi'\end{aligned}$$

because the noise is white. Carrying out one integration gives

$$\begin{aligned}\mathrm{E}[n_1 n_2^*] &= \frac{N_0}{2}\frac{1}{E_p}\int_{-\infty}^{\infty} s_1(-\xi)s_2^*(-\xi)\,\mathrm{d}\xi\\ &= 0\end{aligned}$$

and hence the noise outputs are uncorrelated at $t = 0$. If the noise inputs are gaussian, then the outputs are also gaussian, and therefore independent because uncorrelated gaussian random variables are independent. □

2.8 Resolution and apodization

The study of resolution was introduced by Lord Rayleigh to quantify the way in which an effect such as a filter impulse response (or, in two dimensions, a pointspread function) prevents us from seeing things as they are. A resolution criterion describes the filtered pulse as it is; it does not include any attempt to sharpen the pulse by processing or deconvolution. No single definition of resolution is accepted as the best definition. In different circumstances, different criteria for measuring resolution will be more appropriate. In this section, we discuss resolution in one dimension. We will discuss resolution in two dimensions in Section 3.9.

Let $s(t)$ be a pulse with a maximum at the origin. A pair of identical copies of $s(t)$ separated by ν is the function

$$p(t) = s(t - \nu/2) + s(t + \nu/2).$$

We want to define a measure of the width of $s(t)$ so that, whenever this pair of pulses is separated by a value of ν larger than this width, the pair can be readily distinguished from the single pulse $s(t)$. It is clear that the statement of the problem is too vague to determine a unique definition of resolution. In fact, because we know $s(t)$, we can always distinguish between $s(t)$ and $s(t - \nu/2) + s(t + \nu/2)$ for any nonzero ν. However, this distinction requires close attention to $p(t)$, whereas a discussion of resolution is limited to a more cursory inspection of the pulse. This is because the actual applications will have multiple copies of $s(t)$ attenuated and translated by unspecified amounts and observed in additive noise. Advanced techniques for separating pulses in such an environment are called *superresolution techniques*.

Many criteria for resolution are in use; they amount to different definitions of the width of $s(t)$. One criterion for resolution is the *Rayleigh resolution criterion*, which has long been used in optics. This width of pulse $s(t)$ is defined as the difference between the value of t where the maximum of $s(t)$ occurs and the value of t where the zero of $s(t)$ closest to the maximum occurs. This definition has served well in the field of optics, but it fails if $s(t)$ has no zero, as for a gaussian pulse.

Another way to measure resolution is the *Sparrow resolution criterion*. The Sparrow width is the smallest value of ν for which $p(t)$ has a second derivative at the origin equal to zero. This definition fails for triangular pulses or square pulses. Figure 2.10 shows a comparison of the Rayleigh and Sparrow resolution criteria for the pulses $s(t) = \operatorname{sinc} t$ and $s(t) = \operatorname{sinc}^2 t$.

A third criterion for resolution is the *half-power criterion*. For a pulse whose maximum is at $t = 0$, the half-power width is the smallest positive t for which $s^2(t) = \frac{1}{2}s^2(0)$. This is a well-defined criterion, though the choice of one-half (rather than, say, one-fourth) may be viewed as somewhat arbitrary.

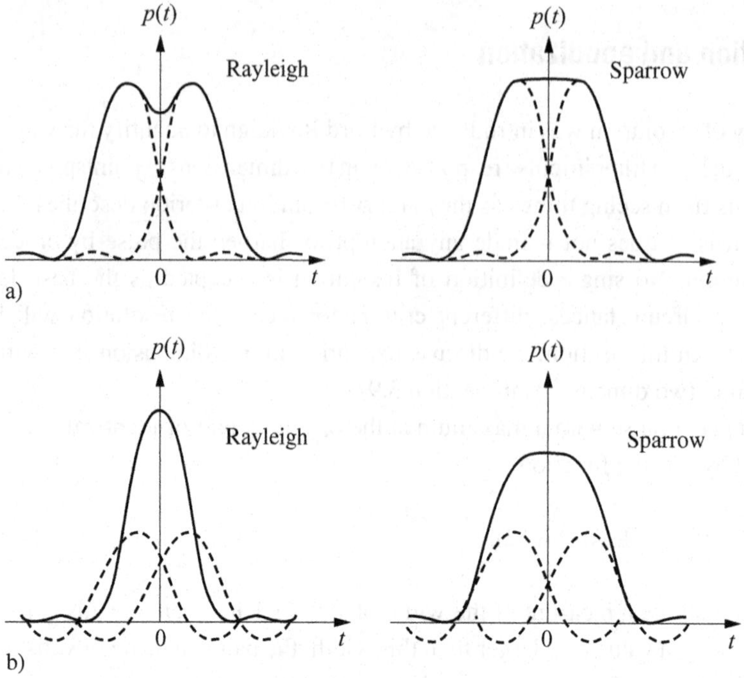

Figure 2.10 A comparison of resolution measures: a) $s(t) = \text{sinc}\, t$; b) $s(t) = \text{sinc}^2 t$

A fourth criterion for resolution is the *Woodward resolution criterion*, defined as

$$\Delta t = \frac{\int_{-\infty}^{\infty} |s(t)|^2 \, dt}{\max_t |s(t)|^2} = \frac{E_p}{|s(0)|^2}$$

for a pulse whose maximum is at the origin. The Woodward timewidth is the width of a square pulse with the same energy and peak value as $s(t)$. Finally, the Gabor timewidth, which was defined in Section 2.6, can be used as a resolution criterion.

We may also measure resolution on the frequency axis. The width of the spectrum $S(f)$ can be measured by the Rayleigh resolution criterion, the Sparrow resolution criterion, or the half-power resolution criterion, which are defined the same way on the frequency axis as on the time axis. Another measure of bandwidth, which is sometimes useful for discussing noise, is the *noise bandwidth* defined as

$$B_N = \frac{\int_{-\infty}^{\infty} |S(f)|^2 \, df}{\max_f |S(f)|^2}.$$

The noise bandwidth is the dual of the Woodward timewidth, and might also be called the *Woodward bandwidth*.

A second issue in distinguishing pulses is the structure of the sidelobes of the pulse $s(t)$ because the sidelobes of one copy of $s(t)$ could mask the main lobe of a faint translated copy of $s(t)$. Given the pulse $s(t)$, the optimum filter to discriminate against noise is the matched filter, but the matched filter may have undesirable sidelobes in the response to the signal. Instead of using the matched filter, one sometimes chooses a filter that suppresses the sidelobes of the filter output even though the noise reduction may be less. Any processing device designed to reduce sidelobes is called *apodization*. It is usually possible to satisfy the requirement for apodization with a filter that differs only slightly from the matched filter, so with only a slight reduction in the output signal-to-noise ratio, and perhaps a slight reduction in resolution, the sidelobes are suppressed.

Problems

2.1 Prove that if $s(t)$ has the Fourier transform $S(f)$, then $s(at)$ has the Fourier transform $|a|^{-1}S(f/a)$.

2.2 Prove that if $s(t)$ is a real-valued pulse, then its Fourier transform $S(f)$ satisfies $S^*(f) = S(-f)$.

2.3 Let s_i for $i = 0, \ldots, n-1$ be a vector of complex numbers. The *discrete Fourier transform* of the vector $\boldsymbol{s}$ is the vector $\boldsymbol{S}$ given by

$$S_k = \sum_{i=0}^{n-1} \mathrm{e}^{-\mathrm{j}2\pi ik/n} s_i.$$

a. Prove that the inverse discrete Fourier transform is

$$s_i = \frac{1}{n}\sum_{k=0}^{n-1} \mathrm{e}^{\mathrm{j}2\pi ik/n} S_k.$$

b. The *cyclic convolution* of two vectors $\boldsymbol{p}$ and $\boldsymbol{q}$, each of blocklength n, is defined as

$$r_i = \sum_{k=0}^{n-1} p_{((i-k))} q_k$$

and denoted by $\boldsymbol{r} = \boldsymbol{p} * \boldsymbol{q}$ where $((i-k))$ denotes $i-k$ modulo n. Prove the convolution theorem for the discrete Fourier transform, that $\boldsymbol{r} = \boldsymbol{p} * \boldsymbol{q}$ if and only if their transforms satisfy $R_k = P_k Q_k$ for $k = 0, \ldots, n-1$.

2.4 Prove that the *autocorrelation function* of a pulse, $s(t)$, defined as

$$\phi(\tau) = \int_{-\infty}^{\infty} s(t)s^*(t-\tau)\,\mathrm{d}t,$$

has the Fourier transform $|S(f)|^2$.

2.5 Let

$$S(f) = 2\,\mathrm{sinc}\,f \cos 4\pi f.$$

Prove that

$$S(f) * S(f) = S(f).$$

2.6 Prove that

$$\text{rect}\left(\frac{t}{T - 2|\tau|}\right) = \text{rect}\left(\frac{\tau}{T - 2|t|}\right).$$

2.7 Pulses exist that are equal to their own Fourier transforms. Verify the following two Fourier transform pairs:

$$e^{-\pi t^2} \leftrightarrow e^{-\pi f^2}$$
$$\frac{1}{\sqrt{|t|}} \leftrightarrow \frac{1}{\sqrt{|f|}}.$$

Consequently, the integral equation

$$s(y) = \int_{-\infty}^{\infty} s(x) e^{-j2\pi xy} \, dx$$

has at least two distinct solutions for $s(x)$.

2.8 **a.** Prove that the delta function satisfies the elementary property

$$\delta\left(\frac{t}{T}\right) = T\delta(t)$$

for $T > 0$.

b. Prove that the comb function satisfies the property

$$\sum_{\ell=-\infty}^{\infty} \delta\left(\frac{t}{T} - \ell\right) = T \sum_{\ell=-\infty}^{\infty} \delta(t - \ell T).$$

c. Prove the Fourier transform pair

$$\sum_{\ell=-\infty}^{\infty} \delta(t - \ell T) \leftrightarrow \frac{1}{T} \sum_{\ell=-\infty}^{\infty} \delta\left(f - \frac{\ell}{T}\right)$$

for $T > 0$.

2.9 Prove the *Poisson sum formula*

$$\sum_{\ell=-\infty}^{\infty} s(t + \ell) = \sum_{\ell=-\infty}^{\infty} e^{j2\pi \ell t} S(\ell).$$

2.10 The Fourier transform of a train of 2^k uniformly spaced, identical pulses can be obtained in two ways: by writing down the dirichlet function, or by a process of successive constructions expressing the train of 2^j pulses as a doublet of trains of 2^{j-1} pulses. Show that both of these approaches give the same result.

2.11 A pulse of width 8τ can be thought of as a train of eight pulses, each of width τ and with pulse spacing τ. Starting with the transform of a pulse of width τ, and using the transform of the array, derive the transform of the wide pulse. How does this compare with the transform of the wide pulse computed directly? Now discuss what happens if the spacing between the eight pulses is slightly larger than τ. (The wide pulse is a little wider, but with holes.)

2.12 Prove that to minimize the Gabor bandwidth of a pulse $s(t) = |s(t)|e^{j\theta(t)}$ over a choice of phase, the phase $\theta(t)$ must be chosen to be of the form $\theta(t) = \theta_0 + \dot{\theta}_0 t$.

2.13 Define the *effective timewidth* of a pulse, $s(t)$, as

$$T_E = \frac{\left|\int_{-\infty}^{\infty} s(t)\,dt\right|}{|s(0)|}.$$

This is the width of a rectangular pulse, $|s(0)|\,\mathrm{rect}(t/T_E)$, that has the same area as $s(t)$. Define the *effective bandwidth* as

$$B_E = \frac{\left|\int_{-\infty}^{\infty} S(f)\,df\right|}{|S(0)|}.$$

Prove that

$$T_E B_E = 1.$$

2.14 A passband signal,

$$s(t) = s_R(t)\cos 2\pi f_0 t + s_I(t)\sin 2\pi f_0 t,$$

can be filtered by convolving it with the passband filter

$$h(t) = h_R(t)\cos 2\pi f_0 t + h_I(t)\sin 2\pi f_0 t.$$

Alternatively, the passband signal can be converted to the complex signal $s_R(t) + js_I(t)$, then convolved with the complex filter $h_R(t) + jh_I(t)$, and then reconverted back to a passband signal. By explicitly writing out the convolutions, show that the result is the same in the two alternatives. Is the exercise a convincing demonstration of the utility of the complex representation?

2.15 Suppose that a signal,

$$s(t) = a(t)\cos(2\pi f_c t + \theta(t)),$$

has an amplitude modulation, $a(t)$, phase modulation, $\theta(t)$, and carrier frequency, f_c. Find the in-phase and quadrature components that are obtained with respect to the reference frequency f_0.

2.16 Using time-domain arguments in place of frequency-domain arguments, show that the complex representation of a passband signal is unique, that is, two different passband signals have different complex representations (and vice versa).

2.17 The passband filter with impulse response

$$h(t) = te^{-at}\sin 2\pi f_0 t \quad t \geq 0,$$

where f_0 is very large compared to a, is excited by a passband pulse,

$$s(t) = \text{rect}(t) \sin 2\pi f_0 t.$$

a. Find the output by a direct convolution.

b. Give a complex representation for the filter and the pulse, and find the complex representation for the filter output by complex convolution.

c. Which is easier?

2.18 Use the Schwarz inequality on the integral

$$\int_{-\infty}^{\infty} [(t - \bar{t})s(t)] \left[\frac{ds(t)}{dt} - 2\pi j \bar{f} s(t) \right] dt$$

to obtain a more direct proof of the uncertainty principle when the time center and the frequency center of $s(t)$ are arbitrary.

2.19 Prove that the Gabor skew parameter satisfies

$$-1 \leq \rho \leq 1,$$

first for the case where $\bar{t} = \bar{f} = 0$, then for the general case.

2.20 Suppose that $g(t)$ has a maximum frequency, B (in hertz). Let $g(t)$ be sampled at a rate of $4B$ samples per second. Define a time-domain interpolation formula based upon the frequency filter.

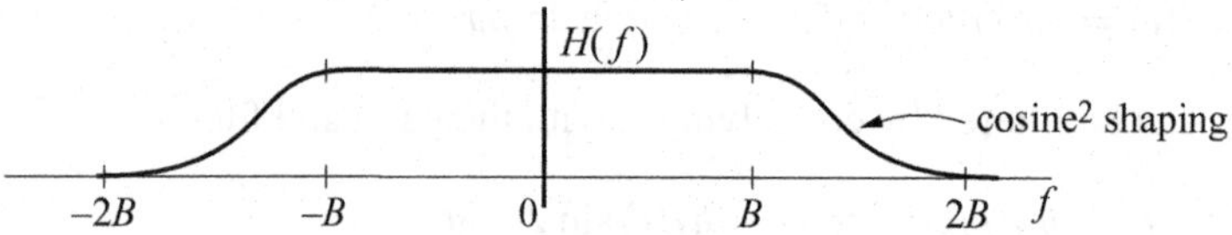

This interpolation filter gives theoretically perfect reconstruction as does sinc interpolation. Why might this interpolation rule be preferred to sinc interpolation?

2.21 A system is to be designed to process signals with spectra confined to the frequency regions shown in the following illustration.

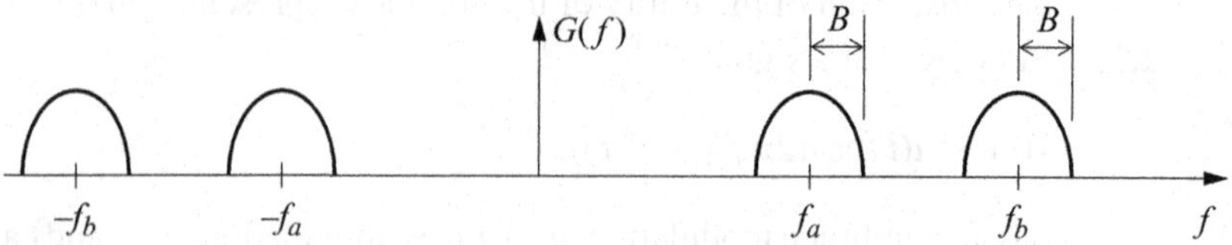

No other prior information is known about the signals.

a. What is the minimum number of samples per second that suffice to specify the signal?

b. Sketch at least three different schemes for sampling the signal.

2.22 **(Chirp Filter)** Show that the Fourier transform

$$S(f) = \int_{-\infty}^{\infty} s(t)e^{-j2\pi ft} dt$$

can be written as

$$S(f) = e^{-j\pi f^2}[e^{j\pi t^2} * (e^{-j\pi t^2} s(t))].$$

Discuss how this expression can be used to rapidly compute a Fourier transform with three "chirp filters" of the form $h(t) = e^{-j\pi t^2}$ and two multipliers. Sketch a functional block diagram of this circuit. Explain the consequence of using a chirp filter of finite duration, such as

$$h(t) = \begin{cases} e^{j\pi t^2} & |t| \leq T/2 \\ 0 & |t| > T/2. \end{cases}$$

With $s(t) = \text{rect}(t/T)$, sketch the waveform at various points of the circuit.

2.23 **(Chirp Filter)** Show that

$$e^{j\pi t^2} * [e^{j2\pi t^2}[e^{j\pi t^2} * s(t)]] = e^{j2\pi t^2} s(-t).$$

This expression is closely related to the lens law of optics.

2.24 **(Pulse Compression)** Show that the filter $g(t) = e^{j\pi\alpha t^2}$, whose impulse response is a quadratic phase pulse, can be implemented as

$$u(t) = e^{j\pi\alpha t^2} \int_{-\infty}^{\infty} e^{-j2\pi\alpha\xi t} \left[v(\xi) e^{j2\pi\alpha\xi^2} \right] d\xi.$$

If $s(t) = e^{j\pi\alpha(t-\tau)^2}\text{rect}(t/T)$, what is $r(t) = s(t)e^{-j\pi\alpha t^2}$? This operation is called *dechirping*. How does the bandwidth of $r(t)$ compare to the bandwidth of $s(t)$? Explain why one might place a sampling operation after a dechirping operation rather than before.

2.25 Suppose that the complex baseband noise process $n(t) = n_R(t) + jn_I(t)$ is passed through the complex baseband filter $g(t) = g_R(t) + jg_I(t)$. Show that if $n_R(t)$ and $n_I(t)$ have the same correlation function and are uncorrelated, then the noise components at the output of the filter have these same properties.

2.26 Derive the whitened matched filter directly as follows. With the aid of Parseval's formula,

$$\int_{-\infty}^{\infty} g(\xi) w^*(\xi)\, d\xi = \int_{-\infty}^{\infty} G(f) W^*(f)\, df,$$

show that the signal power out of the linear filter $g(t)$ at time zero is

$$S = \left| \int_{-\infty}^{\infty} G(f) S(f)\, df \right|^2.$$

Then, with the aid of the Schwarz inequality, show that the maximum signal-to-noise ratio

$$\frac{S}{N} = \int_{-\infty}^{\infty} \frac{|S(f)|^2}{N(f)}\, df$$

is achieved by the matched filter. Finally, specialize this expression to the case of white noise $N(f) = N_0/2$.

Notes

The uncertainty relationship for pulses was first given by Gabor (1946). The notation for the rectangular pulse and the sinc pulse was introduced by Woodward (1953). The matched filter was originally introduced in a laboratory report by North (1943); it is sometimes called the *North filter*.

A form of the sampling theorem was first stated by Whittaker (1915). It reappeared in the engineering literature in papers by Nyquist (1928) and Shannon (1948); and in the Russian literature by Kotel'nikov (1933).

3 Signals in two dimensions

The Fourier transform of a two-dimensional function – or of an n-dimensional function – can be defined by analogy with the Fourier transform of a one-dimensional function. A multidimensional Fourier transform is a mathematical concept. Because many engineering applications of the two-dimensional Fourier transform deal with two-dimensional images, it is common practice, and ours, to refer to the variables of a two-dimensional function as "spatial coordinates" and to the variables of its Fourier transform as "spatial frequencies."

The study of the two-dimensional Fourier transform closely follows the study of the one-dimensional Fourier transform. As the study develops, however, the two-dimensional Fourier transform displays a richness beyond that of the one-dimensional Fourier transform.

3.1 The two-dimensional Fourier transform

A function, $s(x, y)$, possibly complex, of two variables x and y is called a *two-dimensional signal* or a *two-dimensional function* (or, more correctly, a function of a two-dimensional variable). A common example is an image, such as a photographic image, wherein the variables x and y are the coordinates of the image, and $s(x, y)$ is the amplitude. In a photographic image, the amplitude is a nonnegative real number. In other examples, the function $s(x, y)$ may also take on negative or even complex values. Figure 3.1 shows a graphical representation of a two-dimensional complex signal in terms of the real and imaginary parts. Figure 3.2 shows the magnitude of the function depicted in two different ways: one as a three-dimensional graph, and one as a plan view.

The two-dimensional Fourier transform is completely analogous to the one-dimensional Fourier transform. Given the complex function $s(x, y)$ whose energy

$$E_p = \int_{-\infty}^{\infty} \int_{-\infty}^{\infty} |s(x, y)|^2 \, dx \, dy$$

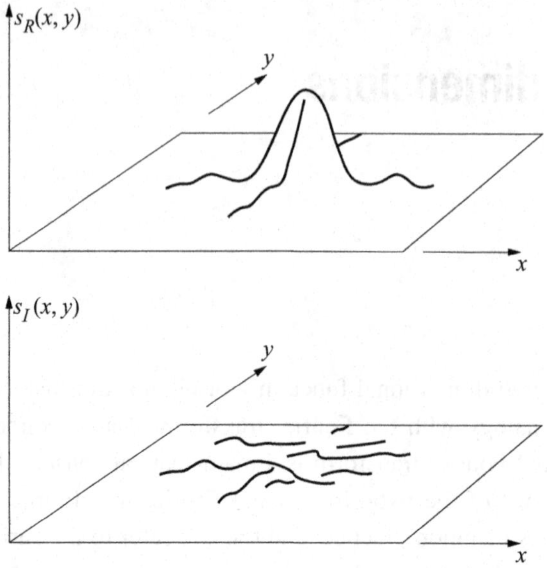

Figure 3.1 Depiction of a two-dimensional complex function

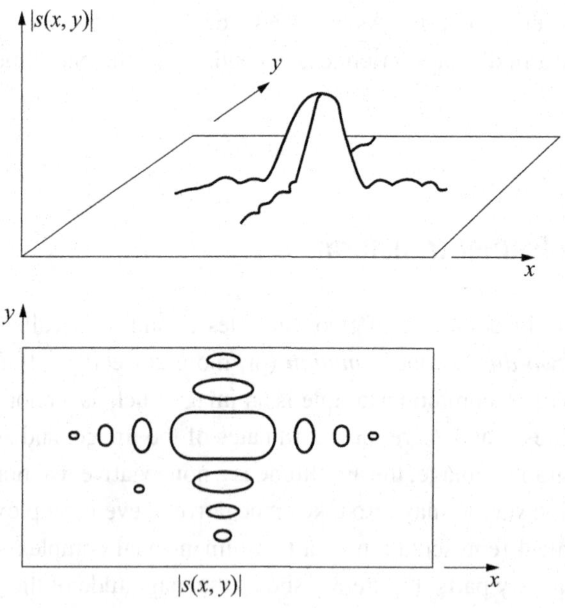

Figure 3.2 Depiction of magnitude of a two-dimensional complex function

is finite, the *two-dimensional Fourier transform* is defined as

$$S(f_x, f_y) = \int_{-\infty}^{\infty} \int_{-\infty}^{\infty} s(x, y) e^{-j2\pi(f_x x + f_y y)} \, dx \, dy.$$

We shall use the term "two-dimensional Fourier transform" to refer to both the function $S(f_x, f_y)$ and the formula that defines $S(f_x, f_y)$ in terms of $s(x, y)$. The transform relationship between the pair of functions $s(x, y)$ and $S(f_x, f_y)$ will be denoted by

$$s(x, y) \Leftrightarrow S(f_x, f_y).$$

Notice that the doubly-shafted arrow $\Leftrightarrow$ is used instead of the singly-shafted arrow $\leftrightarrow$ used with the one-dimensional Fourier transform.

The inverse two-dimensional Fourier transform is given by

$$s(x, y) = \int_{-\infty}^{\infty}\int_{-\infty}^{\infty} S(f_x, f_y)\mathrm{e}^{\mathrm{j}2\pi(f_x x + f_y y)}\,\mathrm{d}f_x\,\mathrm{d}f_y.$$

To show that this is the inverse Fourier transform, we simply view $S(f_x, f_y)$ as formed by a sequence of two one-dimensional Fourier transforms. Let

$$\Gamma(f_x, y) = \int_{-\infty}^{\infty} s(x, y)\mathrm{e}^{-\mathrm{j}2\pi f_x x}\,\mathrm{d}x$$

be the Fourier transform of $s(x, y)$ with respect to x with y held constant. Then

$$S(f_x, f_y) = \int_{-\infty}^{\infty} \Gamma(f_x, y)\mathrm{e}^{-\mathrm{j}2\pi f_y y}\,\mathrm{d}y$$

and

$$s(x, y) \leftrightarrow \Gamma(f_x, y) \leftrightarrow S(f_x, f_y).$$

From the one-dimensional inversion formula, we have

$$s(x, y) = \int_{-\infty}^{\infty} \Gamma(f_x, y)\mathrm{e}^{\mathrm{j}2\pi f_x x}\,\mathrm{d}f_x$$

$$\Gamma(f_x, y) = \int_{-\infty}^{\infty} S(f_x, f_y)\mathrm{e}^{\mathrm{j}2\pi f_y y}\,\mathrm{d}f_y,$$

as was to be proved.

Using similar reasoning, it is a simple exercise to prove many basic properties of the two-dimensional Fourier transform. Often these properties can be deduced simply by twice using the properties of the one-dimensional Fourier transform. The basic properties are as follow:

1. **Linearity:** For any constants, a and b, possibly complex,

$$as_1(x, y) + bs_2(x, y) \Leftrightarrow aS_1(f_x, f_y) + bS_2(f_x, f_y).$$

2. **Sign Reversal:**

$$s(-x, y) \Leftrightarrow S(-f_x, f_y)$$
$$s(x, -y) \Leftrightarrow S(f_x, -f_y).$$

3. **Conjugation:**

$$s^*(x, y) \Leftrightarrow S^*(-f_x, -f_y).$$

If $s(x, y)$ is real, then $S^*(-f_x, -f_y) = S(f_x, f_y)$.

4. **Scaling Property:**[1] For any real nonzero constants, a and b,

$$s(ax, by) \Leftrightarrow \frac{1}{|ab|} S\left(\frac{f_x}{a}, \frac{f_y}{b}\right).$$

5. **Translation:** For any real constants, a and b,

$$s(x - a, y - b) \Leftrightarrow S(f_x, f_y) \mathrm{e}^{-\mathrm{j}2\pi(af_x + bf_y)}.$$

6. **Modulation:** For any real constants, a and b,

$$s(x, y)\mathrm{e}^{\mathrm{j}2\pi(ax+by)} \Leftrightarrow S(f_x - a, f_y - b).$$

7. **Convolution:**

$$g(x, y) ** h(x, y) \Leftrightarrow G(f_x, f_y)H(f_x, f_y)$$

where $**$ denotes a two-dimensional convolution given by

$$g(x, y) ** h(x, y) = \int_{-\infty}^{\infty} \int_{-\infty}^{\infty} g(\xi, \eta)h(x - \xi, y - \eta)\, \mathrm{d}\xi\, \mathrm{d}\eta.$$

8. **Product:**

$$g(x, y)h(x, y) \Leftrightarrow G(f_x, f_y) ** H(f_x, f_y).$$

9. **Spatial Differentiation:**

$$\frac{\partial s(x, y)}{\partial x} \Leftrightarrow \mathrm{j}2\pi f_x S(f_x, f_y)$$

$$\frac{\partial s(x, y)}{\partial y} \Leftrightarrow \mathrm{j}2\pi f_y S(f_x, f_y).$$

10. **Frequency Differentiation:**

$$-\mathrm{j}2\pi x s(x, y) \Leftrightarrow \frac{\partial S(f_x, f_y)}{\partial f_x}$$

$$-\mathrm{j}2\pi y s(x, y) \Leftrightarrow \frac{\partial S(f_x, f_y)}{\partial f_y}.$$

11. **Parseval's Formula:**

$$\int_{-\infty}^{\infty} \int_{-\infty}^{\infty} g(x, y)h^*(x, y)\, \mathrm{d}x\, \mathrm{d}y = \int_{-\infty}^{\infty} \int_{-\infty}^{\infty} G(f_x, f_y)H^*(f_x, f_y)\, \mathrm{d}f_x\, \mathrm{d}f_y.$$

[1] A scale change in two dimensions is sometimes called *magnification*; in one dimension, it is sometimes called *dilation*.

12. Energy Relation:

$$\int_{-\infty}^{\infty}\int_{-\infty}^{\infty}|s(x,y)|^2\,dx\,dy = \int_{-\infty}^{\infty}\int_{-\infty}^{\infty}|S(f_x,f_y)|^2\,df_x\,df_y.$$

13. Coordinate Rotation:

$$s(x\cos\psi - y\sin\psi,\, x\sin\psi + y\cos\psi)$$
$$\Leftrightarrow S(f_x\cos\psi - f_y\sin\psi,\, f_x\sin\psi + f_y\cos\psi).$$

14. Coordinate Transformation:

$$s(a_1x + b_1y, a_2x + b_2y) \Leftrightarrow \frac{1}{|a_1b_2 - a_2b_1|}S(A_1f_x + A_2f_y, B_1f_x + B_2f_y)$$

where

$$\begin{bmatrix} A_1 & B_1 \\ A_2 & B_2 \end{bmatrix} = \begin{bmatrix} a_1 & b_1 \\ a_2 & b_2 \end{bmatrix}^{-1}.$$

The coordinate conversion formula may be stylistically rewritten in the more suggestive form

$$s\left(\begin{bmatrix} a_1 & b_1 \\ a_2 & b_2 \end{bmatrix}\begin{bmatrix} x \\ y \end{bmatrix}\right) \Leftrightarrow \frac{1}{|a_1b_2 - a_2b_1|}S\left(\begin{bmatrix} a_1 & a_2 \\ b_1 & b_2 \end{bmatrix}^{-1}\begin{bmatrix} f_x \\ f_y \end{bmatrix}\right).$$

3.2 Transforms of some useful functions

A list of two-dimensional Fourier transform pairs for some useful functions is given in Table 3.1. Some of these Fourier transform pairs are developed in this section, while others will be presented later in the book.

Whenever $s(x, y)$ factors as

$$s(x, y) = s'(x)s''(y),$$

then the integral defining the two-dimensional Fourier transform separates into a product of integrals. Hence

$$S(f_x, f_y) = S'(f_x)S''(f_y).$$

In this way, many two-dimensional Fourier transforms are easily obtained as products of one-dimensional Fourier transforms.

Perhaps the simplest two-dimensional function is the *two-dimensional rectangle function*, which is defined as a product of one-dimensional rectangle functions:

$$\text{rect}(x, y) = \begin{cases} 1 & \text{if } |x| \le 1/2,\ |y| \le 1/2 \\ 0 & \text{otherwise.} \end{cases}$$

Table 3.1 *A table of two-dimensional Fourier transform pairs*

$s(x, y)$	$S(f_x, f_y)$
$\mathrm{rect}(x, y)$	$\mathrm{sinc}(f_x, f_y)$
$\mathrm{sinc}(x, y)$	$\mathrm{rect}(f_x, f_y)$
$\mathrm{circ}(x, y)$	$\mathrm{jinc}(f_x, f_y)$
$\mathrm{jinc}(x, y)$	$\mathrm{circ}(f_x, f_y)$
$\mathrm{lypd}(x, y)$	$\mathrm{sinc}^2(f_x, f_y)$
$\mathrm{chat}(x, y)$	$\mathrm{jinc}^2(f_x, f_y)$
$\delta(x, y)$	1
1	$\delta(f_x, f_y)$
$\delta(x)$	$\delta(f_y)$
$\mathrm{e}^{-\pi(x^2+y^2)}$	$\mathrm{e}^{-\pi(f_x^2+f_y^2)}$
$\mathrm{e}^{\mathrm{j}\pi(x^2+y^2)}$	$\mathrm{j}\mathrm{e}^{-\mathrm{j}\pi(f_x^2+f_y^2)}$
$\frac{1}{\sqrt{x^2+y^2}}$	$\frac{1}{\sqrt{f_x^2+f_y^2}}$
$\mathrm{comb}(x, y)$	$\mathrm{comb}(f_x, f_y)$
$\mathrm{comb}(x)$	$\mathrm{comb}(f_x)\delta(f_y)$
$\mathrm{comb}_N(x, y)$	$\mathrm{dirc}_N(f_x, f_y)$
$\mathrm{ring}(x, y)$	$\pi J_0\left(\pi\sqrt{f_x^2+f_y^2}\right)$
$\frac{1}{\sqrt{b^2+x^2+y^2}}\mathrm{e}^{-2\pi a\sqrt{b^2+x^2+y^2}}$	$\frac{1}{\sqrt{a^2+f_x^2+f_y^2}}\mathrm{e}^{-2\pi b\sqrt{a^2+f_x^2+f_y^2}}$
$\frac{1}{\sqrt{b^2+x^2+y^2}}\mathrm{e}^{-\mathrm{j}2\pi a\sqrt{b^2+x^2+y^2}}$	$\frac{1}{\mathrm{j}\sqrt{a^2-f_x^2-f_y^2}}\mathrm{e}^{-\mathrm{j}2\pi b\sqrt{a^2-f_x^2-f_y^2}}$
$4\pi(1-4x^2-4y^2)\mathrm{circ}(x, y)$	$J_2\left(\pi\sqrt{f_x^2+f_y^2}\right)/(f_x^2+f_y^2)$
$\sqrt{1-4x^2-4y^2}\,\mathrm{circ}(x, y)$	$\frac{1}{2\pi^2}\frac{\sin\pi\sqrt{f_x^2+f_y^2}-\sqrt{f_x^2+f_y^2}\cos\pi\sqrt{f_x^2+f_y^2}}{\left(\pi\sqrt{f_x^2+f_y^2}\right)^3}$

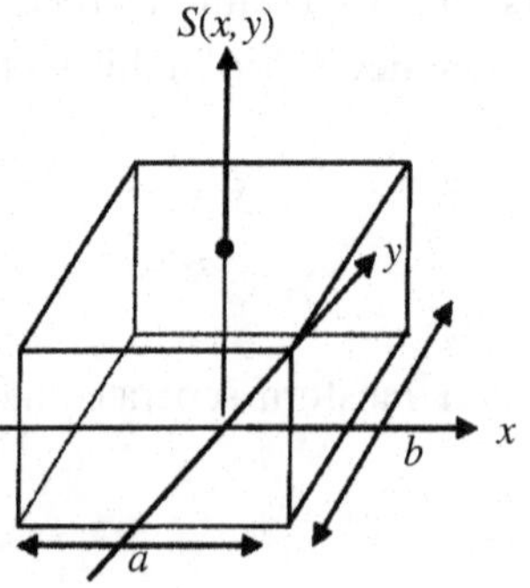

Figure 3.3 The two-dimensional rectangle function

The function $\mathrm{rect}(x, y)$ is shown in Figure 3.3. A simpler way to illustrate the two-dimensional rectangle function is by the plan view shown in Figure 3.4.

The *two-dimensional sinc function* is defined as a product of one-dimensional sinc functions. Thus

$$\mathrm{sinc}(x, y) = \mathrm{sinc}(x)\mathrm{sinc}(y).$$

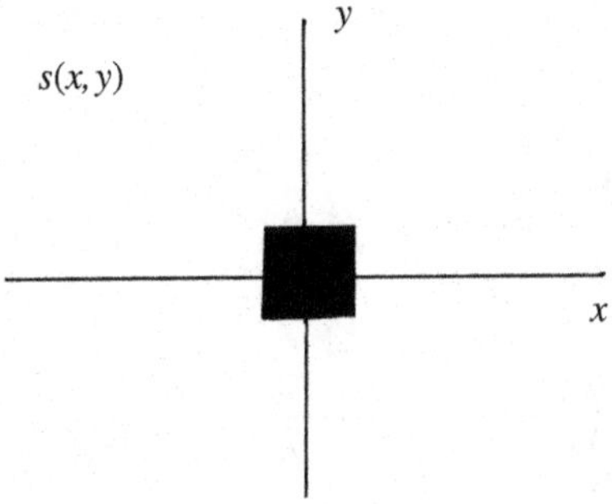

Figure 3.4 Plan view of the two-dimensional rectangle function

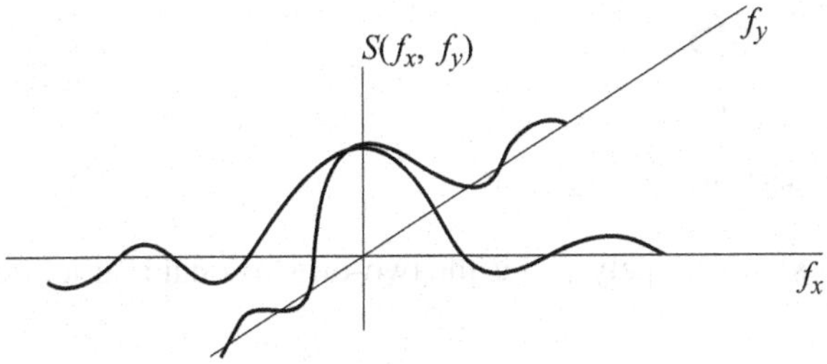

Figure 3.5 The two-dimensional sinc-by-sinc function

Because the two-dimensional rectangle function factors as

$$\mathrm{rect}(x, y) = \mathrm{rect}(x)\mathrm{rect}(y),$$

the appropriate two-dimensional Fourier transform pair is

$$\mathrm{rect}(x, y) \Leftrightarrow \mathrm{sinc}(f_x, f_y)$$

and, by the duality property,

$$\mathrm{sinc}(x, y) \Leftrightarrow \mathrm{rect}(f_x, f_y).$$

The two-dimensional sinc function is illustrated in Figure 3.5. Its plan view is shown in Figure 3.6.

The *two-dimensional impulse function*[2] is defined by

$$\delta(x, y) = \delta(x)\delta(y).$$

[2] Mathematically a *two-dimensional impulse function* (or *delta function*), $\delta(x, y)$, is a generalized function defined by the property that if $f(x, y)$ is any continuous (complex-valued) function at the origin, then

$$\int_{\mathcal{A}} \delta(x, y) f(x, y)\, \mathrm{d}x\, \mathrm{d}y = \begin{cases} f(0, 0) & \text{if } (0, 0) \in \mathcal{A} \\ 0 & \text{otherwise} \end{cases}$$

where $\mathcal{A}$ is any open region in the plane.

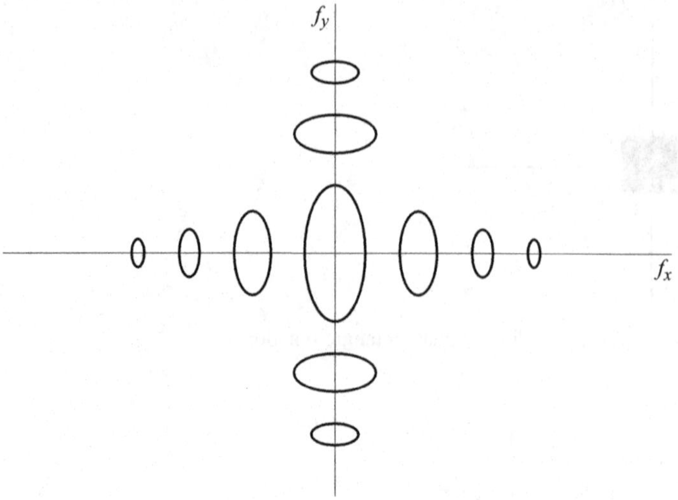

Figure 3.6 Plan view of sinc function

The separation of variables immediately yields the two-dimensional Fourier transform pair

$$\delta(x, y) \Leftrightarrow 1,$$

as well as

$$1 \Leftrightarrow \delta(f_x, f_y).$$

The *two-dimensional gaussian pulse* is defined by

$$e^{-\pi(x^2+y^2)} = e^{-\pi x^2} e^{-\pi y^2}.$$

The separation of variables immediately leads to the two-dimensional Fourier transform pair

$$e^{-\pi(x^2+y^2)} \Leftrightarrow e^{-\pi(f_x^2+f_y^2)}.$$

The one-dimensional impulse function $\delta(x)$ can be considered as a function of two variables that is constant in the y coordinate. In this case, it is called a *line impulse.* The line impulse is actually independent of y and should be visualized as an infinitely thin, infinitely high ridge lying along the y axis. Because the function is constant in y and it is an impulse in x, we have the two-dimensional Fourier transform pair

$$\delta(x) \Leftrightarrow \delta(f_y).$$

Another use of the impulse notation in two dimensions is the *ring impulse* or *ring function*, defined by $\text{ring}(x, y) = \delta\left(\sqrt{x^2 + y^2} - \frac{1}{2}\right)$. The ring impulse should be visualized as an infinitely thin, infinitely high ridge on the circle of unit diameter $\sqrt{x^2 + y^2} = \frac{1}{2}$.

We shall defer the discussion of the two-dimensional Fourier transform of ring(x, y) to Section 3.3.

The *two-dimensional comb function* is defined by

$$\text{comb}(x, y) = \text{comb}(x)\text{comb}(y).$$

The separation of variables immediately leads to the two-dimensional Fourier transform pair

$$\text{comb}(x, y) \Leftrightarrow \text{comb}(f_x, f_y).$$

Similarly, a *two-dimensional finite comb function* is defined by

$$\text{comb}_N(x, y) = \text{comb}_N(x)\text{comb}_N(y),$$

and a *two-dimensional dirichlet function* is defined by

$$\text{dirc}_N(x, y) = \text{dirc}_N(x)\text{dirc}_N(y).$$

This gives the two-dimensional Fourier transform pair

$$\text{comb}_N(x, y) \Leftrightarrow \text{dirc}_N(f_x, f_y).$$

The properties of the two-dimensional Fourier transform allow us to find quickly the two-dimensional Fourier transforms of many other simple functions. We will use the scaling property to find the Fourier transform of a thin strip. Then we will use the modulation property to find the Fourier transforms of simple arrays.

An extreme example of a rectangle is a long, thin strip. If b is small and a is large, then the rectangle function takes the form of a strip along the x axis, as shown in Figure 3.7. By the scaling property of the Fourier transform,

$$\text{rect}\left(\frac{x}{a}, \frac{y}{b}\right) \Leftrightarrow ab\, \text{sinc}(af_x, bf_y).$$

The transform then becomes narrow in f_x and wide in f_y; it is approximately a strip along the f_y axis, but it does exhibit the sidelobes of the sinc function. The long, thin

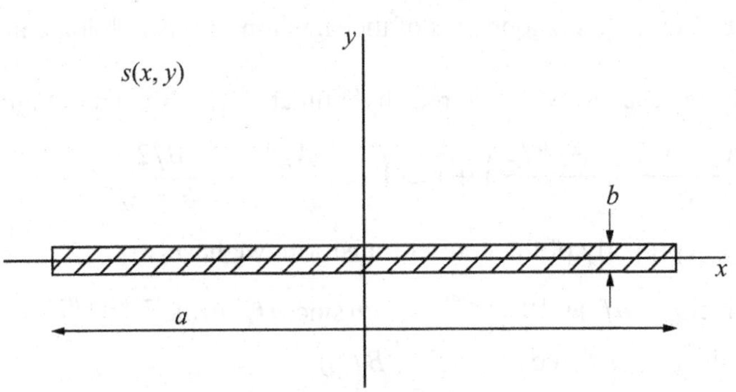

Figure 3.7 A long, thin rectangle function

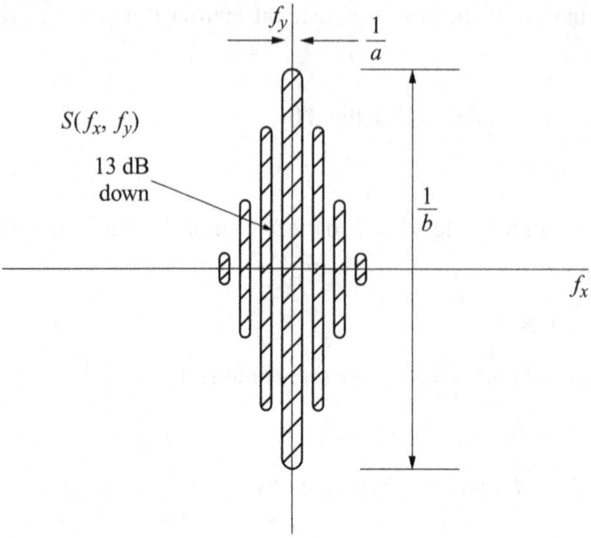

Figure 3.8 Transform of a long, thin rectangle function

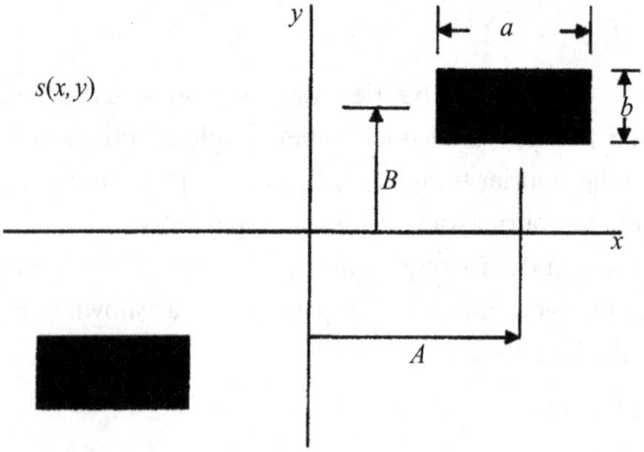

Figure 3.9 A pair of rectangles

sinc function is shown in plan view in Figure 3.8. Notice that the long axis of the Fourier transform is perpendicular to the long axis of the signal in the spatial domain.

Suppose that we have an array of two rectangle functions, as shown in Figure 3.9:

$$p(x, y) = \operatorname{rect}\left(\frac{x - A/2}{a}, \frac{y - B/2}{b}\right) + \operatorname{rect}\left(\frac{x + A/2}{a}, \frac{y + B/2}{b}\right).$$

Using the translation property of the Fourier transform, we have

$$\begin{aligned} P(f_x, f_y) &= ab \operatorname{sinc}(af_x, bf_y)\mathrm{e}^{-\mathrm{j}\pi(Af_x+Bf_y)} + ab \operatorname{sinc}(af_x, bf_y)\mathrm{e}^{\mathrm{j}\pi(Af_x+Bf_y)} \\ &= 2ab \operatorname{sinc}(af_x, bf_y)\cos\left(\pi(Af_x + Bf_y)\right). \end{aligned}$$

The two-dimensional Fourier transform of a pair of rectangles, which is shown in plan

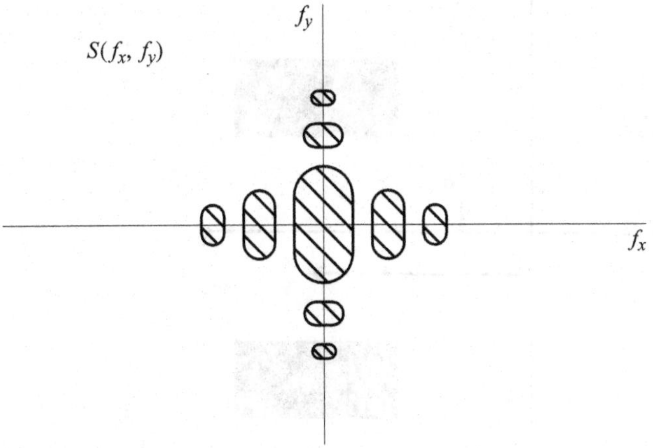

Figure 3.10 Transform of a pair of rectangles

view in Figure 3.10, is the transform of a single rectangle function multiplied by a sinusoid in spatial frequency. This is completely analogous to similar situations for the one-dimensional Fourier transform.

We may refer to the case of a one-dimensional pulse train to recognize the form of the transform of a "train" of rectangles along a common line. Thus

$$p(x, y) = \sum_{\ell=-N}^{N} \operatorname{rect}\left(\frac{x - \ell A/2}{a}, \frac{y - \ell B/2}{b}\right)$$

has the Fourier transform

$$P(f_x, f_y) = ab \operatorname{sinc}(af_x, bf_y) \operatorname{dirc}_{2N+1}(Af_x + Bf_y).$$

This Fourier transform is similar to that shown in Figure 3.10 except that the "cosine ridges" are replaced by the thinner "dirichlet ridges."

Next, consider the array of four rectangles shown in Figure 3.11. The Fourier transform can be computed directly, but it is also instructive to compute it in two steps. First, let

$$p'(x, y) = \operatorname{rect}\left(\frac{x - A/2}{a}, \frac{y}{b}\right) + \operatorname{rect}\left(\frac{x + A/2}{a}, \frac{y}{b}\right),$$

which has the Fourier transform

$$P'(f_x, f_y) = 2ab \operatorname{sinc}(af_x, bf_y) \cos(\pi Af_x).$$

Then the array of four rectangles is

$$p(x, y) = p'(x, y - B/2) + p'(x, y + B/2),$$

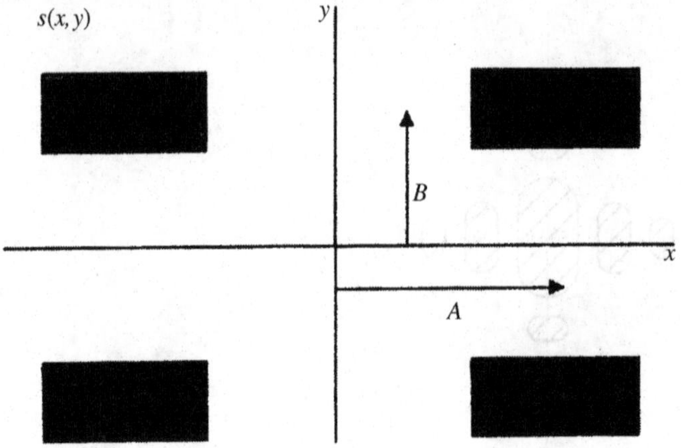

Figure 3.11 An array of rectangles

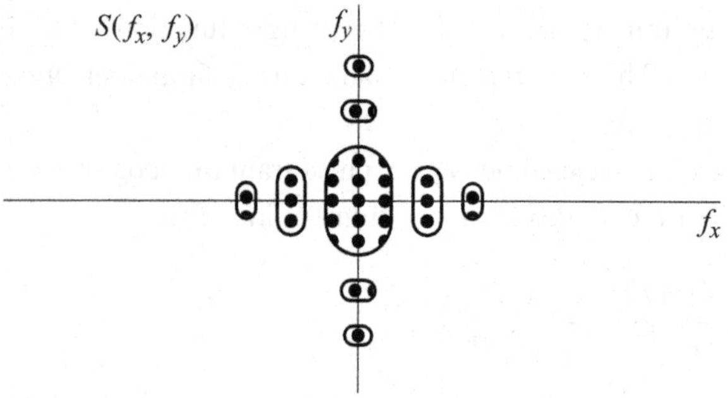

Figure 3.12 Transform of an array of rectangles

and

$$P(f_x, f_y) = 2P'(f_x, f_y)\cos(\pi B f_y).$$

Consequently,

$$P(f_x, f_y) = 4ab\,\mathrm{sinc}(af_x, bf_y)\cos(\pi A f_x)\cos(\pi B f_y),$$

which is illustrated in plan view in Figure 3.12. Now the two-dimensional sinc function is multiplied by an array of positive and negative bumps created by the cosine-by-cosine function. More generally, we may consider a two-dimensional N by N array of copies of the pulse $s(x, y)$ on an A by B grid. In Section 3.6, the Fourier transform is easily found to be

$$\begin{aligned} P(f_x, f_y) &= S(f_x, f_y)\,\mathrm{dirc}_N(Af_x)\mathrm{dirc}_N(Bf_y) \\ &= S(f_x, f_y)\,\mathrm{dirc}_N(Af_x, Bf_y). \end{aligned}$$

The cosine-by-cosine function has been replaced by a two-dimensional dirichlet function.

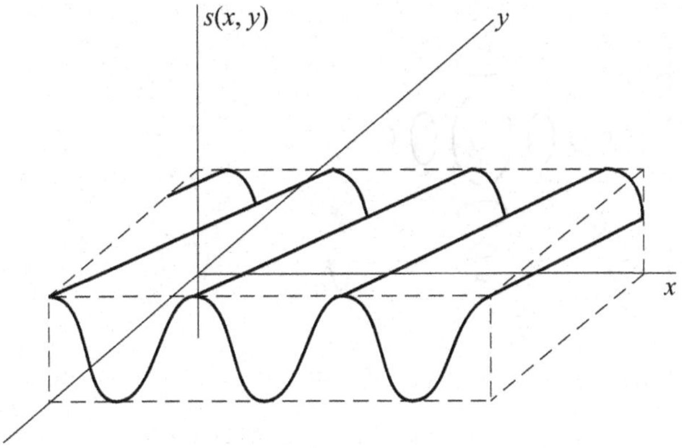

Figure 3.13 A two-dimensional sinusoid

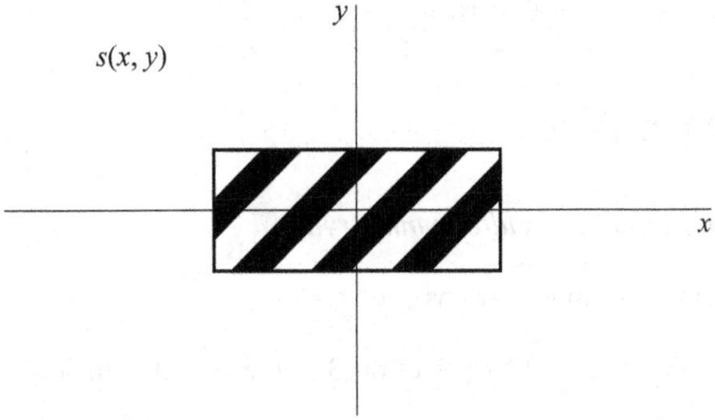

Figure 3.14 Plan view of a two-dimensional sinusoid

Finally, we shall look at some examples in which a rectangle function is sinusoidally modulated. Figure 3.13 shows the function

$$s(x, y) = \text{rect}\left(\frac{x}{a}, \frac{y}{b}\right) \cos(2\pi(Ax + By)).$$

A depiction in plan view is shown in Figure 3.14. The rectangle is modulated by a spatial sinusoid with the frequency $f = \sqrt{A^2 + B^2}$ at an angle of $\tan^{-1}(A/B)$ with respect to the x axis. Using the modulation property of the Fourier transform,

$$\begin{aligned} S(f_x, f_y) &= \tfrac{1}{2}ab\, \text{sinc}(a(f_x - A))\, \text{sinc}(b(f_y - B)) \\ &\quad + \tfrac{1}{2}ab\, \text{sinc}(a(f_x + A))\, \text{sinc}(b(f_y + B)), \end{aligned}$$

as shown in plan view in Figure 3.15. In the terminology of communication theory, the "signal" $s(x, y) = \text{rect}(x/a, y/b)$ has been "modulated" by the carrier $\cos 2\pi(Ax + By)$, thereby producing a sinusoidally modulated signal. The Fourier transform consists of two copies of the two-dimensional sinc function centered at spatial frequencies $\pm(Ax + By)$.

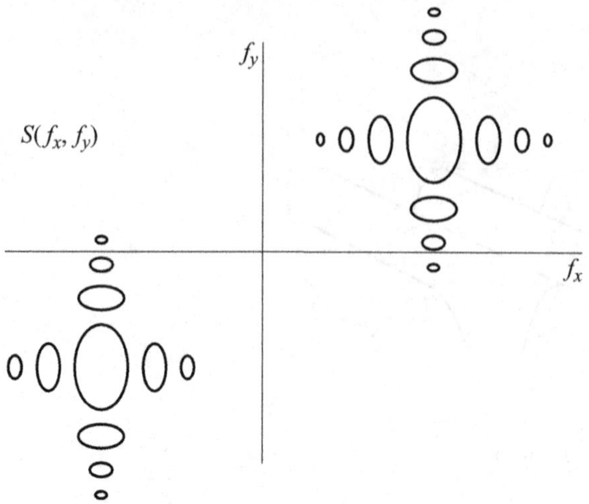

Figure 3.15 Transform of a two-dimensional sinusoid

3.3 Circularly symmetric functions

The function $s(x, y)$ is said to have *circular symmetry* if

$$s(x, y) = s(x \cos \psi - y \sin \psi, x \sin \psi + y \cos \psi)$$

for all ψ. Consequently, by property 13 of Section 3.1, the two-dimensional Fourier transform must satisfy

$$S(f_x, f_y) = S(f_x \cos \psi - f_y \sin \psi, f_x \sin \psi + f_y \cos \psi)$$

for any ψ. Therefore $S(f_x, f_y)$ is circularly symmetric whenever $s(x, y)$ is circularly symmetric.

It is often convenient to restate functions with circular symmetry in polar coordinates. Express the x, y plane and the f_x, f_y plane in polar coordinates by

$$x = r \cos \theta$$
$$y = r \sin \theta$$
$$f_x = \rho \cos \phi$$
$$f_y = \rho \sin \phi.$$

Let $s^\circ(r)$ denote the radial dependence of the circularly symmetric function $s(x, y)$:

$$s^\circ(r) = s(r \cos \theta, r \sin \theta).$$

By assumption, the right side has no θ dependence. Let

$$S^\circ(\rho) = S(\rho \cos \phi, \rho \sin \phi).$$

By the rotation properties of the two-dimensional Fourier transform, the right side has no ϕ dependence. Rewriting the equation

$$S(f_x, f_y) = \int_{-\infty}^{\infty} \int_{-\infty}^{\infty} s(x, y) e^{-j2\pi(f_x x + f_y y)} \, dx \, dy$$

in polar coordinates shows that the functions $s^{\circ}(r)$ and $S^{\circ}(\rho)$ are related by

$$S^{\circ}(\rho) = \int_{-\pi}^{\pi} \left[\int_0^{\infty} \left[r s^{\circ}(r) e^{-j2\pi r\rho(\cos\theta \cos\phi + \sin\theta \sin\phi)} \, dr \right] \right] d\theta$$

$$= \int_0^{\infty} r s^{\circ}(r) \left[\int_{-\pi}^{\pi} e^{-j2\pi r\rho \cos(\theta - \phi)} d\theta \right] dr.$$

But the inner integral is over one period of a periodic function. Hence, as expected, that integral does not depend on ϕ, so ϕ can be set equal to any convenient value. Thus, with $\phi = \pi/2$,

$$S^{\circ}(\rho) = \int_0^{\infty} r s^{\circ}(r) \left[\int_{-\pi}^{\pi} e^{-j2\pi r\rho \sin\theta} d\theta \right] dr.$$

The integral on θ cannot be evaluated in closed form; it must be integrated numerically. This integral occurs often and is widely tabulated: it is called the *zero-order Bessel function of the first kind*,[3] and is defined by

$$J_0(t) = \frac{1}{2\pi} \int_{-\pi}^{\pi} e^{-jt \sin\theta} d\theta.$$

Thus the two-dimensional Fourier transform of a circularly symmetric function, which itself is circularly symmetric, is given in polar coordinates by

$$S^{\circ}(\rho) = 2\pi \int_0^{\infty} r s^{\circ}(r) J_0(2\pi r\rho) \, dr.$$

We can easily write the inverse of this transform as

$$s^{\circ}(r) = 2\pi \int_0^{\infty} \rho S^{\circ}(\rho) J_0(2\pi r\rho) \, d\rho$$

because, except for a sign change, there is no difference between the two-dimensional Fourier transform and the inverse two-dimensional Fourier transform. The sign change drops out in the transformation to polar coordinates.

[3] The integral representation for the nth-order Bessel function of the first kind is

$$J_n(x) = \frac{1}{2\pi} \int_{-\pi}^{\pi} e^{-jx \sin\theta + jn\theta} d\theta.$$

Bessel functions are widely tabulated in mathematical handbooks. Notice that

$$J_{-n}(x) = J_n(-x) = (-1)^n J_n(x).$$

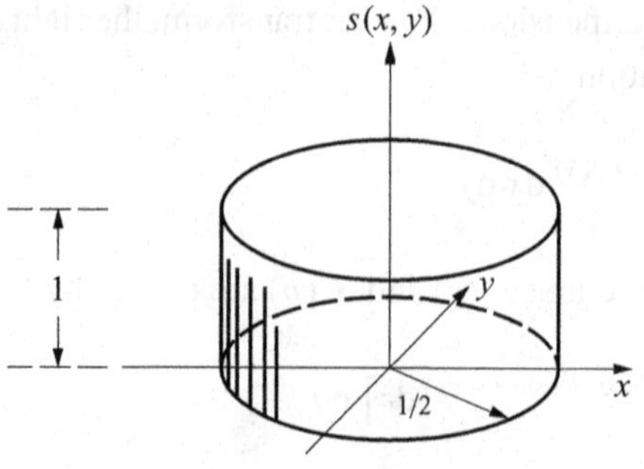

Figure 3.16 The circle function

A simple and important example of a circularly symmetric function is the *circle function*, defined by

$$\begin{aligned}\operatorname{circ}(x, y) &= \begin{cases} 1 & \text{if } \sqrt{x^2+y^2} \le \frac{1}{2} \\ 0 & \text{otherwise} \end{cases} \\ &= \operatorname{rect}\left(\sqrt{x^2+y^2}\right),\end{aligned}$$

and illustrated in Figure 3.16. For the circle function,

$$S^{\circ}(\rho) = 2\pi \int_0^{\frac{1}{2}} r J_0(2\pi r\rho)\, dr.$$

This integration can be evaluated by using the standard identity from the theory of Bessel functions:

$$\int_0^x \xi J_0(\xi)\, d\xi = x J_1(x)$$

where $J_1(x)$ is the first-order Bessel function of the first kind. Then

$$S^{\circ}(\rho) = \frac{J_1(\pi\rho)}{2\rho}.$$

Define the *jinc function* or *jinc pulse* as

$$\operatorname{jinc}(t) = \frac{J_1(\pi t)}{2t}.$$

The function jinc(t) is illustrated in Figure 3.17. Its general appearance is very similar to the appearance of the sinc function, a comparison can be found in Figure 8.1 of Chapter 8. However, the zeros of the jinc function are not spaced periodically, and the magnitudes of the sidelobes of the jinc function fall off more quickly than do those of the sinc function. At $t = 0$, the numerator and denominator in the definition of jinc(t) are both zero. One way to evaluate jinc(0) is to use l'Hôpital's rule. A more immediate way is to note that jinc(f_x, f_y) is the Fourier transform of

Table 3.2 *Locations of zeros and extrema of the jinc function*

	x	Normalized amplitude	Decibels
Central maximum ...	0	1.0	
First zero	1.2197	0	
Second peak	1.6347	−0.1323	−17.6
Second zero	2.2331	0.0	
Third peak	2.6793	0.0644	−23.8
Third zero	3.2383	0.0	
Fourth peak	3.6987	−0.0400	−28.0
Fourth zero	4.2411	0.0	
Fifth peak	4.7097	0.0279	−31.1
Fifth zero	5.2428	0.0	

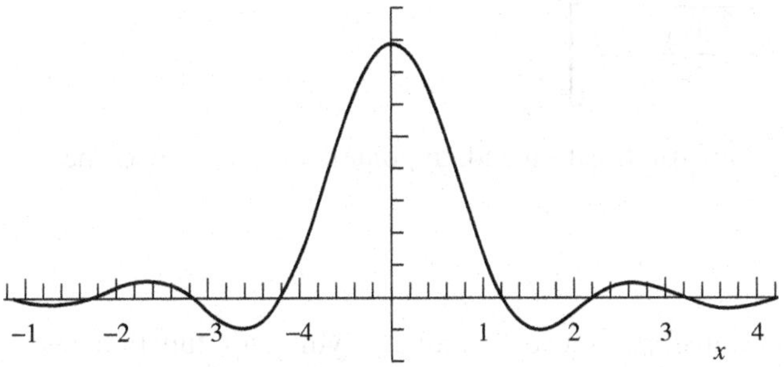

Figure 3.17 The jinc function

circ(x, y), which means that jinc(0) is the area of a circle of unit diameter. Therefore jinc(0) $= \pi/4$. The locations of the zeros and extrema of the jinc function are given in Table 3.2.

In polar coordinates, the two-dimensional Fourier transform of circ(x, y) is given in terms of the jinc function as

$$S^\circ(\rho) = \text{jinc}(\rho).$$

In rectangular coordinates, the Fourier transform of circ(x, y) is

$$S(f_x, f_y) = \text{jinc}\left(\sqrt{f_x^2 + f_y^2}\right).$$

Often, it is natural to express the jinc function as a function of two variables. The *two-dimensional jinc function* is defined as

$$\text{jinc}(x, y) = \text{jinc}\left(\sqrt{x^2 + y^2}\right).$$

We now have the two-dimensional Fourier transform pair

$$\mathrm{circ}(x, y) \Leftrightarrow \mathrm{jinc}(f_x, f_y)$$

and, by duality,

$$\mathrm{jinc}(x, y) \Leftrightarrow \mathrm{circ}(f_x, f_y).$$

The squared magnitude $\mathrm{jinc}^2(x, y)$ is called the *Airy disk* and is denoted $\mathrm{Airy}(x, y)$.

A ring of thickness ϵ and height $1/\epsilon$, centered on a circle of unit diameter, can be written in terms of the difference of a circle function of diameter $1+\epsilon$ and a circle function of diameter $1-\epsilon$:

$$\begin{aligned} s(x, y) &= \frac{1}{\epsilon}\left[\mathrm{circ}\left(\frac{x}{1+\epsilon}, \frac{y}{1+\epsilon}\right) - \mathrm{circ}\left(\frac{x}{1-\epsilon}, \frac{y}{1-\epsilon}\right)\right] \\ &= \frac{1}{\epsilon}\,\mathrm{rect}\left[\frac{2\sqrt{x^2+y^2}-1}{2\epsilon}\right]. \end{aligned}$$

This is a circularly symmetric function and, in polar coordinates, becomes

$$s^{\circ}(r) = \frac{1}{\epsilon}\,\mathrm{rect}\left(\frac{2r-1}{2\epsilon}\right).$$

Therefore the Fourier transform is also a circularly symmetric function, and

$$\begin{aligned} S^{\circ}(\rho) &= 2\pi \int_0^{\infty} r s^{\circ}(r) J_0(2\pi r\rho)\, dr \\ &= \frac{2\pi}{\epsilon} \int_{\frac{1-\epsilon}{2}}^{\frac{1+\epsilon}{2}} r\, J_0(2\pi r\rho)\, dr. \end{aligned}$$

In the limit, as ϵ goes to zero, this becomes

$$S^{\circ}(\rho) = \pi J_0\left(\frac{2\pi\rho}{2}\right),$$

so we conclude that the ring impulse, defined as

$$\mathrm{ring}(x, y) = \lim_{\epsilon\to 0} \frac{1}{\epsilon}\,\mathrm{rect}\left[\frac{2\sqrt{x^2+y^2}-1}{2\epsilon}\right],$$

has the Fourier transform

$$S(f_x, f_y) = \pi J_0\left(\pi\sqrt{f_x^2+f_y^2}\right),$$

which is circularly symmetric, as it must be.

There are two more Fourier transform pairs of circularly symmetric functions that we wish to highlight to complete the section. These are

$$\frac{1}{\sqrt{b^2+x^2+y^2}}\,\mathrm{e}^{-2\pi a\sqrt{b^2+x^2+y^2}} \Leftrightarrow \frac{1}{\sqrt{a^2+f_x^2+f_y^2}}\,\mathrm{e}^{-2\pi b\sqrt{a^2+f_x^2+f_y^2}}$$

$$\frac{1}{\sqrt{b^2+x^2+y^2}}\,\mathrm{e}^{-\mathrm{j}2\pi a\sqrt{b^2+x^2+y^2}} \Leftrightarrow \frac{1}{\mathrm{j}\sqrt{a^2-f_x^2-f_y^2}}\,\mathrm{e}^{-\mathrm{j}2\pi b\sqrt{a^2-f_x^2-f_y^2}}.$$

The second of these is important in the study of wave propagation. Rather than derive these as Fourier transforms here, we will derive them in the guise of Hankel transforms in Section 3.5.

3.4 The projection-slice theorem

The *projection-slice theorem* is a theorem relating a two-dimensional Fourier transform to the one-dimensional Fourier transforms of certain one-dimensional functionals, called *projections*, of the two-dimensional function.

Let $s(x, y)$ be a two-dimensional signal, possibly complex, of finite energy. The projection of $s(x, y)$ onto the x axis is

$$p(x) = \int_{-\infty}^{\infty} s(x, y)\,\mathrm{d}y.$$

More generally, the *projection* of $s(x, y)$ onto an axis at angle θ is defined as

$$p_\theta(t) = \int_{-\infty}^{\infty} s(t\cos\theta - r\sin\theta, t\sin\theta + r\cos\theta)\,\mathrm{d}r.$$

The angle θ specifies a rotation that relates the variables (t, r) to the variables (x, y). Figure 3.18 illustrates the sense in which $s(x, y)$ is "projected" into $p_\theta(t)$ by integrating along lines at an angle of θ from the y axis.

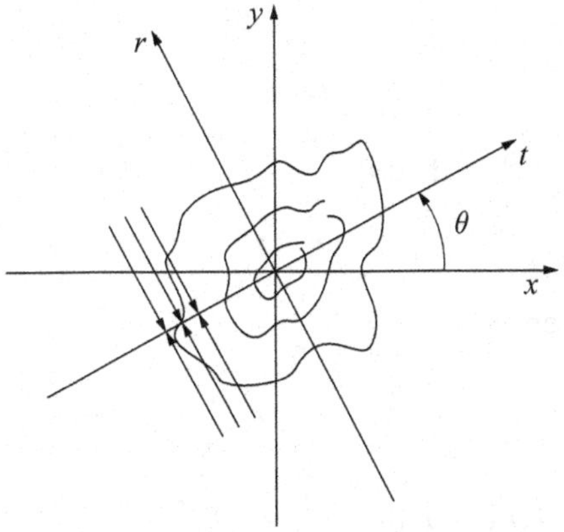

Figure 3.18 Illustrating the projection-slice theorem

Let $S(f_x, f_y)$ be a two-dimensional signal, possibly complex, of finite energy. The *slice* of $S(f_x, f_y)$ along the f_x axis is $S(f, 0)$; the *slice* of $S(f_x, f_y)$ along an axis at angle θ is $S(f\cos\theta, f\sin\theta)$.

The following theorem says that the one-dimensional Fourier transform of a projection of $s(x, y)$ is a slice of $S(f_x, f_y)$.

Theorem 3.4.1 (Projection-Slice Theorem) *If $p_\theta(t)$ is the projection of $s(x, y)$ at angle θ, and $S(f_x, f_y)$ is the Fourier transform of $s(x, y)$, then*

$$P_\theta(f) = S(f\cos\theta, f\sin\theta)$$

is the Fourier transform of $p_\theta(t)$.

Proof:

$$\begin{aligned} P_\theta(f) &= \int_{-\infty}^{\infty} p_\theta(t) e^{-j2\pi ft}\, dt \\ &= \int_{-\infty}^{\infty}\int_{-\infty}^{\infty} s(t\cos\theta - r\sin\theta, t\sin\theta + r\cos\theta) e^{-j2\pi ft}\, dr\, dt. \end{aligned}$$

Now make the change in variables:

$$\begin{aligned} x &= t\cos\theta - r\sin\theta \\ y &= t\sin\theta + r\cos\theta. \end{aligned}$$

Consequently,

$$t = x\cos\theta + y\sin\theta$$

and

$$dr\, dt = dx\, dy.$$

Therefore

$$\begin{aligned} P_\theta(f) &= \int_{-\infty}^{\infty}\int_{-\infty}^{\infty} s(x, y) e^{-j2\pi(f\cos\theta x + f\sin\theta y)}\, dx\, dy \\ &= S(f\cos\theta, f\sin\theta), \end{aligned}$$

which completes the proof of the theorem. □

The projection

$$p_\theta(t) = \int_{-\infty}^{\infty} s(t\cos\theta - r\sin\theta, t\sin\theta + r\cos\theta)\, dr$$

was introduced with θ considered as a single fixed angle. If we define a projection in this way for every value of angle θ, the projection then becomes a function of two variables, t and θ, and is called the *Radon transform*[4].

Definition 3.4.2 *Given a two-dimensional signal, $s(x, y)$, of finite energy, the Radon transform of $s(x, y)$ is the bivariate function*

$$p(t, \theta) = \int_{-\infty}^{\infty} s(t\cos\theta - r\sin\theta, t\sin\theta + r\cos\theta)\, dr.$$

The projection-slice theorem implies that the Radon transform can be inverted. Specifically, if $P(f, \theta)$ denotes the one-dimensional Fourier transform of $p(t, \theta)$ in the variable t, then

$$P(f, \theta) = S(f\cos\theta, f\sin\theta).$$

That is, $P(f, \theta)$ is the two-dimensional Fourier transform of $s(x, y)$ expressed in polar coordinates.

We shall give two simple examples of the Radon transform. Let

$$s(x, y) = \text{circ}(x, y).$$

Because the circle function satisfies

$$\text{circ}(t\cos\theta - r\sin\theta, t\sin\theta + r\cos\theta) = \text{circ}(t, r),$$

the Radon transform is independent of θ. It is easily seen to be

$$p(t, \theta) = \begin{cases} \sqrt{1 - 4t^2} & t \le 1/2 \\ 0 & t \ge 1/2, \end{cases}$$

which is independent of θ.

The two-dimensional gaussian pulse

$$s(x, y) = e^{-\pi(x^2+y^2)}$$

has the Radon transform

$$\begin{aligned} p(t, \theta) &= \int_{-\infty}^{\infty} s(t\cos\theta - r\sin\theta, t\sin\theta + r\cos\theta)\, dr \\ &= \int_{-\infty}^{\infty} e^{-\pi(t^2+r^2)}\, dr \\ &= e^{-\pi t^2}, \end{aligned}$$

which is independent of θ.

[4] The Radon transform can also be defined in multidimensional space. The Radon transform of order (n, k) is defined on the n-dimensional space over the real field or complex field and maps a function onto its integrals on k-dimensional hyperplanes. A Radon transform of order $(n, n-1)$, also called simply the *Radon transform*, maps an n-dimensional function onto its integral on $(n-1)$-dimensional hyperplanes. The Radon transform of order $(n, 1)$, also called the *shadow transform* (or the X-ray transform), maps an n-dimensional function onto its integral on (one-dimensional) lines.

3.5 Hankel transforms

The formulas

$$S(f) = 2\pi \int_0^\infty t s(t) J_0(2\pi f t)\,\mathrm{d}t$$

and

$$s(t) = 2\pi \int_0^\infty f S(f) J_0(2\pi f t)\,\mathrm{d}f$$

arose in the study of the two-dimensional Fourier transform of circularly symmetric functions. When regarded simply as an invertible relationship between the one-dimensional functions $s(t)$ and $S(f)$, these are known as the *Hankel transform*[5] and the *inverse Hankel transform*. The Hankel transform takes a complex function, $s(t)$, defined on the nonnegative real numbers into another complex function, $S(f)$, on the nonnegative real numbers.

Our two examples of Hankel transforms, given next, correspond to the two-dimensional Fourier transforms given at the end of Section 3.3. They have a parallel structure with the real constant a in the first example formally replaced by the imaginary constant ja in the second. The second example is important in the study of propagating waves. The first example is important in the study of diffusing waves.

Proposition 3.5.1 *The following are Hankel transform pairs,*

$$\frac{1}{\sqrt{b^2+r^2}}\,\mathrm{e}^{-2\pi a\sqrt{b^2+r^2}} \overset{h}{\longleftrightarrow} \frac{1}{\sqrt{a^2+\rho^2}}\,\mathrm{e}^{-2\pi b\sqrt{a^2+\rho^2}}$$

$$\frac{1}{\sqrt{b^2+r^2}}\,\mathrm{e}^{-\mathrm{j}2\pi a\sqrt{b^2+r^2}} \overset{h}{\longleftrightarrow} \frac{1}{\mathrm{j}\sqrt{a^2-\rho^2}}\,\mathrm{e}^{-\mathrm{j}2\pi b\sqrt{a^2-\rho^2}},$$

with the understanding that $\mathrm{j}\sqrt{a^2-\rho^2} = \sqrt{\rho^2-a^2}$ *when* ρ *is larger than* a.

Proof: The proof of the first Hankel transform pair begins with the definite integral

$$\frac{1}{\sqrt{\pi}} \int_0^\infty \mathrm{e}^{-A^2t^2-B^2t^{-2}}\,\mathrm{d}t = \frac{1}{2A}\,\mathrm{e}^{-2AB},$$

[5] The Hankel transform of order n and its inverse are given by

$$S(f) = 2\pi \int_0^\infty t s(t) J_n(2\pi f t)\,\mathrm{d}t$$

and

$$s(t) = 2\pi \int_0^\infty f S(f) J_n(2\pi f t)\,\mathrm{d}f.$$

which is developed in Problem 3.15a. By differentiation of both sides with respect to B, we also have the definite integral

$$\frac{1}{\sqrt{\pi}}\int_0^\infty \frac{1}{t^2}\mathrm{e}^{-A^2t^2-B^2t^{-2}}\,\mathrm{d}t = \frac{1}{2B}\mathrm{e}^{-2AB}.$$

We will prove the Hankel transform stated in the theorem by deriving the corresponding two-dimensional Fourier transform. The idea of the proof is to use the first definite integral above to replace an expression in $2A = \sqrt{b^2+x^2+y^2}$ by an expression in $4A^2 = b^2+x^2+y^2$ whose two-dimensional Fourier transform is known. The second definite integral is then used to eliminate the extraneous integration introduced in the first step.

Let $2A = \sqrt{b^2+x^2+y^2}$ and $B = 2\pi a$. Then, using the definite integral above,

$$\begin{aligned}
s(x,y) &= \frac{1}{\sqrt{b^2+x^2+y^2}}\mathrm{e}^{-2\pi a\sqrt{b^2+x^2+y^2}} \\
&= \frac{1}{\sqrt{\pi}}\int_0^\infty \mathrm{e}^{-(b^2+x^2+y^2)(t^2/4)-4\pi^2a^2t^{-2}}\,\mathrm{d}t \\
&= \frac{1}{\sqrt{\pi}}\int_0^\infty \mathrm{e}^{-4\pi^2a^2t^{-2}-b^2t^2/4}[\mathrm{e}^{-t^2(x^2+y^2)/4}]\,\mathrm{d}t.
\end{aligned}$$

The variables x and y only appear in the two-dimensional gaussian pulse highlighted by the brackets. Take the two-dimensional Fourier transform of $s(x,y)$ by moving the integration on x and y inside the integration on t and using the two-dimensional Fourier transform pair

$$\mathrm{e}^{-C(x^2+y^2)} \Leftrightarrow \frac{\pi}{C}\mathrm{e}^{-\pi^2(f_x^2+f_y^2)/C}.$$

This gives

$$\begin{aligned}
S(f_x,f_y) &= \frac{1}{\sqrt{\pi}}\int_{-\infty}^\infty \mathrm{e}^{-4\pi^2a^2t^{-2}-b^2t^2/4}\left[\frac{4\pi}{t^2}\mathrm{e}^{-4\pi^2(f_x^2+f_y^2)t^{-2}}\right]\mathrm{d}t \\
&= 4\sqrt{\pi}\int_0^\infty \frac{1}{t^2}\mathrm{e}^{-(b^2/4)t^2-4\pi^2(a^2+f_x^2+f_y^2)t^{-2}}\,\mathrm{d}t.
\end{aligned}$$

Next, use the second definite integral written above to reduce the right side, now with $A = b/2$ and $B = 2\pi\sqrt{a^2+f_x^2+f_y^2}$. This gives

$$S(f_x,f_y) = \frac{1}{\sqrt{a^2+f_x^2+f_y^2}}\mathrm{e}^{-2\pi b\sqrt{a^2+f_x^2+f_y^2}},$$

which proves the first Hankel transform pair of the theorem.

The proof of the second Hankel transform pair of the theorem is similar. It begins with the definite integral

$$\frac{1-\mathrm{j}}{\sqrt{2\pi}}\int_0^\infty \mathrm{e}^{-\mathrm{j}A^2t^2-\mathrm{j}B^2t^{-2}}\,\mathrm{d}t = \frac{1}{2A}\mathrm{e}^{-\mathrm{j}2AB},$$

which is developed in Problem 3.15b. Differentiation of both sides with respect to B gives

$$\frac{1-\mathrm{j}}{\sqrt{2\pi}}\int_0^\infty \frac{1}{t^2}\mathrm{e}^{-\mathrm{j}A^2t^2-\mathrm{j}B^2t^{-2}}\mathrm{d}t = \frac{1}{2B}\mathrm{e}^{-\mathrm{j}2AB}.$$

Let $2A = \sqrt{b^2+x^2+y^2}$ and $B = 2\pi a$. Then, using the first definite integral above,

$$\begin{aligned}
s(x, y) &= \frac{1}{\sqrt{b^2+x^2+y^2}}\mathrm{e}^{-\mathrm{j}2\pi a\sqrt{b^2+x^2+y^2}} \\
&= \frac{1-\mathrm{j}}{\sqrt{2\pi}}\int_0^\infty \mathrm{e}^{-\mathrm{j}(b^2+x^2+y^2)(t^2/4)-\mathrm{j}4\pi^2a^2t^{-2}}\mathrm{d}t \\
&= \frac{1-\mathrm{j}}{\sqrt{2\pi}}\int_0^\infty \mathrm{e}^{-\mathrm{j}(4\pi^2a^2t^{-2}+b^2t^2/4)}\left[\mathrm{e}^{-\mathrm{j}t^2(x^2+y^2)/4}\right]\mathrm{d}t.
\end{aligned}$$

Take the two-dimensional Fourier transform of $s(x, y)$, recalling that

$$\mathrm{e}^{-\mathrm{j}C(x^2+y^2)} \Leftrightarrow -\mathrm{j}\frac{\pi}{C}\mathrm{e}^{\mathrm{j}\pi^2(f_x^2+f_y^2)/C}.$$

Therefore

$$\begin{aligned}
S(f_x, f_y) &= \frac{1-\mathrm{j}}{\sqrt{2\pi}}\int_0^\infty \mathrm{e}^{-\mathrm{j}(4\pi^2a^2t^{-2}+b^2t^2/4)}\left[\frac{1}{\mathrm{j}}\frac{4\pi}{t^2}\mathrm{e}^{\mathrm{j}4\pi^2(f_x^2+f_y^2)t^{-2}}\right]\mathrm{d}t \\
&= \frac{4\pi}{\mathrm{j}}\frac{(1-\mathrm{j})}{\sqrt{2\pi}}\int_0^\infty \frac{1}{t^2}\mathrm{e}^{-\mathrm{j}(b^2/4)t^2-\mathrm{j}4\pi^2(a^2-f_x^2-f_y^2)t^{-2}}\mathrm{d}t.
\end{aligned}$$

Finally, provided $a^2 - f_x^2 - f_y^2$ is positive, use the second definite integral written above to reduce the right side, now with $A = b/2$ and $B = 2\pi\sqrt{a^2-f_x^2-f_y^2}$. This gives

$$S(f_x, f_y) = \left(\frac{2\pi}{\mathrm{j}}\right)\frac{1}{\sqrt{(4\pi^2)(a^2-f_x^2-f_y^2)}}\mathrm{e}^{-\mathrm{j}2\pi b\sqrt{a^2-f_x^2-f_y^2}},$$

provided $f_x^2 + f_y^2 \le a^2$. For the case that $a^2 - f_x^2 - f_y^2$ is negative, the definite integral given in Problem 3.16c can be used in a similar way to prove that

$$S(f_x, f_y) = \frac{1}{\sqrt{f_x^2+f_y^2-a^2}}\mathrm{e}^{-2\pi b\sqrt{f_x^2+f_y^2-a^2}}.$$

This completes the proof of the proposition. □

3.6 Two-dimensional pulse arrays

A two-dimensional pulse array is a generalization of a one-dimensional pulse train to two dimensions. Let $s(x, y)$ be a finite-energy pulse. Then an N by M rectangular array of pulses is given by

$$p(x, y) = \sum_{m=0}^{M-1} \sum_{n=0}^{N-1} s(x - nA, y - mB).$$

One way of computing the two-dimensional Fourier transform of $p(x, y)$ is to regard the two-dimensional array as a one-dimensional array of one-dimensional arrays. Thus

$$p(x, y) = \sum_{m=0}^{M-1} p'(x, y - mB)$$

where

$$p'(x, y) = \sum_{n=0}^{N-1} s(x - nA, y).$$

Now we can use Theorem 2.3.1 twice to write

$$P(f_x, f_y) = P'(f_x, f_y)\text{dirc}_M(Bf_y)\text{e}^{-\text{j}\pi f_y B(M-1)}$$

where

$$P'(f_x, f_y) = S(f_x, f_y)\text{dirc}_N(Af_x)\text{e}^{-\text{j}\pi f_x A(N-1)}.$$

If the two-dimensional array is centered at the origin, the phase terms drop out and

$$P(f_x, f_y) = S(f_x, f_y)\text{dirc}_N(Af_x)\text{dirc}_M(Bf_y).$$

For each f_x at which Af_x is an integer, $\text{dirc}_N(Af_x)$ has a grating lobe, and for each f_y at which Bf_y is an integer, $\text{dirc}_M(Bf_y)$ has a grating lobe. The two-dimensional function $P(f_x, f_y)$ has a two-dimensional grating lobe whenever both Af_x and Bf_y are integers.

For example, the two-dimensional square array of 81 circle functions,

$$\begin{aligned} p(x, y) &= \sum_{m=-4}^{4} \sum_{n=-4}^{4} \text{circ}(x - n, y - m), \\ &= \text{circ}(x, y) ** \text{comb}_9(x, y), \end{aligned}$$

has the Fourier transform

$$\begin{aligned} P(f_x, f_y) &= \text{jinc}\left(\sqrt{f_x^2 + f_y^2}\right) \text{dirc}_9(f_x)\text{dirc}_9(f_y) \\ &= \text{jinc}\left(\sqrt{f_x^2 + f_y^2}\right) \text{dirc}_9(f_x, f_y) \end{aligned}$$

where $\text{dirc}_N(f_x, f_y)$ is a two-dimensional dirichlet function. The structure of $P(f_x, f_y)$ is an infinite array of grating lobes on a square grid with the amplitude of the grating lobes modulated radially by the jinc function.

It is instructive to also consider a very large number of small circles arranged on a square grid but not in the form of a square. We will fill a large circle with a large number of small circles on a square grid. Place the small circles at each point of the integer lattice, provided that the small circle lies inside the large circle of radius R centered at the origin. This is approximated by

$$p(x, y) = \left(\sum_{m=-N}^{N} \sum_{n=-N}^{N} \operatorname{circ}(x - n, y - m) \right) \operatorname{circ}\left(\frac{x}{R}, \frac{y}{R} \right).$$

This approximation uses a circle of diameter R as a "cookie cutter" to cut a large circle from a square array of small circles. If N is larger than $R/2$, the pattern of small circles will completely fill the large circle. Some of the small circles may be cut by the large circle. The expression is an approximation to the original statement because it includes the pieces of the small circles that have been cut.

The two-dimensional Fourier transform of $p(x, y)$ is

$$P(f_x, f_y) = \left[\operatorname{jinc}(f_x, f_y) \operatorname{dirc}_{2N+1}(f_x, f_y) \right] ** R \operatorname{jinc}(Rf_x, Rf_y).$$

Making N larger does not change $p(x, y)$ because the additional small circles outside the large circle are discarded by the cookie cutter. Therefore, making N larger cannot change $P(f_x, f_y)$ either. Even making N infinite does not change $p(x, y)$. Hence

$$P(f_x, f_y) = \left[\operatorname{jinc}(f_x, f_y) \operatorname{comb}(f_x, f_y) \right] ** R \operatorname{jinc}(Rf_x, Rf_y).$$

Each of the impulses in the comb function is amplitude weighted by $\operatorname{jinc}(f_x, f_y)$, then convolved with $\operatorname{jinc}(Rf_x, Rf_y)$. Thus we have the further approximation

$$P(f_x, f_y) \approx R \operatorname{jinc}(f_x, f_y) \sum_{n=-\infty}^{\infty} \sum_{m=-\infty}^{\infty} \operatorname{jinc}(R(f_x - n), R(f_y - m)).$$

This is an array of thin jinc functions with the amplitude of the array modulated by a wide jinc function.

3.7 Sampling in two dimensions

The sampling theorem in two dimensions states that if the support of $S(f_x, f_y)$ lies inside the unit square, then the sampling operation

$$s'(x, y) = \operatorname{comb}(x, y) s(x, y)$$

can be inverted by

$$s(x, y) = \operatorname{sinc}(x, y) ** s'(x, y).$$

To see how this works in the general case, suppose that we have any signal, $s(x, y)$, in two dimensions that has finite energy and whose transform $S(f_x, f_y)$ is zero except within a finite region, $\mathcal{R}$, of the f_x, f_y plane. Let A and B denote the length and width of any rectangle that is centered at the origin of the f_x, f_y plane and encloses $\mathcal{R}$. The

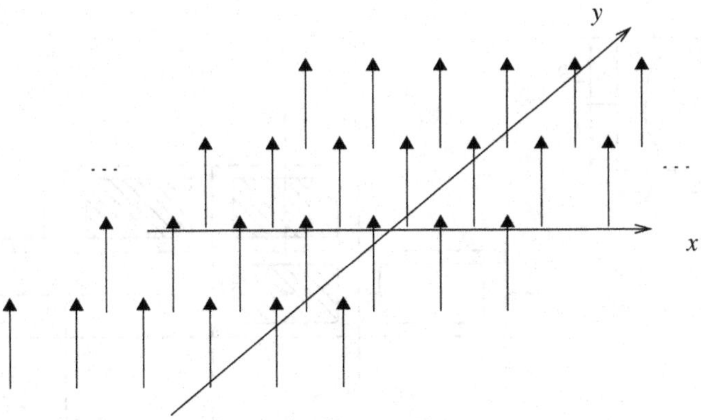

Figure 3.19 A two-dimensional array of impulses

signal $s(x, y)$ is represented by the set of its values, called *two-dimensional Nyquist samples*, on a rectangular grid of points spaced in x by $1/A$ and spaced in y by $1/B$:

$$s_{ii'} = s\left(\frac{i}{A}, \frac{i'}{B}\right).$$

Two-dimensional sampling can be described as a multiplication with the two-dimensional comb function, shown in Figure 3.19. Then the function

$$s'(x, y) = \text{comb}(Ax, By)s(x, y)$$

is a two-dimensional array of impulses. The impulse located at $x = i/A$ and $y = i'/B$ takes an amplitude which is the sample $s_{ii'}$. To transpose this relationship into the frequency domain, we use the Fourier transform relationship

$$AB\,\text{comb}(Ax, By) \Leftrightarrow \text{comb}\left(\frac{f_x}{A}, \frac{f_y}{B}\right).$$

The convolution theorem now allows us to write

$$S'(f_x, f_y) = \text{comb}\left(\frac{f_x}{A}, \frac{f_y}{B}\right) \; ** \; S(f_x, f_y).$$

The convolution with the impulses of the comb function produces images of $S(f_x, f_y)$ translated by jA in f_x and by $j'B$ in f_y. Because of the assumption that $S(f_x, f_y)$ is equal to zero outside of the A by B rectangle, the translated copies of $S(f_x, f_y)$ do not overlap, as shown in Figure 3.20. From this picture, we see that, to recover $S(f_x, f_y)$, we must multiply by a rectangle function that covers only the central rectangle of the frequency plane. That is, define

$$H(f_x, f_y) = \frac{1}{AB}\,\text{rect}\left(\frac{f_x}{A}, \frac{f_y}{B}\right)$$

which corresponds to the pointspread function $h(x, y) = \text{sinc}(Ax, By)$. It is clear that

$$S(f_x, f_y) = H(f_x, f_y)S'(f_x, f_y),$$

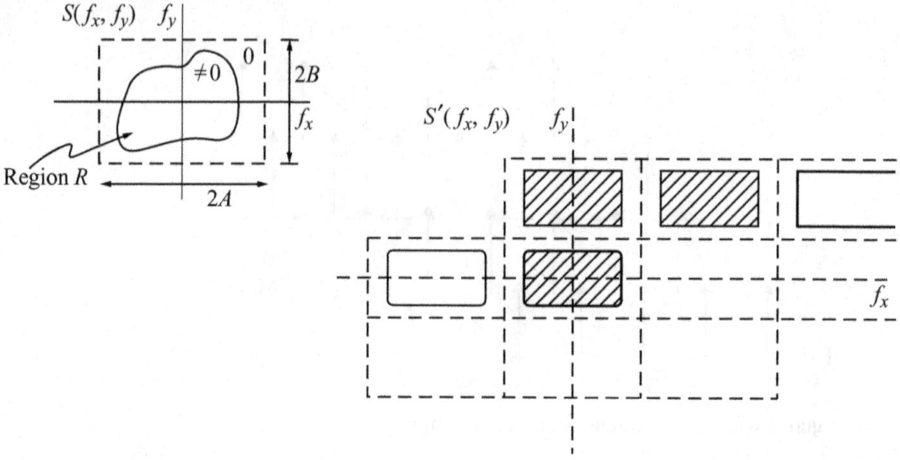

Figure 3.20 Images of a region under sampling

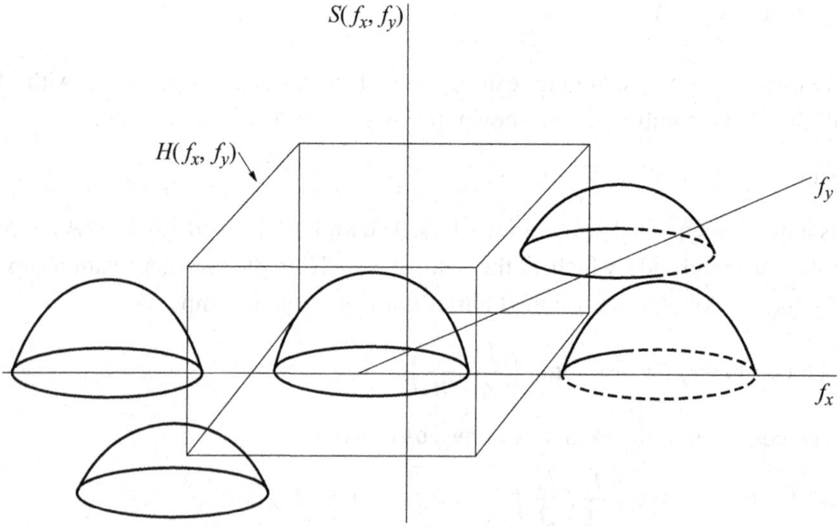

Figure 3.21 Illustrating the derivation of an interpolation formula

as is illustrated in Figure 3.21. In the spatial domain, this is equivalent to a two-dimensional convolution,

$$s(x, y) = h(x, y) ** s'(x, y),$$

which leads to the two-dimensional Nyquist–Shannon interpolation formula

$$s(x, y) = \sum_{i=-\infty}^{\infty} \sum_{i'=-\infty}^{\infty} s_{ii'} \operatorname{sinc}(Ax - i, By - i').$$

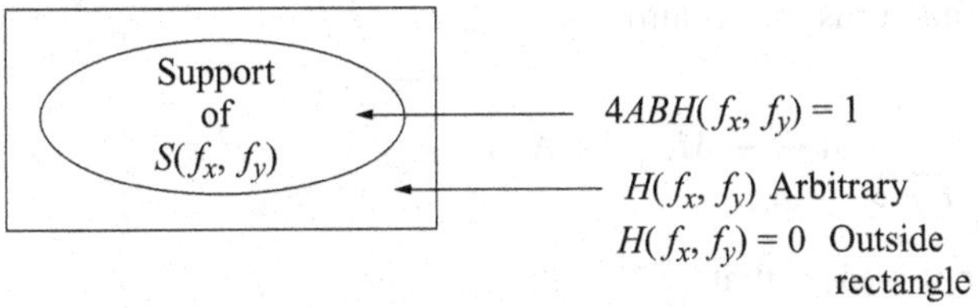

Figure 3.22 Sufficient conditions on $H(f_x, f_y)$

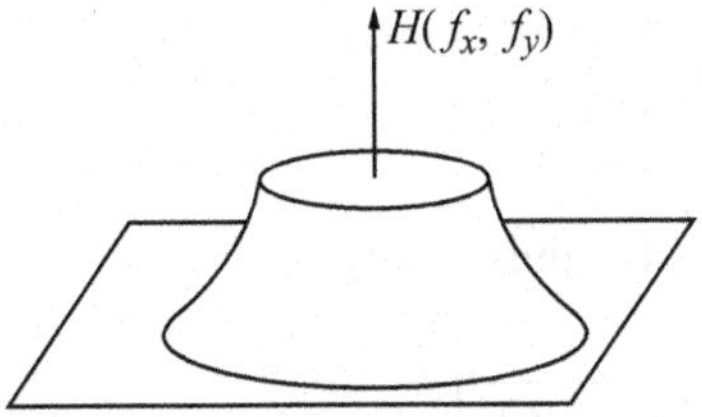

Figure 3.23 A suitable interpolating filter

Two-dimensional sinc interpolation is usually unsatisfactory because the sidelobes of sinc(x, y) fall off too slowly. Many alternative interpolation formulas exist in two dimensions. It is sufficient to choose any interpolation filter, $h(x, y)$, such that

$$H(f_x, f_y) = \begin{cases} 1/AB & \text{if} \quad S(f_x, f_y) \neq 0 \\ 0 & \text{if} \quad |f_x| \geq A/2 \text{ or } |f_y| \geq B/2, \end{cases}$$

as illustrated in Figure 3.22. In the transition region where $|f_x| < A/2$, $|f_y| < B/2$, and $S(f_x, f_y) = 0$, the function $H(f_x, f_y)$ is arbitrary and can be chosen so as to control the sidelobes of the interpolation formula. One possible choice of $H(f_x, f_y)$ is shown in Figure 3.23.

We shall illustrate such alternative interpolation formulas by describing jinc interpolation. Suppose that $s(x, y)$ is equal to zero everywhere outside a region that can be contained within a circle of radius $A/2$. Because this circle is contained in a square of size A, we could recover the signal by the Nyquist–Shannon interpolation formula described before. To develop an alternative interpolation formula, let

$$H(f_x, f_y) = \frac{1}{A^2} \operatorname{circ}\left(\frac{f_x}{A}, \frac{f_y}{A}\right),$$

with the transform

$$h(x, y) = \operatorname{jinc}\left(A\sqrt{x^2 + y^2}\right).$$

Then the supports of the terms in the sum

$$S'(f_x, f_y) = A^2 \sum_{i=-\infty}^{\infty} \sum_{i'=-\infty}^{\infty} S(f_x - Ai, f_y - Ai')$$

are nonintersecting, so it is clear that

$$S(f_x, f_y) = H(f_x, f_y)S'(f_x, f_y).$$

Consequently, in the spatial domain,

$$s(x, y) = h(x, y) ** s'(x, y).$$

Therefore we have the alternative interpolation formula

$$s(x, y) = \sum_{i=-\infty}^{\infty} \sum_{i'=-\infty}^{\infty} s_{ii'} \operatorname{jinc}\left[\sqrt{(Ax - i)^2 + (Ay - i')^2}\right].$$

Although jinc interpolation also has sidelobes, as any interpolation formula must, these sidelobes fall off more quickly than those of sinc interpolation, and so jinc interpolation is less sensitive to error.

Figure 3.24 shows that there is "wasted" space between the circles of the interpolating filter $H(f_x, f_y)$. We may presume from this configuration that if we could rearrange the spatial samples, the images would be placed more closely together in the f_x, f_y plane, and this would place the spatial-domain samples farther apart. In particular, we shall place the spatial samples in a hexagonal grid, and then the sampling images in the frequency domain will also be hexagonally arranged, as shown in Figure 3.25.

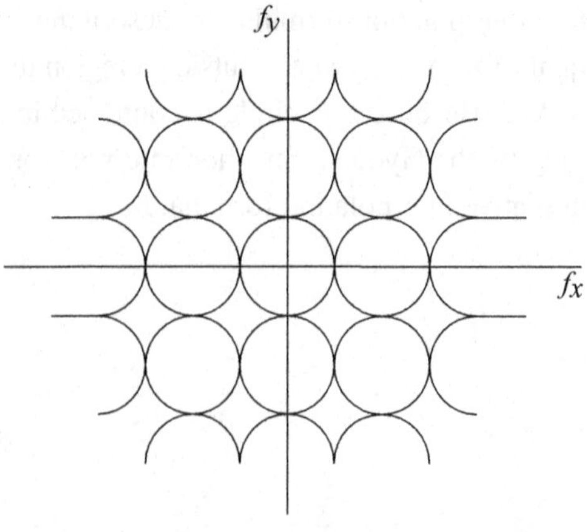

Figure 3.24 Images of a circular region under a rectangular sampling format

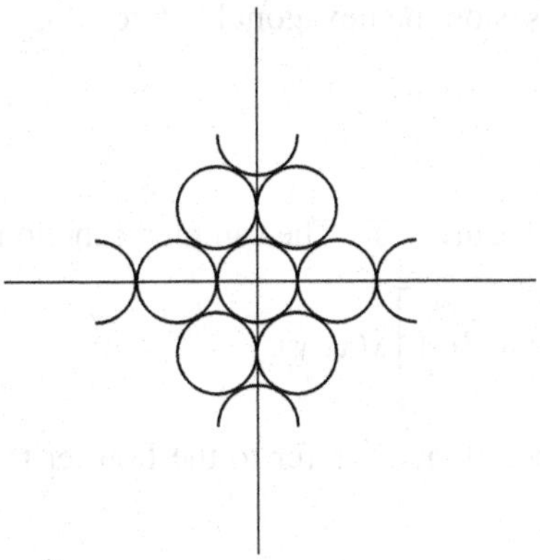

Figure 3.25 Images of a circular region under a hexagonal sampling format

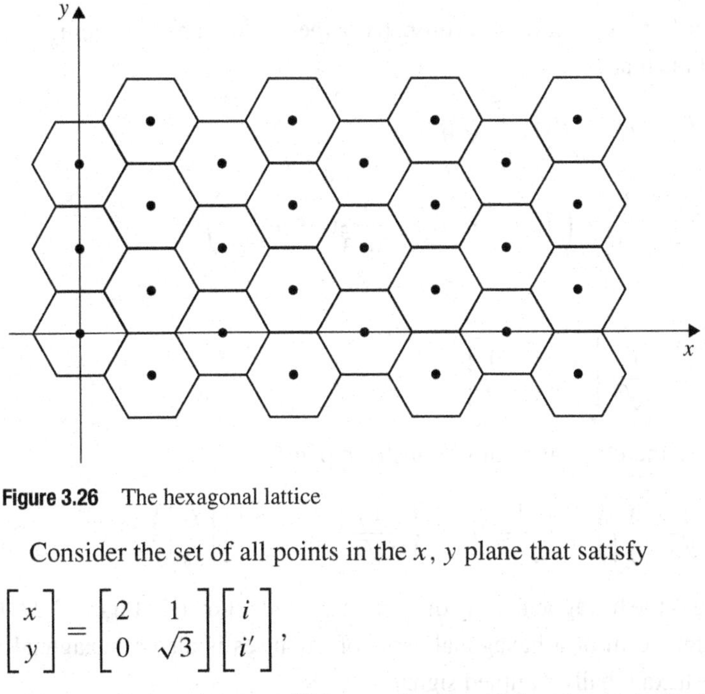

Figure 3.26 The hexagonal lattice

Consider the set of all points in the x, y plane that satisfy

$$\begin{bmatrix} x \\ y \end{bmatrix} = \begin{bmatrix} 2 & 1 \\ 0 & \sqrt{3} \end{bmatrix} \begin{bmatrix} i \\ i' \end{bmatrix},$$

where i and i' are integers. This set of points in the plane is called a *hexagonal lattice*.[6] Figure 3.26 shows the set of points of this lattice; the hexagons are drawn only to show how the lattice derives its name.

[6] In general, a two-dimensional *lattice* is the set of points in the plane that can be written

$$\begin{bmatrix} x \\ y \end{bmatrix} = \boldsymbol{M} \begin{bmatrix} i \\ i' \end{bmatrix}$$

where i and i' are arbitrary integers, and $\boldsymbol{M}$ is any fixed, real-valued, nonsingular, two by two matrix called the *generator matrix* of the lattice. A planar lattice is a regular arrangement of points in the plane. The two lattices formed by a matrix and its inverse are called *reciprocal lattices*. The *fundamental cell* of the lattice is the parallelepiped defined by four points corresponding to $(i, i') = (0, 0), (1, 0), (0, 1)$, and $(1, 1)$.

Define a two-dimensional array of impulses on the hexagonal lattice

$$\sum_{i=-\infty}^{\infty}\sum_{i'=-\infty}^{\infty}\delta\left(x-(2i+i'),\, y-\sqrt{3}i'\right)$$

to form the sampling array that is shown in Figure 3.27. The sampled function is

$$s'(x,y)=\left[\sum_{i=-\infty}^{\infty}\sum_{i'=-\infty}^{\infty}\delta\left(x-(2i+i'),\, y-\sqrt{3}i'\right)\right]s(x,y).$$

To transform this equation into the frequency domain, refer to the Fourier transform pair for an infinite array of impulses:

$$\sum_{i=-\infty}^{\infty}\sum_{i'=-\infty}^{\infty}\delta\left(x-i,\, y-i'\right)\;\Leftrightarrow\;\sum_{\ell=-\infty}^{\infty}\sum_{\ell'=-\infty}^{\infty}\delta(b_x-\ell,\, b_y-\ell').$$

Using the coordinate transformation property of the two-dimensional Fourier transform, we can conclude that

$$\sum_{i=-\infty}^{\infty}\sum_{i'=-\infty}^{\infty}\delta\left(x-(2i+i'),\, y-\sqrt{3}i'\right)$$

$$\Leftrightarrow\frac{1}{2\sqrt{3}}\sum_{\ell=-\infty}^{\infty}\sum_{\ell'=-\infty}^{\infty}\delta\left(f_x-\frac{1}{2}\ell,\, f_y-\frac{1}{2\sqrt{3}}(-\ell+2\ell')\right)$$

because

$$\begin{bmatrix}2 & 1\\ 0 & \sqrt{3}\end{bmatrix}^{-1}=\frac{1}{2\sqrt{3}}\begin{bmatrix}\sqrt{3} & -1\\ 0 & 2\end{bmatrix}.$$

A compact statement of this Fourier transform pair is

$$\mathrm{comb}\left(\begin{vmatrix}2 & 1\\ 0 & \sqrt{3}\end{vmatrix}\begin{vmatrix}x\\ y\end{vmatrix}\right)\Leftrightarrow\frac{1}{2\sqrt{3}}\mathrm{comb}\left(\frac{1}{2\sqrt{3}}\begin{vmatrix}\sqrt{3} & -1\\ 0 & 2\end{vmatrix}\begin{vmatrix}f_x\\ f_y\end{vmatrix}\right).$$

The right side is the hexagonal array of sampling points shown in Figure 3.27. Because the Fourier transform of a hexagonal array of impulses is also a hexagonal array of impulses, the hexagonally sampled signal

$$s'(x,y)=\left[\sum_{i=-\infty}^{\infty}\sum_{i'=-\infty}^{\infty}\delta\left(x-(2i+i'),\, y-\sqrt{3}i'\right)\right]s(x,y)$$

has the Fourier transform

$$\begin{aligned}S'(f_x,f_y)&=\frac{1}{2\sqrt{3}}\left[\sum_{\ell=-\infty}^{\infty}\sum_{\ell'=-\infty}^{\infty}\delta\left(f_x-\frac{1}{2}\ell,\, f_y-\frac{1}{2\sqrt{3}}(-\ell+2\ell')\right)\right]**\,S(f_x,f_y)\\&=\frac{1}{2\sqrt{3}}\sum_{\ell=-\infty}^{\infty}\sum_{\ell'=-\infty}^{\infty}S\left(f_x-\frac{1}{2}\ell,\, f_y-\frac{1}{2\sqrt{3}}(-\ell+2\ell')\right).\end{aligned}$$

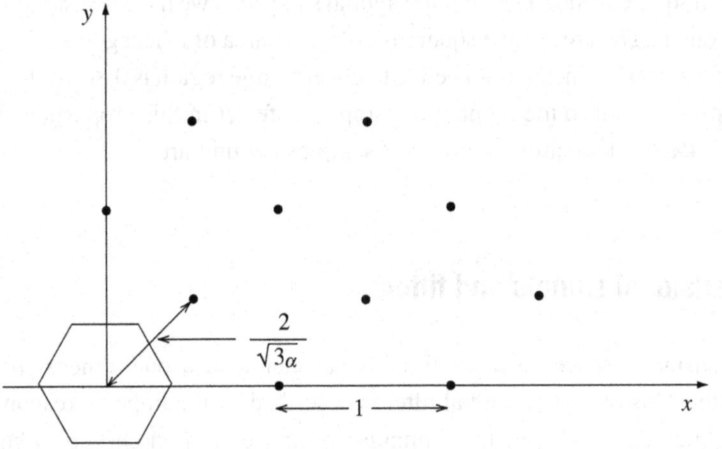

Figure 3.27 The sampling array

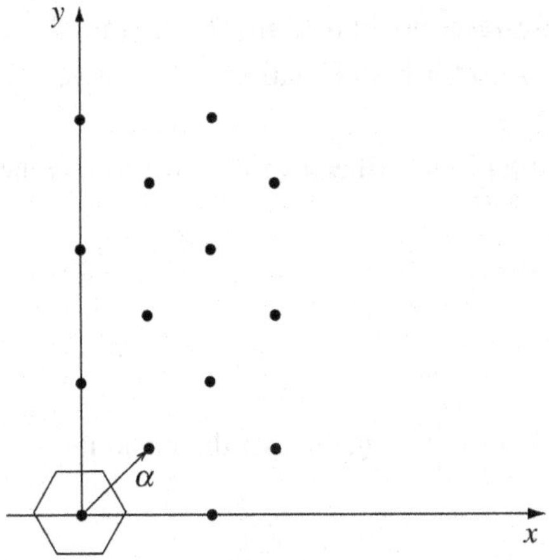

Figure 3.28 Frequency-domain array

An examination of the hexagons in Figure 3.28 shows that, if the x and y axes are scaled by the parameter $A/2$, the images are packed as tightly as possible without overlapping. With this scaling, the spatial-domain samples are $s\left(\frac{4}{A}i + \frac{2}{A}i', \frac{2}{A}\sqrt{3}i'\right)$.

The earlier development of jinc interpolation easily carries over to hexagonal sampling. Consequently, we have the interpolation formula

$$s(x, y) = \sum_{i=-\infty}^{\infty} \sum_{i'=-\infty}^{\infty} s\left((4i + 2i')/A, 2\sqrt{3}i'/A\right) \operatorname{jinc}\left[2\pi\sqrt{(Ax - 4i)^2 + (Ay - 2i')^2}\right].$$

The only remaining task is to compare the density of sampling points for hexagonal sampling and for rectangular sampling. For rectangular sampling, we have one sampling point for each square of side $1/A$. For hexagonal sampling, we have one sampling point for each hexagon. The area of the square is A^{-2}. The area of a hexagon is $(2/\sqrt{3})A^{-2}$. Therefore the number of hexagons needed to cover a large region is 0.866 of the number of squares needed, and so the number of samples is fewer in this proportion. This is a saving of 13.4% of the required number of samples per unit area.

3.8 Two-dimensional signals and filters

A two-dimensional, space-invariant filter is an analog of a one-dimensional, time-invariant filter. This one-dimensional filter is described by the impulse-response function of the filter. A one-dimensional impulse response is often chosen to be causal, but in two dimensions there is no analog of this realizability condition. It is possible for a two-dimensional spatial filter to be nonzero for all x and y. Consequently, the two-dimensional function $h(x, y)$ is usually called a *pointspread function* or a *spread function*, rather than an impulse-response function. This terminology carries the connotation that $h(x, y)$ can be nonzero for both negative and positive values of x and y.

The operation of a two-dimensional filter is described by the convolution relationship

$$\begin{aligned} v(x, y) &= \int_{-\infty}^{\infty}\int_{-\infty}^{\infty} h(x-\xi, y-\eta)s(\xi, \eta)\, d\xi\, d\eta \\ &= h(x, y) ** s(x, y). \end{aligned}$$

After the change in variables $\xi' = x - \xi$, $\eta' = y - \eta$, the two-dimensional convolution becomes

$$v(x, y) = \int_{-\infty}^{\infty}\int_{-\infty}^{\infty} s(x-\xi', y-\eta')h(\xi', \eta')\, d\xi'\, d\eta'.$$

Thus two-dimensional convolution is a commutative operation,

$$h(x, y) ** s(x, y) = s(x, y) ** h(x, y).$$

The role of the pointspread function in forming $v(x, y)$ from $s(x, y)$ can be depicted by a block diagram as shown in Figure 3.29. The operation of the filter as a convolution

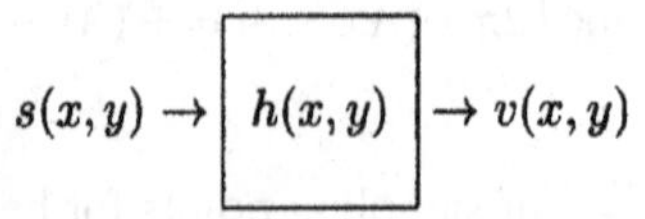

Figure 3.29 Two-dimensional, pointspread function as a filter

can be illustrated by taking $s(x, y) = \delta(x, y)$. Then the filtering operation, given by

$$\begin{aligned} v(x, y) &= \int_{-\infty}^{\infty}\int_{-\infty}^{\infty} h(x-\xi, y-\eta)\delta(\xi, \eta)\, d\xi\, d\eta \\ &= h(x, y), \end{aligned}$$

replaces the impulse $\delta(x, y)$ at its input by the pointspread function $h(x, y)$ at its output; the "point" $\delta(x, y)$ is spread into $h(x, y)$. Similarly, the impulse $\delta(x - x_o, y - y_o)$ at the input is replaced by $h(x - x_o, y - y_o)$. This, again, is the pointspread function but now centered at (x_o, y_o). Consequently, the infinitesimal impulse $s(\xi, \eta)\, d\xi\, d\eta\, \delta(x - \xi, y - \eta)$ of weight $s(\xi, \eta)\, d\xi\, d\eta$ located at (ξ, η), when passed through the filter, produces the output $s(\xi, \eta)h(x - \xi, y - \eta)\, d\xi\, d\eta$. The convolution equation is then the superposition of the response to all such inputs, as expressed by the convolution integral.

The two-dimensional filter is expressed in the frequency domain by using the two-dimensional convolution theorem. In the frequency domain, the filter becomes

$$V(f_x, f_y) = H(f_x, f_y)S(f_x, f_y).$$

In the frequency domain, the two-dimensional filter becomes a multiplication of two functions of two-dimensional frequency (f_x, f_y).

For example, the filter with the two-dimensional transfer function

$$H(f_x, f_y) = \text{circ}\left(\frac{f_x}{A}, \frac{f_y}{A}\right)$$

will reject all frequency components with $f_x^2 + f_y^2 > A^2$, and will pass all other frequencies without change. The two-dimensional impulse response of this filter is

$$h(x, y) = A \text{ jinc}\left(A\sqrt{x^2 + y^2}\right).$$

Because the circle function has a sharp cutoff in the frequency domain, impulses or edges in the space-domain input signal $s(x, y)$ are filtered by $h(x, y)$ to produce the sidelobes of the jinc function in the filtered output signal $v(x, y)$. A softer cutoff can be obtained by choosing the two-dimensional filter with the response

$$H_n(f_x, f_y) = \frac{1}{1 + (f_x^2 + f_y^2)^n}.$$

This filter, $H_n(f_x, f_y)$, is known as a *two-dimensional Butterworth filter of order n*. Because $H_n(f_x, f_y)$ approaches circ$(f_x/2, f_y/2)$ as n goes to infinity, the response $h_n(x, y)$ can be described as a softened version of a jinc function, and with smaller sidelobes. A Butterworth filter might be used to suppress high-frequency noise from a two-dimensional image without introducing significant sidelobes as unwanted artifacts.

3.9 Resolution and apodization

The quality of an imaging system is judged, in part, by the resolution of the system. For a linear system, the scene $s(x, y)$ is spread by $h(x, y)$, the pointspread function of the imaging system, to produce the image $v(x, y)$. The term "resolution" then refers to the sharpness of the main lobe of $h(x, y)$. The resolution refers to the ability to distinguish two points, $\delta(x, y)$ and $\delta(x - a, y - b)$, after the sum $\delta(x, y) + \delta(x - a, y - b)$ is convolved with $h(x, y)$.

A plan view of the main lobe of a typical pointspread function with a simple shape is shown in Figure 3.30. Figure 3.31 shows a plan view of the output after a pair of impulses is convolved with $h(x, y)$.

The many criteria for quantifying resolution that were discussed for one-dimensional signals in Section 2.8 can also be used for two-dimensional signals. These include the

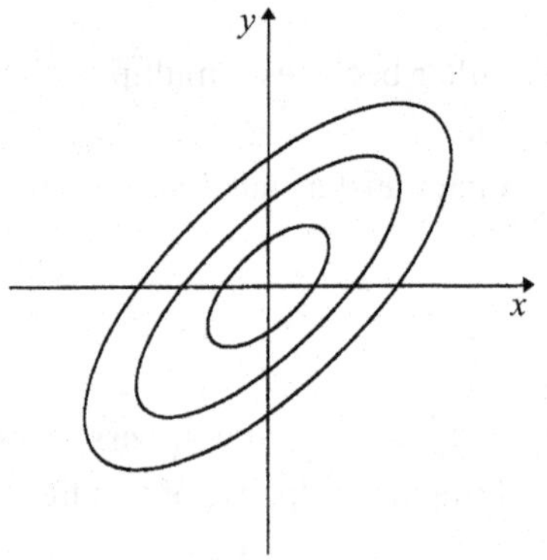

Figure 3.30 Plan view of a pointspread function

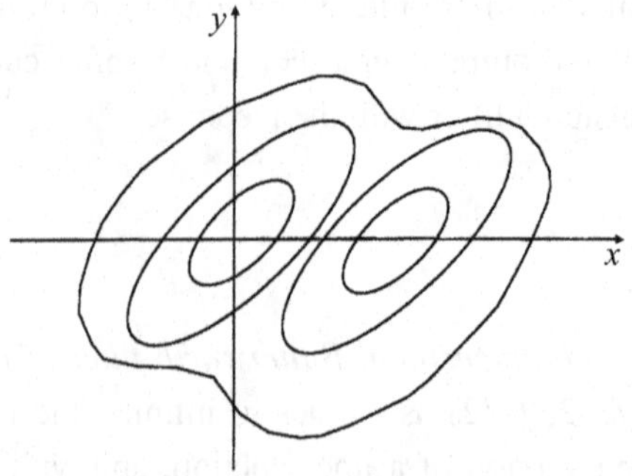

Figure 3.31 Convolution of a pointspread function with two impulses

Rayleigh resolution criteria, the Gabor resolution criteria, the Sparrow resolution criteria, and the Woodward resolution criteria. Of these, the Gabor resolution criteria have the cleanest mathematical properties, but the Rayleigh resolution criteria is convenient and the most popular.

The Rayleigh resolution of the jinc function, $\text{jinc}\left(\sqrt{x^2+y^2}/\text{d}\right)$, is defined as the value of the separation between two copies of the function at which the main lobe of the translated copy falls on the first zero of the nontranslated copy. This occurs where $2\pi\sqrt{x^2+y^2}/\text{d} = 0.610$. Consequently, the Rayleigh resolution of the jinc function is $0.61d/2\pi$. The Rayleigh resolution can be defined for any function, $h(x, y)$, for which the first zero contour $h(x, y) = 0$ is a circle surrounding the maximum. If the first zero contour is not a circle, then one might use the maximum radial distance of that contour as a measure of resolution. If there is no such closed contour, the Rayleigh resolution is not meaningful.

To define the half-power resolution, consider the set of points (x, y) forming the outermost closed contour, if it exists, enclosing the peak of $h(x, y)$ and satisfying

$$|h(x, y)|^2 = \tfrac{1}{2} \max |h(x, y)|^2.$$

The half-power resolution is defined as the maximum distance between this contour and the origin.

Resolution is not the only criterion for judging a pointspread function. Sidelobes of the pointspread function may cause artifacts in the output that are more serious than the lack of resolution. One may wish to suppress these sidelobes by apodization. Apodization of two-dimensional signals is the same as apodization of one-dimensional signals. A pointspread function $h(x, y)$ with unacceptable sidelobes is replaced by a pointspread function with acceptable sidelobes. For example, a jinc function might be replaced by something looking more like a gaussian pulse, reducing sidelobes at the cost of a wider main lobe. Because a jinc function maximally occupies spatial bandwidth, apodization always requires additional spatial bandwidth or a loss of resolution.

Problems

3.1 Find the two-dimensional Fourier transform of the two-dimensional gaussian pulse

$$s(x, y) = \mathrm{e}^{-(ax^2+by^2)}.$$

3.2 **a.** Prove that

$$\text{jinc}\left(\sqrt{x^2+y^2}\right) ** \text{jinc}\left(\sqrt{x^2+y^2}\right) = \text{jinc}\left(\sqrt{x^2+y^2}\right)$$

where

$$\text{jinc}(r) = \frac{J_1(\pi r)}{2r}.$$

b. Prove that

$$\text{jinc}(x, y) \ {**}\ \text{sinc}(x, y) = \text{jinc}(x, y)$$

where

$$\text{jinc}(x, y) = \text{jinc}\left(\sqrt{x^2 + y^2}\right).$$

3.3 What is the two-dimensional Fourier transform of the two-dimensional Fourier transform of $s(x, y)$?

3.4 **a.** Let $s(x, y)$ be a function that is zero outside of the unit square $-\frac{1}{2} \leq x \leq \frac{1}{2}$, $-\frac{1}{2} \leq y \leq \frac{1}{2}$, and, inside this square, takes the shape of a pyramid with a square base (see illustration). Find the two-dimensional Fourier transform of $s(x, y)$.

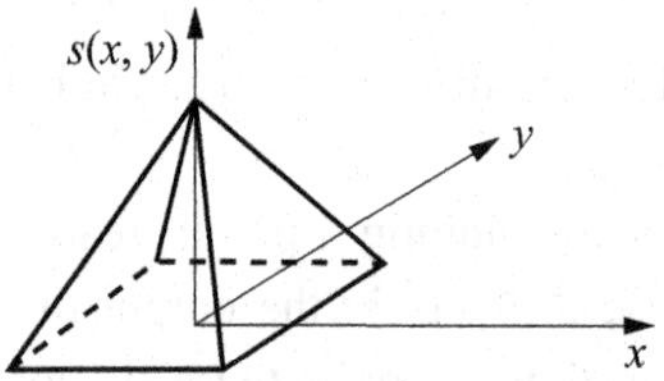

b. Define the lazy pyramid function, denoted $\text{lzpd}(x, y)$, by

$$\text{lzpd}(x, y) = \text{rect}(x, y) ** \text{rect}(x, y).$$

Find $\text{lzpd}(x, y)$ and its two-dimensional Fourier transform.

3.5 **a.** The one-dimensional convolution of two rectangle functions is called the one-dimensional *triangle function* and is denoted $\text{trng}(t)$,

$$\text{trng}(t) = \text{rect}(t) * \text{rect}(t).$$

Give an expression for $\text{trng}(t)$.

b. Given two overlapping circles of unit radius with centers at distance d, find the area of the lens-shaped intersection.

c. The two-dimensional convolution of two circle functions is called the two-dimensional *hat function* (or the *Chinese hat function*) and is denoted $\text{chat}(x, y)$,

$$\text{chat}(x, y) = \text{circ}(x, y) ** \text{circ}(x, y).$$

Give a closed-form expression for $\text{chat}(x, y)$.

d. Give an expression for circ$(x, y) ** $circ$(ax, ay)$. This function may be regarded as a two-dimensional generalization of a one-dimensional trapezoid function, rect$(t) *$ rect(at).

e. What is the Fourier transform of chat(x, y)?

3.6 Let $H(f_x, f_y)$ be an ideal lowpass, two-dimensional filter, given by

$$H(f_x, f_y) = \begin{cases} 1 & \text{if } \sqrt{f_x^2 + f_y^2} \leq 1 \\ 0 & \text{otherwise.} \end{cases}$$

Let the filter input be

$$s(x, y) = \text{sinc}(2x)\,\text{sinc}(2y).$$

Find the filter output.

3.7 Prove that if

$$s(x, y) \Leftrightarrow S(f_x, f_y),$$

then

$$s(x\cos\psi - y\sin\psi, x\sin\psi + y\cos\psi)$$
$$\Leftrightarrow S(f_x\cos\psi - f_y\sin\psi, f_x\sin\psi + f_y\cos\psi),$$

and more generally,

$$s(a_1x + b_1y, a_2x + b_2y) \Leftrightarrow \frac{1}{|a_1b_2 - a_2b_1|} S(A_1f_x + A_2f_y, B_1f_x + B_2f_y)$$

where

$$\begin{bmatrix} A_1 & B_1 \\ A_2 & B_2 \end{bmatrix} = \begin{bmatrix} a_1 & b_1 \\ a_2 & b_2 \end{bmatrix}^{-1}.$$

3.8 Find the two-dimensional Fourier transform of $s(x, y) = xy\,e^{-\pi(x^2+y^2)}$.

3.9 Prove the following two-dimensional Fourier transform relationships:

a.

$$(a - 2|x|)\text{sinc}[y(a - 2|x|)] \Leftrightarrow (a - 2|f_y|)\text{sinc}[f_x(a - 2|f_y|)].$$

b.

$$(a - 2|x|)\text{sinc}[(y - bx)(a - 2|x|) \Leftrightarrow (a - 2|f_y|)\text{sinc}(f_x + bf_y)(a - 2|f_y|)].$$

3.10 Prove the following one-dimensional Fourier transform pairs:

a.

$$J_0(2\pi t) \leftrightarrow \frac{\text{rect}(f/2)}{\pi(1 - f^2)^{1/2}}.$$

b.

$$\mathrm{jinc}(t) \leftrightarrow (1 - 4f^2)^{1/2}\mathrm{rect}(f).$$

3.11 An "elliptical disc" is given by

$$s(x, y) = \mathrm{circ}(x, ay)$$

for $a \neq 1$. Find the two-dimensional Fourier transform in terms of the jinc function.

3.12 Suppose that $g(x, y)$ is a two-valued function, assuming only the values zero and one. The region of the x, y plane where $g(x, y)$ equals one is called the *aperture*. One wishes to design an aperture whose total area is small but whose Fourier transform has a narrow (two-dimensional) main lobe. A cross configuration has been suggested with the argument that the wide width in x should give a narrow transform in the f_x direction, and the wide width in y should give a narrow transform in the f_y direction. Calculate the Fourier transform and decide whether the suggestion is sound.

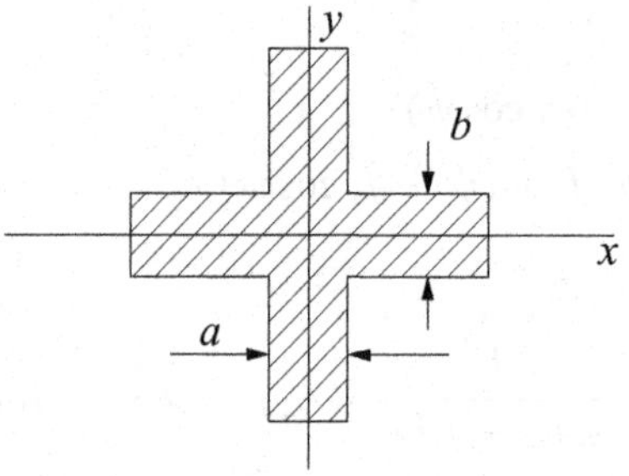

3.13 Using simple graphical sketches, describe the general nature of the two-dimensional Fourier transform of the thin ring

$$s(x, y) = \mathrm{circ}(x, y) - \mathrm{circ}\left(\frac{x}{1-\epsilon}, \frac{y}{1-\epsilon}\right)$$

where ϵ is very small.

3.14 Prove the following equality:

$$\int_0^x \xi J_0(\xi)\,\mathrm{d}\xi = xJ_1(x)$$

where

$$J_n(x) = \frac{1}{2\pi}\int_{-\pi}^{\pi} \mathrm{e}^{-\mathrm{j}x\sin\theta + \mathrm{j}n\theta}\,\mathrm{d}\theta.$$

3.15 An n-dimensional signal, $s(\boldsymbol{x})$, where $\boldsymbol{x} = (x_1, \ldots, x_n)$ has an n-dimensional Fourier transform, $S(\boldsymbol{f})$, where $\boldsymbol{f} = (f_1, \ldots, f_n)$, given by

$$S(\boldsymbol{f}) = \int_{-\infty}^{\infty} \cdots \int_{-\infty}^{\infty} s(\boldsymbol{x})\mathrm{e}^{-\mathrm{j}2\pi(\boldsymbol{x}\cdot\boldsymbol{f})}\mathrm{d}x_1 \cdots \mathrm{d}x_n.$$

a. State and prove the inverse multidimensional Fourier transform.
b. State and prove the multidimensional convolution theorem.
c. Prove that if $\boldsymbol{A}$ is an orthogonal matrix (that is, $\boldsymbol{A}^T = \boldsymbol{A}^{-1}$), then $s(\boldsymbol{x}\boldsymbol{A})$ and $S(\boldsymbol{f}\boldsymbol{A})$ are a Fourier transform pair.

3.16 Consider the lattice formed by the matrix

$$\boldsymbol{M} = \begin{bmatrix} 1 & 0 \\ a & b \end{bmatrix}.$$

Find the reciprocal lattice. Sketch the lattice and its reciprocal if $a = b = \sqrt{2}$.

3.17 Generalize the two-dimensional Fourier series to an arbitrary lattice, that is, let

$$S_{ii'} = \int_{\mathcal{A}}\int s(x, y)\mathrm{e}^{-\mathrm{j}2\pi(ix+i'y)}\,\mathrm{d}x\,\mathrm{d}y$$

where $\mathcal{A}$ is the aperture of the fundamental cell. Then prove that

$$s(x, y) = \frac{1}{|\mathcal{A}|}\sum_{i}\sum_{i'} S_{ii'}\mathrm{e}^{\mathrm{j}2\pi(ix+i'y)}.$$

3.18 **a.** Prove that

$$\int_0^\infty \mathrm{e}^{-a^2t^2-b^2t^{-2}}\mathrm{d}t = \frac{\sqrt{\pi}}{2a}\mathrm{e}^{-2ab}$$

where a and b are both positive. **Hint:** Let $I(b)$ denote the integral on the left, considered as a function of b, and show that I satisfies the first-order differential equation

$$\frac{\mathrm{d}I}{\mathrm{d}b} + 2aI = 0.$$

Find $I(b)$ by solving this differential equation.

b. Prove that

$$\int_0^\infty \mathrm{e}^{-\mathrm{j}a^2t^2-\mathrm{j}b^2t^{-2}}\mathrm{d}t = \frac{1-\mathrm{j}}{\sqrt{2}}\frac{\sqrt{\pi}}{2a}\mathrm{e}^{-\mathrm{j}2ab}.$$

c. Prove that

$$\int_0^\infty \mathrm{e}^{\mathrm{j}a^2t^2-\mathrm{j}b^2t^{-2}}\mathrm{d}t = \frac{1-\mathrm{j}}{\sqrt{2}}\frac{\sqrt{\pi}}{2a}\mathrm{e}^{-2ab}.$$

3.19 **a.** The two-dimensional Fourier transform of the unit square $\mathrm{rect}(x, y)$ is $\mathrm{sinc}(f_x, f_y)$. Let $s(t)$ be a triangle pulse

$$s(t) = \begin{cases} 1 - |t| & |t| \le 1 \\ 0 & \text{otherwise.} \end{cases}$$

Use the rotation properties of the two-dimensional Fourier transform applied to $\mathrm{rect}(x, y)$ and the projection-slice theorem to derive the Fourier transform of $s(t)$. Compare this to the Fourier transform of $s(t)$ derived directly.

b. Let $s(t)$ be the pulse

$$s(t) = \begin{cases} \sqrt{1-t^2} & |t| \leq 1 \\ 0 & \text{otherwise.} \end{cases}$$

Derive the Fourier transform of $s(t)$ by using the two-dimensional Fourier transform of the unit circle and the projection-slice theorem.

3.20 **a.** Using spherical coordinates, find the three-dimensional Fourier transform of the sphere function, defined as

$$\text{sphr}(x, y, z) = \begin{cases} 1 & \text{if } \sqrt{x^2+y^2+z^2} \leq \frac{1}{2} \\ 0 & \text{otherwise.} \end{cases}$$

Hint: By symmetry, it is enough to compute $S(0, 0, f)$. The integral $\int x \cos x \, \mathrm{d}x = \cos x + x \sin x$ may be helpful. (Whereas the two-dimensional version of this problem requires Bessel functions, the three-dimensional version requires only trigonometric functions. More generally, the n-dimensional version of the problem requires Bessel functions if n is even, but only trigonometric functions if n is odd.)

b. Using a three-dimensional generalization of the projection-slice theorem, find the Fourier transform of

$$s(x, y) = \begin{cases} \sqrt{1-x^2-y^2} & x^2+y^2 \leq 1 \\ 0 & x^2+y^2 \geq 1. \end{cases}$$

3.21 State and prove a two-dimensional version of the uncertainty principle.

3.22 Establish the following useful properties of Bessel functions:

a. Prove that

$$J_\nu(a)J_\nu(b) = \frac{1}{2\pi}\int_{-\pi}^{\pi} J_0\left(\sqrt{a^2+b^2+2ab\cos\alpha}\right) \mathrm{e}^{-\mathrm{j}\nu\alpha} \, \mathrm{d}\alpha.$$

b. Prove that

$$J_0\left(\sqrt{a^2+b^2-2ab\cos\alpha}\right) = \sum_{\ell=-\infty}^{\infty} J_\ell(a)J_\ell(b)\cos\ell\alpha.$$

c. Prove that

$$J_n(a+b) = \sum_{\ell=-\infty}^{\infty} J_\ell(a)J_{n-\ell}(b).$$

3.23 Given the infinite array of finite sampling apertures

$$\delta_\epsilon(x, y) = \sum_{\ell''=-\infty}^{\infty}\sum_{\ell'=-\infty}^{\infty} \text{circ}\left(\frac{x-\ell'\Delta}{\epsilon}, \frac{y-\ell''\Delta}{\epsilon}\right):$$

a. Describe the sampled signal

$$s_\epsilon(x, y) = \delta_\epsilon(x, y)s(x, y)$$

in the space domain and in the spatial frequency domain where $s(x, y)$ is a large, slowly varying image.

b. What happens as ϵ goes to zero?

c. State a condition under which there is no "aliasing."

3.24 A charge-coupled device (CCD) is a modern light-density to electron-density converter that is widely used as an optical sensor in devices such as camcorders. A CCD is divided into nonoverlapping cells called "pixels." In each cell, the total incident light intensity received in time Δt over the area of that cell is converted to a voltage measurement.

a. Prove that this device can be modeled as

$$s'(x, y) = h(-x, -y) * * s(x, y)$$

where $h(x, y)$ describes the shape of a pixel, followed by a perfect sampler with samples centered on pixels.

b. If $s(x, y) = \text{circ}\left(\frac{x}{a}, \frac{y}{a}\right)$ and the pixel is also $\text{circ}\left(\frac{x}{a}, \frac{y}{a}\right)$, describe how the CCD degrades the image.

3.25 Suppose $S(f_x, f_y) = \text{sinc}\sqrt{f_x^2 + f_y^2}$. Is the inverse Fourier transform $s(x, y)$ a cone? What are the projections of $s(x, y)$? What are the projections of a cone? A cone is described by

$$\text{cone}(x, y) = \begin{cases} 1 - \sqrt{x^2 + y^2} & x^2 + y^2 \le 1 \\ 0 & \text{otherwise.} \end{cases}$$

3.26 Suppose that $s(x, y)$ has the property that $S(f_x, f_y) = s(-f_x, f_y)$. (Except for a sign change, $s(x, y)$ is its own two-dimensional Fourier transform.) Use the projection-slice theorem to derive a constraint on the shape of $s(x, y)$. Specifically, for $s(x, y)$ to have a narrow projection at angle θ, it must be wide at a slice $90°$ from θ.

Notes

The two-dimensional Fourier transform and many of its properties form an obvious generalization of the theory of the one-dimensional Fourier transform. It plays an important role in image formation algorithms, Fourier optics, and Fourier antenna theory. The Hankel transform is closely related to the two-dimensional Fourier transform.

The square of the two-dimensional Fourier transform of the circle function is widely used in connection with problems in optical astronomy and was studied by Airy in 1835. Rayleigh, in 1879, defined his resolution limit in terms of the Airy disk to show that optical resolution by conventional methods using monochromatic light is of the order of the wavelength of the light.

The Radon transform was introduced by Johann Radon in a 1917 paper that went unnoticed for many years. The transform and its properties were rediscovered through the years by many people in many fields and in varying degrees of explicitness. The Radon transform and its relationship with the Fourier transform were explicitly stated and formalized in papers by Bracewell (1956, 1958) and by Lighthill (1958). The projection-slice theorem was first stated in the context of Fourier transform theory by Bracewell (1956), although it was implicit earlier in many applications. The shadow transform was discussed by Solomon (1976).

The sampling theorem does not have a completely satisfactory statement in polar coordinates. This topic was studied by Stark (1979), with comments by Fan and Sanz (1985). Two-dimensional sampling on a nonrectangular array was discussed by Petersen and Middleton (1962). Two-dimensional sampling on a hexagonal grid was further studied by Mersereau (1979). Indeed, the sampling theorem can be generalized to any lattice in n-dimensional space.

4 Optical imaging systems

The earliest imaging systems were optical imaging systems, and optical systems are still the most common imaging systems. Optical imaging systems that employ the simple lens are widespread; they are found both in biological organisms and in man-made devices. Much of optics, including the properties of the ideal lens, can be understood in the language of signal processing in terms of pointspread functions, convolutions, and Fourier transforms. More generally, we shall describe the propagation and diffraction of waves based on a two-dimensional pointspread function. In this setting, the Huygens–Fresnel principle of optics will be presented simply as a special case of the convolution theorem of the two-dimensional Fourier transform.

In principle, the diffraction of electromagnetic waves should be explained directly from Maxwell's equations, which give a complete description of electromagnetic fields. However, there may be mathematical difficulties when starting from first principles because there may be concerns about how to model a given problem, or how to specify a consistent and accurate set of boundary conditions. It may be difficult to formulate the boundary conditions at a level of detail needed to apply Maxwell's equations, while the weaker conditions needed for diffraction theory may be readily apparent. This is why we formulate the theories of diffraction as distinct from, but subservient to, electromagnetic theory.

We choose to describe diffraction simply as a property of wave propagation in a homogeneous medium. The Huygens–Fresnel principle then becomes a description of a signal when that signal has the form of a propagating wave of constant velocity. As such the Huygens–Fresnel principle applies to many subjects, including optics and infrared optics, antenna theory (both for radar and for communications), sonar hydrophones, acoustics (including speaker design), ultrasonics, and seismics. It is unnecessary to present the mathematical study of diffraction as restricted to any particular instance of wave propagation.

4.1 Scalar diffraction

Diffraction is the study of the relationship between the complex amplitude distribution of a propagating wave at one plane that the wavefront passes through and the complex amplitude distribution of that same wavefront at a subsequent plane. The terms *diffraction* and *interference* refer to the same basic phenomenon. The term interference is usually chosen for the interaction of a finite number of wavefronts such as when the source consists of a finite number of point sources or line sources. The term diffraction is preferred for the general situation such as when there is an uncountable infinity of interacting wavefronts.

Scalar diffraction theory is a mathematical formulation of diffraction that provides a satisfactory description of the physical phenomenon of diffraction for scalar waves such as acoustic waves. Scalar diffraction theory usually gives satisfactory results even for vector waves such as electromagnetic waves, especially if the significant dimensions in the problem are large in comparison to a wavelength. There are rare cases, however, where scalar diffraction theory is inadequate and fails to describe the physical phenomenon. Then a richer theory, known as the *vector diffraction theory*, must be used.

Scalar diffraction is precisely described by the *Huygens–Fresnel principle*, which describes a wavefront, $s(x, y, z)$, on the x, y plane at $z = d$ in terms of that wavefront on a previous x, y plane at $z = 0$. With $s_0(x, y) = s(x, y, 0)$ and $s_d(x, y) = s(x, y, d)$, the Huygens–Fresnel principle, which will be derived in Section 4.2, is

$$s_d(x, y) = h(x, y) ** s_0(x, y)$$

where $h(x, y)$ is a function known as the *Huygens–Fresnel pointspread function*[1] or the *pointspread function of free space*, and is given by

$$h(x, y) = \left[\frac{-\mathrm{j}d/\lambda}{(d^2 + x^2 + y^2)} + \frac{d/2\pi}{(d^2 + x^2 + y^2)^{3/2}} \right] e^{\mathrm{j}2\pi\sqrt{d^2+x^2+y^2}/\lambda}.$$

The Huygens–Fresnel pointspread function describes the relationship between a wavefront in the x, y planes at two different values of z. There need not be any special significance to the planes at $z = 0$ and $z = d$. The wave need not be generated nor sensed at either of these planes. Indeed, since the coordinate origin is arbitrary, the Huygens–Fresnel principle describes the relationship between the complex amplitude of the wavefront in any two parallel planes. It says nothing about the

[1] The signal-processing term *pointspread function* may be called a *Green's function* in the study of wave propagation, though the term Green's function usually refers to the kernel of a three-dimensional convolution rather than that of a two-dimensional convolution.

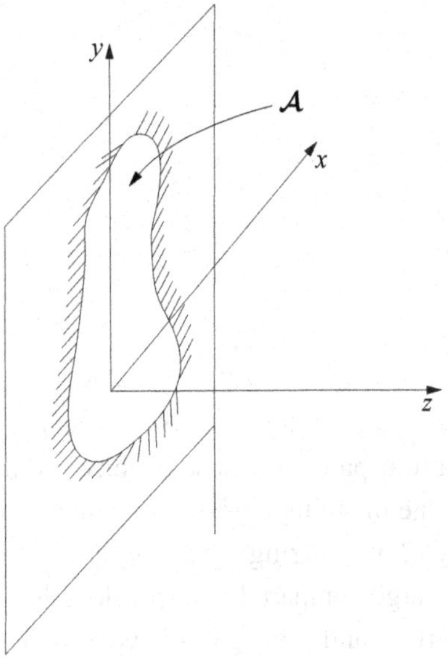

Figure 4.1 An aperture

realizability of the physical boundary conditions necessary to generate a particular wavefront.

We will presume that we can generate any desired wavefront, $s_0(x, y)$, leaving the plane with $z = 0$ as by passing a monodirectional plane wave through a "transparency" in the plane with $z = 0$. Actually, it may be physically very difficult (or impossible) to generate a wave with a specified complex amplitude in the plane $z = 0$. For example, details smaller than the wavelength of light may not be easy, or even possible, to generate by passing light through a transparency. Nevertheless, we will consider the propagation of all such waves. Mathematically, there is no reason to exclude them. Physically, if such waves can be generated, they will propagate, and our only goal is to describe the waves during propagation.

Consider an opaque screen in the x, y plane at $z = 0$ with an aperture, $\mathcal{A}$, cut out of it, as shown in Figure 4.1. The aperture can be described as a closed set of points $\mathcal{A}$ such that the screen at (x, y) is removed if $(x, y) \in \mathcal{A}$. Passing a plane wave through the aperture can be described as multiplying a two-dimensional signal by a two-dimensional function, called a *transmittance function*, defined by

$$t(x, y) = \begin{cases} 1 & (x, y) \in \mathcal{A} \\ 0 & (x, y) \notin \mathcal{A} \end{cases}.$$

If the incoming spatial signal incident on the aperture is $c(x, y)$, then the outgoing spatial signal $s_0(x, y)$, called the *illumination function* (at the output of the

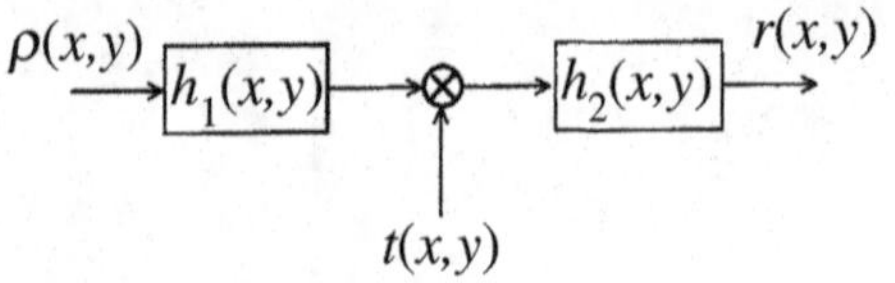

Figure 4.2 Functional model of an optical system

aperture), is

$$s_0(x, y) = t(x, y)c(x, y).$$

In the general case, we shall regard $t(x, y)$ as any complex function of x and y. We may visualize this transmittance physically as a transparency whose opacity and thickness vary with x and y. The opacity determines the magnitude of $t(x, y)$, and the thickness determines the phase of $t(x, y)$. This way of visualizing $t(x, y)$ is quite reasonable physically as long as the detail in $t(x, y)$ is large compared to a wavelength.[2]

The signal $s_0(x, y)$ leaves the planar aperture and propagates beyond it. The signal arriving at a plane located at distance d beyond the aperture is described by the Huygens–Fresnel principle

$$\begin{aligned} s_d(x, y) &= h(x, y) ** s_0(x, y) \\ &= h(x, y) ** t(x, y)c(x, y) \end{aligned}$$

where $h(x, y)$ is the Huygens–Fresnel pointspread function at distance d. If $c(x, y)$ represents a plane wave with an amplitude equal to one moving along the z axis, then $c(x, y) = 1$ and

$$s_d(x, y) = h(x, y) ** t(x, y).$$

A more complicated situation is shown in Figure 4.2. The transparency $t(x, y)$ is illuminated by an incident wavefront that has the complex amplitude $\rho(x, y)$ at distance d_1 prior to the transparency. This might have been formed by illuminating another transparency, $\rho(x, y)$, with a plane wave, $c(x, y) = 1$, to produce the signal $\rho(x, y)$ at its output. This produces the signal $h_1(x, y) ** \rho(x, y)$ at the distance d_1, which passes through a second transparency $t(x, y)$. At the distance d_2 from the second transparency, the signal is

$$r(x, y) = h_2(x, y) ** [t(x, y)[h_1(x, y) ** \rho(x, y)]]$$

where $h_1(x, y)$ and $h_2(x, y)$ denote the Huygens–Fresnel pointspread function with distances d_1 and d_2, respectively.

[2] The branch of optics that assumes the existence of complex transmittance functions and studies their interaction with the Huygens–Fresnel principle is called *Fourier optics*.

4.2 The Huygens–Fresnel principle

A monochromatic plane wave satisfying the wave equation and moving entirely in the z direction cannot have a spatially varying amplitude, as was discussed in Section 1.3. Conversely, if a monochromatic wave passing through the x, y plane at $z = 0$ has a spatially varying amplitude across that plane, then the wave must be a composite of plane waves moving in different directions. This simple fact underlies diffraction and the Huygens–Fresnel principle. We shall develop these consequences in detail.

As was discussed in Section 1.7, a monochromatic, monodirectional plane wave of amplitude A and phase θ has the complex baseband representation

$$s(x, y, z) = A e^{-j\theta} e^{j2\pi f_0(\alpha x + \beta y + \gamma z)/c}$$

where α, β, and γ are the direction cosines and $\gamma = \sqrt{1 - \alpha^2 - \beta^2}$. Then, as was discussed in Section 1.8, a general monochromatic wave is composed of a linear superposition of such monodirectional waves:

$$s(x, y, z) = \int_{-1}^{1} \int_{-1}^{1} a(\alpha, \beta) e^{j2\pi f_0 \sqrt{1-\alpha^2-\beta^2} z/c} e^{j2\pi f_0(\alpha x + \beta y)/c} \, d\alpha \, d\beta$$

where $a(\alpha, \beta)$ is known as the *angular spectrum* of the wave.

At this point, it is convenient to replace the limits of integration by $\pm\infty$. Then

$$s(x, y, z) = \int_{-\infty}^{\infty} \int_{-\infty}^{\infty} a(\alpha, \beta) e^{j2\pi f_0 \sqrt{1-\alpha^2-\beta^2} z/c} e^{j2\pi f_0(\alpha x + \beta y)/c} \, d\alpha \, d\beta.$$

This puts the expression in the form of an inverse Fourier transform. If we simply require that $a(\alpha, \beta) = 0$ for $|\alpha| \geq 1$ or $|\beta| \geq 1$, we really change nothing while introducing the standard form of the Fourier transform. However, there is good reason to allow the direction cosines to range from $-\infty$ to ∞ without constraining $a(\alpha, \beta)$. As discussed in Section 1.8, waves known as *evanescent waves*, do exist with $a(\alpha, \beta)$ nonzero for $|\alpha| \geq 1$ or $|\beta| \geq 1$.

Let $s_0(x, y) = s(x, y, 0)$ denote the wave $s(x, y, z)$ restricted to the x, y plane at $z = 0$,

$$s_0(x, y) = \int_{-\infty}^{\infty} \int_{-\infty}^{\infty} a(\alpha, \beta) e^{j2\pi \left(\frac{\alpha}{\lambda} x + \frac{\beta}{\lambda} y\right)} \, d\alpha \, d\beta.$$

This equation has the form of the inverse two-dimensional Fourier transform

$$s_0(x, y) = \int_{-\infty}^{\infty} \int_{-\infty}^{\infty} S_0(f_x, f_y) e^{j2\pi (f_x x + f_y y)} \, df_x \, df_y$$

where $S_0(f_x, f_y)$ is the Fourier transform of $s_0(x, y)$. Therefore with $f_x = \alpha/\lambda$ and

$f_y = \beta/\lambda$, the angular spectrum can be interpreted as

$$\lambda^2 a(\alpha, \beta) = S_0\left(\frac{\alpha}{\lambda}, \frac{\beta}{\lambda}\right).$$

Thus, the angular spectrum is essentially the Fourier transform of the wave as it appears in the x, y plane with $z = 0$.

Now consider the wavefront as it appears in the x, y plane with $z = d$, letting $s_d(x, y) = s(x, y, d)$. Then, with the same reasoning,

$$s_d(x, y) = \int_{-\infty}^{\infty}\int_{-\infty}^{\infty}\left[S_0(f_x, f_y)e^{j2\pi\frac{d}{\lambda}\sqrt{1-\lambda^2 f_x^2-\lambda^2 f_y^2}}\right]e^{j2\pi(f_x x+f_y y)}\,df_x\,df_y.$$

This is just the inverse Fourier transform of the product

$$S_d(f_x, f_y) = S_0(f_x, f_y)e^{j2\pi d\sqrt{\lambda^{-2}-f_x^2-f_y^2}}.$$

By the convolution theorem,

$$s_d(x, y) = h(x, y) ** s_0(x, y)$$

where $h(x, y)$ is defined by the Fourier transform relationship

$$h(x, y) \Leftrightarrow e^{j2\pi d\sqrt{\lambda^{-2}-f_x^2-f_y^2}}.$$

Our last task of this development is to find $h(x, y)$. To find $h(x, y)$, refer to Proposition 3.5.1, which, with the conjugation property, gives the Fourier transform pair

$$\frac{-1}{\sqrt{b^2+x^2+y^2}}e^{j2\pi a\sqrt{b^2+x^2+y^2}} \Leftrightarrow \frac{1}{j\sqrt{a^2-f_x^2-f_y^2}}e^{j2\pi b\sqrt{a^2-f_x^2-f_y^2}}.$$

Take the derivative of both sides with respect to b to obtain

$$\left[\frac{-j2\pi ab}{(b^2+x^2+y^2)}+\frac{b}{(b^2+x^2+y^2)^{3/2}}\right]e^{j2\pi a\sqrt{b^2+x^2+y^2}} \Leftrightarrow 2\pi e^{j2\pi b\sqrt{a^2-f_x^2-f_y^2}}.$$

Consequently, by setting $b = d$ and $a = \lambda^{-1}$, we have $h(x, y)$, as stated in the following definition:

Definition 4.2.1 *The Huygens–Fresnel pointspread function is*

$$h(x, y) = \left[\frac{-jd/\lambda}{(d^2+x^2+y^2)}+\frac{d/2\pi}{(d^2+x^2+y^2)^{3/2}}\right]e^{j2\pi\sqrt{d^2+x^2+y^2}/\lambda}.$$

Theorem 4.2.2 (Huygens–Fresnel Principle) *If a monochromatic wave, $s(x, y, z)$, takes the values $s_0(x, y)$ in the x, y plane at $z = 0$, then in the x, y plane at $z = d$, it takes the values*

$$s_d(x, y) = h(x, y) ** s_0(x, y)$$

where $h(x, y)$ is the Huygens–Fresnel pointspread function.

Proof: The proof is implicit in the discussion prior to the definition. □

The mathematically complete form of $h(x, y)$, which is given in Definition 4.2.1, has two terms: one term decreases as the inverse square of $\sqrt{d^2 + x^2 + y^2}$; the other decreases as the inverse cube. The second term, sometimes called the *reactive term*, can be important in situations where d is only a few wavelengths. In more traditional situations, d is large, and the reactive term can be neglected. For such cases, the Huygens–Fresnel pointspread function is commonly written as

$$h(x, y) = \frac{-\mathrm{j}d/\lambda}{d^2 + x^2 + y^2} e^{\mathrm{j}2\pi\sqrt{d^2+x^2+y^2}/\lambda}.$$

This approximate form, now known more simply as the *Huygens pointspread function*, is entirely adequate for most conventional applications.

The Huygens pointspread function has a satisfying intuitive interpretation. First, consider a monochromatic, complex passband, spherical wave radiating omnidirectionally from the origin. It is given by

$$\widetilde{c}(t, x, y, z) = \frac{e^{-\mathrm{j}2\pi f_0(t-\sqrt{x^2+y^2+z^2}/c)}}{\sqrt{x^2 + y^2 + z^2}}.$$

The attenuation term in the denominator assures the conservation of energy. Each sphere about the origin will have the same energy passing through it per unit time. The complex baseband form of this wave, crossing the x, y plane at $z = d$, is

$$c(x, y, d) = \frac{1}{\sqrt{x^2 + y^2 + d^2}} e^{\mathrm{j}2\pi\sqrt{x^2+y^2+d^2}/\lambda}.$$

The Huygens pointspread function $h(x, y)$ can now be redeveloped in a qualitative way, known as the *Huygens principle*, which states that each point of an advancing wavefront can be regarded as the source of an infinitesimal secondary spherical wavelet; the subsequent advancing wavefront is then the cumulative effect of the secondary wavelets. The secondary wavelets are precisely described by the Huygens pointspread function, which we now rewrite as follows:

$$h(x, y) = -\mathrm{j}\frac{1}{\lambda}\left[\frac{1}{\sqrt{x^2 + y^2 + d^2}} e^{\mathrm{j}2\pi\sqrt{x^2+y^2+d^2}/\lambda}\right]\frac{d}{\sqrt{x^2 + y^2 + d^2}}.$$

Thus the radiated Huygens wavelet is (1) a spherical wavelet, (2) phase shifted by 90° because of the term $-\mathrm{j}$, (3) attenuated by the wavelength described by the term $1/\lambda$, and (4) attenuated by the "obliquity factor" $d/\sqrt{x^2 + y^2 + d^2}$, which is the cosine of the angle between the spherical wavefront at (x, y, d) and the x, y plane passing through the point (x, y, d). We can think of the obliquity factor as due to the projection of the spherical wavefront onto that x, y plane.

This intuitive interpretation is often used as the basis of a physical "derivation" of Huygens' principle. Each point of an advancing wavefront becomes the source of a secondary Huygens wavelet that radiates spherically. The wave at a subsequent surface is then the integral of these secondary wavelets. This argument is a useful heuristic derivation, but it is flawed because it is "intuitively obvious" only in retrospect or to a willing listener. It does not readily explain the term j/λ, nor does it explain why the secondary Huygens wavelets do not propagate in the backwards direction. Most important, it gives the wrong answer; the complete form of $h(x, y)$, given in Definition 4.2.1, is not obtained.

4.3 Fresnel and Fraunhofer approximations

The general theory of scalar diffraction is based on the Huygens–Fresnel principle, which was derived in Section 4.2. In its exact form, the Huygens–Fresnel principle is usually considered needlessly complicated for practical applications. Even the Huygens principle is usually too complicated. Various approximations are in common use, primarily *Fresnel diffraction* or *near-field diffraction*, and *Fraunhofer diffraction* or *far-field diffraction*. We shall see that the term "Fraunhofer diffraction pattern" can be translated into the term "two-dimensional Fourier transform," thereby changing the terminology from that of optics to that of signal processing, but without otherwise changing the facts.

Huygens' principle

$$s_d(x, y) = h(x, y) ** s_0(x, y)$$

has the form of the convolution integral

$$s_d(x, y) = \int_{-\infty}^{\infty} \int_{-\infty}^{\infty} s_0(\xi, \eta) \frac{d/j\lambda}{(x-\xi)^2 + (y-\eta)^2 + d^2} e^{j(2\pi/\lambda)\sqrt{(x-\xi)^2+(y-\eta)^2+d^2}} \, d\xi \, d\eta.$$

This integral expression is rather inscrutable as it stands, especially because of the square root in the exponent. In common practice, one or several standard simplifying approximations make the expression much more useful. First, we assume that the range of x, y, d, ξ, and η are such that the denominator of the pointspread function $h(x, y)$ is adequately approximated by replacing $\sqrt{x^2 + y^2 + d^2}$ by the constant d. This is implied by the statement that d is large compared to x, y, ξ, and η. Then the pointspread function

$$h(x, y) = \frac{1}{j\lambda d} e^{j(2\pi/\lambda)\sqrt{x^2+y^2+d^2}}$$

can be used. Now the amplitude term of $1/j\lambda d$ is a constant, which is not of immediate interest. By redefining $s_0(\xi, \eta)$, the amplitude term $1/j\lambda d$ can be absorbed into $s_0(\xi, \eta)$ and is not explicitly written. Then the Huygens principle is written in the approximated form

$$s_d(x, y) = \int_{-\infty}^{\infty} \int_{-\infty}^{\infty} s_0(\xi, \eta) \, e^{j(2\pi/\lambda)\sqrt{(x-\xi)^2+(y-\eta)^2+d^2}} \, d\xi \, d\eta,$$

now corresponding to the simplified pointspread function

$$h(x, y) = e^{j(2\pi/\lambda)\sqrt{x^2+y^2+d^2}}.$$

Next, we turn our attention to the square root in the exponent. The square root makes the expression unwieldy and almost useless for an elementary analysis of optical systems. There are two standard approximations to the exponent, the *Fresnel approximation* and the *Fraunhofer approximation*, that may be used in appropriate situations.

In Chapter 15, we will study various criteria for judging the quality of phase approximations. For the present, we will make an approximation – without further comment – that it is accurate enough if the exponent is approximated by an expression that is in error by a phase angle not larger than some constant such as $\pi/4$.

Fresnel approximation

To obtain the Fresnel approximation, the exponent in the Huygens pointspread function is expanded as the series

$$\sqrt{(x-\xi)^2+(y-\eta)^2+d^2}$$
$$\approx d\left[1+\frac{1}{2}\left(\frac{(x-\xi)^2+(y-\eta)^2}{d^2}\right)-\frac{1}{8}\left(\frac{(x-\xi)^2+(y-\eta)^2}{d^2}\right)^2+\cdots\right],$$

and, for large d, approximated by terms up to quadratic in x and y. Therefore, in the Fresnel approximation,

$$h(x, y) = e^{jkd}e^{j\frac{k}{2d}(x^2+y^2)}$$

where $k = 2\pi/\lambda$. In the Fresnel approximation, $s_d(x, y)$ has the form of a two-dimensional convolution of $s_0(x, y)$ with a two-dimensional, quadratic-phase pulse,

$$s_d(x, y) = e^{jkd}\left[s_0(x, y) ** e^{j(2\pi/\lambda)(x^2+y^2)/2d}\right].$$

The term e^{jkd} simply describes a phase shift between a plane wave at $z = 0$ and that same wave at $z = d$. It is common to suppress this term as uninteresting, though it is understood to be there. The pointspread function

$$h(x, y) = e^{j(2\pi/\lambda)(x^2+y^2)/2d},$$

or, with the amplitude term explicit,

$$h(x, y) = \frac{1}{j\lambda d}e^{j(2\pi/\lambda)(x^2+y^2)/2d},$$

is called the *pointspread function of free space in the Fresnel approximation*, or the *Fresnel pointspread function*.

To determine a sufficient condition on the region in which the Fresnel approximation is adequate, we compare the largest neglected term of the Taylor series expansion to $\pi/4$

$$\frac{2\pi}{\lambda} d \frac{1}{8} \left[\frac{(x-\xi)^2 + (y-\eta)^2}{d^2} \right]^2 \leq \frac{\pi}{4}.$$

From this, we conclude that the Fresnel approximation is satisfactory if

$$[(x-\xi)^2 + (y-\eta)^2]^2 \leq \lambda d^3.$$

This is a sufficient, but not necessary, condition. The Fresnel approximation may sometimes be satisfactory even under weaker conditions.

Fraunhofer approximation

The Fraunhofer approximation is an alternative approximation to the Huygens convolution equation. Unlike the Fresnel approximation, the Fraunhofer approximation is not an approximation to the pointspread function $h(x, y)$ alone. Rather, it is an approximation to the entire convolution integral in which $h(x, y)$ appears. The approximation replaces the exponent within the convolution equation

$$s_d(x, y) = \int_{-\infty}^{\infty} \int_{-\infty}^{\infty} s_0(\xi, \eta) e^{j(2\pi/\lambda)\sqrt{(x-\xi)^2 + (y-\eta)^2 + d^2}} \, d\xi \, d\eta$$

by a term that is linear in x and y. Let $R^2 = x^2 + y^2 + z^2$, then

$$\begin{aligned} \sqrt{(x-\xi)^2 + (y-\eta)^2 + z^2} &= R\sqrt{1 - \frac{2x\xi}{R^2} - \frac{2y\eta}{R^2} + \frac{\xi^2 + \eta^2}{R^2}} \\ &= R\left[1 - \frac{1}{2}\left(\frac{2x\xi + 2y\eta - \xi^2 - \eta^2}{R^2}\right)\right. \\ &\quad \left. - \frac{1}{8}\left(\frac{2x\xi + 2y\eta - \xi^2 - \eta^2}{R^2}\right)^2 + \cdots\right]. \end{aligned}$$

We can then make the approximation

$$\sqrt{(x-\xi)^2 + (y-\eta)^2 + z^2} \approx R - \frac{x}{R}\xi - \frac{y}{R}\eta,$$

provided R is sufficiently large. Therefore, in the Fraunhofer approximation,

$$\begin{aligned} s(x, y) &= e^{j2\pi R/\lambda} \int_{-\infty}^{\infty} \int_{-\infty}^{\infty} s_0(\xi, \eta) e^{-j(2\pi/\lambda)(\alpha\xi + \beta\eta)} \, d\xi \, d\eta \\ &= e^{j2\pi R/\lambda} S_0\left(\frac{\alpha}{\lambda}, \frac{\beta}{\lambda}\right) \end{aligned}$$

where $\alpha = x/R$ and $\beta = y/R$ are the direction cosines. Thus, in the Fraunhofer approximation, the Huygens–Fresnel convolution integral is replaced with a Fourier transform.

The region in which the Fraunhofer approximation is adequate is the region where the quadratic term in the preceding Taylor series approximation can be neglected. We will specify that the quadratic term in the series expansion can be neglected if it is not more than $\pi/4$. Thus examining those terms that are quadratic in ξ and η gives

$$\left|\frac{2\pi R}{\lambda}\left[\frac{1}{2}\left(\frac{\xi^2+\eta^2}{R^2}\right)-\frac{1}{2}\left(\frac{x\xi+y\eta}{R^2}\right)^2\right]\right| \leq \frac{\pi}{4},$$

which can be rewritten as

$$\left|\xi^2+\eta^2-\left(\frac{x}{R}\xi+\frac{y}{R}\eta\right)^2\right| \leq \frac{\lambda R}{4}.$$

For x and y small compared to R, this becomes

$$\xi^2+\eta^2 \leq \frac{\lambda R}{4}.$$

To summarize this conclusion, let r be the radius of a circle enclosing the aperture. The Fraunhofer approximation is usually satisfactory if

$$r^2 \leq \frac{\lambda R}{4}.$$

In a typical optics application, we may have $r = 0.03$ meter and $\lambda = 0.4 \times 10^{-6}$ meter. This requires that R be greater than 9000 meters. This shows that the Fraunhofer approximation is not valid in a typical optics application.

In a typical radar application, we may have $r = 1$ meter and $\lambda = 0.1$ meter. The above condition requires that R be greater than 40 meters. This suggests that the Fraunhofer approximation is valid in many radar applications. It would fail, however, with $r = 100$ meters, $\lambda = 0.01$ meter, and $R = 10^6$ meters.

4.4 The geometrical optics approximation

The Fresnel and the Fraunhofer approximations to the Huygens–Fresnel principle are each valid for λ larger than a certain value, which depends on the distances within the geometrical situation. An approximation of a much different kind, called the *geometrical optics approximation*, results if λ is taken instead to be vanishingly small. This approximation is usually well justified in many everyday optical situations where optical wavelengths are smaller than 10^{-6} meter. Indeed, the geometrical optics approximation, leading to the field of geometrical optics and ray tracing, is the justification for much of practical optics.

The geometrical optics approximation says that, for sufficiently small λ, the Huygens–Fresnel pointspread function can be approximated as

$$h(x, y) \approx \delta(x, y)e^{j2\pi d/\lambda}.$$

The impulse function $\delta(x, y)$ in the approximation is to be understood, as usual, in terms of its behavior as an operator under an integral sign.

The geometrical optics approximation follows from the Huygens–Fresnel principle by using the *principle of stationary phase*. This principle makes the observation that, in an integral of the form

$$I = \int_{-\infty}^{\infty} a(t)e^{jb(t)}\,dt,$$

there will be little contribution to the integral in regions of the time axis where $a(t)$ is changing slowly and $b(t)$ is large. If $b(t)$ is a smooth function with a minimum, and $b(t)$ grows large compared to 2π on either side of this minimum, then the primary contributions to the integral come from an interval surrounding the minimum of $b(t)$, provided $a(t)$ is varying sufficiently slowly with t. For many integrals of this form, it is possible to obtain reasonably good approximations by using the principle of stationary phase.

Suppose that $b(t)$ takes its minimum at $t = t_o$, and expand $b(t)$ in a series about this point,

$$b(t) = b(t_o) + \tfrac{1}{2}b''(t_o)(t - t_o)^2 + \tfrac{1}{6}b'''(t_o)(t - t_o)^3 + \cdots .$$

Then

$$\int_{-\infty}^{\infty} a(t)e^{jb(t)}\,dt = e^{jb(t_o)} \int_{-\infty}^{\infty} a(t)e^{j\left[\frac{1}{2}b''(t_o)(t-t_o)^2+\cdots\right]}\,dt.$$

The principle of stationary phase makes the approximation that if $a(t)$ is slowly varying, then

$$\int_{-\infty}^{\infty} a(t)e^{jb(t)}\,dt \approx e^{jb(t_o)}a(t_o) \int_{-\infty}^{\infty} e^{j\frac{1}{2}b''(t_o)(t-t_o)^2}\,dt.$$

We now use the principle of stationary phase to develop the geometrical optics approximation, starting with the Huygens–Fresnel principle

$$s_d(x, y) = \int_{-\infty}^{\infty}\int_{-\infty}^{\infty} s_0(\xi, \eta)a(x - \xi, y - \eta)e^{j(2\pi/\lambda)\sqrt{(x-\xi)^2+(y-\eta)^2+d^2}}\,d\xi\,d\eta$$

where

$$a(x, y) = \frac{-jd/\lambda}{(x^2 + y^2 + d^2)} + \frac{d/2\pi}{(x^2 + y^2 + d^2)^{3/2}}.$$

The exponent can be expanded in a series as

$$e^{j(2\pi/\lambda)\sqrt{(x-\xi)^2+(y-\eta)^2+d^2}} = e^{j(2\pi/\lambda)\left[d+(x-\xi)^2/2d+(y-\eta)^2/2d+\cdots\right]}.$$

By the principle of stationary phase,

$$s_d(x, y) \approx e^{j2\pi d/\lambda}s_0(x, y)a(0, 0) \int_{-\infty}^{\infty}\int_{-\infty}^{\infty} e^{j(\pi/\lambda d)[(x-\xi)^2+(y-\eta)^2]}\,d\xi\,d\eta.$$

Make the appropriate changes in the variables of integration so that $(\xi - x)$ and $(\eta - y)$ become replaced by ξ and η. Then

$$s_d(x, y) \approx e^{j2\pi d/\lambda} s_0(x, y) \left[\frac{-jd/\lambda}{d^2} + \frac{d/2\pi}{d^3} \right] \int_{-\infty}^{\infty} e^{j(\pi/\lambda d)\xi^2} \, d\xi \int_{-\infty}^{\infty} e^{j(\pi/\lambda d)\eta^2} \, d\eta.$$

Next, use the widely-tabulated definite integrals

$$\int_{-\infty}^{\infty} \cos x^2 \, dx = \int_{-\infty}^{\infty} \sin x^2 \, dx = \sqrt{\frac{\pi}{2}}$$

to write

$$\int_{-\infty}^{\infty} e^{j(\pi/\lambda d)\xi^2} \, d\xi = \int_{-\infty}^{\infty} e^{j(\pi/\lambda d)\eta^2} \, d\eta = \sqrt{\frac{\lambda d}{2}}(1 + j).$$

Then

$$\begin{aligned} s_d(x, y) &\approx e^{j2\pi d/\lambda} s_0(x, y) \left[\frac{-j}{\lambda d} + \frac{1}{2\pi d^2} \right] \frac{\lambda d}{2}(1 + j)^2 \\ &\approx e^{j2\pi d/\lambda} s_0(x, y) \left[1 + \frac{j\lambda}{2\pi d} \right]. \end{aligned}$$

For λ small, we have

$$s_d(x, y) \approx e^{j2\pi d/\lambda} s_0(x, y).$$

This is the response of the pointspread function

$$h(x, y) = e^{j2\pi d/\lambda} \delta(x, y).$$

In the geometrical optics approximation, the wave $s_0(x, y)$ propagates without diffraction and with a phase shift due to the propagation distance. In this approximation, the complex passband wave will be described as

$$s(t, x, y, z) = s_0(x, y) e^{j2\pi f_0(t - z/c)}.$$

4.5 The ideal lens

The design of high-quality lenses is a sophisticated subject. We shall be interested only in the mathematical description of the input/output relationship of the lens. For our purposes, the important processing properties of the simple lens are adequately explained by using a simple mathematical model called the *ideal lens*. The ideal lens can be used to process a waveform in certain ways, and thus is an example of a processor.

An ideal lens, depicted symbolically in Figure 4.3, is a mathematical function that is an idealized model of certain physical lenses. An ideal lens is a transmittance function, $t(x, y)$, nonzero on a finite aperture, $\mathcal{A}$, that introduces a phase factor across the

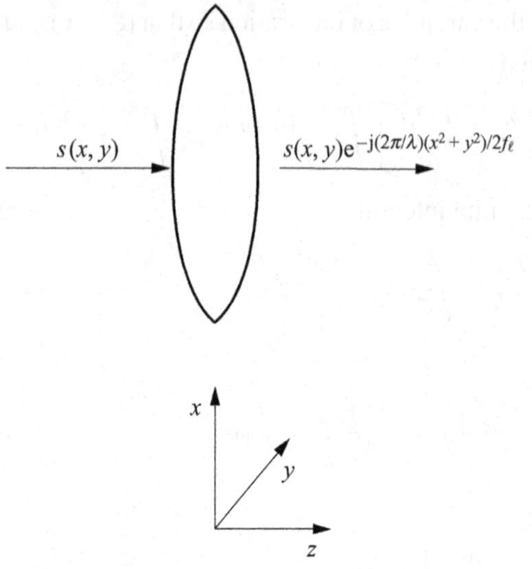

Figure 4.3 The ideal lens

aperture which varies quadratically in the radial direction. Specifically, on an aperture, $\mathcal{A}$, described by the aperture function

$$A(x, y) = \begin{cases} 1 & (x, y) \in \mathcal{A} \\ 0 & (x, y) \notin \mathcal{A}, \end{cases}$$

the transmittance is given by

$$t(x, y) = e^{-j2\pi(x^2+y^2)/2\lambda f_\ell} A(x, y)$$

where f_ℓ is a constant known as the *focal length*[3] of the lens. If f_ℓ is positive, the lens is called a *positive lens*, while if f_ℓ is negative, the lens is called a *negative lens*. A plane wave traveling in the z-direction, when multiplied by the transmittance $t(x, y)$, has a phase change at coordinate (x, y) that is proportional to $x^2 + y^2$.

Most commonly, the aperture $\mathcal{A}$ is a circle. For a circle of radius r,

$$t(x, y) = e^{-j(2\pi/\lambda)(x^2+y^2)/2f_\ell} \operatorname{circ}\left(\frac{x}{2r}, \frac{y}{2r}\right).$$

Notice that the phase of the ideal lens depends only on the single constant λf_ℓ.

In spite of its apparent simplicity, the processing power of a lens can be enormous if the processing task is exactly the right one. A lens can perform either of two processing tasks, shown in Figure 4.4. It can form an image, or it can form a two-dimensional

[3] The focal length f_ℓ should not be confused with frequency, which is also denoted by a subscripted f. The use of f for focal length has too long a precedence to ignore. This leads to the bizarre and perhaps confusing notational coincidence that the spatial coordinates in the focal plane, which we might denote as (x_f, y_f), represent the spatial frequencies (f_x, f_y) because of the Fourier transforming property of the lens.

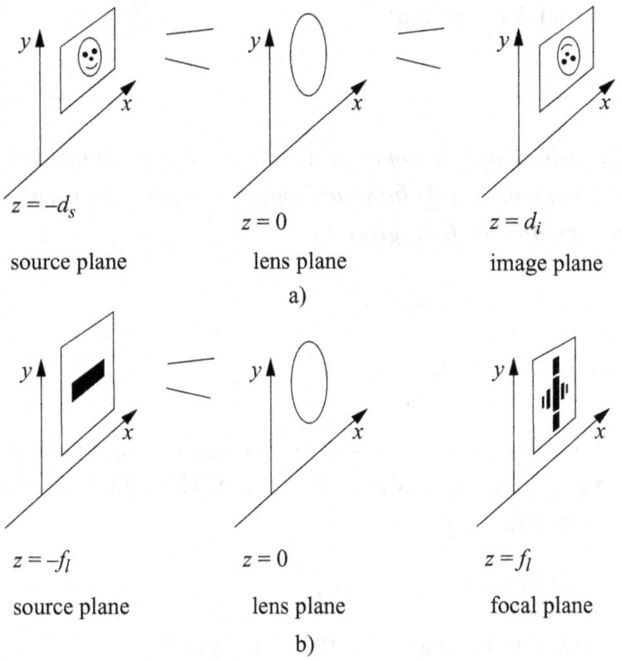

Figure 4.4 Two functions of an ideal lens: a) lens forms an image; b) lens forms a Fourier transform

Fourier transform. A lens can be cascaded with other optical elements to create a processor of enormous throughput, but of quite limited flexibility.

At the end of this section, we shall discuss the relationship between $t(x, y)$ and the geometrical shape of a practical lens. But first, as our primary interest, we will study the properties of the ideal lens as a circuit element.

Lens as an imager

The ideal lens can be used to form images either with coherent light, which we will study in this section, or with noncoherent light, which we will study in the next section. A *geometrical image* of the function $\rho(x, y)$ is a function $r(x, y) = Be^{j\theta(x,y)}\rho(\pm x/M, \pm y/M)$. The image is a magnified and attenuated version of $\rho(x, y)$, possibly with a position-dependent phase shift. The *magnification* M may be larger than one or smaller than one. The image is inverted if there is a sign reversal on x and y. The quantity $|r(x, y)|^2$ is called the *intensity* of the image of $\rho(x, y)$.

The formation of an image by a lens is described by the lens law. First, we shall give this law for the case in which the aperture is infinitely large[4] (or is so large that it may be treated as infinitely large).

[4] Of course, the Fresnel approximation is not valid for an infinitely large lens. The lens law for an infinite aperture in the Fresnel approximation is a mathematical statement, not a physical statement.

Theorem 4.5.1 (Lens Law) *Suppose that*

$$\frac{1}{d_s} + \frac{1}{d_i} = \frac{1}{f_\ell}.$$

Within the Fresnel approximation, the complex wavefront $\rho(x, y)$ situated at the distance d_s before an ideal lens with an infinitely large aperture will produce a geometrical image at the distance d_i behind the lens, given by

$$r(x, y) = \frac{-1}{M} e^{j(\pi/\lambda d_i^2)(d_s+d_i)(x^2+y^2)} \rho\left(-\frac{x}{M}, -\frac{y}{M}\right),$$

where the magnification is $M = d_i/d_s$.

Proof: A coherent monochromatic wavefront with complex amplitude $\rho(x, y)$ at $z = -d_s$ propagates toward an ideal lens placed at $z = 0$. The signal incident on the lens, in the Fresnel approximation, is

$$\sigma(x, y) = e^{j(\pi/\lambda d_s)(x^2+y^2)} ** \rho(x, y).$$

The output of the lens (leaving the plane $z = 0$) is $\sigma(x, y)e^{-j(\pi/\lambda f_\ell)(x^2+y^2)}$. Using the Fresnel approximation again to propagate the signal to the plane $z = d_i$ gives

$$r(x, y) = \left\{\left[\rho(x, y) ** e^{j(\pi/\lambda d_s)(x^2+y^2)}\right] e^{-j(\pi/\lambda f_\ell)(x^2+y^2)}\right\} ** e^{j(\pi/\lambda d_i)(x^2+y^2)}.$$

To find the image, we must expand the two convolutions in this equation. For simplicity of exposition, we will expand instead the equation in one variable,

$$r(x) = \left\{\left[\rho(x) * e^{j(\pi/\lambda d_s)x^2}\right] e^{-j(\pi/\lambda f_\ell)x^2}\right\} * e^{j(\pi/\lambda d_i)x^2}.$$

The original two-dimensional equation expands in the same way. Writing out the two convolutions gives

$$\begin{aligned} r(x) &= \int_{-\infty}^{\infty} e^{j(\pi/\lambda d_i)(x-\xi)^2} \left[e^{-j(\pi/\lambda f_\ell)\xi^2} \int_{-\infty}^{\infty} \rho(\eta) e^{j(\pi/\lambda d_s)(\eta-\xi)^2} \, d\eta\right] d\xi \\ &= e^{j(\pi/\lambda d_i)x^2} \int_{-\infty}^{\infty} e^{j(\pi/\lambda)(d_i^{-1}-f_\ell^{-1}+d_s^{-1})\xi^2} e^{-j2\pi x\xi/\lambda d_i} \left[\int_{-\infty}^{\infty} \rho(\eta) e^{j(\pi/\lambda d_s)\eta^2} e^{-j2\pi\eta\xi/\lambda d_s} \, d\eta\right] d\xi. \end{aligned}$$

In the image plane, $z = d_i$ and, by assumption, $1/d_s + 1/d_i = 1/f_\ell$. Then

$$r(x) = e^{j(\pi/\lambda d_i)x^2} \int_{-\infty}^{\infty} e^{-j2\pi x\xi/\lambda d_i} \left[\int_{-\infty}^{\infty} \left[\rho(\eta) e^{j(\pi/\lambda d_s)\eta^2}\right] e^{-j2\pi\eta\xi/\lambda d_s} \, d\eta\right] d\xi.$$

We can regard the inner integral as a Fourier transform of $\rho(x)e^{j\pi x^2/\lambda d_s}$. To be more explicit, make the change in variables so that $\eta = -\frac{d_s}{d_i}\tau$ and $\xi = -\lambda d_i f_x$. Then

$$r(x) = \lambda d_s e^{j(\pi/\lambda d_i)x^2} \int_{-\infty}^{\infty} e^{j2\pi f_x x} \int_{-\infty}^{\infty} \left[\rho\left(-\frac{d_s}{d_i}\tau\right) e^{j(\pi d_s/d_i^2\lambda)\tau^2}\right] e^{-j2\pi f_x\tau} \, d\tau \, df_x.$$

The equation now has the form of an inverse Fourier transform of a Fourier transform. Because these two operations cancel, we can conclude that

$$r(x) = \lambda d_s e^{j(\pi/\lambda d_i)x^2}\left[\rho\left(-\frac{d_s}{d_i}x\right)e^{j(\pi d_s/\lambda d_i^2)x^2}\right].$$

But for a magnification and sign reversal, the image $r(x)$ is a complex multiple of $\rho(x)$.

The equation for the two-dimensional signal $r(x, y)$ expands in the same way as the equation for $r(x)$. This completes the proof of the theorem. □

Thus the object $\rho(x, y)$ in the object plane $z = d_s$ will produce an image, $r(x, y)$, in the image plane $z = d_i$ that is a magnified, inverted, and phase-shifted copy of $\rho(x, y)$. Moreover, the image intensity satisfies

$$\begin{aligned} i(x, y) &= |r(x, y)|^2 \\ &= \lambda^2 d_s^2 |\rho(-x/M, -y/M)|^2, \end{aligned}$$

which is an inverted and magnified copy of $|\rho(x, y)|^2$.

Theorem 4.5.1 ignores the diffraction caused by a finite aperture. Every lens has a finite aperture, so diffraction must always occur. To include the effect of the finite aperture, write

$$t(x, y) = e^{-j(\pi/\lambda f_\ell)(x^2+y^2)} A(x, y)$$

where

$$A(x, y) = \begin{cases} 1 & (x, y) \in \mathcal{A} \\ 0 & (x, y) \notin \mathcal{A}. \end{cases}$$

Let $s(x, y)$ denote the image in the presence of the finite aperture $\mathcal{A}$. We will relate $s(x, y)$ to the undiffracted image $r(x, y)$, as given in Theorem 4.5.1.

If we examine the proof of Theorem 4.5.1, we will see that, in the equation for $r(x)$, the term $e^{-j(\pi/\lambda f_\ell)\xi^2}$ must be replaced by the term $A(\xi)e^{-j(\pi/\lambda f_\ell)\xi^2}$. Then, in one dimension, the diffracted image is

$$s(x) = e^{j(\pi/\lambda d_i)x^2} \int_{-\infty}^{\infty} A(\xi) e^{j(\pi/\lambda)(d_i^{-1} - f_\ell^{-1} + d_s^{-1})\xi^2} e^{-j2\pi x\xi/\lambda d_i} \left[\int_{-\infty}^{\infty} \rho(\eta) e^{j(\pi/\lambda d_s)\eta^2} e^{-j2\pi\eta\xi/\lambda d_s}\, d\eta\right] d\xi.$$

In the image plane, $1/d_s + 1/d_i = 1/f_\ell$. Hence

$$s(x) = e^{j(\pi/\lambda d_i)x^2} \int_{-\infty}^{\infty} e^{-j2\pi x\xi/\lambda d_i} \left[A(\xi) \int_{-\infty}^{\infty} \rho(\eta) e^{j(\pi/\lambda d_s)\eta^2} e^{-j2\pi\eta\xi/\lambda d_s}\, d\eta\right] d\xi.$$

As before, make the change of variables $\eta = -\frac{d_s}{d_i}\tau$ and $\xi = -\lambda d_i f_x$,

$$s(x) = \lambda d_s e^{j(\pi/\lambda d_i)x^2} \int_{-\infty}^{\infty} e^{j2\pi f_x x} \left[A(-\lambda d_i f_x) \int_{-\infty}^{\infty} \left[\rho\left(-\frac{d_s}{d_i}\tau\right) e^{j(\pi d_s/\lambda d_i^2)\tau^2}\right] e^{-j2\pi f_x \tau}\, d\tau\right] df_x.$$

To recognize the role that the aperture function plays in this equation, notice that the outer integral is an inverse Fourier transform. Thus

$$\int_{-\infty}^{\infty}\left[s(x)e^{-j(\pi/\lambda d_i)x^2}\right]e^{-j2\pi f_x x}\,dx = \lambda d_s A(-\lambda d_i f_x)\int_{-\infty}^{\infty}\left[\rho\left(-\frac{d_s}{d_i}\tau\right)e^{j(\pi d_s/\lambda d_i^2)\tau^2}\right]e^{-j2\pi f_x\tau}\,d\tau$$
$$= A(-\lambda d_i f_x)\int_{-\infty}^{\infty}\left[r(\tau)e^{-j(\pi/\lambda d_i)\tau^2}\right]e^{-j2\pi f_x\tau}\,d\tau.$$

The second line follows because the undiffracted image is

$$r(x) = \lambda d_s e^{j(\pi/\lambda d_i)x^2}\left[\rho\left(-\frac{d_s}{d_i}x\right)e^{j(\pi d_s/\lambda d_i^2)x^2}\right].$$

In two dimensions, we conclude that the aperture, in the form $A(-\lambda d_i f_x, -\lambda d_i f_y)$, multiplies the Fourier transform of the geometrical (undiffracted) image $r(x, y)e^{-j(\pi/\lambda d_i)(x^2+y^2)}$. Therefore the image itself is convolved with the appropriate inverse Fourier transform of the aperture. Let $a(x, y)$ be the inverse Fourier transform of the aperture function $A(f_x, f_y)$, and define the *coherent optical pointspread function*

$$g_c(x, y) = a\left(\frac{-x}{\lambda d_i}, \frac{-y}{\lambda d_i}\right).$$

The filter $g_c(x, y)$ has the Fourier transform $G_c(f_x, f_y)$, known as the *coherent optical transfer function*, given by

$$G_c(f_x, f_y) = (\lambda d_i)^2 A(-\lambda d_i f_x, -\lambda d_i f_y).$$

Proposition 4.5.2 *Within the Fresnel approximation, a finite aperture will produce a diffracted image, $s(x, y)$, given by*

$$s(x, y) = e^{j(\pi/\lambda d_i)(x^2+y^2)}[g_c(x, y) ** r(x, y)e^{-j(\pi/\lambda d_i)(x^2+y^2)}],$$

where the geometrical image is

$$r(x, y) = -\frac{1}{M}e^{j(\pi/\lambda d_i^2)(d_s+d_i)(x^2+y^2)}\rho\left(-\frac{x}{M}, -\frac{y}{M}\right),$$

and $g_c(x, y)$ is the coherent optical pointspread function.

Proof: The geometrical image $r(x, y)$ was given in Theorem 4.5.1. The modification needed to describe the diffracted image follows from the discussion. □

Whenever $g_c(x, y)$ is sufficiently like an impulse so that one phase term in Proposition 4.5.2 can be (approximately) pulled through the convolution to cancel the other, as is often the case, the equation can be replaced by the convenient approximation

$$s(x, y) \approx g_c(x, y) ** r(x, y).$$

In the language of signal theory, an ideal lens rescales and inverts the signal $\rho(x, y)$ to form $r(x, y)$, then passes $r(x, y)$ through the two-dimensional filter $g_c(x, y)$. The output of this filtering operation is the diffracted image $s(x, y)$. Consequently, we see that coherent imaging is actually the convolution of the true image with the inverse Fourier transform of the system aperture $A(x, y)$. The effect of diffraction is to limit the resolution of an image because of the finite size of the aperture.

Lens as a Fourier transformer

We have seen that the ideal lens can be used to form images. The ideal lens can also be used to perform another useful function, that is, it can form a two-dimensional Fourier transform. Figure 4.4 shows the ideal lens in its two major roles, imaging and transforming. In both cases, the ideal lens consists of a perfect transparency (infinitely thin) with a quadratic phase across an aperture, and the source is modeled as a transparency illuminated by a plane wave. The lens lies in an x, y plane at $z = 0$, and the source lies in an x, y plane at $z = -d_s$. As we saw in Theorem 4.5.1, an image will form in the x, y plane at the value $z = d_i$ satisfying the lens law

$$\frac{1}{d_i} = \frac{1}{f_\ell} - \frac{1}{d_s}.$$

There are two other planes, one on each side of the lens at $z = \pm f_\ell$, called the *focal planes*. We will show that the focal plane to the right of the lens (at $z = f_\ell$) contains the two-dimensional Fourier transform $P(f_x, f_y)$ of the wave amplitude $\rho(x, y)$ that lies in the focal plane to the left of the lens (at $z = -f_\ell$). More fundamentally, the spatial signals in the planes at distance $\pm f_\ell$ at each side of an ideal lens are Fourier transforms of each other (in the Fresnel approximation).

To make the derivation as intuitive as possible, we will first treat the case in which the source plane is at $z = 0$, and so the source plane is coincident with the lens plane.[5] Later, we will move the source plane to $z = -f_\ell$.

The output of the lens with the source at $z = 0$ is $\rho(x, y)A(x, y)e^{-j(\pi/\lambda f_\ell)(x^2+y^2)}$ for all x, y, where $A(x, y)$ is the aperture function. If the aperture is large enough to encompass the support of $\rho(x, y)$, then $\rho(x, y)A(x, y) = \rho(x, y)$. Using the Fresnel approximation, the signal $r(x, y)$ at distance $z = d$ is

$$r(x, y) = \left[\rho(x, y)e^{-j(\pi/\lambda f_\ell)(x^2+y^2)}\right] ** e^{j(\pi/\lambda d)(x^2+y^2)}.$$

Expanding the convolution gives

$$r(x, y) = \int_{-\infty}^{\infty}\int_{-\infty}^{\infty} \rho(\xi, \eta)e^{-j(\pi/\lambda f_\ell)(\xi^2+\eta^2)}e^{j(\pi/\lambda d)[(x-\xi)^2+(y-\eta)^2]}\, d\xi\, d\eta.$$

[5] This requires the lens to be infinitely thin, as is the case with an ideal lens. This physically unreasonable assumption is only introduced temporarily to partition the explanation into two steps. It will disappear when the source is moved away from the lens.

If $d = f_\ell$, then this reduces to

$$r(x, y) = \mathrm{e}^{\mathrm{j}(\pi/\lambda f_\ell)(x^2+y^2)} \int_{-\infty}^{\infty}\int_{-\infty}^{\infty} \rho(\xi, \eta)\mathrm{e}^{-\mathrm{j}(2\pi/\lambda f_\ell)(x\xi+y\eta)}\,\mathrm{d}\xi\,\mathrm{d}\eta$$

$$r(x, y) = \mathrm{e}^{\mathrm{j}(\pi/\lambda f_\ell)(x^2+y^2)} P\left(\frac{x}{\lambda f_\ell}, \frac{y}{\lambda f_\ell}\right).$$

Thus $r(x, y)$ is the Fourier transform of $\rho(x, y)$, rescaled and multiplied by the extraneous phase term $\mathrm{e}^{\mathrm{j}(\pi/\lambda f_\ell)(x^2+y^2)}$.

Now we move the source from the x, y plane at $z = 0$ to the x, y plane at $z = -f_\ell$. This means that $\rho(x, y)$ is convolved with the pointspread function (in the Fresnel approximation)

$$h(x, y) = \mathrm{e}^{\mathrm{j}(\pi/\lambda f_\ell)(x^2+y^2)}$$

and the signal that is incident on the lens is $\rho'(x, y) = h(x, y) ** \rho(x, y)$. The pointspread function $h(x, y)$ has the Fourier transform

$$H(f_x, f_y) = \mathrm{j}\lambda f_\ell \mathrm{e}^{-\mathrm{j}(\pi\lambda f_\ell)(f_x^2+f_y^2)}.$$

By the convolution theorem, if the source is at $z = -f_\ell$, then the Fourier transform of the signal that is incident on the lens is

$$P'(f_x, f_y) = \mathrm{j}\lambda f_\ell \mathrm{e}^{-\mathrm{j}(\pi\lambda f_\ell)(f_x^2+f_y^2)} P(f_x, f_y).$$

Therefore $P'(f_x, f_y)$ must be used in place of $P(f_x, f_y)$ in the equation for $r(x, y)$,

$$r(x, y) = \mathrm{e}^{\mathrm{j}(\pi/\lambda f_\ell)(x^2+y^2)} P'\left(\frac{x}{\lambda f_\ell}, \frac{y}{\lambda f_\ell}\right)$$

$$= \mathrm{j}\lambda f_\ell P\left(\frac{x}{\lambda f_\ell}, \frac{y}{\lambda f_\ell}\right).$$

Thus placing the source at $z = -f_\ell$ creates a phase term that cancels the extraneous spatially-varying phase term that arises when the source is at $z = 0$, leaving only the constant $\mathrm{j}\lambda f_\ell$.

In summary, suppose that a partially transparent object at $z = -f_\ell$ has the complex transmittance $\rho(x, y)$. A plane wave passing through the object along the z axis takes on the complex amplitude $\rho(x, y)$ as a function of x and y. When this plane wave is incident on the lens, we obtain a wave at the second focal plane with the complex amplitude (proportional to)

$$P(f_x, f_y) = \int_{-\infty}^{\infty}\int_{-\infty}^{\infty} \rho(x, y)\mathrm{e}^{-\mathrm{j}\frac{k}{f_\ell}(xf_x+yf_y)}\,\mathrm{d}x\,\mathrm{d}y$$

where $k = 2\pi/\lambda$. By placing an appropriate intensity sensor, such as photographic film, in the focal plane, one can record the intensity $|P(f_x, f_y)|^2$ of the Fourier transform.

There is one detail that has been overlooked in this discussion. Moving the source plane from $z = 0$ to $z = -f_\ell$ means that a finite aperture, $A(x, y)$, at $z = 0$ will fail to

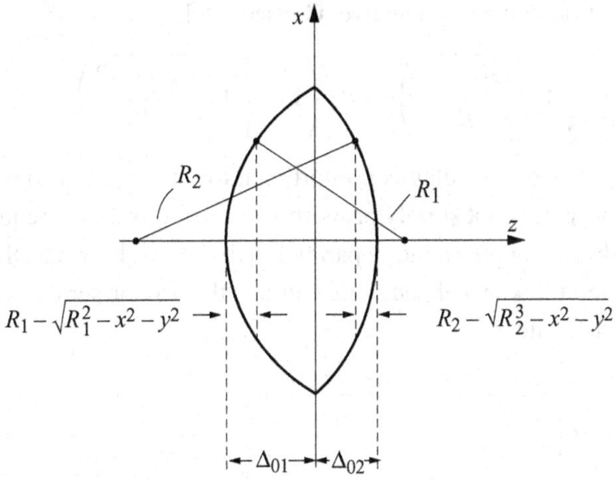

Figure 4.5 A lens

capture some of the signal because $\rho(x, y) ** h(x, y)$ has infinite support. This means that there will always be some error in the Fourier transform because of the truncation of the propagating signal due to the finite aperture.

Lens construction

To conclude this section, we shall describe briefly how the ideal lens can be closely approximated by the shaping of a piece of glass. The design of lenses is a highly advanced branch of the subject of optics, which we can touch on only briefly. Figure 4.5 shows the cross section of a lens, often made of glass, for which the lens material has a propagation velocity different than the propagation velocity of the material outside of the lens, typically air. The velocity of the wave within the glass is $(n - 1)c$ where n is the *refractive index* of the glass. We will consider the case where the lens completely fills the aperture $A(x, y)$. The lens is to be shaped so that its transmittance approximates that of the ideal lens.

A ray entering a glass lens will move with a smaller velocity and may also change direction within the lens. We shall make the *thin lens approximation.* A physical lens is called a *thin lens* if a ray entering approximately along the z axis at one face at coordinates (x, y) emerges from the lens at approximately the same coordinates and the same direction at the other face. There is a negligible translation of the ray within the thin lens. Consequently, the phase delay in the lens at position x, y is simply proportional to the thickness of the lens at the point x, y.

The lens in Figure 4.5 has faces that are segments of spheres with radii R_1 and R_2. The thickness $\Delta(x, y)$ is the sum of two terms, $\Delta_1(x, y)$ and $\Delta_2(x, y)$, given by

$$\Delta_1(x, y) = \Delta_{01} - \left(R_1 - \sqrt{R_1^2 - x^2 - y^2}\right)$$

$$\Delta_2(x, y) = \Delta_{02} - \left(R_2 - \sqrt{R_2^2 - x^2 - y^2}\right)$$

where we use the convention that R_2 is negative. Consequently,

$$\Delta(x, y) = \Delta_0 - R_1\left(1 - \sqrt{1 - \frac{x^2+y^2}{R_1^2}}\right) - R_2\left(1 - \sqrt{1 - \frac{x^2+y^2}{R_2^2}}\right)$$

where $\Delta_0 = \Delta_{01} + \Delta_{02}$. The phase delay is proportional to $\Delta(x, y)$, and so is a complicated function of x and y. It is not simply proportional to $x^2 + y^2$, as is required for the ideal lens. For rays that are approximately parallel to the z axis, however, the phase delay depends only on the thickness; hence one can use the Taylor series expansion $\sqrt{1-t} = 1 - \frac{1}{2}t + \cdots$ to write

$$1 - \sqrt{1 - \frac{x^2+y^2}{R_1^2}} \approx \frac{x^2+y^2}{2R_1^2}$$

$$1 - \sqrt{1 - \frac{x^2+y^2}{R_2^2}} \approx \frac{x^2+y^2}{2R_2^2}.$$

With this approximation, known as the *paraxial approximation*, the thickness is

$$\Delta(x, y) = \Delta_0 - \frac{x^2+y^2}{2}\left(\frac{1}{R_1} + \frac{1}{R_2}\right),$$

and the phase delay is $2\pi f_0(n-1)\Delta(x, y)/c$. In the absence of the lens, in a distance, Δ_0, the phase delay would be $2\pi f_0 \Delta_0/c$. Consequently, because $k = 2\pi f_0/c$, the transmittance is

$$t(x, y) = \mathrm{e}^{\mathrm{j}kn\Delta_0}\mathrm{e}^{-\mathrm{j}k(n-1)\frac{x^2+y^2}{2}\left(\frac{1}{R_1}+\frac{1}{R_2}\right)},$$

and the focal length is

$$\frac{1}{f_\ell} = (n-1)\left(\frac{1}{R_1} + \frac{1}{R_2}\right).$$

Then

$$t(x, y) = \mathrm{e}^{\mathrm{j}kn\Delta_0}\mathrm{e}^{-\mathrm{j}\frac{k}{2f_\ell}(x^2+y^2)}.$$

Therefore, within the limitations of the thin lens approximation and the paraxial approximation, the thin lens with spherical surfaces has the transmittance of the ideal lens.

4.6 Noncoherent imaging

We have seen that the ideal lens will process a spatially coherent wave originating at the object $\rho(x, y)$ to produce an image of the object. However, in everyday circumstances, light from a source is both spatially noncoherent and temporally noncoherent. In this section, we shall show that, for spatially noncoherent illumination, the ideal lens still forms images, but only in intensity, not in amplitude.

A *spatially noncoherent wave* originating from the source $\rho(x, y)$ in the plane $z = 0$ is a complex wavefront of the form

$$v(x, y, t) = \rho(x, y)u(x, y, t)$$

where $u(x, y, t)$ is modeled as a spatially white random process defined by the correlation function

$$\mathrm{E}[u(x, y, t)u^*(x', y', t)] = \delta(x - x', y - y').$$

Consequently,

$$\mathrm{E}[v(x, y, t)v^*(x', y', t)] = |\rho(x, y)|^2\delta(x - x', y - y').$$

Because the time variable t has the same value in both copies of $v(x, y, t)$ appearing within the expectation, the definition of spatial noncoherence is silent with respect to randomness on the time axis.

A *temporally noncoherent* wave is a time-stationary process defined by the correlation function

$$\mathrm{E}[v(x, y, t)v^*(x, y, t')] = |\rho(x, y)|^2\delta(t - t').$$

The definition of temporal noncoherence is silent with regard to randomness in the space coordinates. Of course, a process can satisfy both definitions and so be both temporally and spatially noncoherent.

Intermediate between the cases of (spatially) coherent imaging and (spatially) noncoherent imaging is the case of partially coherent imaging. A partially spatially coherent waveform satisfies

$$\begin{aligned}\mathrm{E}\left[v(x, y, t)v^*(x', y', t)\right] &= |\rho(x, y)|^2\mathrm{E}\left[u(x, y, t)u^*(x', y', t)\right]\\ &= |\rho(x, y)|^2\phi(x - x', y - y')\end{aligned}$$

where $\phi(x, y)$ is a two-dimensional correlation function that is not an impulse. We shall not deal with the difficult case of imaging partially coherent scenes.

An image of a spatially noncoherent object that is formed with temporally coherent illumination, such as laser or ultrasound illumination, can have a granular appearance called *speckle*. Speckle is a statistical fluctuation of the image intensity in space caused by unresolved roughness in the surface of the imaged object, or caused by irregularities in the propagation medium. Because the illumination is a passband waveform, each reflecting element returns an echo of that passband waveform altered both in amplitude and phase. If the many unresolved complex reflecting elements within one resolution cell are random and independent, the central limit theorem of probability theory implies that the in-phase and quadrature components of the received signal from one resolution cell are gaussian random variables. Consequently, the amplitude is a Rayleigh random variable. Speckle is usually considered to be an impairment in the quality of an image, and the image may be further processed to smooth the speckle. Sometimes, instead, the statistics of the speckle may be estimated as a useful way to characterize the smoothness and structure of a reflector at a level finer than the resolution of the image.

The next proposition says that the expected value of a noncoherent image, which does not depend on time, is described as a convolution of intensities[6] rather than as a convolution of amplitudes.

Proposition 4.6.1 *For noncoherent imaging in the Fresnel approximation, the intensity*

$$i(x, y) = \mathrm{E}|r(x, y, t)|^2$$

is given by the intensity convolution equation

$$i(x, y) = |g_c(x, y)|^2 ** |\rho\left(-\frac{x}{M}, -\frac{y}{M}\right)|^2$$

where $g_c(x, y)$ is the coherent optical pointspread function.

Proof: The coherent imaging equation

$$s(x, y)\mathrm{e}^{-\mathrm{j}(\pi/\lambda d_i)(x^2+y^2)} = g_c(x, y) ** r(x, y)\mathrm{e}^{-\mathrm{j}(\pi/\lambda d_i)(x^2+y^2)}$$

was given in Proposition 4.5.2 where

$$r(x, y) = -\frac{1}{M}\mathrm{e}^{\mathrm{j}(\pi/\lambda d_i^2)(d_s+d_i)(x^2+y^2)}\rho\left(-\frac{x}{M}, -\frac{y}{M}\right),$$

and $g_c(x, y)$ has the Fourier transform

$$G_c(f_x, f_y) = (\lambda d_i)^2 A(-\lambda d_i f_x, -\lambda d_i f_y).$$

The imaging equation is now replaced with

$$s(x, y)\mathrm{e}^{-\mathrm{j}(\pi/\lambda d_i)(x^2+y^2)} = g_c(x, y) ** v'(x, y, t)\mathrm{e}^{-\mathrm{j}(\pi/\lambda d_i)(x^2+y^2)}$$

where

$$v'(x, y, t) = -\frac{1}{M}\mathrm{e}^{\mathrm{j}(\pi/\lambda d_i^2)(d_s+d_i)(x^2+y^2)}v\left(-\frac{x}{M}, -\frac{y}{M}, t\right),$$

[6] The intensity recorded in a stationary system is actually the sample time average

$$i(x, y) = \frac{1}{T}\int_0^T |r(x, y, t)|^2\, dt,$$

whereas we are dealing with the ensemble average

$$i(x, y) = \mathrm{E}|r(x, y, t)|^2.$$

If the illumination is temporally coherent, as with a laser source, the time average and the ensemble average need not be equal or even approximately equal. This causes *speckle* for temporally coherent light. Speckle is an important difference between imaging with temporally coherent light and temporally noncoherent light.

and $v(x, y, t) = \rho(x, y)u(x, y, t)$. For the noncoherent case, the image intensity is given by the expected value of the squared magnitude of this expression:

$$i(x, y) = \mathrm{E}\left[|s(x, y)|^2\right].$$

Therefore

$$i(x, y) = \int_{-\infty}^{\infty}\int_{-\infty}^{\infty}\int_{-\infty}^{\infty}\int_{-\infty}^{\infty} g_c(x-\xi, y-\eta)g_c^*(x-\xi', y-\eta')\phi(\xi, \eta, \xi', \eta')\,\mathrm{d}\xi\,\mathrm{d}\eta\,\mathrm{d}\xi'\,\mathrm{d}\eta'$$

where

$$\phi(\xi, \eta, \xi', \eta') = \mathrm{E}[v'(\xi, \eta, t)\mathrm{e}^{-\mathrm{j}(\pi/\lambda d_i)(\xi^2+\eta^2)}v'^*(\xi', \eta', t)\,\mathrm{e}^{\mathrm{j}(\pi/\lambda d_i)(\xi'^2+\eta'^2)}].$$

To evaluate this expectation, recall that

$$\mathrm{E}[v(x, y, t)v^*(x', y', t)] = |\rho(x, y)|^2\delta(x-x', y-y'),$$

so

$$\phi(\xi, \eta, \xi', \eta') = \left|\rho\left(-\frac{\xi}{M}, -\frac{\eta}{M}\right)\right|^2 \delta(\xi-\xi', \eta-\eta').$$

Then

$$i(x, y) = \frac{1}{M^2}\int_{-\infty}^{\infty}\int_{-\infty}^{\infty} |g_c(x-\xi, y-\eta)|^2 \left|\rho\left(-\frac{\xi}{M}, -\frac{\eta}{M}\right)\right|^2 \mathrm{d}\xi\,\mathrm{d}\eta,$$

which completes the proof of the theorem. □

The term

$$g_n(x, y) = \frac{|g_c(x, y)|^2}{\int_{-\infty}^{\infty}\int_{-\infty}^{\infty} |g_c(x, y)|^2\,\mathrm{d}x\,\mathrm{d}y}$$

is called the (normalized) *noncoherent optical pointspread function*. For example, if $A(x, y)$ is a circle function, then $g_n(x, y)$ is an Airy disk.

The *noncoherent optical transfer function* is defined as the Fourier transform of $g_n(x, y)$:

$$G_n(f_x, f_y) = \frac{\int_{-\infty}^{\infty}\int_{-\infty}^{\infty} |g_c(x, y)|^2\mathrm{e}^{-\mathrm{j}2\pi(f_x x + f_y y)}\,\mathrm{d}x\,\mathrm{d}y}{\int_{-\infty}^{\infty}\int_{-\infty}^{\infty} |g_c(x, y)|^2\,\mathrm{d}x\,\mathrm{d}y}.$$

The magnitude $|G_n(f_x, f_y)|$ is called the *modulation transfer function*. The relationship between the noncoherent optical transfer function $G_n(f_x, f_y)$ and the aperture transmittance function $A(x, y)$ is easily found to be

$$G_n(f_x, f_y) = \frac{\int_{-\infty}^{\infty}\int_{-\infty}^{\infty} A(\xi, \eta)A^*(\xi + \lambda d_i f_x, \eta + \lambda d_i f_y)\,\mathrm{d}\xi\,\mathrm{d}\eta}{\int_{-\infty}^{\infty}\int_{-\infty}^{\infty} |a(x, y)|^2\,\mathrm{d}x\,\mathrm{d}y}.$$

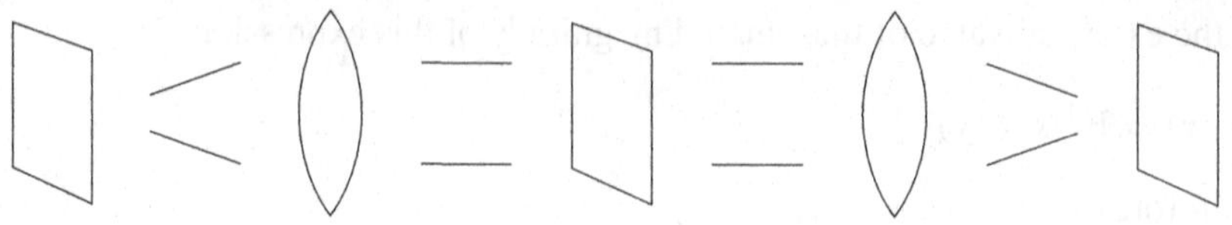

Figure 4.6 An optical processor for spatial filtering

4.7 Optical filtering

The mathematical operation of two-dimensional filtering may be performed by any convenient processing technology. We shall describe how a two-dimensional filter can be implemented optically by using the ideal lens as an optical processor, which is one important application of the lens. A two-dimensional filtering operation performed by a pair of lenses uses the convolution theorem to multiply in the Fourier transform domain, as shown in Figure 4.6. One lens is used to compute a two-dimensional Fourier transform of the source plane $s(x, y)$, and a second lens is used to compute an inverse Fourier transform. Between them is a transparency with the transmittance $G(f_x, f_y)$. The resulting approximate relationship between the image plane and the source plane is

$$r(x, y) = g(x, y) ** s(x, y)$$

if the system is coherent, and

$$|r(x, y)|^2 = |g(x, y)|^2 ** |s(x, y)|^2$$

if the system is noncoherent. For the noncoherent case, this means that only spatial filters that have the form of a squared magnitude of another function can be realized.

Spatial filtering of an image can be used to suppress the effects of an unwanted pointspread function and to sharpen the image. Image detail is enhanced by de-emphasizing the low spatial frequencies. Spatial filtering can also be used to suppress the effects of noise, usually by de-emphasizing the high spatial frequencies. For example, the artifacts introduced by half-tone modulation can be suppressed by using a spatial lowpass filter. When the filtering operation is used to remove the effect of an unwanted pointspread function, such as a blur on the image, it is called *spatial equalization* or *deconvolution*. The general topic of deconvolution is studied in Chapter 9.

Many egregious impairments in photographs can be modeled as blurs arising in the system that produced the photographs. Blurring can be caused by poor focusing or even by camera movement. A two-dimensional signal of interest, $s(x, y)$, received as a blurred signal is

$$v(x, y) = h(x, y) ** s(x, y)$$

where $h(x, y)$ is a pointspread function that blurs the signal $s(x, y)$. Such a relationship might be an adequate model of the diffraction of a lens, for example. If the pointspread function were an impulse there would be no impairments.

To improve a blurred image, one may pass it through a compensating filter and rerecord it. In general, ignoring noise, we may compute

$$\begin{aligned} u(x, y) &= g(x, y) ** v(x, y) \\ &= [g(x, y) ** h(x, y)] ** s(x, y) \end{aligned}$$

in the absence of noise. This process will recover $s(x, y)$ exactly if

$$g(x, y) ** h(x, y) = \delta(x, y).$$

In the frequency domain, one choice for the *equalization filter* is

$$G(f_x, f_y) = \frac{1}{H(f_x, f_y)}.$$

This ideal form for $G(f_x, f_y)$ will be impossible to realize exactly because $H(f_x, f_y)$ will be zero, or very small, for some values of f_x and f_y. Then $G(f_x, f_y)$ would be very large at those frequencies, causing imperfections in the observed image to be magnified. In general, the received signal is contaminated by additive noise. Then

$$v(x, y) = h(x, y) ** s(x, y) + n(x, y)$$

where $n(x, y)$ is the additive noise. In the language of information theory, the signal $v(x, y)$ has been passed through a noisy channel with the transfer function $H(f_x, f_y)$. The equalization filter $1/H(f_x, f_y)$ is often unsatisfactory in the presence of noise because of noise amplification. Then one may use the alternative filter

$$G(f_x, f_y) = \frac{H^*(f_x, f_y)}{|H(f_x, f_y)|^2 + C},$$

for which $H(f_x, f_y)G(f_x, f_y)$ is not larger than C^{-1}. This is a form of the *Wiener filter*, which will be studied in Section 9.1. The construction of an equalization filter requires that the pointspread function of the channel be known.

Figure 4.7 shows a pointspread function, $h(x, y)$, that may be used to model streaking due to camera motion. The motion may cause a point object to be reproduced by a streak of very small width at the angle θ. The streak can be represented by a two-dimensional rectangle function of width b and length a. If the coordinate system is chosen so that the streak is along the x axis, then

$$H(f_x, f_y) = ab \operatorname{sinc}(af_x) \operatorname{sinc}(bf_y),$$

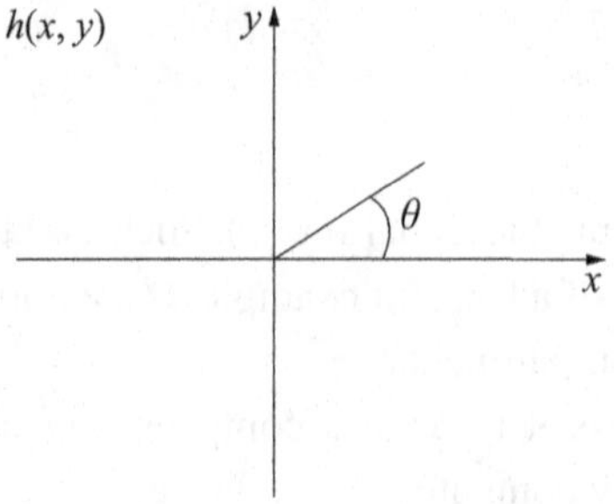

Figure 4.7 Pointspread function for camera movement

which is zero whenever af_x is a nonzero integer. If $s(x, y)$ is periodic in the x direction with period a^{-1}, then $S(f_x, f_y)$ is nonzero only where af_x is an integer, so $H(f_x, f_y)S(f_x, f_y)$ is zero everywhere unless $f_x = 0$. It is clear that equalization in this case is not possible. In the space domain, if $s(x, y)$ is periodic, then the convolution $h(x, y) ** s(x, y)$ is a constant, and $s(x, y)$ cannot be recovered. The Wiener equalization filter for this example is

$$G(f_x, f_y) = \frac{ab\,\mathrm{sinc}(af_x)\,\mathrm{sinc}(bf_y)}{|ab\,\mathrm{sinc}(af_x)\,\mathrm{sinc}(bf_y)|^2 + C}.$$

The values of the magnitude of $G(f_x, f_y)$ near the zeros of the two-dimensional sinc function can be kept small by the choice of the constant C.

Another common image impairment is blurring caused by focusing error (see Problem 4.10). A simple model of blurring is the pointspread function

$$h(x, y) = \mathrm{circ}\left(\frac{x}{a}, \frac{y}{a}\right).$$

In the frequency domain,

$$H(f_x, f_y) = \mathrm{jinc}\left(a\sqrt{f_x^2 + f_y^2}\right),$$

and the Wiener equalization filter is

$$G(f_x, f_y) = \frac{\mathrm{jinc}\left(a\sqrt{f_x^2 + f_y^2}\right)}{\mathrm{jinc}^2\left(a\sqrt{f_x^2 + f_y^2}\right) + C}.$$

For a third example, suppose that the signal $s(x, y)$ is transmitted or recorded by some process that inserts false lines into the image or inserts multiple streaks parallel to the x axis. This is an additive noise process. It is different from the pointspread function due to camera movement because the horizontal lines are not a result of details of the image. One instance is an image of a large section of the earth that is formed by mosaicking a number of small photographs. When this is done, there may be visible lines in the mosaic where the small pictures abut. These lines may be modeled as additive noise and removed by filtering.

4.8 Phase-contrast imaging

An early application of spatial filtering arose in the microscopic imaging of transparent objects (such as organisms) that affect light, not by absorption, but by a spatially dependent delay, which appears in the light signal as a spatially dependent phase shift. Because either a recording film or the eye would detect only the intensity of light but not its phase, objects that are completely transparent are impossible to see directly. However, they can be seen if a device is placed in the optical path that will convert phase modulation to amplitude modulation. A simple technique is the method known as the *central dark ground method* in which a small spot in the frequency domain is used to modify the components of the Fourier transform near the origin of the f_x, f_y plane.

Suppose that a transparent object has a complex transmittance consisting of a phase variation only

$$t(x, y) = e^{j\phi(x,y)},$$

and that it is coherently illuminated in an image forming system. The intensity image of $t(x, y)$ produced by a conventional optical imaging system would have the intensity

$$|t(x, y)|^2 = 1,$$

which is independent of the signal, so the signal is lost. To preserve the desired signal, $t(x, y)$ is first passed through a *Zernike filter* prior to computing the magnitude. The Zernike filter is defined in the frequency domain as

$$H(f_x, f_y) = \begin{cases} ja & \text{if } \sqrt{f_x^2 + f_y^2} \le \epsilon \\ 1 & \text{otherwise} \end{cases}$$

where ϵ is a small number and a is a real constant. Imaging a transparent object with the Zernike filter is shown in Figure 4.8.

The Zernike filter can be implemented optically as a glass slide on which a small transparent dot of dielectric is placed at the origin $(f_x, f_y) = (0, 0)$. The glass slide is

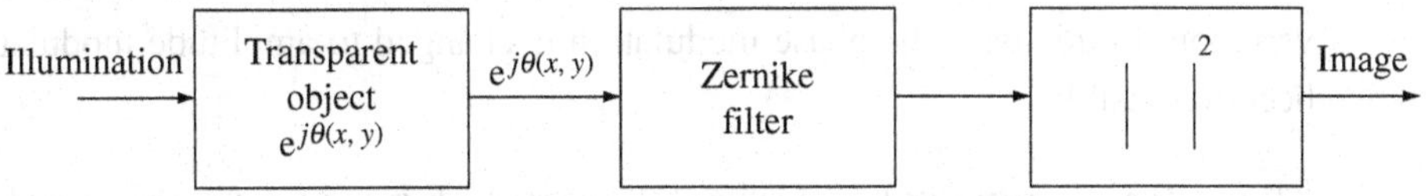

Figure 4.8 Imaging a transparent object

placed in the Fourier transform plane between two lenses. If the thickness of the dot corresponds to an optical phase shift of $\pi/2$ radians, then the filter is a Zernike filter.

The output of the filter is $h(x, y) ** t(x, y)$. If $\phi(x, y)$ is less than one radian, then an adequate approximation is

$$t(x, y) \approx 1 + \mathrm{j}\phi(x, y),$$

and, in the transform domain,

$$T(f_x, f_y) = \delta(f_x, f_y) + \mathrm{j}\Phi(f_x, f_y).$$

Because $\Phi(f_x, f_y)$ is negligible compared to $\delta(f_x, f_y)$ at the origin, the filter output in the transform domain is approximately

$$H(f_x, f_y)T(f_x, f_y) = \mathrm{j}a\delta(f_x, f_y) + \mathrm{j}\Phi(f_x, f_y).$$

Consequently,

$$h(x, y) ** t(x, y) \approx \mathrm{j}(a + \phi(x, y)),$$

which has the intensity

$$|a + \phi(x, y)|^2 \approx a^2 + 2\phi(x, y).$$

Whereas the intensity of the unfiltered signal is independent of the object $\phi(x, y)$, the intensity of the filtered signal depends on the object, although seen against a bright background denoted by a^2.

Another method of phase-contrast imaging can be used in applications such as wind tunnel photographs in which the air stream has a nonuniform index of refraction, which may be caused by pressure differences or temperature differences due to the details of the airflow. The nonuniform index of refraction introduces phase modulation into the optical field. The phase modulation can be changed into amplitude modulation by the use of a *knife-edge filter*.[7] This knife-edge filter is defined in the frequency domain as

$$H(f_x, f_y) = \tfrac{1}{2}(1 + \mathrm{sgn}(f_x)).$$

This is a two-dimensional version of the *Hilbert filter*. The knife-edge filter can be constructed optically by placing a knife edge in the Fourier transform plane between two lenses. The output of the knife-edge filter is

$$I(x, y) = \frac{1}{4}\left[1 - \frac{2}{\pi}\int_{-\infty}^{\infty} \frac{\phi(x', y')}{x - x'}\,\mathrm{d}x'\right],$$

which is the Hilbert transform of the phase variation translated into a real intensity variation. In this way, the phase modulation is changed to amplitude modulation and becomes visible.

[7] The use of a knife-edge filter for converting phase into amplitude is known as the *schlieren method* from a German word meaning streak or striation.

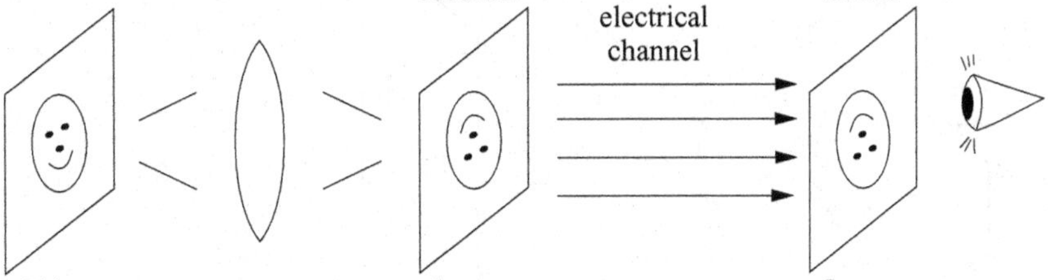

Figure 4.9 Transferring a recorded image

4.9 Wavefront reconstruction

Given the spatial signal $s(x, y)$, represented in the form of the passband optical signal $s(x, y)e^{j2\pi f_0 t}$, a camera can record this signal by using a lens to form an image and placing a recording medium in the image plane of the lens. A recording medium, such as a photographic film or an array of optoelectronic photosensors, is suitable. The recorded image can then be scanned and transferred to another place, as by a communication system, and displayed. Figure 4.9 shows the situation.

Even if the image in the image plane were recorded perfectly, the redisplayed image would fail to have all the properties of the original unrecorded image. The unrecorded image in the image plane has depth. A viewer can change the viewing angle slightly to change the view of the scene. In contrast, the reconstructed displayed image lacks this depth. Whereas the actual visual field is three dimensional, the displayed image is only two dimensional. What happened to the missing information about the image?

The explanation for this apparent inconsistency is that present-day recorders record only the intensity of the signal in the image plane. The signal in the image plane also has a meaningful phase angle at each point, and this phase angle is what allows the propagating light wave to reconstruct itself in different ways as it travels in different directions; this is where the three-dimensional effect arises.

A more ambitious goal is to record both the magnitude and the phase of the advancing wavefront $s(x, y)$ because there is useful information in the phase. However, a medium that directly records optical phase is difficult to develop, in part because the wavelength is so small.

One way of recording phase is to turn the phase information into amplitude modulation of another wavefront and then record the intensity of that wavefront. This is called *holography*. In the simplest approach, the signal and a coherent reference are added, and the sum intensity is recorded, as described below. Usually, the recording medium is photographic film, which is developed as a transparency called a *hologram*.

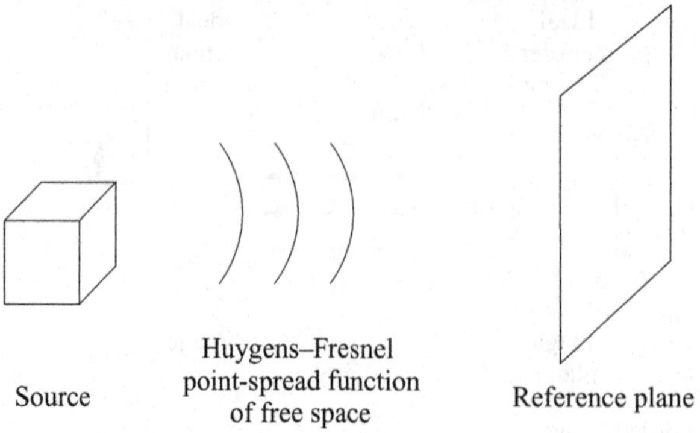

Figure 4.10 Free-space propagation

The hologram does not contain an image; the image is formed by illuminating the developed hologram with the coherent reference and reconstructing the optical field.

Let $\rho(x, y)$ be the source of a waveform. The waveform propagates to an arbitrary plane at distance z from the source where it appears, after diffraction, as the spatial signal $s(x, y)$ in the recording plane, as shown in Figure 4.10. That is

$$s(x, y) = |s(x, y)|e^{j\theta(x,y)}$$

is the signal as it appears in the recording plane. If it were allowed to pass undisturbed through that plane, it could be made to form an image at a subsequent plane by a suitable lens. Let

$$a(x, y) = |a(x, y)|e^{j\psi(x,y)}$$

be a fixed reference signal as it appears in the recording plane. A similar reference must be regenerated at a later time when the hologram is to be read. Define

$$v(x, y) = s(x, y) + a(x, y),$$

and record the magnitude $|v(x, y)|^2$. This consists of four terms:

$$\begin{aligned} |v(x, y)|^2 &= |s(x, y) + a(x, y)|^2 \\ &= |s(x, y)|^2 + a^*(x, y)s(x, y) + a(x, y)s^*(x, y) + |a(x, y)|^2. \end{aligned}$$

We shall see that we can recover $s(x, y)$ from either the second term or the third term. In either case, the other three terms of the four will create spatial interference in the regenerated spatial signal.

To view the hologram, illuminate it with a reference waveform, $b(x, y)$, to produce

$$\begin{aligned} r(x, y) &= b(x, y)|v(x, y)|^2 \\ &= b(x, y)|s(x, y)|^2 + b(x, y)a^*(x, y)s(x, y) \\ &\quad + b(x, y)a(x, y)s^*(x, y) + b(x, y)|a(x, y)|^2. \end{aligned}$$

Now specify that $b(x, y) = a(x, y)$ and choose $a(x, y)$ to be a monochromatic plane wave such that $|a(x, y)| = 1$. Then

$$r(x, y) = a(x, y)|s(x, y)|^2 + s(x, y) + (a(x, y))^2 s^*(x, y) + a(x, y).$$

The second term is the desired reconstructed signal. An observer of $r(x, y)$ would see the signal $s(x, y)$ in $r(x, y)$ just as if the original source of $s(x, y)$ were still there. Thus we say that the waveform $r(x, y)$ contains a copy of $s(x, y)$, diverging from a *virtual image*, $\rho(x, y)$, that would appear quite real to the viewer. The other three terms of $r(x, y)$ are recorded as interference appearing as artifacts in the visual field, and we shall want to suppress them.

Alternatively, again specify that $|a(x, y)| = 1$, but now $b(x, y) = a^*(x, y)$. Then

$$r(x, y) = a^*(x, y)|s(x, y)|^2 + (a^*(x, y))^2 s(x, y) + s^*(x, y) + a^*(x, y).$$

In this case, the third term is the conjugate of signal $s(x, y)$. Therefore this gives a method for reversing the sign of the phase, which leads to some interesting properties. As before, the three remaining terms are to be regarded as interference in the visual field, which we shall want to suppress.

We shall describe two variations of holography: the *Gabor hologram*, which was the first to be proposed, and the *Leith–Upatnieks hologram*, which has better properties and is more practical. The difference between the two is the way in which the reference is combined with the signal. The Gabor hologram relies on the existence of an ambient illuminating waveform that leaks past the object $\rho(x, y)$. The Leith–Upatnieks hologram explicitly captures the reference signal by a system of mirrors.

Let the coherent, monochromatic, monodirectional wave given by $a(x, y) = 1$ be used to illuminate the object $\rho(x, y)$. The signal in the secondary plane is

$$v(x, y) = s(x, y) + 1$$

where $s(x, y)$ is the signal, originating as $\rho(x, y)$, as it appears in the recording plane. The signal $s(x, y)$ may be complex and $|s(x, y)| \ll 1$. (Notice the similarity of this model and the approximation used earlier while developing the Zernike filter.) The signal intensity recorded at the photographic film is

$$\begin{aligned} |v(x, y)|^2 &= |s(x, y) + 1|^2 \\ &= 1 + s(x, y) + s^*(x, y) + |s(x, y)|^2. \end{aligned}$$

Consequently, the recorded intensity consists of a strong, uniform plane wave; two signal terms; and a small term, $|s(x, y)|^2$, which is ignored under the assumption that $|s(x, y)| \ll 1$. If the latter term cannot be neglected, then the Gabor hologram may be unsatisfactory. The two signal terms correspond to "twin images": one real image and one virtual image. We can choose to focus either image, but that image then will be superimposed on the other image, which is far out of focus. The out-of-focus image contaminating the in-focus image is a disadvantage of the Gabor hologram.

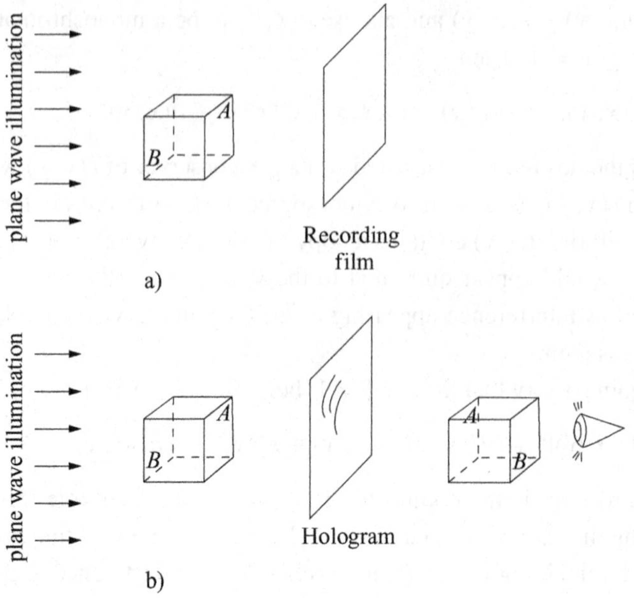

Figure 4.11 The Gabor (transmission) hologram: a) recording the hologram; b) viewing the hologram

The transparency $|v(x, y)|^2$ made from the recorded image is viewed by passing a uniform plane wave of amplitude B through the transparency, as shown in Figure 4.11. Then the viewed signal is

$$B|v(x, y)|^2 = B[s(x, y) + s^*(x, y)]$$

where the constant term and the small quadratic term have been ignored as uninteresting. Within the plane of the hologram, the signal $s(x, y)$ is just as it appeared when the hologram was made. The signal reaching the viewer due to $s(x, y)$ is just as it would appear if the original subject were still behind the hologram. A viewer focusing on this point sees a virtual image of $\rho(x, y)$.

The second term $s^*(x, y)$ has the sign of its phase angle reversed. Thus the term appears as it would appear if it came from an object located at the opposite side of the hologram. The illumination causes a signal that converges to a *real image* to the right of the hologram. The viewer to the right of this point who focuses on it will see the real image.

The Gabor hologram is usually unsatisfactory because of the four waves reaching the viewer, corresponding to the four complex amplitudes 1, $|s(x, y)|^2$, $s(x, y)$, and $s^*(x, y)$. Of these, if only the wave $s(x, y)$ is of interest, it can be isolated by focusing. The other three terms are interference. This limitation of the Gabor hologram can be circumvented by introducing the reference waveform from another direction. This is the Leith–Upatnieks hologram, shown in Figure 4.12.

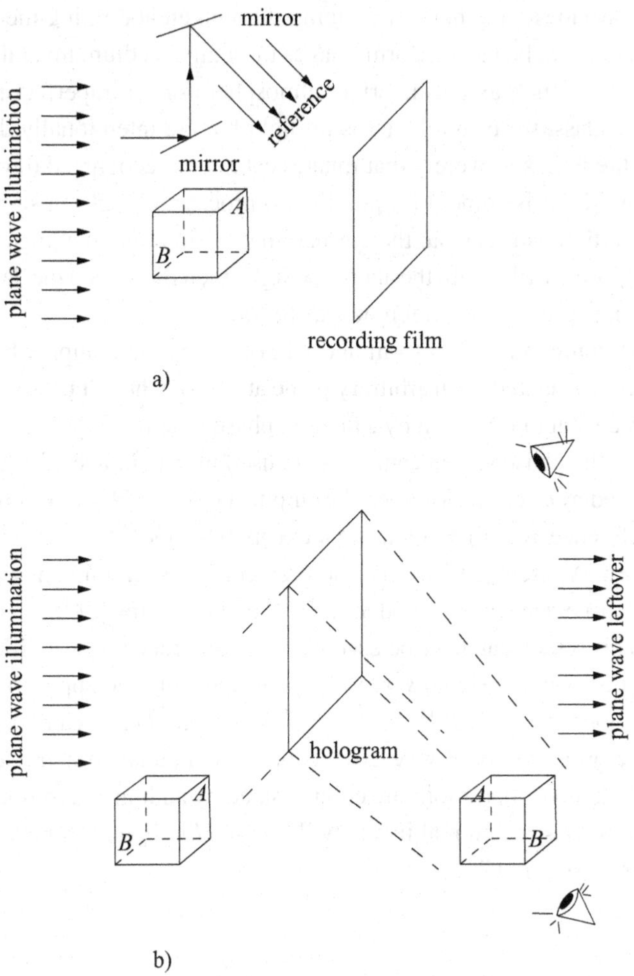

Figure 4.12 The Leith–Upatnieks (transmission) hologram: a) recording the hologram; b) viewing the hologram

We close the chapter with the remark that the methods of holography can be used (at least in principle) to form images of objects that cannot be viewed directly by the eye. The two principal examples of this are the imaging of transparent objects and the imaging of opaque objects through an aberrating medium. In each case, the received optical signal in a recording plane is a scrambled function of the desired image, but not an understandable function and, if recorded, not an invertible function because the phase is lost. In principle, holography can make the recording invertible.

Holographic imaging of transparent objects (such as bacteria or air-flow patterns) is an alternative to phase-contrast imaging. An air flow may be recorded by exposing the hologram film in two stages: once without the air flow and a second time with the air flow. The illuminated hologram will show the air flow as an intensity variation.

Holography also provides a method for imaging through an aberrating medium, at least in principle. This could be used to form images through a medium that introduces uncontrolled aberrations, such as an imperfect atmosphere or an imperfect piece of glass. Some kinds of glass, for example, transmit light but are intentionally designed to distort or diffuse the light so severely that images cannot be recognized through the glass. In theory, however, it is sometimes possible to process the light in such a way as to cancel the aberration and recreate the correct image – in effect making the glass transparent. First, expose the film with the aberrated light, then illuminate the developed film with an appropriate prerecorded reference, as follows.

Let $\rho(x, y)$ be the source of a waveform and let $s(x, y)$ be the complex-baseband waveform after it has propagated to an arbitrary plane at $z = 0$, where it passes through a thin aberrating screen that multiplies it by a random phase function $e^{j\theta(x,y)}$. The signal $s(x, y)e^{j\theta(x,y)}$ leaving the phase screen cannot be focused into an image of $\rho(x, y)$ because it is contaminated by $\theta(x, y)$. However, the output $s(x, y)e^{j\theta(x,y)}$ can be combined with a plane wave reference $a(x, y) = 1$ and the intensity $|s(x, y)e^{j\theta(x,y)} + 1|^2$ recorded as a magnitude screen. A reference wave is also prerecorded by illuminating the random phase screen with a plane wave to produce $e^{j\theta(x,y)}$ and recording $|e^{j\theta(x,y)} + 1|^2$ as a magnitude screen. This screen can later be used to form a reference wave to illuminate the other screen. In this way the signal $s(x, y)$ is recovered from the appropriate term of the product $|s(x, y)e^{j\theta(x,y)} + 1|^2|e^{j\theta(x,y)} + 1|^2$. This signal $s(x, y)$ can be focused into an image of $\rho(x, y)$ in the usual way. By this process, an unaberrated real image will form because the known aberrations are compensated by adding an arbitrary phase to its conjugate. The reconstructed real image will be *pseudoscopic*, meaning that the contours will appear "inside-out."

Problems

4.1 Prove that the Huygens–Fresnel pointspread function is consistent under convolution in the following sense. If $s_0(x, y) = s(x, y, 0)$ is a signal in a plane at $z = 0$ that propagates into $s_{d_1}(x, y) = s(x, y, d_1)$ in a plane at $z = d_1$, and $s_{d_1}(x, y)$ is a signal in a plane at $z = d_1$ that propagates into $s_{d_2}(x, y) = s(x, y, d_2)$ in a plane at $z = d_2 > d_1$, prove that $s_{d_2}(x, y)$ is the same as the signal obtained by propagating $s_0(x, y)$ to the plane at $z = d_2$ directly. In particular, the signal in any plane at any distance between a source and an observation can be considered as an alternative source.

4.2 (Conjugation) Suppose that $s_0(x, y) = s(x, y, 0)$ is a real signal at $z = 0$. Let $s(x, y, d_1)$ be the signal in a plane at $z = d_1$ obtained by propagating $s_0(x, y)$. What signal is received in a plane at $z = 2d_1$ if the conjugate signal $s^*(x, y, z)$ is propagated from the plane at $z = d_1$? (A hypothetical device in the plane at $z = d_1$ conjugates the complex baseband optical signal.)

What signal will be observed in a plane at $z = 3d_1$?

4.3 **(Very Near Field)**

a. Show that, for large λ, the Huygens–Fresnel pointspread function reduces to

$$h(x, y) = \frac{d/2\pi}{(x^2 + y^2 + d^2)^{3/2}}.$$

Specifically, show that this approximation holds for λ much larger than $2\pi\sqrt{x^2 + y^2 + d^2}$.

b. Show that, for large λ, the Fourier transform of the Huygens–Fresnel pointspread function reduces to

$$H(f_x, f_y) = -e^{-2\pi d\sqrt{f_x^2+f_y^2}}.$$

c. Conclude that the secondary term of the Huygens–Fresnel pointspread function is primarily associated with evanescent waves. Can this correspondence be made exact?

4.4 A transparency consists of opaque squares and transparent squares in a checkerboard pattern. What is the far-field (Fraunhofer) diffraction pattern?

4.5 Prove the lens law in the Fresnel approximation by operations in the transform domain.

4.6 An ideal lens in the zero plane $z = 0$ is used to form an image of the object $s_0(x, y)$ in the plane $z = -d_0$. Prove that, for any x, y, the line defined by $(x, y, -d_0)$ and $(0, 0, 0)$, and the line defined by $(x, y, 0)$ and $(0, 0, f_\ell)$, cross in the image plane and moreover cross at the image of $s_0(x, y)$. Give a sketch of a simple ray-tracing method of locating the image of an object. Give a "derivation" of the lens law from the ray-tracing sketch. Give an explanation of the magnification. Describe how to decompose the analysis of an imaging system into the computation of a diffraction-free image by ray tracing and the computation of a pointspread function by Fourier optics.

4.7 The *f-number* (or *relative aperture*) of a simple lens is the ratio of the focal length to the diameter of the aperture. A *fast lens* is an ideal lens whose f-number is smaller than one. Will the Fresnel approximation be adaquate for analyzing a fast lens?

4.8 A *diffraction grating* consists of an aperture with narrow slits ruled across it. Suppose that a diffraction grating is 4 centimeters wide and has 10 000 slits ruled across it. What is the wavelength of the light whose first-order grating lobe is 45° when the diffraction grating is illuminated from far behind? What is the wavelength of light whose second-order and third-order grating lobes are at 45°? Are any of these wavelengths visible light? If so, what is the color?

4.9 Derive the Fraunhofer approximation by rewriting the Fresnel pointspread function in terms of chirp filters. Under what condition can the quadratic-phase term inside the integration be dropped?

4.10 **(Focusing Error)** Instead of the lens law,

$$\frac{1}{d_s} + \frac{1}{d_i} - \frac{1}{f_\ell} = 0,$$

suppose that a noncoherent optical-imaging system is "out-of-focus," satisfying the expression

$$\frac{1}{d_s} + \frac{1}{d_i} - \frac{1}{f_\ell} = \Delta.$$

Show that the effect of this can be described by cascading the ideal lens with the two-dimensional function

$$e^{jkW(f_x, f_y)} = e^{j\pi(\Delta/\lambda)(f_x^2 + f_y^2)}.$$

4.11 **(Focusing Error)** A one-dimensional ideal lens,

$$t(x) = e^{-j(\pi/\lambda f_\ell)x^2} \operatorname{rect}\left(\frac{x}{a}\right),$$

is used to form a simple image. Suppose that the image is out of focus with

$$\frac{1}{d_s} + \frac{1}{d_i} - \frac{1}{f_\ell} = \Delta.$$

a. Find the noncoherent optical transfer function by deriving and evaluating the following expression,

$$G_n\left(\frac{x}{-\lambda d_i}\right) = \int_{-\infty}^{\infty} A\left(\xi + \frac{x}{2}\right) A^*\left(\xi - \frac{x}{2}\right) d\xi,$$

for an appropriate choice of the "aperture function" $A(x)$.

b. Find the two-dimensional, noncoherent, optical transfer function for an out-of-focus image formed by an ideal lens with a square aperture.

c. Show that if

$$\frac{1}{8}\left(\frac{\Delta}{\lambda}\right) \gg \left(\frac{\lambda}{a}\right)^2,$$

the noncoherent, optical pointspread function reduces to the geometrical optics approximation obtained by ray tracing.

d. What is the noncoherent, optical pointspread function in the geometrical optics approximation of an out-of-focus image formed by an ideal lens with a circular aperture?

4.12 Based on the Rayleigh resolution criterion, what is the angular resolution at wavelength λ of a simple telescope with a uniformly illuminated circular aperture of diameter a?

4.13 A pinhole can be used to produce images of low intensity. Using geometrical optics and ray tracing, explain how the set of rays originating on the source

$\rho(x, y)$ at $z = -d$, and passing through the point (0, 0, 0), will produce an image at $z = d'$. Why are pinhole cameras not in common use?

4.14 A *Fresnel zone plate* is a two-dimensional weighting function,

$$t(x, y) = \tfrac{1}{2}\left[1 + \mathrm{sgn}(\cos \alpha(x^2 + y^2))\right] \mathrm{circ}(x/a, y/a).$$

By using the Fourier series expansion,

$$1 + \mathrm{sgn}(\cos 2\pi t) = \sum_{n=-\infty}^{\infty} \frac{\sin(\pi n/2)}{\pi n} \mathrm{e}^{\mathrm{j}2\pi nt},$$

show that $t(x, y)$ acts like a composite of ideal lenses with different focal lengths. What are the focal lengths? Describe the output of the Fresnel zone plate when the input is the plane wave $s(x, y)$. Describe the appearance of the Fresnel zone plate. Give a formula for the area of each ring of the Fresnel zone plate.

4.15 A *Fresnel phase plate* is a two-dimensional weighting function,

$$t(x, y) = \mathrm{e}^{\mathrm{j}(\pi/2)[1+\mathrm{sgn}(\cos \alpha(x^2+y^2))]} \mathrm{circ}(x/a, y/a).$$

Compare the Fresnel phase plate with the Fresnel zone plate. In particular, compare the intensity of the corresponding images.

4.16 A thin lens, here called a thin *refractive lens*, has relative phase shift $(2\pi/\lambda)(x^2 + y^2)/2f_\ell$ at position x, y. A *diffractive lens* has relative phase shift $(2\pi/\lambda)(x^2 + y^2)/2f_\ell (\mathrm{mod}\ 2\pi)$ at position x, y. Sketch a thin refractive lens and a thin diffractive lens in cross section. Do a thin refractive lens and a thin diffractive lens both provide the same approximation to an ideal lens? Do they behave the same under a wavelength perturbation?

4.17 One version of *Babinet's principle* says that, if $t_1(x, y)$ and $t_2(x, y)$ sum to a circle function,

$$t_1(x, y) + t_2(x, y) = \mathrm{circ}(x, y),$$

then their diffraction patterns sum to the diffraction pattern of a circle function.

a. Prove Babinet's principle and state its form in the Fraunhofer region.

b. Use Babinet's principle to state the behavior in the Fresnel region of the weighting function

$$t(x, y) = \tfrac{1}{2}[1 - \mathrm{sgn}(\cos \alpha(x^2 + y^2))]\mathrm{circ}(x/a, y/a).$$

4.18 **(Gaussian Beams)** A spatially coherent, monochromatic wave propagating nominally in the z direction has a gaussian amplitude distribution in the x, y plane at $z = 0$

$$s(x, y, 0) = \mathrm{e}^{-(x^2+y^2)/w^2}.$$

What is the intensity distribution $|s(x, y, z)|^2$ in the plane $z = d$? The parameter w is called the "waist" of the beam. Can the intensity at $z = d$ be characterized by a waist?

4.19 Is the following statement correct? A gaussian beam remains gaussian as it propagates, and a nongaussian beam tends to become more gaussian as it propagates.

4.20 The *Talbot effect* is the following self-imaging property of a periodic transparency. A transparency at $z = 0$, periodic in the x direction with period Δ and illuminated by a normally incident wave of wavelength λ, reproduces itself periodically in planes at $z = l z_T$ where l is an integer and $z_T = 2\Delta^2/\lambda$ is a constant called the *Talbot distance*.

Derive the Talbot self-imaging effect from the Fresnel approximation for an infinitely large aperture. Is any further approximation necessary? Describe how the effect changes for a finite aperture.

4.21 Given the transmittance $t(x, y)$ in the plane at $z = 0$, let $\rho(x, y)$ be the complex amplitude of a monochromatic wave at $z = -d_s$. This wave passes through the transmittance and is observed in the plane at $z = d_i$. Show that, in the Fresnel approximation, the complex amplitude $r(x, y)$ in this plane is described as a scaled and magnified copy of $\rho(x, y)$ passed through a pointspread function, given by the Fourier transform of $t(x, y)e^{j2\pi(x^2+y^2)/2\lambda f_\ell}$ where f_ℓ is the constant

$$\frac{1}{f_\ell} = \frac{1}{d_s} + \frac{1}{d_i}.$$

4.22 **a.** A transmittance function,

$$t(x, y) = e^{-j(2\pi/\lambda)(x^2+y^2)/2f_\ell} e^{-j2\pi(a + b \cos 2\pi x/\Delta)} \operatorname{circ}(x/r, y/r),$$

is placed at the origin. An object, $\rho(x, y)$, is placed at $z = -d_s$. Under the Fresnel approximation, describe the noncoherent image observed in the plane at $z = d_i$ where d_i satisfies

$$\frac{1}{d_i} = \frac{1}{f_\ell} - \frac{1}{d_s}.$$

b. Describe the noncoherent image in the same situation if $t(x, y)$ is arbitrary. Specifically, what is the pointspread function through which the magnified image is observed?

4.23 A transmittance function,

$$t(x, y) = e^{-j(2\pi/\lambda)(x^2+y^2)/2f_\ell} \operatorname{sinc}(ax, ay),$$

is placed at the origin. An object, $\rho(x, y)$, is placed at $z = -d_s$. Under the Fresnel approximation, describe the noncoherent image observed in the plane at $z = d_i$ where d_i satisfies

$$\frac{1}{d_i} = \frac{1}{f_\ell} - \frac{1}{d_s}.$$

4.24 Explain the nature of a "pseudoscopic image" that may occur in certain forms of holographic aberration correction.

Notes

In 1678, Christian Huygens first put forth the view that each point on an advancing wavefront could be regarded as the source of a secondary wavefront. Huygens stated his principle only in qualitative terms. Much later, in 1818, Augustin Fresnel stated this principle in precise mathematics, explaining diffraction as an interference phenomenon although his formulation was incorrect. Because of continuing controversy about the proper statement of consistent boundary conditions, there have been many subsequent derivations of the Huygens–Fresnel principle, by Kirchoff, by Rayleigh, and by Sommerfeld. To avoid this discussion, we have sidestepped questions of boundary conditions by stating the Huygens–Fresnel principle as a relationship between a wave in two planes, and deriving it as a simple consequence of a certain two-dimensional Fourier transform pair. This gives a precise mathematical relationship between a wavefront passing through one plane and that wavefront passing through a subsequent plane without regard for the source of the wavefront.

The use of the lens for magnification, as in microscopes and telescopes, has been widespread for some time as has been the use of the lens for readjusting focal length, for example, in eyeglasses. The first study of the lens in the spirit of this chapter was the Abbe–Porter experiment begun by Abbe (1873) and continued by Porter (1906). They experimented with placing simple obstructions in the Fourier transform plane of a lens to modify the image. Their experiments became the basis for many later developments. The phase-contrast microscope was proposed by Zernike (1935). From a mathematical point of view, the Zernike filter is the two-dimensional version of the FM radio demodulator proposed by Armstrong (1936). The schlieren method had been proposed much earlier by Foucault (1858).

The development of optics as a branch of information theory that we have adopted here emphasizes the role of the Fourier transform in optics, which was recognized early on by Duffieux (1946). This development began with Elias (1953), who was motivated by his earlier paper with Grey and Robinson (1952). The suggestion was developed considerably by O'Neill and others (1956, 1962). Vander Lugt (1964) proposed the use of an optical processor to calculate a two-dimensional convolution as a product in the Fourier transform domain, thereby providing a two-dimensional filter. Lohmann (1977) described a method to bypass the limitations of noncoherent optics used for filtering. At the same time that this restatement of the theory of optics was underway, applications emerged for the use of optics to perform certain two-dimensional Fourier transforms of microwave signals, as discussed by Cutrona, Leith, Palermo, and Porcello (1960). The use of the Fourier transform to derive the Fresnel diffraction formula was discussed by

Banerjee (1985). The use of the ambiguity function to design cubic-phase apertures for large depth-of-field imaging was proposed by Dowski and Cathey (1995).

Systems for reconstructing images from Fourier transform data greatly influenced Gabor (1948), who, in turn, worked out the ideas of holography as a lensless imaging process for electron microscopy. Sometime later, holography was seen to be an important method for optical signals. Various versions of holography have been formulated since Gabor's work; the most important advance was by Leith and Upatnieks (1962, 1964).

5 Antenna systems

An antenna (or hydrophone) is a linear device that forms the interface between free propagation of electromagnetic waves (or pressure waves) and guided propagation of electromagnetic signals. An antenna can be used either to transmit an electromagnetic signal or to receive an electromagnetic signal. During transmission, the function of the antenna is to concentrate the electromagnetic wave into a beam that points in the desired spatial direction. During reception, the function of the antenna is to collect the incident signal and deliver it to the receiver. An important theorem of antenna theory, known as the *reciprocity theorem*, allows us to deal with the antenna either as a transmitting device or as a receiving device, depending on which is more convenient for a particular discussion.

The only aspect of antennas that we shall study is the propagation and diffraction of waves traveling between the antenna aperture and points far away. During transmission, an antenna creates a time-varying and spatially distributed signal across its aperture to form a wave that will propagate as desired through free space. The spatial distribution of the signal across the antenna aperture is called the *illumination function*. The distribution in the far field of the waveform amplitude and phase over the spherical coordinate angles is called the *antenna radiation pattern* or the *antenna pattern*. The relationship between the antenna pattern and the aperture illumination function can be described with the aid of the two-dimensional Fourier transform.

The wavefront launched by an illumination function is completely described by the Huygens–Fresnel principle, which was discussed in detail in Chapter 4. In this chapter, we shall give an alternative and much looser development that is suitable only in the (Fraunhofer) far-field diffraction region. This heuristic approach to the Huygens principle leads to a simple and intuitive method of deriving radiation patterns. It complements and reinforces the formal approach of Chapter 4.

Large antennas can be formed by combining the signals from the elements of an array of simpler antenna elements, just as a pulse train can be formed by combining many copies of a simple pulse. We shall study some of the processing techniques needed to form antenna arrays. Again, our concern is only with the mathematical relationship between the illumination function and the antenna pattern. The many physical effects associated with real antennas and arrays will not be considered.

5.1 Aperture and pattern

A physical antenna is the interface between guided wave propagation and free space. As far as its effect on the signal is concerned, an antenna is described by its antenna radiation pattern. The antenna radiation pattern is a function, $E(\phi, \psi)$, of the spherical coordinates (ϕ, ψ). The function $E(\phi, \psi)$ is a complex function that gives the magnitude and phase of the signal radiated in direction ϕ, ψ. An *omnidirectional antenna*, which does not exist, is one for which $E(\phi, \psi)$ is a constant. The squared magnitude of $E(\phi, \psi)$, denoted $P(\phi, \psi) = |E(\phi, \psi)|^2$, is also called the antenna radiation pattern. To distinguish $E(\phi, \psi)$ from $P(\phi, \psi)$, we may speak of the antenna signal radiation pattern or the antenna power radiation pattern.

An antenna can be described as a device for setting up an illumination function, $s_0(x, y)$, possibly complex, on a region of the x, y plane called the *aperture*; the signal radiated into each spatial direction by this illumination function then constitutes the antenna radiation pattern. In general, the aperture need not be planar, but we shall study only planar apertures and their radiation patterns.

The province of the antenna designer is to configure an arrangement of conducting objects and their electrical feeds, as shown in Figure 5.1, to generate the illumination function $s_0(x, y)$ within the aperture. We will not study the design of antennas; we will study only the relationship between the illumination function $s_0(x, y)$ and the antenna radiation pattern.

Just as we can speak of the mathematical response of a filter without discussing the physical phenomena that are used to implement the filter, so, too, we can speak of

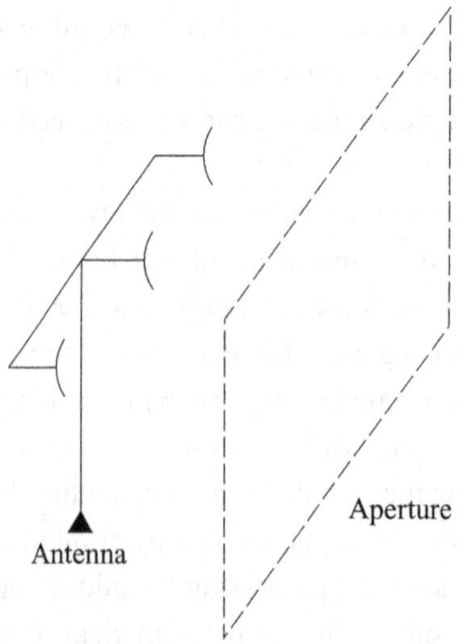

Figure 5.1 Portraying an antenna by an aperture

the mathematical response of an antenna without examining the underlying physical phenomena, as are described by Maxwell's equations. The mathematical analysis does not specify the nature of the radiation, only that the radiation from each point source be a spherical wave of frequency f_0 and velocity c.

At each point (x, y) of the aperture, a spherical Huygens wavelet, $s_0(x, y)e^{-j2\pi f_0 t}dx\,dy$, is launched. The magnitude and phase of $s_0(x, y)$ describe the magnitude and phase of the infinitesimal wave launched from this (x, y). One may wish to visualize the aperture as a screen with a hole in it that is illuminated from behind by a plane wave of frequency f_0. By visualizing the hole filled with a semitransparent medium with the attenuation and phase described by $s_0(x, y)$, we complete the model of the aperture. This visualization helps to explain the role of the illumination function $s_0(x, y)$ in general, though perhaps it is more suggestive of an optical aperture. In the microwave bands of the electromagnetic spectrum, the aperture and its illumination function would normally refer to a specified region of space in an arbitrary plane lying in front of the physical antenna.

A one-dimensional radiation pattern can be related to a one-dimensional illumination function, $s_0(x)$, by integrating all signals propagating in a given direction; a signal arriving from an infinitesimal element of the aperture has a weight given by the illumination function and is delayed by an amount depending on the path length. At a point very far from the aperture, the paths from points on the aperture are essentially parallel lines, and the path from the point x is shorter than the path from the origin by the amount $x \sin\phi$ as shown in Figure 5.2. Let τ denote the time that it takes a signal from the origin to reach a point in the far field. The differential time delay of the signal coming from an infinitesimal element at x, with respect to the delay from the infinitesimal element

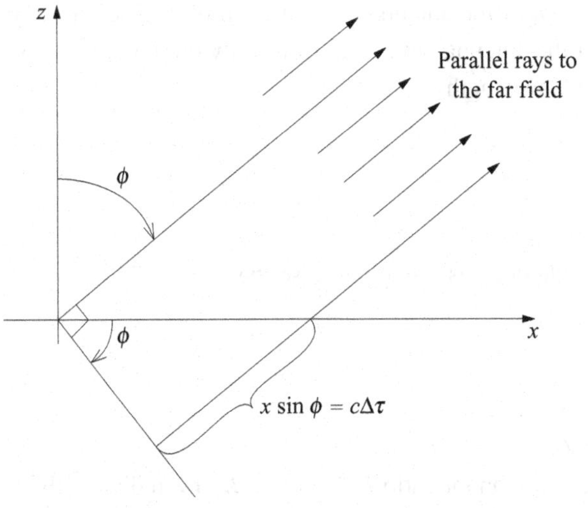

Figure 5.2 Incident wave in two dimensions

at the origin, is

$$\Delta\tau = -\frac{x}{c}\sin\phi$$

because it travels a path that is longer by the distance $x\sin\phi$. Then the signal received at this distant point is

$$v(t,\phi)\mathrm{e}^{-\mathrm{j}2\pi f_0(t-\tau)} = \int_{-\infty}^{\infty} s_0(x)\mathrm{e}^{-\mathrm{j}2\pi f_0\left(t-\tau+\frac{x}{c}\sin\phi\right)}\,\mathrm{d}x.$$

More generally, a time-varying signal, $c(t)$, may modulate the carrier. The waveform $c(t)$ at a carrier frequency of f_0 gives rise to a spatially distributed signal within the aperture, given by $c(t)\mathrm{e}^{-\mathrm{j}2\pi f_0 t}s_0(x)$. This leads to a composite signal radiated at angle ϕ, given by

$$v(t,\phi)\mathrm{e}^{-\mathrm{j}2\pi f_0(t-\tau)} = \int_{-\infty}^{\infty} s_0(x)c\left(t-\tau+\frac{x}{c}\sin\phi\right)\mathrm{e}^{-\mathrm{j}2\pi f_0\left(t-\tau+\frac{x}{c}\sin\phi\right)}\,\mathrm{d}x.$$

In this section we require that the bandwidth of $c(t)$ is very small compared to f_0. This allows us to make the narrowband approximation, which is that $c\left(t+\frac{x}{c}\sin\phi\right)$ is equal to $c(t)$. Then

$$\begin{aligned} v(t,\phi) &= c(t-\tau)\int_{-\infty}^{\infty} s_0(x)\mathrm{e}^{-\mathrm{j}2\pi\frac{x}{\lambda}\sin\phi}\,\mathrm{d}x \\ &= c(t-\tau)S_0\left(\frac{\sin\phi}{\lambda}\right) \end{aligned}$$

where $S_0(f)$ is the Fourier transform of $s_0(x)$. The function $E(\phi)$, defined as

$$E(\phi) = \frac{1}{\lambda}S_0\left(\frac{\sin\phi}{\lambda}\right),$$

is called the *antenna pattern* of the antenna as a function of ϕ. The multiplying term $1/\lambda$ is introduced so that the antenna pattern depends only on the ratio of the antenna spatial dimensions to the wavelength.

For example, let

$$s_0(x) = \begin{cases} 1 & |x| \le \frac{L}{2} \\ 0 & \text{otherwise.} \end{cases}$$

The transform of a rectangle pulse is a sinc pulse, so that

$$\begin{aligned} E(\phi) &= \frac{1}{\lambda}S_0\left(\frac{\sin\phi}{\lambda}\right) \\ &= \frac{L}{\lambda}\,\mathrm{sinc}\left(\frac{L}{\lambda}\sin\phi\right). \end{aligned}$$

Notice that $E(\phi)$ depends only on the ratio L/λ, not on L or λ individually.

The antenna pattern of the rectangle illumination function has the sidelobes of a sinc function. However, the zeros of the antenna pattern are uniformly spaced, not in ϕ but in $\sin\phi$. A zero occurs when $(L/\lambda)\sin\phi$ is a nonzero integer. If $L < \lambda$, there is no zero. If L is only a little bigger than λ, there will be only one zero, or perhaps several. Finally, if L/λ is large, there will be many zeros and the antenna pattern will be small, except near $\sin\phi = 0$. In this case, we can make the approximation

$$E(\phi) = \frac{L}{\lambda}\,\mathrm{sinc}\left(\frac{L}{\lambda}\phi\right)$$

by setting $\sin\phi = \phi$.

The antenna pattern is meaningful only for $|\sin\phi| \le 1$. Consequently, the *visible region* of the frequency domain is defined as those spatial frequencies satisfying $|f_x| \le \lambda^{-1}$. In particular, energy incident on the aperture that is not in the visible region does not radiate. This flaw in the representation of the antenna pattern as a Fourier transform, which is not significant if the aperture is large compared to a wavelength, is related to the Fraunhofer approximation. This consideration is accommodated within the Fraunhofer approximation by the maxim that the invisible energy is "stored" in the antenna and not transmitted. To quantify this effect, the *antenna quality factor* is

$$Q = \frac{\displaystyle\int_{-\infty}^{\infty} |S_0(f)|^2\,\mathrm{d}f}{\displaystyle\int_{-\lambda^{-1}}^{\lambda^{-1}} |S_0(f)|^2\,\mathrm{d}f}.$$

For antennas larger than a wavelength, the antenna quality factor is nearly equal to 1.

A two-dimensional radiation pattern is related to a two-dimensional illumination function in the same way that a one-dimensional radiation pattern is related to a one-dimensional illumination function. The differential delay in the signal transmitted from the point (x, y) relative to the signal transmitted from the origin is $\frac{1}{c}(x\cos\psi + y\sin\psi)\sin\phi$. To see this more clearly, temporarily rotate the coordinate system, shown in Figure 5.3, by the angle ψ about the z axis into a new (x', y', z) coordinate system such that the far-field point is in the (x', z) plane. The differential delay then is $\frac{1}{c}(x'\sin\phi)$. Replacing x' by $x\cos\psi + y\sin\psi$ gives the delay. Consequently, as before,

$$\begin{aligned} v(t, \phi, \psi) &= c(t-\tau)\int_{-\infty}^{\infty}\int_{-\infty}^{\infty} s_0(x, y)e^{-j2\pi\left(\sin\phi\cos\psi\frac{x}{\lambda} + \sin\phi\sin\psi\frac{y}{\lambda}\right)}\,\mathrm{d}x\,\mathrm{d}y \\ &= c(t-\tau)S_0\left(\frac{\sin\phi\cos\psi}{\lambda}, \frac{\sin\phi\sin\psi}{\lambda}\right) \end{aligned}$$

where $S_0(f_x, f_y)$ is the two-dimensional Fourier transform of $s_0(x, y)$. The *two-dimensional antenna pattern* is defined as

$$E(\phi, \psi) = \frac{1}{\lambda^2}\,S_0\left(\frac{\sin\phi\cos\psi}{\lambda}, \frac{\sin\phi\sin\psi}{\lambda}\right).$$

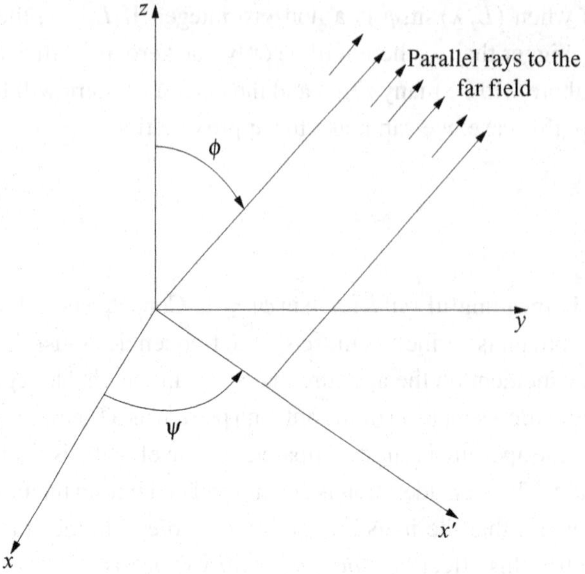

Figure 5.3 Incident wave in three dimensions

Recall that the trigonometric quantities $\sin\phi\cos\psi$ and $\sin\phi\sin\psi$ are two of the direction cosines. They are the cosines, respectively, of the angles between the direction of propagation and the x and y axes. The terms

$$\cos\phi_x = \sin\phi\cos\psi$$
$$\cos\phi_y = \sin\phi\sin\psi$$

are sometimes used to express the direction cosines. Then with

$$\sin\overline{\phi}_x = \sin\phi\cos\psi$$
$$\sin\overline{\phi}_y = \sin\phi\sin\psi$$

we can write

$$E'(\overline{\phi}_x, \overline{\phi}_y) = \frac{1}{\lambda^2} S_0\left(\frac{\sin\overline{\phi}_x}{\lambda}, \frac{\sin\overline{\phi}_y}{\lambda}\right),$$

which sometimes is also called the antenna pattern, though it is not the same function as $E(\phi, \psi)$. The angles $\overline{\phi}_x$ and $\overline{\phi}_y$ are the complements of the angles between the direction of propagation and the x and y axes. For directions close to the z axis, the small-angle

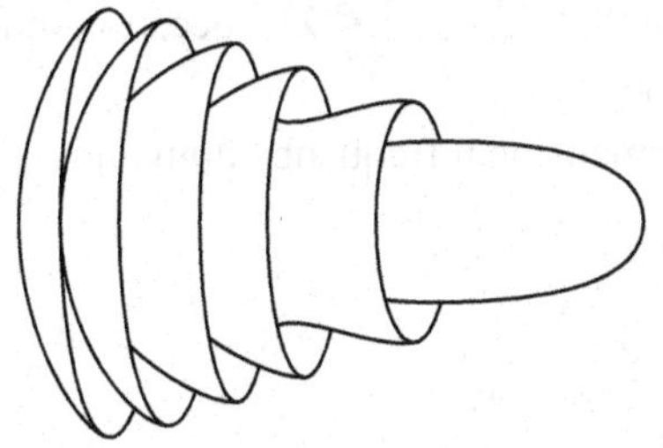

Figure 5.4 A circularly symmetric antenna pattern

approximation to the sine gives

$$\overline{\phi}_x \approx \phi \cos \psi$$
$$\overline{\phi}_y \approx \phi \sin \psi.$$

With these approximations, a highly directive antenna pattern can be expressed as

$$E'(\overline{\phi}_x, \overline{\phi}_y) = \frac{1}{\lambda^2} S_0 \left(\frac{\phi \cos \psi}{\lambda}, \frac{\phi \sin \psi}{\lambda} \right),$$

which is a small-angle approximation to the correct spherical trigonometry.

Figure 5.4 shows a two-dimensional antenna pattern formed by an illumination function that is circularly symmetric. If the illumination function is a uniformly illuminated circle of diameter $2r$,

$$s_0(x, y) = \operatorname{circ}\left(\frac{x}{2r}, \frac{y}{2r} \right),$$

then the antenna pattern is

$$E(\phi, \psi) = \frac{4r^2}{\lambda^2} \operatorname{jinc}\left(\frac{2r}{\lambda} \sin \phi \right).$$

In this case, the sidelobes shown in Figure 5.4 are the sidelobes of the jinc function as a function of $\sin \phi$. Because $\sin \phi \le 1$, there are only a finite number of sidelobes depending on the ratio of r/λ. The first zero of the jinc function occurs when $(2r/\lambda) \sin \phi = 1.22$, or $\phi = \sin^{-1}(0.61\lambda/r)$. The first zero of the jinc function will not be seen for any value of ϕ if

$$\frac{r}{\lambda} \le 0.61.$$

Note in the example that $E(\phi, \psi)$ does not display all of $S_0(f_x, f_y)$. It only displays

$S_0(f_x, f_y)$ over the range of frequencies such that $f_x^2 + f_y^2 \leq \lambda^{-2}$ because, otherwise, the direction cosines would be larger than one.

In general, the visible region of the two-dimensional frequency domain is

$$\mathcal{V} = \{(f_x, f_y) : f_x^2 + f_y^2 \leq \lambda^{-2}\}.$$

The radiated energy is the energy in the visible region

$$E_v = \int\int_{\mathcal{V}} |S_0(f_x, f_y)|^2 \, \mathrm{d}f_x \, \mathrm{d}f_y.$$

Definition 5.1.1 *The directivity (or gain pattern) of an antenna in the direction with the spherical coordinates ϕ, ψ is defined as*

$$G(\phi, \psi) = \frac{|E(\phi, \psi)|^2}{\frac{1}{4\pi}\int\int_A |s_0(x, y)|^2 \, \mathrm{d}x \, \mathrm{d}y}.$$

The peak-of-beam gain (or gain) is defined as $G = \max_{\phi,\psi} G(\phi, \psi)$.

If the aperture is large in comparison with λ, then the antenna quantity factor is nearly one. If the aperture is not large in comparison with λ, the alternative definition

$$G(\phi, \psi) = \frac{|E(\phi, \psi)|^2}{\frac{1}{4\pi}\int\int_{\mathcal{V}} |S_0(f_x, f_y)|^2 \, \mathrm{d}f_x \, \mathrm{d}f_y}$$

may be preferred because it reduces the gain by the antenna quality factor.

The *resolution* of an antenna is determined by the width of the main beam, which varies inversely with the peak-of-beam gain of the antenna. The directivity, or gain, of an antenna is defined in a normalized form by dividing out the total energy in the illumination function $s_0(x, y)$. This means that the antenna gain cannot be increased simply by scaling the amplitude of the illumination function.

The directivity of an antenna is sometimes displayed in a spherical coordinate system, as shown in Figure 5.5. Each contour line on the sphere denotes the locus of points for which $G(\phi, \psi)$ is equal to a given constant.

Theorem 5.1.2 *The peak-of-beam gain, G, of an aperture of area A satisfies*

$$G \leq 4\pi \frac{A}{\lambda^2}$$

with equality if and only if the aperture illumination function is of the form

$$s_0(x, y) = \mathrm{e}^{\mathrm{j}2\pi(x \sin\phi_0 \cos\psi_0 + y \sin\phi_0 \sin\psi_0)/\lambda}.$$

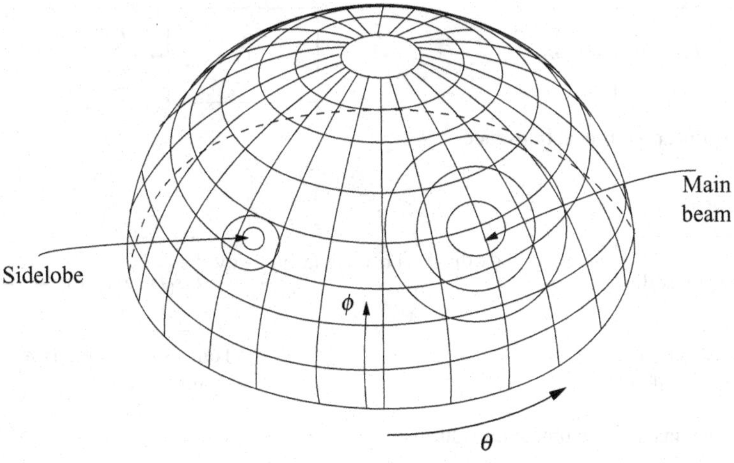

Figure 5.5 A gain pattern in spherical coordinates

Proof: The proof uses the Schwarz inequality. The function $s_0(x, y)$ is nonzero only for x, y in the region $\mathcal{A}$. Hence

$$\begin{aligned}|\lambda E(\phi, \psi)|^2 &= \left| \int\!\!\int_{\mathcal{A}} s_0(x, y) e^{j2\pi(x \sin\phi \cos\psi + y \sin\phi \sin\psi)/\lambda} \, dx \, dy \right|^2 \\ &\leq \int\!\!\int_{\mathcal{A}} |s_0(x, y)|^2 \, dx \, dy \int\!\!\int_{\mathcal{A}} dx \, dy,\end{aligned}$$

from which the inequality of the theorem follows. The Schwarz inequality is satisfied with equality if and only if $s_0(x, y)$ has the form

$$s_0(x, y) = e^{j2\pi(x \sin\phi_0 \cos\psi_0 + y \sin\phi_0 \sin\psi_0)/\lambda},$$

and $(\phi, \psi) = (\phi_0, \psi_0)$. □

Definition 5.1.3 *The effective area of an antenna aperture is that value A_e satisfying*

$$G = 4\pi \frac{A_e}{\lambda^2}.$$

The effective area of an antenna is the area of a uniformly illuminated aperture that has the same gain as the antenna. Theorem 5.1.2 says that the effective area is never larger than the actual area.

5.2 Reciprocity

The gain of an antenna has been defined for the case where the antenna is used to transmit waves, transforming a signal into an illumination function that launches a propagating wave. An antenna may also be used to intercept a propagating wave to

Figure 5.6 A linear reciprocal transducer

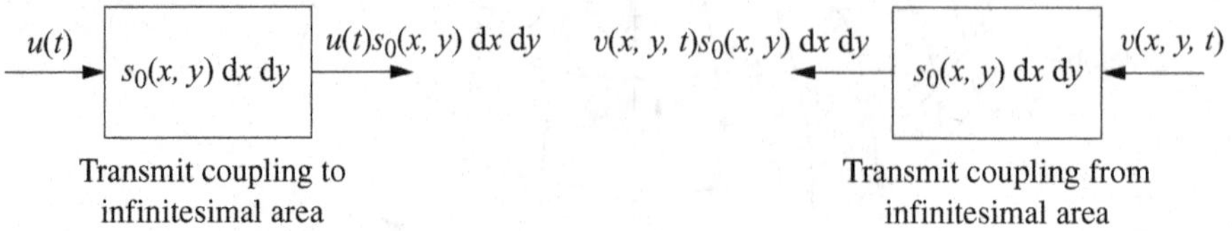

Figure 5.7 Antenna as a reciprocal transducer

form the received signal. The reciprocity theorem of antenna theory, loosely stated, says that the properties of an antenna are the same whether the antenna is used for transmission or reception. Specifically, the gain pattern $G(\phi, \psi)$ of an antenna that is used to transmit is equal to the gain pattern when that antenna is used to receive. The relationship between the antenna gain function and the illumination function within the aperture is the same for reception as for transmission even though the wave-front in the two cases is distinctly different. At a deeper level, and really beyond our scope, is the reciprocity between the illumination function and the transmitter/receiver circuitry.

The reciprocity theorem for antennas requires that the relationship between the antenna transmitter/receiver circuitry and the antenna illumination function is that of a linear reciprocal transducer. A reciprocal transducer, shown symbolically in Figure 5.6, is one with two ports such that when the signal s is the input at the first port, then Gs is the output at the second port; while if the signal r is the input at the second port, then Gr is the output at the first port.

A linear reciprocal relationship between a signal and an illumination function is one in which the signal $u(t)$ feeding the antenna generates the illumination function $u(t)s_0(x, y)$, and the input illumination function $v(t, x, y)$ incident on the aperture generates the signal

$$r(t) = \int_{-\infty}^{\infty}\int_{-\infty}^{\infty} v(x, y, t)s_0(x, y)\,dx\,dy$$

captured by the antenna. More specifically, as shown in Figure 5.7, if the input at the signal port is $u(t)$, the output at antenna position (x, y) is $u(t)s_0(x, y)\,dx\,dy$; while if the intercepted signal at antenna position (x, y) is $v(x, y, t)$, its contribution to the signal port is $v(x, y, t)s_0(x, y)\,dx\,dy$.

Because it is not our purpose to study the internal structure of antennas, we will simply define an antenna as a device with the appropriate reciprocity properties for which superposition applies.

The following theorem says that this assumption of reciprocity within the antenna structure implies a reciprocity in the antenna pattern.

Theorem 5.2.1 (Antenna Reciprocity Theorem) *The gain pattern of an antenna is the same for reception as for transmission.*

Proof: The proof consists of comparing the two cases. When transmitting, as we have already seen, the signal $s(t)$ reaching a point in the far field with spherical coordinates ϕ, ψ is

$$v(t, \phi, \psi)e^{-j2\pi f_0 t} = c(t-\tau)e^{-j2\pi f_0 t} S_0\left(\frac{\sin\phi\cos\psi}{\lambda}, \frac{\sin\phi\sin\psi}{\lambda}\right).$$

When receiving, the signal $c(t)e^{-j2\pi f_0 t}$ is transmitted from a distant point with the spherical coordinates ϕ, ψ and is seen by the aperture at point (x, y) with a relative delay, $\tau = -\frac{1}{c}(x\sin\phi\cos\psi + y\sin\phi\sin\psi)$, as referenced to the origin. Consequently, the signal intercepted at point (x, y) of the aperture is

$$\begin{aligned} v(x, y, t)e^{-j2\pi f_0 t} &= c(t-\tau)e^{-j2\pi f_0(t-\tau)} \\ &= c(t-\tau)e^{-j2\pi f_0 t}e^{-j2\pi(x\sin\phi\cos\psi + y\sin\phi\sin\psi)/\lambda}, \end{aligned}$$

and the contribution to the received signal is $v(x, y, t)s_0(x, y)\,dx\,dy$. But the reciprocal relationship between the antenna illumination function and the transmitter circuitry gives

$$r(t) = \int_{-\infty}^{\infty}\int_{-\infty}^{\infty} v(x, y, t)s_0(x, y)\,dx\,dy.$$

From this we can immediately write

$$\begin{aligned} r(t)e^{-j2\pi f_0 t} &= c(t-\tau)e^{-j2\pi f_0 t}\int_{-\infty}^{\infty}\int_{-\infty}^{\infty} s_0(x, y)e^{-j2\pi(x\sin\phi\cos\psi + y\sin\phi\sin\psi)/\lambda}\,dx\,dy \\ &= c(t-\tau)e^{-j2\pi f_0 t} S_0\left(\frac{\sin\phi\cos\psi}{\lambda}, \frac{\sin\phi\sin\psi}{\lambda}\right), \end{aligned}$$

as was to be proved. □

5.3 Antenna arrays

A large antenna may be constructed by arranging a number of simple antennas into a suitable pattern. In this context, the simple antenna is called an *antenna element*, and the large antenna is called an *antenna array*. The relationship between the element

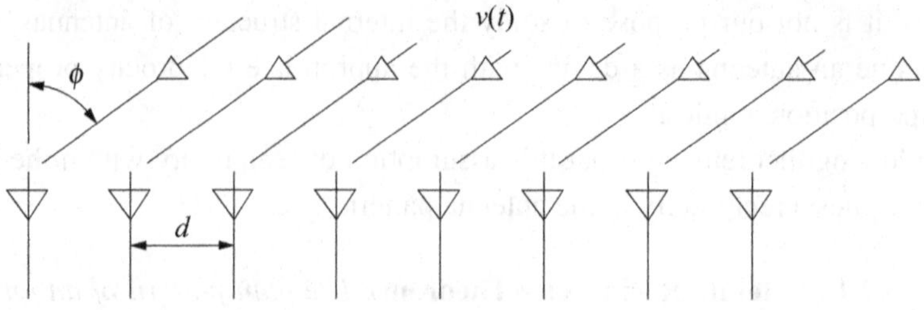

Figure 5.8 Antenna array

and the array is analogous to the relationship between a pulse and a pulse train. The larger array has a narrower Fourier transform and hence a narrower main beam. For this reason, combining the signal to and from the elements is called *beamforming*. Ideally, each element of the array radiates with the same pattern as it does when it is physically isolated; the total radiation pattern is the sum of the individual patterns, as adjusted for their individual placement. In practice, there will be blocking and coupling effects, which can modify the results of a superposition calculation, but these effects can be kept small by the careful design of the physical elements. We shall ignore these effects and study only the idealized model in which the antennas do not affect each other and superposition applies.

First, we shall study one-dimensional arrays. The antenna pattern for a two-dimensional array will be an obvious generalization that we can write down by inspection of the formula for the one-dimensional array.

Figure 5.8 shows a one-dimensional antenna array. Denote the radiation pattern of the element by $E_e(\phi)$. The radiation pattern of the array, denoted $E(\phi)$, is the superposition of all of the element radiation patterns.

Theorem 5.3.1 *The linear antenna pattern of N equispaced antenna elements, centered at the origin, is given by*

$$E(\phi) = E_e(\phi)\text{dirc}_N\left(\frac{d}{\lambda}\sin\phi\right)$$

where d is the element spacing.

Proof: Suppose that the plane wave $v(t)$ at frequency f_0 and wavelength $\lambda = c/f_0$ is incident on the array at angle ϕ. The output of the nth antenna element is a signal at this frequency but at a phase angle based on the path delay $\tau = nd\sin\phi/c$. Hence the output of the nth antenna element is

$$g_n(t) = v(t)E_e(\phi)e^{j[2\pi(d/\lambda)\sin\phi]n}.$$

These N signals are combined to produce the signal

$$\begin{aligned} g(t) &= \sum_{n=0}^{N-1} g_n(t) \\ &= v(t)E_e(\phi)\sum_{n=0}^{N-1} e^{j[2\pi(d/\lambda)\sin\phi]n}. \end{aligned}$$

Consequently, the array radiation pattern is

$$E(\phi) = E_e(\phi)\sum_{n=0}^{N-1} e^{j[2\pi(d/\lambda)\sin\phi]n}.$$

The sum has been encountered previously in connection with the definition of the dirichlet function. Thus

$$\begin{aligned} E(\phi) &= E_e(\phi)e^{j(N-1)\pi(d/\lambda)\sin\phi}\text{dirc}_N\left(\frac{d}{\lambda}\sin\phi\right) \\ &= E_e(\phi)e^{j(N-1)\pi(d/\lambda)\sin\phi}\frac{\sin[N\pi(d/\lambda)\sin\phi]}{\sin[\pi(d/\lambda)\sin\phi]}. \end{aligned}$$

Centering the array at the origin eliminates the phase term, so the proof is complete.

□

The squared magnitude of the array radiation pattern is the power radiation pattern

$$P(\phi) = |E(\phi)|^2\text{dirc}_N^2\left(\frac{d}{\lambda}\sin\phi\right).$$

The square of the dirichlet function determines the main beam of the array. This beam is broadside to the array and has its first null at the angle ϕ satisfying

$$N(d/\lambda)\sin\phi = 1,$$

or

$$\phi = \sin^{-1}[\lambda/(Nd)],$$

from which we see that the width of the main beam depends inversely on the length of the array Nd.

The dirichlet function has its first grating lobe at the angle ϕ satisfying

$$(d/\lambda)\sin\phi = 1,$$

or

$$\phi = \sin^{-1}\left(\frac{\lambda}{d}\right).$$

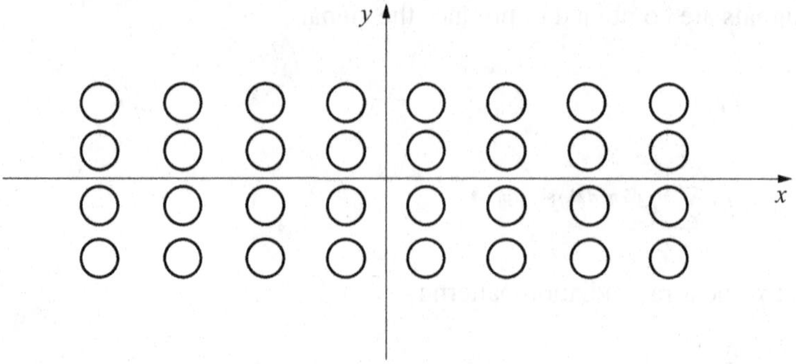

Figure 5.9 A rectangular array of antennas

One would like to choose the element illumination function so that the element pattern $E_e(\phi)$ is small at the grating lobes. However, this will require the element size to be comparable (or equal) to the spacing d, which may lead to practical difficulties.

Figure 5.9 shows a two-dimensional K by N array of antennas placed on a rectangular grid. We can regard this grid as a one-dimensional array of antennas along the y axis, each of which is an array of antenna elements along the x axis. Consequently, with the origin at the center of the array so that the main beam is real, the antenna pattern for the two-dimensional array is

$$E(\phi, \psi) = E_e(\phi, \psi)\text{dirc}_N\left(\frac{d_1}{\lambda}\sin\phi\cos\psi\right)\text{dirc}_K\left(\frac{d_2}{\lambda}\sin\phi\sin\psi\right).$$

This pattern has a two-dimensional grating lobe whenever each dirichlet function has a one-dimensional grating lobe. This occurs whenever both sines in the denominator have arguments that are integer multiples of π. Equivalently, a grating lobe occurs at angular coordinates ϕ, ψ whenever

$$\sin\phi\cos\psi = \ell_1\frac{\lambda}{d_1}$$
$$\sin\phi\sin\psi = \ell_2\frac{\lambda}{d_2}$$

where ℓ_1 and ℓ_2 are arbitrary integers. These conditions are a form of the *Bragg-Laue equations* of crystallography.

Instead of a rectangular array of antenna elements, one can use a two-dimensional array of antennas based on the hexagonal lattice, shown in Figure 5.10. Because the calculation of an antenna pattern from an illumination function is essentially the calculation of a two-dimensional Fourier transform, the calculation of the pattern of a hexagonal array is straightforward and proceeds in a way that is similar to the analysis of hexagonal sampling in Section 3.7. A hexagonal antenna array has the

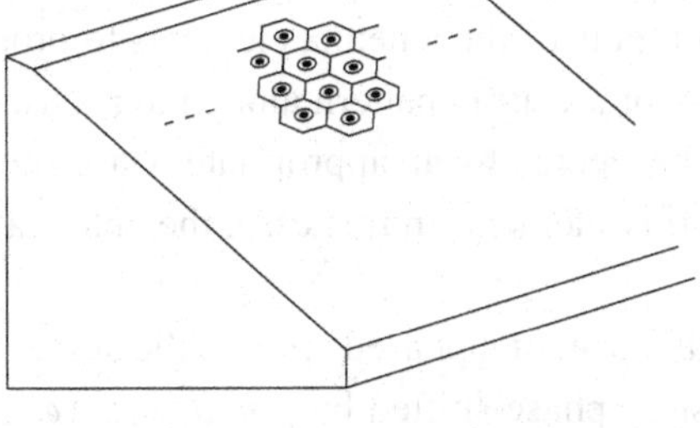

Figure 5.10 Hexagonal antenna array

advantage that, for a fixed total aperture, which determines the width of the main lobe, fewer antenna elements are needed to satisfy a given angular spread of grating lobes.

An antenna used both to transmit a signal and to receive an echo of that signal affects the signal twice: once when transmitting and once when receiving. Thus the received signal (not the received power) is proportional to the square of the antenna radiation pattern. For an array centered at the origin, this squared pattern is

$$E^2(\phi) = E_e^2(\phi)\mathrm{dirc}_N^2\left(\frac{d}{\lambda}\sin\phi\right).$$

It is instructive to contrast this with an alternative way of using the array. Rather than using all N antenna elements simultaneously to transmit a single pulse, N identical copies of a pulse are transmitted, one pulse at a time. The nth copy of the pulse is transmitted from the nth antenna element and received only at the nth antenna. Each return is recorded as a complex signal as it is received; subsequently, all recorded returns are coherently added together. We may refer to this as a *sequentially formed array*. The nth sequential return is proportional to the square of $E_e(\phi)\mathrm{e}^{\mathrm{j}[2\pi(d/\lambda)\sin\phi]n}$. Thus the sum of the received echoes gives

$$\mathrm{e}^{-\mathrm{j}\theta}E_e^2(\phi)\sum_{n=0}^{N-1}\mathrm{e}^{2\mathrm{j}[2\pi(d/\lambda)\sin\phi]n} = E_e^2(\phi)\mathrm{dirc}_N\left(\frac{2d}{\lambda}\sin\phi\right)$$

where $\theta = 2(N-1)(d/\lambda)\sin\phi$. This expression differs from the earlier expression in that d is replaced by $2d$, and the dirichlet function is not squared. Because d is replaced by $2d$, the main beam is half as wide so that the resolution is twice as good.[1] At the same time, the grating lobes become more closely spaced. Because the dirichlet function is not squared, the peak value of the beam is N rather than N^2. This loss can be attributed to the fact that each reflected signal is incident on all N antennas but is recorded on only one, so only one Nth of the energy is used.

[1] This way to improve resolution can be important in applications, such as ultrasound transducers, in which the array aperture is constrained in size.

Steered arrays

Thus far we have simply summed the outputs of the antenna elements to produce the array pattern. This places the main beam of the array pattern along the z axis. Instead, the main beam of an antenna array can be steered by an appropriate phase shift on the signal received or transmitted at each array element. In this case, the antenna array is called a *phased array*.

Suppose that we wish to steer the main beam of a phased array to the angle ϕ_0. Then the signal at the nth array element must by phase-shifted by $[2\pi(d/\lambda)\sin\phi_0]n$. For a receiving array, the beamsteering is described by the equation

$$G(t) = \sum_{n=0}^{N-1} g_n(t)\mathrm{e}^{-\mathrm{j}[2\pi(d/\lambda)\sin\phi_0]n}$$

where $g_n(t)$ is the signal received at the nth array element, and $G(t)$ is the final received signal. This leads to the new radiation pattern, denoted $E_{\phi_0}(\phi)$, for the steered array, given by

$$\begin{aligned} E_{\phi_0}(\phi) &= E_e(\phi)\sum_{n=0}^{N-1} \mathrm{e}^{\mathrm{j}[2\pi(d/\lambda)(\sin\phi - \sin\phi_0)]n} \\ &= E_e(\phi)\mathrm{e}^{\mathrm{j}(N-1)\pi(d/\lambda)(\sin\phi-\sin\phi_0)}\mathrm{dirc}_N\left(\frac{d}{\lambda}(\sin\phi - \sin\phi_0)\right), \end{aligned}$$

and

$$|E_{\phi_0}(\phi)|^2 = |E_e(\phi)|^2\mathrm{dirc}_N^2\left(\frac{d}{\lambda}(\sin\phi - \sin\phi_0)\right).$$

The main lobe of the dirichlet function now occurs at $\phi = \phi_0$. The first null occurs at

$$\phi = \sin^{-1}[\lambda/(Nd) + \sin\phi_0].$$

Because of the nonlinearity of the sine function, the angle from the main beam to the first null is greater for the steered array when ϕ_0 is nonzero than when ϕ_0 is zero. As the main beam is steered farther off the boresight ($\phi_0 = 0$), the width of the main beam increases.

A useful property of a phased array is that a number of steered beams can be created simultaneously from the same aperture. One simply processes the element signals of the array simultaneously through several sets of phase angles, thereby forming several steered beams simultaneously, as shown in Figure 5.11.

Let $\phi_0, \phi_1, \phi_2, \ldots, \phi_{N'-1}$ denote a set of steering offset angles. The set of antenna radiation patterns

$$|E_{\phi_\ell}(\phi)|^2 = |E_e(\phi)|^2\mathrm{dirc}_N^2\left(\frac{d}{\lambda}(\sin\phi - \sin\phi_\ell)\right) \qquad \ell = 0, \ldots, N'-1$$

describes N' simultaneous beams, which can be formed by the set of beamforming

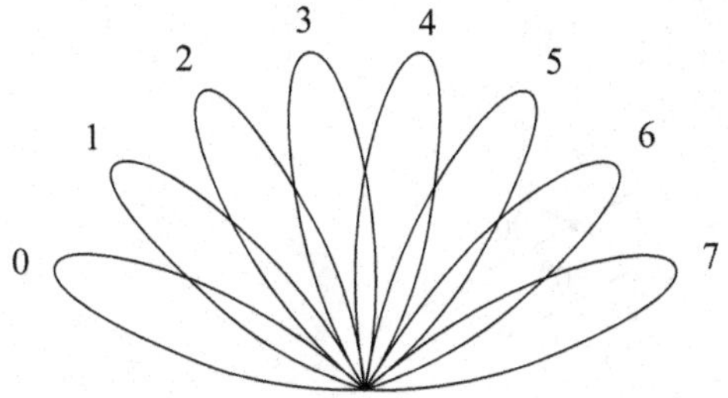

Figure 5.11 Simultaneous beams of an array antenna

equations

$$G_\ell(t) = \sum_{n=0}^{N-1} g_n(t)e^{-j[2\pi(d/\lambda)\sin\phi_\ell]n} \quad \ell = 0, \ldots, N' - 1.$$

This set of equations has the form of a Fourier transform and, by choosing the steering angles appropriately, we can obtain the form of a discrete Fourier transform. Choose $N' = N$ and choose ϕ_ℓ to satisfy

$$(d/\lambda)\sin\phi_\ell = \frac{\ell}{N} \quad \ell = 0, \ldots, N - 1.$$

Then N beams, at the angles

$$\phi_\ell = \sin^{-1}\left(\frac{\lambda}{d}\frac{\ell}{N}\right) \quad \ell = 0, \ldots, N - 1,$$

are formed simultaneously by the equation

$$G_\ell = \sum_{n=0}^{N-1} g_n e^{-j2\pi n\ell/N} \quad \ell = 0, \ldots, N - 1.$$

This equation has the form of a discrete Fourier transform.

The Butler array

The discrete Fourier transform

$$G_\ell = \sum_{n=0}^{N-1} g_n e^{-j2\pi N^{-1}n\ell} \quad \ell = 0, \ldots, N - 1$$

is a mathematical transformation that often appears in the processing of signals. As the equation is written, it appears that N^2 complex multiplications are necessary to compute the transform because there are N complex multiplications for each value of ℓ. When N is large, this will be a formidable task that may be impractical for some applications. By clever factoring of the equation, however, it is possible to use far fewer multiplications in the calculations.

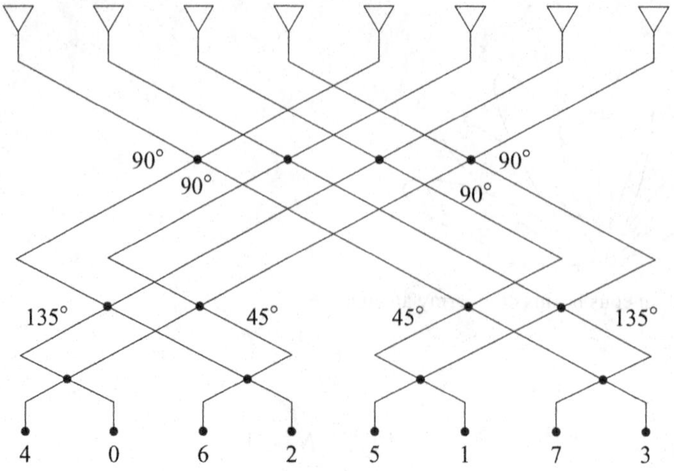

Figure 5.12 An eight-element Butler matrix

This fact has been discovered independently in many situations. In the study of signal processing, it is known as the *Cooley–Tukey fast Fourier transform* (FFT). In the study of antenna arrays, it is known as a *Butler array*. An eight-element Butler array, shown in Figure 5.12, is an example of a *decimation algorithm*.

Suppose that $n = 0, \ldots, N-1$ is an index, and N factors as $N = KI$. The index n can be broken down into a coarse index, i, and a vernier index, k just as a long time interval is broken into hours and minutes. We count with the index k from 0 to $K-1$. When k overflows, the index i is increased by one and k returns to zero. Then the original index n can be written

$$n = k + Ki \qquad \begin{aligned} i &= 0, \ldots, I-1 \\ k &= 0, \ldots, K-1 \\ n &= 0, \ldots, IK-1. \end{aligned}$$

We count the same components with the double indices k and i as we did with the single index n. But, by converting the discrete Fourier transform into this more complicated indexing scheme, we find that the computations become simpler. Therefore we trade the original formulation for one that is more difficult to understand, but is computationally more efficient. Of course, we will only make this exchange when we wish to do the numerical computations – otherwise, we prefer the original form.

Now let us see how this happens. We shall express the output index by

$$\ell = i' + Ik' \quad \begin{aligned} i' &= 0, \ldots, I-1 \\ k' &= 0, \ldots, K-1 \\ \ell &= 0, \ldots, IK-1. \end{aligned}$$

Then

$$G_\ell = \sum_{n=0}^{N-1} \mathrm{e}^{-\mathrm{j}2\pi N^{-1} n\ell} g_n$$

for $\ell = 0, \ldots, N-1$, becomes

$$G_{i'+Ik'} = \sum_{k=0}^{K-1} \sum_{i=0}^{I-1} \mathrm{e}^{-\mathrm{j}2\pi (IK)^{-1}(k+Ki)(i'+Ik')} g_{k+Ki}$$

for $i' = 0, \ldots, I-1$ and $k' = 0, \ldots, K-1$. Expand the product term in the exponent. The term $\exp(-\mathrm{j}2\pi ik')$ equals one, so it can be dropped. Then we have the key decimation formula:

$$G_{i'+Ik'} = \sum_{k=0}^{K-1} \mathrm{e}^{-\mathrm{j}2\pi K^{-1}kk'} \left[\mathrm{e}^{-\mathrm{j}2\pi I^{-1}K^{-1}ki'} \sum_{i=0}^{I-1} \mathrm{e}^{-\mathrm{j}2\pi I^{-1}ii'} g_{k+Ki} \right].$$

Notice that the inner sum is an I-point discrete Fourier transform for each value of k, and the outer sum is a K-point discrete Fourier transform for each value of i'.

The original expression has N^2 complex multiplications. (To be more efficient, we could discount multiplication by 1, but we shall not bother with such fine tuning.) The new expression has an inner sum with KI^2 multiplications, producing N terms. Each of the N sums is multiplied by a complex number that requires N more complex multiplications. Finally, the outer sum requires IK^2 complex multiplications. Altogether, $KI^2 + N + IK^2 = N(K + I + 1)$ multiplications are required. But $(K + I + 1)N$ is less than N^2, so fewer complex multiplications are needed by the decimation formula.

If either K or I can be factored, then the idea can be repeated again to reduce the sum on i or the sum on k, whereby the number of multiplications is reduced even further.

The most important instance of the Cooley–Tukey algorithm is when $N = 2^m$ for some integer m. Then one usually applies the key decimation algorithm in one of two ways: with $K = 2$ and $I = N/2$, which is called *decimation in time*; or with $I = 2$ and $K = N/2$, which is called *decimation in frequency*.

First, consider decimation in time. Let

$$H_k(i') = \sum_{i=0}^{I-1} \mathrm{e}^{-\mathrm{j}2\pi I^{-1}ii'} g_{k+Ki}.$$

Then the key decimation algorithm reduces to

$$G_{i'+0} = H_{0i'} + \mathrm{e}^{\mathrm{j}\pi I^{-1}i'} H_{1i'} \quad i' = 0, \ldots, I-1$$

$$G_{i'+I} = H_{0i'} - \mathrm{e}^{\mathrm{j}\pi I^{-1}i'} H_{1i'} \quad i' = 0, \ldots, I-1.$$

These N equations require a total of $I = N/2$ complex multiplications to compute $G(i)$ from $H_0(i')$ and $H_1(i')$. The same decimation can be applied to compute both

H_0 and H_1 and, in fact, can be repeated $m-1$ times. Altogether, $\frac{N}{2}(\log N)$ complex multiplications are required.

Alternatively, decimation in frequency chooses $I=2$, $K=N/2$, and the key decimation algorithm reduces to

$$G_{0+2k'} = \sum_{k=0}^{K-1} \mathrm{e}^{-\mathrm{j}2\pi K^{-1}kk'}[g_{k+0}+g_{k+K}]$$

$$G_{1+2k'} = \sum_{k=0}^{K-1} \mathrm{e}^{-\mathrm{j}2\pi K^{-1}kk'}[\mathrm{e}^{-\mathrm{j}\pi K^{-1}k}[g_{k+0}-g_{k+K}]].$$

As before, this decimation reduces an N-point discrete Fourier transform to two $N/2$-point discrete Fourier transforms plus an additional N multiplications. Both $N/2$-point transforms can then be reduced in the same way to two $N/4$-point transforms plus $N/2$ multiplications.

5.4 Focused antennas and arrays

The antenna pattern for a single aperture is defined for the far-field region and is a consequence of the Fraunhofer approximation. The antenna pattern of an array of apertures is also defined in the far-field region since the array can be regarded as one large aperture. The phase angles at the individual elements of a phased array aim the antenna beam as seen in the far field. We may refer to such a beam as an unfocused beam or as a beam in focus at infinity.

It is also possible to choose the phase angles at the array elements (or even across an individual aperture) so that the beam is focused at a finite range. We refer to this technique as beamforming in the near-field region. Whereas, for a fixed illumination function of magnitude $|s_0(x,y)|$, the beam is sharpest at infinite range if the phase is a linear function of x and y across the aperture, we shall see that, at a fixed finite range, the beam is sharpest (under a near-field approximation) if the phase is a quadratic function of x and y across the aperture. It is very suggestive to refer to the computation that provides the focusing at each element of a phased array as a "processing lens." A different quadratic function is required for each value of range and for each value of angle. A phased array with variable phase compensation within the processing can be used to focus at a variable range and direction. Indeed, by processing multiple copies of the received signal with different quadratic-phase compensations, a receiving array can be used to focus simultaneously on many ranges and directions.

Suppose that the magnitude $|s_0(\xi,\eta)|$ of the illumination function is specified. Let

$$s_0(\xi,\eta) = |s_0(\xi,\eta)|\mathrm{e}^{\mathrm{j}\theta(\xi,\eta)}$$

where $\theta(\xi,\eta)$ is to be specified to achieve a desired focus. A Huygens wavelet, originating at the point ξ, η, reaches the point x, y, z with the infinitesimal value

$$dc(x, y, z) = s_0(\xi, \eta)\mathrm{e}^{-\mathrm{j}2\pi(f_0/c)\sqrt{(x-\xi)^2+(y-\eta)^2+z^2}}\, \mathrm{d}\xi\, \mathrm{d}\eta.$$

The total signal at x, y, z is

$$c(x, y, z) = \int_{-\infty}^{\infty}\int_{-\infty}^{\infty} s_0(\xi, \eta)\mathrm{e}^{-\mathrm{j}2\pi(f_0/c)\sqrt{(x-\xi)^2+(y-\eta)^2+z^2}}\, \mathrm{d}\xi\, \mathrm{d}\eta.$$

In the Fresnel approximation,

$$c(x, y, z) = \mathrm{e}^{\mathrm{j}kz}\int_{-\infty}^{\infty}\int_{-\infty}^{\infty} |s_0(\xi, \eta)|\, \mathrm{e}^{\mathrm{j}\theta(\xi,\eta)}\mathrm{e}^{-\mathrm{j}\frac{k}{2z}[(x-\xi)^2+(y-\eta)^2]}\, \mathrm{d}\xi\, \mathrm{d}\eta.$$

The integral will be largest if all contributions add in phase. This means that we should choose

$$\theta(\xi, \eta) = \frac{k}{2z}[(x-\xi)^2 + (y-\eta)^2]$$

as the phase function across the aperture to focus the wavefront at the point (x, y, z). Because the phase function depends on (x, y, z), the phase term cannot be expressed in spherical coordinates as simply a function of angle that is independent of range. Each value of range requires a different quadratic-phase function for focusing.

5.5 Nondiffracting beams

A nondiffracting beam does not change its cross section as it propagates. Every finite aperture must launch a diffracting beam because the far-field pattern is the Fourier transform of the aperture, and a signal and its Fourier transform cannot both be confined. However, an infinite aperture can launch a nondiffracting beam with an illumination function that has most of its signal concentrated near the origin. This suggests that it may be possible to approximate this infinite-aperture illumination function to form a beam that nearly approximates a nondiffracting beam well into the near field. In some applications, a nondiffracting beam deep into the near field can be desirable.

The Huygens–Fresnel pointspread function was developed from the wave equation, and any wave that satisfies the wave equation also satisfies the Huygens–Fresnel pointspread function. For the infinite aperture, it is more convenient to work with the wave equation. In cylindrical coordinates, the wave equation is

$$\left[\frac{1}{r}\frac{\partial}{\partial r}\left(r\frac{\partial}{\partial r}\right) + \frac{1}{r^2}\frac{\partial^2}{\partial\phi^2} + \frac{\partial^2}{\partial z^2} - \frac{1}{c^2}\frac{\partial^2}{\partial t^2}\right]\widetilde{s}(x, y, z, t) = 0.$$

The zero-order Bessel function solves the differential equation

$$\frac{1}{r}\frac{\mathrm{d}}{\mathrm{d}r}\left(r\frac{\mathrm{d}J_0(r)}{\mathrm{d}r}\right) + J_0(r) = 0.$$

By a simple change of variables, this becomes

$$\frac{1}{r}\frac{\mathrm{d}}{\mathrm{d}r}\left(r\frac{\mathrm{d}J_0(2\pi\alpha r/\lambda)}{\mathrm{d}r}\right)+(2\pi\alpha/\lambda)^2 J_0(2\pi\alpha r/\lambda)=0.$$

Therefore the function

$$\widetilde{s}(x, y, z, t) = J_0(2\pi\alpha r/\lambda)\mathrm{e}^{-\mathrm{j}2\pi f_0(t-\gamma z/c)}$$

satisfies the wave equation, provided that

$$\alpha^2 + \gamma^2 = 1.$$

Because the x, y dependence of this wavefront is independent of z, the illumination function $J_0(2\pi\alpha r/\lambda)$ will launch the nondiffracting beam,

$$s(x, y, z) = J_0(2\pi\alpha r/\lambda)\,\mathrm{e}^{\mathrm{j}2\pi\gamma z/\lambda}$$

provided that α is smaller than one. Of course, the illumination function $J_0(2\pi\alpha r/\lambda)$ requires an infinite aperture and therefore cannot be realized. However, the alternative illumination function

$$s_0(x, y) = J_0(2\pi\alpha r/\lambda)\mathrm{circ}\left(\frac{r}{R}\right)$$

for large R will be a good approximation in the vicinity of the origin and will produce a nearly nondiffracting beam in the near field.

There are many other examples of nondiffracting beams, all of which must have an infinite aperture because a signal and its Fourier transform cannot both be confined. Since we are interested only in confined beams, it seems appropriate to consider the wave equation in cylindrical coordinates, which implies that all such beams are best described in terms of Bessel functions or sums of Bessel functions. The derivation of these beams is the topic of Problem 5.14.

5.6 Interferometry

An antenna aperture can be used as a linear component to collect all of the signal that is incident on it. The response of the antenna, then, can be described as a beam, as was studied in Sections 5.1 and 5.2. Beamforming is a straightforward and linear way to process an incoming signal across the aperture.

There are other ways to process the distribution of the phase of the signal across the aperture to estimate the direction of arrival. Many devices are in use that depend on the relationship between the angle of arrival of a wavefront at several points, the wavelength of the wavefront, and the relative phase angle of the signal at the several points. These devices can be used to measure the direction of arrival, the wavelength, or the frequency.

Such devices use nonlinear methods of processing and are referred to under the general term *interferometer*. We shall define an interferometer as any measurement device based on the fact that radiation from a point source arrives at different points of the aperture with a phase difference dependent on the angle of arrival.

Definition 5.6.1 *Any device for measurement based on the relative phase of a wavefront across an aperture that cannot be described simply as a spatial integral of the incident signal is called an interferometer.*

The usual interferometer breaks a larger aperture into two or more subapertures, receives a signal in each subaperture, and compares phase. The phase comparison is a nonlinear operation. Consequently, an interferometer is distinctly different from a beamforming system.

The simplest and most widely used interferometer in radar applications uses two identical subapertures separated by the distance d. The two subapertures are regarded as individual antennas. The incoming signal from the far field is received by the two antennas, shown in Figure 5.13. From a radiation source that is far from the aperture in comparison to the separation d, the difference in path length is $d \sin \phi$. Hence the

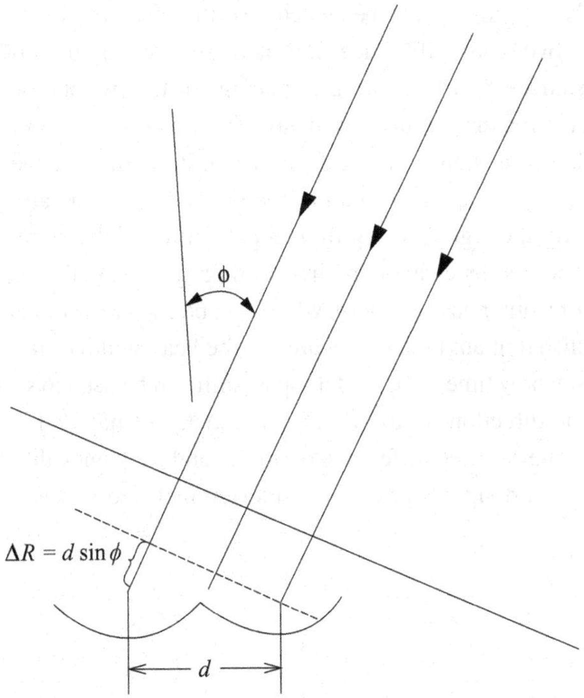

Figure 5.13 An elementary interferometer

phase difference in the signal, as received at the two antennas, is

$$\Delta\theta = 2\pi \frac{d}{\lambda} \sin\phi \quad \text{(radians)}.$$

The phase difference $\Delta\theta$ is measured modulo 2π, so whenever d/λ is larger than 1, there will be more than one value of the angle ϕ corresponding to the given value of $\Delta\theta$.

The equation for $\Delta\theta$ can be inverted to give

$$\phi = \sin^{-1}\left(\frac{\lambda}{d}\frac{\Delta\theta}{2\pi}\right),$$

from which the direction of arrival ϕ can be computed from the phase difference $\Delta\theta$. There will be ambiguities whenever d/λ is larger than 1 because there will be more than one value of ϕ solving the equation. Ambiguities in an interferometric, angle-of-arrival measurement are quite routine and are regularly resolved by using a second interferometer with a different value of the spacing d. Now the calculation is

$$\phi = \sin^{-1}\left(\frac{\lambda}{d_1}\frac{\Delta\theta_1 + 2\pi\ell_1}{2\pi}\right)$$
$$\phi = \sin^{-1}\left(\frac{\lambda}{d_2}\frac{\Delta\theta_2 + 2\pi\ell_2}{2\pi}\right)$$

where ℓ_1 and ℓ_2 are unknown integers to be selected so that the two equations yield the same value of ϕ. The two phase differences $\Delta\theta_1$ and $\Delta\theta_2$ are measured by the two interferometers. The distances d_1 and d_2 are the spacings of the two interferometers and can be chosen so that the pair of equations always has a unique solution.

When multiple simultaneous signals arrive from a multitude of sources, interferometry will break down unless the signals can first be sorted from one another. The simplest way of separating the signals is by frequency sorting, if the signals do not overlap significantly on the frequency axis; by time sorting, if the signals normally arrive at different times; or by antenna directivity, which can be used to eliminate signals that arrive from widely different angles as compared to the beamwidth of the antenna. In Section 7.7, we shall see how time delay and doppler shift can be used to sort signals prior to an interferometric direction-of-arrival measurement. In Chapter 13, we shall also see that, in passive systems, time difference of arrival and frequency difference of arrival can also be used to sort signals prior to an interferometric direction-of-arrival measurement.

5.7 Vector diffraction

A propagating wave may have scalar values or vector values. In Chapter 4, we considered extensively the diffraction of scalar-valued waves. The diffraction of scalar waves is completely described by the Huygens–Fresnel principle and is approximately described by various approximations to that principle.

A vector-valued wave may be regarded as three scalar-valued waves comprising the three components of the vector in a suitable coordinate system. If these three scalar components can be independently specified, then such a wave amounts to nothing more than three independent scalar waves, and no new theory is needed. However, if the three scalar components are interdependent in some way, then this constraint may force a restatement of the Huygens–Fresnel principle.

A *transverse wave* is a vector wave that only takes values perpendicular to its direction of propagation. This perpendicular direction of the wave amplitude is called the *polarization* of the wave. We will begin with a monochromatic, monodirectional transverse vector wave. This transverse vector wave has complex-baseband components and a complex-baseband vector structure

$$\boldsymbol{s}(x, y, z) = [a_x \boldsymbol{i}_x + a_y \boldsymbol{i}_y + a_z \boldsymbol{i}_z] e^{j2\pi f_0(\alpha x + \beta y + \gamma z)/c}$$

where $(\boldsymbol{i}_x, \boldsymbol{i}_y, \boldsymbol{i}_z)$ is a triad of orthogonal unit vectors along the three axes of the coordinate system.

For the wave to be a transverse wave, the vector direction must be perpendicular to the direction of propagation. This means that for a monodirectional transverse wave traveling in the direction

$$\alpha \boldsymbol{i}_x + \beta \boldsymbol{i}_y + \gamma \boldsymbol{i}_z$$

the side condition

$$a_x \alpha + a_y \beta + a_z \gamma = 0$$

must be satisfied. This side condition creates a linkage in the components, which is the reason that vector diffraction is different from scalar diffraction.

A general, monochromatic, transverse vector wave consists of a composite of vector waves traveling in all directions. This composite is given by the vector integral

$$\boldsymbol{s}_0(x, y, z) = \int_{-\infty}^{\infty} \int_{-\infty}^{\infty} [a_x(\alpha, \beta) \boldsymbol{i}_x + a_y(\alpha, \beta) \boldsymbol{i}_y + a_z(\alpha, \beta) \boldsymbol{i}_z] e^{j2\pi f_0(\alpha x + \beta y + \gamma z)/c} \, d\alpha \, d\beta,$$

where $\gamma = \sqrt{1 - \alpha^2 - \beta^2}$ and, for each α and β, the constraining equation

$$\alpha a_x(\alpha, \beta) + \beta a_y(\alpha, \beta) + \gamma a_z(\alpha, \beta) = 0$$

is satisfied. Without the constraining equation, the problem of vector diffraction would separate into three uncoupled problems of scalar diffraction.

Let

$$\boldsymbol{s}_0(x, y) = \boldsymbol{s}(x, y, 0).$$

Then

$$s_0(x, y) = \int_{-\infty}^{\infty}\int_{-\infty}^{\infty} [a_x(\alpha, \beta)\boldsymbol{i}_x + a_y(\alpha, \beta)\boldsymbol{i}_y + a_z(\alpha, \beta)\boldsymbol{i}_z]\mathrm{e}^{\mathrm{j}2\pi f_0(\alpha x+\beta y)/c}\, \mathrm{d}\alpha\, \mathrm{d}\beta,$$

and the constraint is

$$\alpha a_x(\alpha, \beta) + \beta a_y(\alpha, \beta) + \sqrt{1-\alpha^2-\beta^2}a_z(\alpha, \beta) = 0.$$

To interpret this as a two-dimensional Fourier transform, let $f_x = \alpha/\lambda$ and $f_y = \beta/\lambda$. Let

$$S_x(f_x, f_y) = \lambda^2 a_x(\alpha, \beta)$$
$$S_y(f_x, f_y) = \lambda^2 a_y(\alpha, \beta)$$
$$S_z(f_x, f_y) = \lambda^2 a_z(\alpha, \beta).$$

The vector spectrum is then

$$\boldsymbol{S}_0(f_x, f_y) = S_x(f_x, f_y)\boldsymbol{i}_x + S_y(f_x, f_y)\boldsymbol{i}_y + S_z(f_x, f_y)\boldsymbol{i}_z.$$

Now we have

$$s_0(x, y) = \int_{-\infty}^{\infty}\int_{-\infty}^{\infty} \boldsymbol{S}_0(f_x, f_y)\,\mathrm{e}^{\mathrm{j}2\pi(f_x x+f_y y)}\,\mathrm{d}f_x\,\mathrm{d}f_y,$$

and the constraint is

$$f_x S_x(f_x, f_y) + f_y S_y(f_x, f_y) + \sqrt{\lambda^{-2} - f_x^2 - f_y^2}S_z(f_x, f_y) = 0.$$

In the spatial domain, this constraining equation becomes a constraint that interrelates the three components of $s_0(x, y)$. The inverse Fourier transform of this constraint equation is

$$\frac{1}{\mathrm{j}2\pi}\frac{\partial s_x(x, y)}{\partial x} + \frac{1}{\mathrm{j}2\pi}\frac{\partial s_y(x, y)}{\partial y} + g(x, y) ** s_z(x, y) = 0$$

where

$$g(x, y) \Leftrightarrow \sqrt{\lambda^{-2} - f_x^2 - f_y^2}$$

with $f_x^2 + f_y^2 \leq \lambda^{-2}$. Referring to the last entry of Table 3.1 (and Problem 3.20b), we have

$$g(x, y) = \lambda^{-3}2\pi\frac{\sin(2\pi\sqrt{x^2+y^2}/\lambda) - (2\pi\sqrt{x^2+y^2}/\lambda)\cos(2\pi\sqrt{x^2+y^2}/\lambda)}{(2\pi\sqrt{x^2+y^2}/\lambda)^3}.$$

We can now summarize these results to present a full description of vector diffraction. Let

$$\boldsymbol{s}(x, y, z) = s_x(x, y, z)\boldsymbol{i}_x + s_y(x, y, z)\boldsymbol{i}_y + s_z(x, y, z)\boldsymbol{i}_z$$

be a complex vector function (in three-dimensional space). This function can be the complex-baseband representation of a monochromatic transverse wave of wavelength λ if and only if

$$\frac{1}{\mathrm{j}2\pi}\frac{\partial s_{0x}(x, y)}{\partial x} + \frac{1}{\mathrm{j}2\pi}\frac{\partial s_{0y}(x, y)}{\partial y} + g(x, y) ** s_{0z}(x, y) = 0$$

where $s_0(x, y) = s(x, y, 0)$, and $g(x, y)$ is as above. Each component of $s(x, y, z)$ propagates as described by the Huygens–Fresnel pointspread function of free space.

5.8 Scanning antenna patterns

The radar antenna pattern or sonar hydrophone pattern, as studied in Section 5.1, had no dependence on time. It is a straightforward generalization to introduce time dependence into the antenna pattern. Simply visualize an aperture that is moving, either translating in the x, y plane or rotating about an axis in the x, y plane. The wavefront illuminating the aperture is then imagined as translating or rotating with the aperture. Then the antenna radiation pattern will translate or rotate in lock step with the aperture. Such a situation occurs when an antenna is mechanically rotated or carried on a moving vehicle. An antenna aperture that is rotating is called a *scanning aperture*. In contrast, an antenna aperture that is not rotating is called a *staring aperture*.

A radar antenna pattern or sonar hydrophone pattern can also be scanned by changing the phase distribution of the illumination function. There are various ways of doing this. One method is by moving the feed element that is illuminating the aperture to change the illumination function. If the aperture is a phased array, then one may appropriately phase shift the signal fed to each element. This will steer the beam, as discussed in Section 5.3. By varying the phase shifts with time, the direction of the beam can change with time. By using phase shifts of the form

$$\theta_n(t) = 2\pi n \left(\frac{d}{\lambda}\right) \sin\phi_0(t),$$

it is straightforward to form an antenna pattern of the form

$$E_{\phi_0(t)}(\phi) = E_e(\phi)\mathrm{dirc}_N\left(\frac{d}{\lambda}(\sin\phi - \sin\phi_0(t))\right).$$

The main beam of this pattern is at the time-varying angle $\phi_0(t)$. This is called an "electronically" scanned beam, though perhaps a more modern term would be a "computationally" scanned beam. Because of the principle of superposition, one can even use a single array to simultaneously form multiple scanning beams, with an arbitrary time-varying trajectory for each beam. Each time-varying beam can carry its own time-varying signal.

5.9 Wideband radiation patterns

We have studied antenna patterns in some detail under the narrowband approximation. This amounts to replacing time delay by a phase shift at the carrier frequency. Let us now drop the approximation that $c(t)$ is a narrowband transmitted signal; every useful signal has some nonzero bandwidth. We will allow the Fourier transform $C(f)$ to be nonzero in a range of frequencies. To understand the effect of bandwidth, we can imagine taking a thin slice of $C(f)$ near frequency f. Modulating this slice onto a carrier of frequency f_0 gives a signal slice near the frequency $f + f_0$. The wavelength of this signal at the propagation velocity c is $c/(f + f_0)$. This frequency slice, according to the narrowband analysis of Section 5.1, sees an antenna pattern,

$$E(\phi, \psi) = \left(\frac{f+f_0}{c}\right)^2 S\left(\frac{f+f_0}{c}\sin\phi\cos\psi, \frac{f+f_0}{c}\sin\phi\sin\psi\right),$$

where $s(x, y)$ is the aperture illumination function, and $S(f_x, f_y)$ is its Fourier transform. The contribution to the signal $c(t, \phi, \psi)$ in the far field at angular coordinates ϕ, ψ is given in the frequency domain by

$$C(f, \phi, \psi) = C(f)S\left(\frac{f+f_0}{c}\sin\phi\cos\psi, \frac{f+f_0}{c}\sin\phi\sin\psi\right).$$

Thus the signal $c(t, \phi, \psi)$ radiated into direction ϕ, ψ is obtained as an inverse Fourier transform of $C(f, \phi, \psi)$,

$$c(t, \phi, \psi) = \int_{-\infty}^{\infty} C(f, \phi, \psi)\,\mathrm{e}^{\mathrm{j}2\pi ft}\,\mathrm{d}f.$$

This same expression will now be derived in a more formal way in one dimension. The two-dimensional case goes through in the same way. For a one-dimensional antenna, the signal at angle ϕ is

$$\begin{aligned}
c(t, \phi)\mathrm{e}^{\mathrm{j}2\pi f_0 t} &= \int_{-\infty}^{\infty} s(x)c\left(t - \frac{x}{c}\sin\phi\right)\mathrm{e}^{\mathrm{j}2\pi f_0(t-\frac{x}{c}\sin\phi)}\mathrm{d}x \\
&= \int_{-\infty}^{\infty} s(x)\left[\int_{-\infty}^{\infty} \mathrm{e}^{-\mathrm{j}2\pi f\frac{x}{c}\sin\phi}C(f)\mathrm{e}^{\mathrm{j}2\pi ft}\mathrm{d}f\right]\mathrm{e}^{\mathrm{j}2\pi f_0(t-\frac{x}{c}\sin\phi)}\mathrm{d}x \\
&= \int_{-\infty}^{\infty} C(f)\mathrm{e}^{\mathrm{j}2\pi(f_0+f)t}\left[\int_{-\infty}^{\infty} s(x)\mathrm{e}^{-\mathrm{j}2\pi(f_0+f)\frac{x}{c}\sin\phi}\,\mathrm{d}x\right] \\
&= \mathrm{e}^{\mathrm{j}2\pi f_0 t}\int_{-\infty}^{\infty} C(f)S\left(\frac{f_0+f}{c}\sin\phi\right)\mathrm{e}^{\mathrm{j}2\pi ft}\,\mathrm{d}f.
\end{aligned}$$

This means that the Fourier transform of the signal at angle ϕ can be written

$$C(f, \phi) = C(f)S\left(\frac{f_0+f}{c}\sin\phi\right).$$

A similar analysis of the two-dimensional aperture will give

$$C(f, \phi, \psi) = C(f) S\left(\frac{f + f_0}{c} \sin\phi \cos\psi, \frac{f + f_0}{c} \sin\phi \sin\psi\right)$$

and

$$c(t, \phi, \psi) = \int_{-\infty}^{\infty} C(f, \phi, \psi) \mathrm{e}^{\mathrm{j}2\pi f t} \, \mathrm{d}f$$

as we have seen earlier.

For example, if a one-dimensional aperture of length L is uniformly illuminated so that

$$s(x) = \operatorname{rect}\left(\frac{x}{L}\right),$$

then

$$S\left(\frac{f_0 + f}{c} \sin\phi\right) = L \operatorname{sinc}\left(L \frac{f_0 + f}{c} \sin\phi\right),$$

and

$$C(f, \phi) = C(f) L \operatorname{sinc}\left(L \frac{f_0 + f}{c} \sin\phi\right).$$

At every value of ϕ, the spectrum of $C(f)$ will be changed differently to form $C(f, \phi)$. By the convolution theorem

$$c(t, \phi) = c(t) * h(t)$$

where

$$h(t) \leftrightarrow L \operatorname{sinc}\left(L \frac{f_0 + f}{c} \sin\phi\right).$$

Therefore

$$h(t) = \frac{c}{\sin\phi} \operatorname{rect}\left(\frac{ct}{L \sin\phi}\right) \mathrm{e}^{-\mathrm{j}2\phi f_0 t}.$$

Problems

5.1 **a.** The illumination function of one element of a given antenna array is equal to 1 on a circle. That is,

$$s_0(x, y) = \begin{cases} 1 & \text{if } \sqrt{x^2 + y^2} \leq 1 \\ 0 & \text{otherwise.} \end{cases}$$

What is the element pattern as a function of ϕ and ψ?

b. Consider the four-element array with the combined illumination function

$$p_0(x, y) = s_0(x - \Delta/2, y - \Delta/2) + s_0(x - \Delta/2, y + \Delta/2) + s_0(x + \Delta/2, y - \Delta/2) + s_0(x + \Delta/2, y + \Delta/2).$$

What is the antenna pattern?

c. Consider the four-element "phased" array

$$p_0(x, y) = e^{j\theta_{00}} s_0(x - \Delta/2, y - \Delta/2) + e^{j\theta_{01}} s_0(x - \Delta/2, y + \Delta/2) + e^{j\theta_{10}} s_0(x + \Delta/2, y - \Delta/2) + e^{j\theta_{11}} s_0(x + \Delta/2, y + \Delta/2).$$

How should the "electrical" phases $\theta_{00}, \theta_{01}, \theta_{10}, \theta_{11}$ be chosen to steer the peak of the beam into the desired spatial direction, defined by ϕ_0 and ψ_0?

5.2 An antenna pattern is formed by placing sixteen antenna elements in the pattern

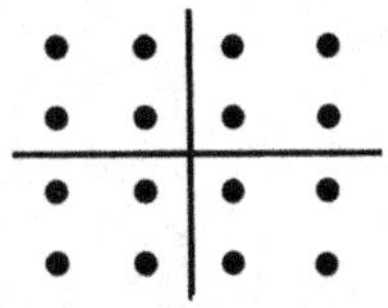

Find the array antenna pattern by considering the array as a two by two array of two by two arrays.

5.3 Because of the requirements imposed by mechanical support, or because of the position of waveguides or feeds, a radar antenna may have part of its aperture blocked. To analyze the effect of blockage, one may simply subtract the "antenna pattern" of the blockage from the antenna pattern of the unblocked aperture.

Find the antenna pattern of a uniformly illuminated, rectangular aperture of dimension A by B that is blocked at the center by an obstruction of dimension a by b.

5.4 A uniformly illuminated circular aperture of diameter D forms a jinc-shaped beam. In order to enlarge the aperture, it is replaced by a D by D square aperture. Does this improve the beam, degrade the beam, or is the answer more complex?

5.5 **a.** Determine the antenna pattern of the "cross antenna" with uniform illumination over the aperture, in the illustration:

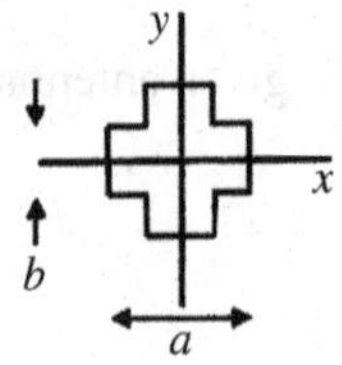

b. Instead of the procedure in part **a**, consider the vertical strip and the horizontal strip as two separate apertures:

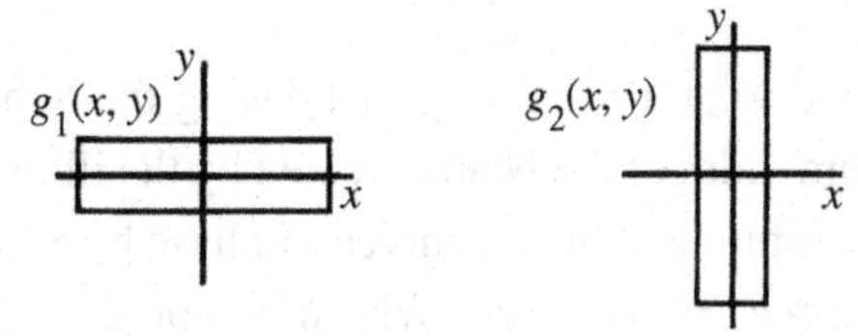

and define

$$M(f_x, f_y) = G_1(f_x, f_y)G_2(f_x, f_y).$$

An antenna array in this shape that uses this kind of multiplicative processing is known as a *Mills cross*. Describe the main lobe of $M(f_x, f_y)$ and describe the sidelobes. Describe the output of the Mills cross if a single signal arrives from a direction with the spherical coordinates (ϕ, ψ). Describe the output if two signals arrive simultaneously from directions (ϕ, ψ) and (ϕ', ψ').

5.6 An interferometer is used to measure the direction of arrival of a signal by the computation

$$\phi = \sin^{-1}\left(\frac{\lambda}{d}\frac{\Delta\theta}{2\pi}\right).$$

Suppose that the received signal is actually the sum of two signals:

$$v_1(t) = A_1 \sin 2\pi\left(f_0 t + \frac{d}{2\lambda}\sin\phi_1\right) + A_2 \sin 2\pi\left(f_0 t + \frac{d}{2\lambda}\sin\phi_2\right)$$

$$v_2(t) = A_1 \sin 2\pi\left(f_0 t - \frac{d}{2\lambda}\sin\phi_1\right) + A_2 \sin 2\pi\left(f_0 t - \frac{d}{2\lambda}\sin\phi_2\right).$$

Let $A_2/A_1 = 0.1$ and $d/\lambda = 10$. By using the dominant terms in a linearized analysis, find an approximate expression for the error in an estimate of ϕ_1 because of the interfering signal.

5.7 A point of a one-dimensional illumination function on a bounded aperture is called the *phase center* of the antenna if it is that point nearest the center of the antenna for which the boresight antenna pattern is real when that point is chosen as the origin. Show that every one-dimensional illumination function has a phase center. Does this generalize to two dimensions?

5.8 The RMS beamwidth of the antenna pattern $E(\phi)$ is defined as

$$B_\phi^2 = \frac{\int_{-\infty}^{\infty}(\phi - \overline{\phi})^2|E(\phi)|^2\, d\phi}{\int_{-\infty}^{\infty}|E(\phi)|^2\, d\phi}$$

where

$$\overline{\phi} = \frac{\int_{-\infty}^{\infty} \phi |E(\phi)|^2 \, d\phi}{\int_{-\infty}^{\infty} |E(\phi)|^2 \, d\phi},$$

and where $E(\phi) = P(\sin\phi)$ and $p(u) = s_0(\lambda u)$. Using the Schwarz inequality, prove that the RMS beamwidth of the beam, created by the illumination function $s_0(u) = |s_0(u)|e^{j\theta(u)}$, is minimized over a choice of phase by a linear phase function, which has the form $\theta(u) = \theta_0 + \theta_0' u$. (Within the approximation $\sin\phi = \phi$.)

5.9 **a.** Use the theory of the Fourier series to prove the relationship

$$e^{-jat \sin\psi} = \sum_{n=-\infty}^{\infty} J_n(a) e^{-jn\psi}.$$

b. A circular antenna array consists of N identical antenna elements equispaced on the circumference of a circle of radius a (and oriented in the same direction). Find the array factor of the antenna pattern expressed as an infinite sum of Bessel functions.

c. Discuss which terms of the sum are significant for moderate to large N.

5.10 An antenna illumination function is given by

$$s(x, y) = \text{circ}\left(\frac{x}{D_1}, \frac{y}{D_1}\right) ** \text{circ}\left(\frac{x}{D_2}, \frac{y}{D_2}\right).$$

a. What is the antenna pattern in the far field? What is the area of the active aperture of the illumination function? What is the effective area?

b. Assuming λ is small, what is the asymptotic decay of the sidelobe magnitudes?

c. If the active area is held fixed, how should D_1 and D_2 be chosen to maximize antenna gain? What happens to the sidelobes?

5.11 **a.** Graph the antenna quality factor as a function of D/λ for a uniformly illuminated, one-dimensional aperture of width D.

b. Graph the antenna quality factor as a function of D/λ for a uniformly illuminated, two-dimensional circular aperture of diameter D.

5.12 What is the far-field (Fraunhofer) diffraction pattern for the illumination function

$$s_0(x, y) = J_0(2\pi a r/\lambda)\text{circ}\left(\frac{r}{R}\right)?$$

5.13 Calculate the energy of the two-dimensional function

$$s(x, y) = J_0\left(2\pi\alpha\sqrt{x^2 + y^2}/\lambda\right).$$

5.14 Show that the illumination function

$$s_0(x, y) = J_n\left(2\pi\alpha\sqrt{x^2 + y^2}/\lambda\right) e^{jn\phi}$$

produces a nondiffracting beam, as does the linear combination of such illumination functions.

Notes

The first antenna was designed around 1887 by the German physicist Heinrich Hertz as part of his work demonstrating the validity of the electromagnetic theory of the British physicist James Clerk Maxwell. Heuristic methods of antenna design were developed by Guglielmo Marconi. At that time, the choice of carrier frequency was largely determined by the consideration of antenna design. Now the antenna considerations are usually subservient to other system considerations in choosing the carrier frequency.

The first array antennas were introduced in the 1920s. The Butler matrix (1961) has the same decimation structure as the Cooley–Tukey (1965) fast Fourier transform, but preceded that work by several years. The Cooley–Tukey paper, however, is credited with first annunciating the decimation structure as a general mathematical identity, unencumbered by a discussion within the context of a single physical application. It seems, however, that the idea had been in use implicitly for many years, for example by radio astronomers, and as far back as Gauss.

Radiation patterns of ultrasound transducers, sonar hydrophones, and arrays of sonar hydrophones can be treated with the same methods used to study antennas if the propagation medium is homogeneous and isotropic. Murino and Trucco (2000) discussed imaging using acoustic waves. However, in sonar applications, these propagation assumptions are at best only first approximations. A better performance can be obtained by incorporating a model of the propagation medium into the processing equations, a technique known as *matched-field processing*, as discussed by Baggeroer, Kuperman, and Mikhalevsky (1993).

The uncertainty principle for pulses also pertains to the relationship between the width of an aperture and the width of its antenna pattern. Rhodes (1974) discussed the role in antenna theory of prolate spheroidal wave functions, first studied by Slepian and Pollak (1961). McEwan and Goldsmith (1989) described the theory of illuminating small reflectors with gaussian beams to obtain very high efficiency. In ultrasound applications, gaussian beams are formed in the near field to sharpen resolution, but with limited depth of focus. To counter this limitation, Durnin (1987) introduced a nondiffracting solution to the scalar wave equation, given an infinite aperture, which is attractive because it is in the same focus at all depths.

Sonar beamforming is a critical technology in submarine warfare. Because the narrowband approximation is often not appropriate, the literature of wideband beamformers in sonar applications is extensive.

6 The ambiguity function

A two-dimensional radar can be described as a device for forming a two-dimensional convolution of the reflectivity density function of an illuminated scene with a two-dimensional function, called an *ambiguity function*, that is associated with the radar waveform. A radar uses the delay or the doppler of the received waveform as a means of obtaining surveillance information, and requires the use of waveforms that are carefully designed to provide adequate resolution and avoid ambiguity. The major analytical tool used to design such waveforms is the ambiguity function. The ambiguity function is a two-dimensional function defined as a functional of the one-dimensional waveform. Every one-dimensional waveform of energy E_p is associated with a two-dimensional ambiguity function of energy E_p^2, which provides a surprising amount of insight into the performance of the waveform.

We shall introduce the ambiguity function formally here, proving a number of its mathematical properties. We will then study the ambiguity functions of some interesting waveforms. Later, in Chapter 7, we shall study the performance of imaging radars from the point of view of the ambiguity function, identifying the coordinates of the ambiguity function with the delay and the doppler of an echo. In Chapter 12, we shall study the performance of search radars from the point of view of the ambiguity function.

6.1 Theory of the ambiguity function

Every finite energy pulse or waveform, whether real or complex, is associated with a complex function of two variables called the ambiguity function. The ambiguity function captures many of the performance properties of any radar or sonar that uses that pulse. It is commonly used as a means of evaluating the usefulness of that pulse for a particular radar or sonar application.

Definition 6.1.1 *The ambiguity function of the finite energy pulse $s(t)$ is the complex function of two variables:*

$$\chi(\tau, \nu) = \int_{-\infty}^{\infty} s(t + \tau/2)s^*(t - \tau/2)e^{-j2\pi\nu t}\,dt.$$

The ambiguity function provides a mapping from the set of complex-valued, finite-energy functions of one variable into the set of complex-valued, finite-energy functions of two variables. This mapping is denoted:

$$s(t) \to \chi(\tau, \nu).$$

The ambiguity function is a complex function with a real part and an imaginary part:

$$\chi(\tau, \nu) = \chi_R(\tau, \nu) + \mathrm{j}\chi_I(\tau, \nu).$$

The magnitude of the ambiguity function $|\chi(\tau, \nu)|$ is called the *ambiguity surface*.

An equivalent form of the definition is

$$\chi(\tau, \nu) = \int_{-\infty}^{\infty} [s(t + \tau/2)\mathrm{e}^{-\mathrm{j}\pi\nu t}][s(t - \tau/2)\mathrm{e}^{\mathrm{j}\pi\nu t}]^* \,\mathrm{d}t.$$

The latter form shows that $\chi(\tau, \nu)$ can be thought of as the correlation between the waveform $s(t)$ shifted in time and in frequency with the same waveform $s(t)$ shifted in the opposite direction in time and in frequency.

A slightly different form of the ambiguity function, which is asymmetric but essentially equivalent, is given by the definition

$$\chi'(\tau, \nu) = \int_{-\infty}^{\infty} s(t)s^*(t - \tau)\mathrm{e}^{-\mathrm{j}2\pi\nu t} \,\mathrm{d}t.$$

While the asymmetric form appears simpler at the onset, the symmetrical form is chosen for theoretical studies because it simplifies the appearance of many later results. In engineering practice, either form is used without comment depending on its convenience. The two forms are related by a linear phase term. This can be seen by a change in variables,

$$\begin{aligned}
\chi(\tau, \nu) &= \int_{-\infty}^{\infty} s(t + \tau/2)s^*(t - \tau/2)\mathrm{e}^{-\mathrm{j}2\pi\nu t} \,\mathrm{d}t \\
&= \mathrm{e}^{\mathrm{j}\pi\tau\nu} \int_{-\infty}^{\infty} s(t + \tau/2)s^*(t - \tau/2)\mathrm{e}^{-\mathrm{j}2\pi\nu(t+\tau/2)} \,\mathrm{d}t \\
&= \mathrm{e}^{\mathrm{j}\pi\tau\nu} \int_{-\infty}^{\infty} s(t')s^*(t' - \tau)\mathrm{e}^{-\mathrm{j}2\pi\nu t'} \,\mathrm{d}t' \\
&= \mathrm{e}^{\mathrm{j}\pi\tau\nu}\chi'(\tau, \nu).
\end{aligned}$$

In particular, notice that

$$|\chi(\tau, \nu)| = |\chi'(\tau, \nu)|.$$

By setting $\nu = 0$ or $\tau = 0$, the ambiguity function is reduced to other well-known functions. By setting $\nu = 0$ in $\chi(\tau, \nu)$, we have the autocorrelation function of $s(t)$,

$$\chi(\tau, 0) = \phi(\tau).$$

By setting $\tau = 0$ in $\chi(\tau, \nu)$, we have the Fourier transform of the square of the pulse

$$\chi(0, \nu) = \int_{-\infty}^{\infty} |s(t)|^2 e^{-j2\pi\nu t}\, dt.$$

By setting both $\tau = 0$ and $\nu = 0$, we have

$$\begin{aligned}\chi(0, 0) &= \int_{-\infty}^{\infty} |s(t)|^2\, dt \\ &= E_p.\end{aligned}$$

Thus the value of the ambiguity function at the origin is equal to the energy in the pulse.

Theorem 6.1.2 *Suppose that*

$$s(t) \to \chi(\tau, \nu).$$

Then

$$\begin{aligned}&\textit{i)}\ s(t - \Delta) \to e^{-j2\pi\nu\Delta}\chi(\tau, \nu) \\ &\textit{ii)}\ s(t)e^{j2\pi f t} \to e^{-j2\pi f\tau}\chi(\tau, \nu) \\ &\textit{iii)}\quad s(at) \to |a|^{-1}\chi(a\tau, \nu/a) \\ &\textit{iv)}\ s(t)e^{j\pi\alpha t^2} \to \chi(\tau, \nu - \alpha\tau).\end{aligned}$$

Proof: We provide only a proof of part iv (which is known as the *quadratic-phase property*). Let $s'(t) = s(t)e^{j\pi\alpha t^2}$,

$$\begin{aligned}\chi_{s'}(\tau, \nu) &= \int_{-\infty}^{\infty} s'(t + \tau/2)s'^*(t - \tau/2)e^{-j2\pi\nu t}\, dt \\ &= \int_{-\infty}^{\infty} s(t + \tau/2)s^*(t - \tau/2)e^{j2\pi\alpha\tau t}e^{-j2\pi\nu t}\, dt \\ &= \chi_s(\tau, \nu - \alpha\tau).\end{aligned}$$

□

Theorem 6.1.3 *Let $S(f)$ be the Fourier transform of $s(t)$. The ambiguity function can be written*

$$\chi(\tau, \nu) = \int_{-\infty}^{\infty} S(f + \nu/2)S^*(f - \nu/2)e^{j2\pi f\tau}\, df.$$

Proof:

$$\chi(\tau, \nu) = \int_{-\infty}^{\infty} [s(t + \tau/2)e^{-j\pi\nu t}][s(t - \tau/2)e^{j\pi\nu t}]^*\, dt.$$

The first term has the transform $S(f + \nu/2)e^{j\pi(f+\nu/2)\tau}$, and the second term has the transform $S(f - \nu/2)e^{-j\pi(f-\nu/2)\tau}$. Then, by Parseval's formula,

$$\chi(\tau, \nu) = \int_{-\infty}^{\infty} [S(f + \nu/2)e^{j\pi(f+\nu/2)\tau}][S(f - \nu/2)e^{-j\pi(f-\nu/2)\tau}]^* \, df$$
$$= \int_{-\infty}^{\infty} [S(f + \nu/2)e^{j\pi f\tau}][S(f - \nu/2)e^{-j\pi f\tau}]^* \, df,$$

as was to proved. □

It is immediately apparent from Theorem 6.1.3 that the ambiguity function can also be written

$$\chi(\tau, \nu) = e^{j\pi\nu\tau} \int_{-\infty}^{\infty} S(f + \nu)S^*(f)e^{j2\pi f\tau} \, df.$$

Corollary 6.1.4 (Duality) *Suppose that $s(t)$, which has the Fourier transform $S(f)$, has the ambiguity function $\chi(\tau, \nu)$. Then $S(t)$, which has the Fourier transform $s(-f)$, has the ambiguity function $\chi(-\nu, \tau)$.*

Proof: The proof follows directly from Theorem 6.1.3. □

Theorem 6.1.5

$$\int_{-\infty}^{\infty} \chi(\tau, \nu)e^{j2\pi\nu t} \, d\nu = s(t + \tau/2)s^*(t - \tau/2)$$
$$\int_{-\infty}^{\infty} \chi(\tau, \nu)e^{-j2\pi f\tau} \, d\tau = S(f + \nu/2)S^*(f - \nu/2).$$

Proof: The first expression is the inverse Fourier transform of the defining equation for $\chi(\tau, \nu)$ with τ fixed. The second expression is the inverse Fourier transform of the expression in Theorem 6.1.3 with ν fixed. □

The theorem suggests that ambiguity functions are scarce in the space of two-dimensional functions because most two-dimensional functions, $f(\tau, t)$, cannot be factored as $f(\tau, t) = s(t + \tau/2)s^*(t - \tau/2)$.

From the first line of Theorem 6.1.5, we can immediately write the following expression,

$$s(\tau) = \frac{1}{s^*(0)} \int_{-\infty}^{\infty} \chi(\tau, \nu)e^{j\pi\nu\tau} \, d\nu,$$

simply by replacing t by $\tau/2$. Thus, in principle, $s(t)$ can be recovered from $\chi(\tau, \nu)$. In practice, the inverse formula can give a poorly behaved numerical computation because small changes in the ambiguity function $\chi(\tau, \nu)$ can cause very large changes in $s(\tau)$.

Table 6.1 *A table of ambiguity functions*

$s(t)$	$\chi(\tau, \nu)$
$\text{rect}(t/T)$	$(T - \lvert\tau\rvert)\text{sinc}[\nu(T - \lvert\tau\rvert)]\text{rect}(\tau/2T)$
$\text{sinc}(t/T)$	$T(1 - T\lvert\nu\rvert)\text{sinc}[(1 - T\lvert\nu\rvert)(\tau/T)]\text{rect}(T\nu/2)$
$\text{chrp}_\alpha(t)\text{rect}\left(\frac{t}{T}\right)$	$(T - \lvert\tau\rvert)\text{sinc}[(\nu - \alpha\tau)(T - \lvert\tau\rvert)]\text{rect}(\tau/2T)$
$e^{-\pi t^2}$	$\sqrt{\frac{1}{2}}e^{-\pi(\tau^2+\nu^2)/2}$
$e^{-\pi(t/T)^2}$	$\sqrt{\frac{1}{2}}Te^{-\pi((\tau/T)^2+(T\nu)^2)/2}$

The two-dimensional inverse Fourier transform of $\chi(\tau, \nu)$ is

$$\Xi(f, t) = \int_{-\infty}^{\infty}\int_{-\infty}^{\infty} \chi(\tau, \nu)e^{j2\pi\nu t}e^{j2\pi\tau f}\, d\nu\, d\tau,$$

which reduces to

$$\Xi(f, t) = \int_{-\infty}^{\infty} s(t + \tau/2)s^*(t - \tau/2)e^{j2\pi\tau f}\, d\tau.$$

This expression superficially resembles the expression defining $\chi(\tau, \nu)$, but it is actually quite different because the integration is in the variable τ. The function $\Xi(f, t)$ is called the *Wigner distribution* of the pulse $s(t)$.

There are a few simple pulses whose ambiguity functions are convenient to compute in closed form. Some of these functions are listed in Table 6.1. For example, the ambiguity function of the simple rectangular pulse $s(t) = \text{rect}(t/T)$ is easy to compute. Because the term $s(t + \tau/2)s^*(t - \tau/2)$ equals 1 for $|t| \leq (T - |\tau|)/2$, and otherwise equals zero, we have

$$s(t + \tau/2)s^*(t - \tau/2) = \text{rect}\left(\frac{t}{T - |\tau|}\right),$$

which is zero if $|\tau| > T$. Consequently,

$$\chi(\tau, \nu) = \begin{cases} (T - |\tau|)\text{sinc}[\nu(T - |\tau|)] & \text{if } |\tau| \leq T \\ 0 & \text{if } |\tau| > T. \end{cases}$$

This function is illustrated in Figure 6.1. It is instructive to notice that for $\nu = 0$, $\chi(\tau, 0)$ takes the form of a triangular pulse in τ,

$$\chi(\tau, 0) = \begin{cases} T - |\tau| & |\tau| \leq T \\ 0 & |\tau| > T, \end{cases}$$

and for $\tau = 0$, $\chi(0, \nu)$ takes the form of a sinc pulse in ν,

$$\chi(0, \nu) = T\,\text{sinc}(\nu T).$$

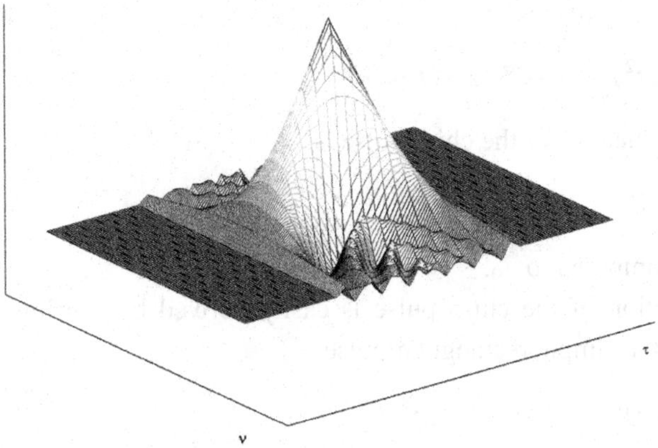

Figure 6.1 Ambiguity function of a square pulse

The zeros of $\chi(\tau, \nu)$ occur along the curves $\nu(T - |\tau|) = k$ where $k = \pm 1, \pm 2, \ldots$.

The ambiguity function of a gaussian pulse

$$s(t) = e^{-\pi t^2}.$$

is also easy to compute. The ambiguity function is

$$\begin{aligned}\chi(\tau, \nu) &= \int_{-\infty}^{\infty} s(t + \tau/2)s^*(t - \tau/2)e^{-j2\pi\nu t}\, dt \\ &= \int_{-\infty}^{\infty} e^{-\pi(t+\tau/2)^2} e^{-\pi(t-\tau/2)^2} e^{-j2\pi\nu t}\, dt \\ &= e^{-\pi\tau^2/2} \int_{-\infty}^{\infty} e^{-2\pi t^2} e^{-j2\pi\nu t}\, dt.\end{aligned}$$

The integration now has the form of a Fourier transform of a gaussian pulse. Therefore

$$\begin{aligned}\chi(\tau, \nu) &= e^{-\pi\tau^2/2}\sqrt{\tfrac{1}{2}}e^{-\pi\nu^2/2} \\ &= \sqrt{\tfrac{1}{2}}e^{-\pi(\tau^2+\nu^2)/2}.\end{aligned}$$

Thus the ambiguity function of a gaussian pulse is a two-dimensional gaussian pulse.

Another important pulse is a chirp pulse. Any pulse whose instantaneous frequency varies linearly across the duration of the pulse is called a *linear frequency-modulated pulse*. If the amplitude of the pulse is constant during its duration, then the linear pulse is also called a *chirp pulse* or a *quadratic-phase pulse*. The chirp pulse has strong properties that are useful for a variety of purposes.

The *finite chirp pulse*, denoted $\text{chrp}_\alpha(\tau/T)\text{rect}(t/T)$, is written in the complex baseband form as

$$s(t) = e^{j\pi\alpha t^2}\text{rect}(t/T).$$

The passband form of this pulse is

$$\widetilde{s}(t) = \cos(2\pi f_0 t + \pi\alpha t^2) \qquad -T/2 \le t \le T/2.$$

The "instantaneous frequency" of the chirp pulse is

$$f(t) = f_0 + \alpha t.$$

Usually, αT is small compared to f_0.

The ambiguity function of the chirp pulse is easily derived by starting with the ambiguity function of the simple rectangular pulse

$$\chi(\tau, \nu) = [T - |\tau|]\mathrm{sinc}(\nu[T - |\tau|]) \quad |\tau| \le T.$$

The quadratic-phase property for ambiguity functions, given in Theorem 6.1.2, says that, if

$$s(t) \to \chi(\tau, \nu),$$

then

$$s(t)e^{j\pi\alpha t^2} \to \chi(\tau, \nu - \alpha\tau).$$

Therefore, for the chirp pulse,

$$\chi(\tau, \nu) = \begin{cases} (T - |\tau|)\mathrm{sinc}[(\nu - \alpha\tau)(T - |\tau|)| & |\tau| \le T \\ 0 & |\tau| > T. \end{cases}$$

The ambiguity function of a chirp pulse is shown in Figure 6.2. The correlation function of the chirp pulse is now easily written by setting $\nu = 0$,

$$\begin{aligned}\phi(\tau) &= \chi(\tau, 0) \\ &= [T - |\tau|]\mathrm{sinc}[\alpha\tau(T - |\tau|)] \quad |\tau| \le T.\end{aligned}$$

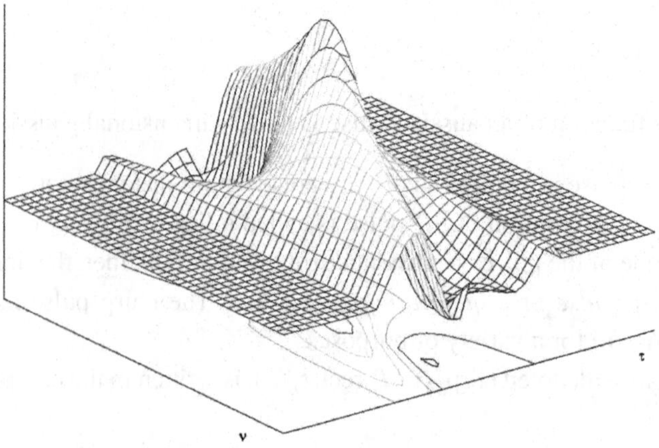

Figure 6.2 Ambiguity function of a chirp pulse

6.2 Properties of the ambiguity function

The ambiguity function has a tidy set of interlocking properties. It is unusual for an arbitrary two-dimensional function to satisfy these properties, and so, in the set of all possible two-dimensional functions, ambiguity functions are rare. Usually, in radar design problems, one has a pretty good idea of the desired ambiguity surface $|\chi(\tau, \nu)|$ or ambiguity function $\chi(\tau, \nu)$, and one wishes to work backwards to find a corresponding waveform. However, this tends to be an ill-posed problem: No satisfactory techniques are available for finding the waveform corresponding to a desired ambiguity surface, nor is a satisfactory set of rules known for determining whether a desired ambiguity surface is, in fact, an ambiguity surface. A waveform, $s(t)$, that gives rise to a desired ambiguity surface need not exist.

The ambiguity function $\chi(\tau, \nu)$ can be tested, in principle, by Corollary 6.1.4 to see whether its inverse Fourier transform has the factored form $s(t + \tau/2)s^*(t - \tau/2)$ required by Theorem 6.1.5. For most $\chi(\tau, \nu)$, the inverse Fourier transform can only be performed numerically and is subject to numerical loss of precision, so the factorization test is tainted.

We shall develop the main properties of the ambiguity function. Some of these properties are similar to properties of the Fourier transform. Those proofs that follow easily from the definition will be omitted.

Property 1 (Symmetry)

$$\chi(\tau, \nu) = \chi^*(-\tau, -\nu).$$

Property 2 (Maximum) The ambiguity function is real and positive at the origin. The largest value of the ambiguity surface is always at the origin,

$$|\chi(\tau, \nu)| \leq \chi(0, 0) = E_p,$$

with strict inequality if $(\tau, \nu) \neq (0, 0)$.

Proof: A straightforward application of the Schwarz inequality gives

$$\begin{aligned}
|\chi(\tau, \nu)|^2 &= \left| \int_{-\infty}^{\infty} [s(t + \tau/2)\mathrm{e}^{-\mathrm{j}\pi\nu t}][s(t - \tau/2)\mathrm{e}^{\mathrm{j}\pi\nu t}]^* \, \mathrm{d}t \right|^2 \\
&\leq \int_{-\infty}^{\infty} \left| s(t + \tau/2)\mathrm{e}^{-\mathrm{j}\pi\nu t} \right|^2 \mathrm{d}t \int_{-\infty}^{\infty} \left| s(t - \tau/2)\mathrm{e}^{\mathrm{j}\pi\nu t} \right|^2 \mathrm{d}t \\
&= \int_{-\infty}^{\infty} |s(t)|^2 \, \mathrm{d}t \int_{-\infty}^{\infty} |s(t)|^2 \, \mathrm{d}t \;=\; E_p^2.
\end{aligned}$$

□

Property 3 (Volume Property)

$$\int_{-\infty}^{\infty}\int_{-\infty}^{\infty} |\chi(\tau, \nu)|^2 \, d\tau \, d\nu = |\chi(0, 0)|^2 = E_p^2.$$

Proof: Recall that $\chi(\tau, \nu)$ can be expressed in either of the following two ways:

$$\chi(\tau, \nu) = e^{j\pi\tau\nu} \int_{-\infty}^{\infty} s(t)s^*(t-\tau)e^{-j2\pi\nu t} \, dt,$$

or

$$\chi(\tau, \nu) = e^{j\pi\tau\nu} \int_{-\infty}^{\infty} S(f+\nu)S^*(f)e^{j2\pi f\tau} \, df.$$

Therefore we have the fourfold integral

$$\int_{-\infty}^{\infty}\int_{-\infty}^{\infty} |\chi(\tau, \nu)|^2 \, d\tau \, d\nu$$
$$= \int_{-\infty}^{\infty}\int_{-\infty}^{\infty}\int_{-\infty}^{\infty}\int_{-\infty}^{\infty} s(t)s^*(t-\tau)S^*(f+\nu)S(f)e^{-j2\pi(\nu t+f\tau)} \, dt \, df \, d\tau \, d\nu.$$

Now identify groups of terms as Fourier transforms:

$$\int_{-\infty}^{\infty} s^*(t-\tau)e^{-j2\pi f\tau} \, d\tau = S^*(f)e^{-j2\pi ft}$$

and

$$\int_{-\infty}^{\infty} S^*(f+\nu)e^{-j2\pi\nu t} \, d\nu = s^*(t)e^{j2\pi ft}.$$

This gives

$$\int_{-\infty}^{\infty}\int_{-\infty}^{\infty} |\chi(\tau, \nu)|^2 \, d\tau \, d\nu = \int_{-\infty}^{\infty}\int_{-\infty}^{\infty} |s(t)|^2|S(f)|^2 \, dt \, df = [\chi(0, 0)]^2,$$

as was to be proved. □

Property 3 gives a strong and important condition necessary for a function to be an ambiguity function. The volume under the surface $|\chi(\tau, \nu)|^2$ is equal to the value of $|\chi(\tau, \nu)|^2$ at the origin. If one wishes for $|\chi(\tau, \nu)|$ to have a narrow "main lobe" at the origin, then there must be an excess of volume that cannot appear under this main lobe, and so must appear somewhere else in the τ, ν plane. Perhaps, away from the origin, $|\chi(\tau, \nu)|$ is small and the excess volume is more or less uniformly distributed across a large region of the τ, ν plane, or perhaps the excess volume is concentrated in one or more large lobes at places other than the origin. Such extraneous lobes are sometimes called "ambiguities."[1]

[1] Hence the name "ambiguity function."

To conclude this section, we shall give several properties that are somewhat more specialized than those already given and perhaps less important. The volume property is a special case of an even stronger property: the squared ambiguity surface equals its own two-dimensional Fourier transform, but with a sign reversal.

Property 4:

$$\int_{-\infty}^{\infty}\int_{-\infty}^{\infty} |\chi(\tau,\nu)|^2 e^{-j2\pi\xi\tau} e^{-j2\pi\eta\nu}\, d\tau\, d\nu = |\chi(-\eta,\xi)|^2.$$

Proof: The proof is similar to the proof of Property 3. Again, using the relationships

$$\chi(\tau,\nu) = e^{j\pi\tau\nu} \int_{-\infty}^{\infty} s(t)s^*(t-\tau)e^{-j2\pi\nu t}\, dt$$

$$\chi(\tau,\nu) = e^{j\pi\tau\nu} \int_{-\infty}^{\infty} S(f+\nu)S^*(f)e^{j2\pi f\tau}\, df,$$

one obtains the fourfold integral

$$\int_{-\infty}^{\infty}\int_{-\infty}^{\infty} |\chi(\tau,\nu)|^2 e^{-j2\pi(\xi\tau+\eta\nu)}\, d\tau\, d\nu$$
$$= \int_{-\infty}^{\infty}\int_{-\infty}^{\infty}\int_{-\infty}^{\infty}\int_{-\infty}^{\infty} s(t)s^*(t-\tau)S^*(f+\nu)S(f)e^{-j2\pi(\tau(f+\xi))+\nu(t+\eta))}\, dt\, df\, d\tau\, d\nu.$$

As in the proof of Property 3, identify the τ integral and the ν integral as Fourier transforms to evaluate two of the integrations. This gives

$$\int_{-\infty}^{\infty}\int_{-\infty}^{\infty} |\chi(\tau,\nu)|^2 e^{-j2\pi(\xi\tau+\eta\nu)}\, d\tau\, d\nu$$
$$= \int_{-\infty}^{\infty} s(t)s^*(t+\eta)e^{-j2\pi\xi t}\, dt \int_{-\infty}^{\infty} S(f)S^*(f+\xi)e^{j2\pi f\eta}\, df$$
$$= \int_{-\infty}^{\infty} s(t)s^*(t-\eta')e^{-j2\pi\xi t}\, dt \int_{-\infty}^{\infty} S(f)S^*(f+\xi)e^{-j2\pi f\eta'}\, df$$

where the second line follows by setting $\eta' = -\eta$. The first integral on the right yields $\chi(\eta',\xi)e^{-j\pi\eta'\xi}$. The second integral yields $\chi^*(\eta',\xi)e^{j\pi\eta'\xi}$. Therefore

$$\int_{-\infty}^{\infty}\int_{-\infty}^{\infty} |\chi(\tau,\nu)|^2 e^{-j2\pi(\xi\tau+\eta\nu)}\, d\tau\, d\nu = |\chi(-\eta,\xi)|^2,$$

as was to be proved. □

By replacing η by $-\eta$, Property 4 can be stated in the alternative form

$$\int_{-\infty}^{\infty}\int_{-\infty}^{\infty} |\chi(\tau,\nu)|^2 e^{-j2\pi(\xi\tau-\eta\nu)}\, d\tau\, d\nu = |\chi(\eta,\xi)|^2,$$

which implies that

$$\int_{-\infty}^{\infty} |\chi(\tau,\nu)|^2 e^{-j2\pi\xi\tau}\, d\tau = \int_{-\infty}^{\infty} |\chi(\eta,\xi)|^2 e^{-j2\pi\eta\nu}\, d\eta.$$

When we combine it with the projection-slice theorem, Property 4 becomes a statement that an ambiguity surface cannot be narrow on both of two orthogonal axes. The next property is a simplified version of this statement.

Property 5:

$$\int_{-\infty}^{\infty} |\chi(\tau,\nu)|^2\, d\tau = \int_{-\infty}^{\infty} |\chi(\tau,0)|^2 e^{-j2\pi\nu\tau}\, d\tau$$
$$\int_{-\infty}^{\infty} |\chi(\tau,\nu)|^2\, d\nu = \int_{-\infty}^{\infty} |\chi(0,\nu)|^2 e^{-j2\pi\nu\tau}\, d\nu.$$

Proof: To prove the first expression, let $\xi = 0$ and replace η by τ on the right side of the preceding equation. The second equation is obtained in a similar way. □

6.3 Shape and resolution parameters

The shape of the main lobe of the ambiguity surface of $s(t)$ near the origin determines much of the performance of a radar that uses the waveform $s(t)$. The size of the main lobe determines the ability of the radar to resolve two closely spaced targets. In this section, we shall relate the shape of the main lobe of $\chi(\tau,\nu)$ and of $|\chi(\tau,\nu)|$ to the properties of the waveform $s(t)$.

The *shape parameters*, describing the shape of the main lobe of $\chi(\tau,\nu)$, are the coefficients of a quadratic surface fit to $\chi(\tau,\nu)$ near the origin.

Theorem 6.3.1 *If $\chi(\tau,\nu)$ has first and second derivatives at the origin, then, near the origin,*

$$|\chi(\tau,\nu)| = E_p(1 - 2\pi^2(T_G^2\nu^2 + 2T_G B_G \rho\tau\nu + B_G^2\tau^2))$$

up to the terms of second order in τ and ν, where T_G, B_G, and ρ are the Gabor parameters of pulse $s(t)$.

Proof: Expand $\chi(\tau,\nu)$ in a Taylor series as follows:

$$\chi(\tau,\nu) = \chi(0,0) + \tau\frac{\partial\chi}{\partial\tau} + \nu\frac{\partial\chi}{\partial\nu} + \tfrac{1}{2}\tau^2\frac{\partial^2\chi}{\partial\tau^2} + \tau\nu\frac{\partial^2\chi}{\partial\tau\partial\nu} + \tfrac{1}{2}\nu^2\frac{\partial^2\chi}{\partial\nu^2} + \cdots.$$

This expansion exists whenever the partial derivatives exist. Notice, however, that the ambiguity function of a square pulse fails to have a first partial derivative with respect to τ.

We want to relate the partial derivatives of $\chi(\tau, \nu)$ to the properties of the pulse $s(t)$. It will be more convenient to work with the alternative version of the ambiguity function given by

$$\chi(\tau, \nu) = \int_{-\infty}^{\infty} s(t)s^*(t-\tau)e^{j2\pi\nu t}\,dt.$$

Now substitute the Taylor series

$$s(t-\tau) = s(t) - \tau\dot{s}(t) + \tfrac{1}{2}\tau^2\ddot{s}(t) + \cdots,$$

and

$$e^{j2\pi\nu t} = 1 + j2\pi\nu t - 2\pi^2\nu^2t^2 + \cdots.$$

Then, up to terms of second order,

$$\begin{aligned}
\chi(\tau, \nu) &= \int_{-\infty}^{\infty} s(t)s^*(t-\tau)e^{j2\pi\nu t}\,dt \\
&\approx \int_{-\infty}^{\infty} [|s(t)|^2 - \tau s(t)\dot{s}^*(t) + \frac{\tau^2}{2}s(t)\ddot{s}^*(t)][1 + j2\pi\nu t - 2\pi^2\nu^2t^2]\,dt \\
&= \int_{-\infty}^{\infty} [|s(t)|^2 + j2\pi\nu t|s(t)|^2 - 2\pi^2\nu^2t^2|s(t)|^2 - \tau s(t)\dot{s}^*(t) \\
&\qquad -j\tau 2\pi\nu t s(t)\dot{s}^*(t) + \frac{\tau^2}{2}s(t)\ddot{s}^*(t)]\,dt \\
&= \chi(0,0)[1 + j2\pi\nu\overline{t} - 2\pi^2\nu^2\overline{t^2} + j\tau 2\pi\overline{f} - 2\pi^2\tau^2\overline{f^2} - 4\pi^2\tau\nu\overline{tf}]
\end{aligned}$$

as was to be proved. □

The proof made use of the definitions

$$\begin{aligned}
\overline{t} &= \frac{1}{E_p}\int_{-\infty}^{\infty} t|s(t)|^2\,dt \\
\overline{t^2} &= \frac{1}{E_p}\int_{-\infty}^{\infty} t^2|s(t)|^2\,dt \\
\overline{tf} &= \frac{1}{E_p}\frac{j}{2\pi}\int_{-\infty}^{\infty} ts(t)\dot{s}^*(t)\,dt \\
\overline{f} &= \frac{1}{E_p}\frac{j}{2\pi}\int_{-\infty}^{\infty} s(t)\dot{s}^*(t)\,dt = \frac{1}{E_p}\int_{-\infty}^{\infty} f|S(f)|^2\,df \\
\overline{f^2} &= \frac{1}{E_p}\frac{-1}{4\pi^2}\int_{-\infty}^{\infty} s(t)\ddot{s}^*(t)\,dt = \frac{1}{E_p}\int_{-\infty}^{\infty} f^2|S(f)|^2\,df = \frac{1}{E_p}\frac{-1}{4\pi^2}\int_{-\infty}^{\infty} |\dot{s}(t)|^2\,dt
\end{aligned}$$

which were introduced in Section 2.6.

The shape of the ambiguity surface $|\chi(\tau, \nu)|$ is closely related to the shape of $\chi(\tau, \nu)$. The shape of the squared ambiguity surface near the origin is given by

$$|\chi(\tau, \nu)|^2 = \chi(0,0)^2\left|1 + j2\pi\nu\overline{t} - 2\pi^2\nu^2\overline{t^2} + j2\pi\tau\overline{f} - 2\pi^2\tau\overline{f^2} - 4\pi^2\tau\nu\overline{tf}\right|^2.$$

Up to the terms of second order, this becomes

$$|\chi(\tau,\nu)| = E_p[1 - 4\pi^2\nu^2(\overline{t^2} - \overline{t}^2) - 4\pi^2\tau^2(\overline{f^2} - \overline{f}^2) - 8\pi^2\tau\nu\mathrm{Re}[\overline{tf} - \overline{t}\,\overline{f}]]^{1/2}.$$

Consequently, because $\sqrt{1+2x^2} = 1 + x^2$ up to the terms of second order, we have

$$\begin{aligned}|\chi(\tau,\nu)| &= E_p[1 - 2\pi^2\nu^2(\overline{t^2} - \overline{t}^2) - 2\pi^2\tau^2(\overline{f^2} - \overline{f}^2) - 4\pi^2\tau\nu\mathrm{Re}[\overline{tf} - \overline{t}\overline{f}]] \\ &= E_p[1 - 2\pi^2\nu^2 T_G^2 - 2\pi^2\tau^2 B_G^2 - 4\pi^2\tau\nu T_G B_G \rho]\end{aligned}$$

up to the terms of second order, where T_G is the Gabor timewidth of the pulse $s(t)$, B_G is the Gabor bandwidth of the pulse $s(t)$, and

$$T_G B_G \rho = \mathrm{Re}[\overline{tf} - \overline{t}\,\overline{f}]$$

is the Gabor skew parameter of the pulse $s(t)$. The Gabor parameters are external descriptors of the waveform $s(t)$ in the sense that they can be measured from $|\chi(\tau,\nu)|$ without knowledge of the details of $s(t)$.

The reciprocal of the Gabor timewidth of the pulse measures the width of the main lobe of the ambiguity surface in the ν direction, and the reciprocal of the Gabor bandwidth measures the width of the main lobe of the ambiguity surface in the τ direction. The shape of the main lobe can be seen more completely by describing the intersection of the main lobe with a horizontal plane. If the plane is high enough, it slices through the main lobe just below the peak, and the shape of the cut describes the shape of the main lobe near the peak. Specifically, choose a convenient constant, C, and set

$$|\chi(\tau,\nu)| = E_p(1 - 2\pi^2 C).$$

Within the quadratic approximation to $\chi(\tau,\nu)$, this becomes the equation of an ellipse in τ and ν,

$$\tau^2 B_G^2 + 2\tau\nu T_G B_G \rho + \nu^2 T_G^2 = C,$$

which is known as the *uncertainty ellipse*. The uncertainly ellipse provides a summary description of the main lobe of the ambiguity function. The constant C has no special importance if we are interested only in the shape of the uncertainty ellipse.

Suppose we are given the pulse $s(t)$ whose uncertainty ellipse is given by

$$B_G^2\tau^2 + T_G^2\nu^2 = C.$$

Then, by the quadratic-phase property, the chirp pulse $s(t)\mathrm{e}^{\mathrm{j}\pi\alpha t^2}$ has the uncertainty ellipse

$$B_G^2\tau^2 + T_G^2(\nu - \alpha\tau)^2 = C,$$

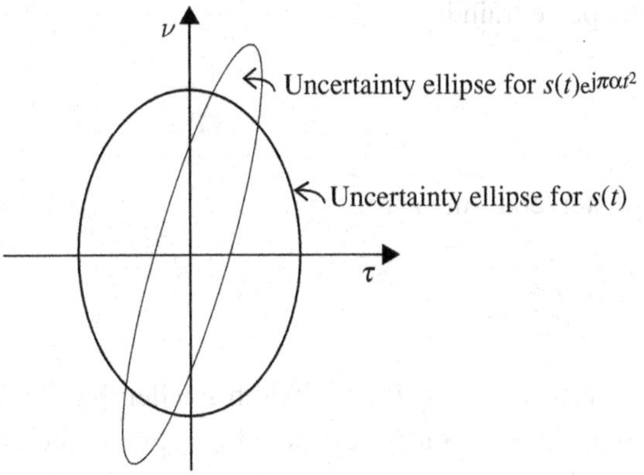

Figure 6.3 Some uncertainty ellipses

which can be rewritten as

$$(B_G^2 + \alpha^2 T_G^2)\tau^2 - 2\alpha T_G^2 \nu\tau + T_G^2 \nu^2 = C.$$

Consequently, if $s(t)$ has the Gabor bandwidth B_G, then $s(t)e^{j\pi\alpha t^2}$ has the Gabor bandwidth $\sqrt{B_G^2 + \alpha^2 T_G^2}$ and has a nonzero skew parameter. The uncertainty ellipses for the pulses $s(t)$ and $s(t)e^{j\pi\alpha t^2}$ are shown in Figure 6.3.

The Gabor bandwidth, the Gabor timewidth, and the Gabor skew parameter ρ describe the shape and width of the main lobe of the ambiguity function near the maximum. These parameters can be used as a measure of resolution, although the curvature at the peak may be inadequate as a description of the width of the main lobe. Other resolution criteria may measure the width of the main lobe in a more appropriate way. For example, the ambiguity function $\chi(\tau, \nu)$ has a Woodward resolution in the τ direction given by

$$\Delta\tau = \frac{\int_{-\infty}^{\infty} |\chi(\tau, 0)|^2 \, d\tau}{|\chi(0, 0)|^2} = \frac{1}{E_p^2} \int_{-\infty}^{\infty} |\phi(\tau)|^2 \, d\tau,$$

where $\phi(\tau) = \chi(\tau, 0)$.

6.4 Ambiguity function of a pulse train

A long-duration waveform can be generated by periodically repeating a suitable short-duration waveform. In this context, the basic waveform is called a *pulse*, although that waveform itself may be very complicated. A pulse train, then, is a waveform consisting of a finite or infinite number of nonoverlapping pulses. In this section, we will study pulse trains with uniformly spaced, identical copies of pulse $s(t)$. Then, with the first pulse centered at time zero, the pulse train is

$$p(t) = \sum_{n=0}^{N-1} s(t - nT_r)$$

or, with the pulse train itself centered at time zero,

$$p(t) = \sum_{n=0}^{N-1} s(t - nT_r + \tfrac{1}{2}(N-1)T_r)$$

where T_r is the pulse repetition interval, and $p(t)$ has a width smaller than T_r. We shall relate the ambiguity function of the pulse train $p(t)$, denoted $\chi_p(\tau, \nu)$, to the ambiguity function of the pulse $s(t)$, denoted $\chi_s(\tau, \nu)$.

Theorem 6.4.1 *Let*

$$p(t) = \sum_{n=0}^{N-1} s(t - nT_r + \tfrac{1}{2}(N-1)T_r).$$

Then

$$\chi_p(\tau, \nu) = \sum_{n=-(N-1)}^{N-1} \chi_s(\tau - nT_r, \nu)\mathrm{dirc}_{N-|n|}\nu T_r.$$

Proof: We will work with the pulse train in the form

$$p(t) = \sum_{n=0}^{N-1} s(t - nT_r),$$

and later slide it to the left. By definition,

$$\begin{aligned}
\chi_p(\tau, \nu) &= \int_{-\infty}^{\infty} \left(\sum_{n=0}^{N-1} s\left(t - nT_r + \tfrac{1}{2}\tau\right)\right)\left(\sum_{m=0}^{N-1} s\left(t - mT_r - \tfrac{1}{2}\tau\right)\right)^{*} \mathrm{e}^{-\mathrm{j}2\pi\nu t}\,\mathrm{d}t \\
&= \sum_{n=0}^{N-1}\sum_{m=0}^{N-1} \int_{-\infty}^{\infty} s\left(t + \tfrac{1}{2}\tau - nT_r\right) s^{*}\left(t - \tfrac{1}{2}\tau - mT_r\right) \mathrm{e}^{-\mathrm{j}2\pi\nu t}\,\mathrm{d}t.
\end{aligned}$$

Replace t by $t + \frac{1}{2}(m+n)T_r$

$$\begin{aligned}
\chi_p(\tau, \nu) &= \sum_{n=0}^{N-1}\sum_{m=0}^{N-1} \mathrm{e}^{-\mathrm{j}2\pi\frac{m+n}{2}\nu T_r} \int_{-\infty}^{\infty} s\left(t + \frac{\tau}{2} + \frac{m-n}{2}T_r\right) \\
&\quad \times s^{*}\left(t - \frac{\tau}{2} - \frac{m-n}{2}T_r\right) \mathrm{e}^{-\mathrm{j}2\pi\nu t}\,\mathrm{d}t \\
&= \sum_{n=0}^{N-1}\sum_{m=0}^{N-1} \mathrm{e}^{-\mathrm{j}2\pi\frac{m+n}{2}\nu T_r} \chi_s(\tau + (m-n)T_r, \nu).
\end{aligned}$$

The sum over the N by N array is indexed along rows and columns by n and m. The sum is now rearranged by indexing along the subdiagonals. Given any N by N matrix, $\boldsymbol{A}$, we can sum the N^2 elements by first summing down the main diagonal, then summing all subdiagonals above the main diagonal and all subdiagonals below the main diagonal. This leads to the identity

$$\sum_{n=0}^{N-1}\sum_{m=0}^{N-1} A_{nm} = \sum_{n=0}^{N-1} A_{nn} + \sum_{n=1}^{N-1}\sum_{k=0}^{N-1-n} A_{k(k+n)} + \sum_{n=1}^{N-1}\sum_{k=0}^{N-1-n} A_{(k+n)k}.$$

The first term is a sum down the main diagonal, the second is the sum of all minor diagonals in the upper triangular matrix, and the third is the sum of all minor diagonals in the lower triangular matrix. The first two terms can be combined to give

$$\sum_{n=0}^{N-1}\sum_{m=0}^{N-1} A_{nm} = \sum_{n=0}^{N-1}\sum_{k=0}^{N-1-n} A_{k(k+n)} + \sum_{n=1}^{N-1}\sum_{k=0}^{N-1-n} A_{(k+n)k}.$$

Replace n by $-n$ in the second term

$$\sum_{n=0}^{N-1}\sum_{m=0}^{N-1} A_{nm} = \sum_{n=0}^{N-1}\sum_{k=0}^{N-1-n} A_{k(k+n)} + \sum_{n=-1}^{-N+1}\sum_{k=0}^{N-1-n} A_{(k-n)k}.$$

In our case,

$$A_{nm} = e^{-j2\pi \frac{m+n}{2} \nu T_r} \chi_s(\tau + (m-n)T_r, \nu).$$

Then

$$A_{k(k+n)} = e^{-j2\pi \frac{2k+n}{2} \nu T_r} \chi_s(\tau + nT_r, \nu),$$

and

$$A_{(k-n)k} = e^{-j2\pi \frac{2k+|n|}{2} \nu T_r} \chi_s(\tau + nT_r, \nu).$$

We now combine the two sums as follows:

$$\begin{aligned}
\chi_p(\tau, \nu) &= \sum_{n=0}^{N-1}\sum_{m=0}^{N-1} A_{nm} \\
&= \sum_{n=-(N-1)}^{N-1}\ \sum_{k=0}^{N-1-|n|} e^{-j2\pi \frac{2k+|n|}{2} \nu T_r} \chi_s(\tau + nT_r, \nu) \\
&= \sum_{n=-(N-1)}^{N-1} e^{-j\pi |n| \nu T_r} \chi_s(\tau + nT_r, \nu) \sum_{k=0}^{N-1-|n|} e^{-j2\pi k \nu T_r}.
\end{aligned}$$

Finally, we will execute the sum on k using the relationship

$$\sum_{k=0}^{N-1} e^{-j2\pi Ak} = e^{-j\pi(N-1)A} \operatorname{dirc}_N A$$

to obtain

$$\begin{aligned}\chi_p(\tau,\nu) &= \sum_{n=-(N-1)}^{N-1} e^{-j\pi|n|\nu T_r}\chi_s(\tau+nT_r,\nu)e^{-j\pi(N-1-|n|)\nu T_r}\,\mathrm{dirc}_{N-|n|}\nu T_r \\ &= e^{-j\pi\nu T_r(N-1)}\sum_{n=-(N-1)}^{N-1}\chi_s(\tau-nT_r,\nu)\mathrm{dirc}_{N-|n|}\nu T_r.\end{aligned}$$

When the pulse train is centered by redefining the time origin, the phase term drops out, and so the proof is complete. □

The statement of Theorem 6.4.1 does not presume any simple structure for the pulse $s(t)$. The pulse may itself be a complex structure. For example, $s(t)$ may itself be a train of pulses with its own pulse repetition interval. The formula for the pulse train ambiguity function may then be embedded into itself to describe a pulse train of pulse trains.

Theorem 6.4.1 is basic to understanding the performance of any pulse train. The summands are a product of two terms. For each n, the first term is the ambiguity function of a single pulse delayed by n multiples of the pulse repetition interval. The second term is a dirichlet function due to the pulse train. For fixed n, the dirichlet function has grating lobes at those values of ν that are integer multiples of $1/T_r$.

Let $s(t)$ be a pulse whose ambiguity function main lobe is described by an uncertainty ellipse, as shown in Figure 6.4. The structure near the origin of the ambiguity function of the pulse train $p(t)$ can be portrayed by combining this uncertainty ellipse with the grating lobes of the dirichlet function for each value of the delay offset nT_r. Then one obtains the depiction of $\chi_p(\tau,\nu)$ near the origin shown in Figure 6.4.

The main lobe of $s(t)$ repeats in $\chi_p(\tau,\nu)$ whenever τ is a multiple of T_r, and so T_r is called the *delay ambiguity*. Similarly, the dirichlet function has grating lobes in the ν

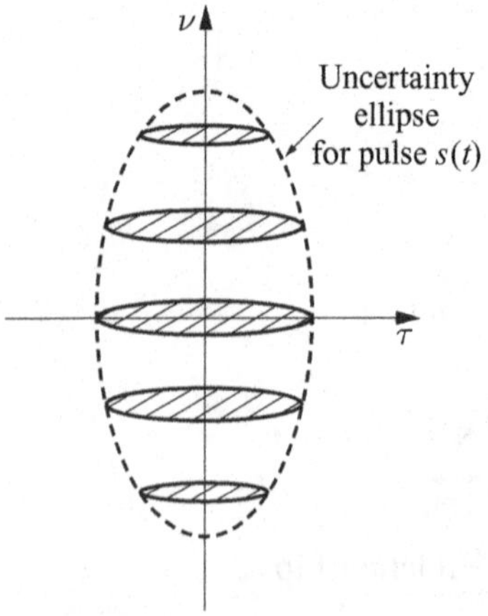

Figure 6.4 Illustrating the formation of doppler grating lobes for a simple pulse shape

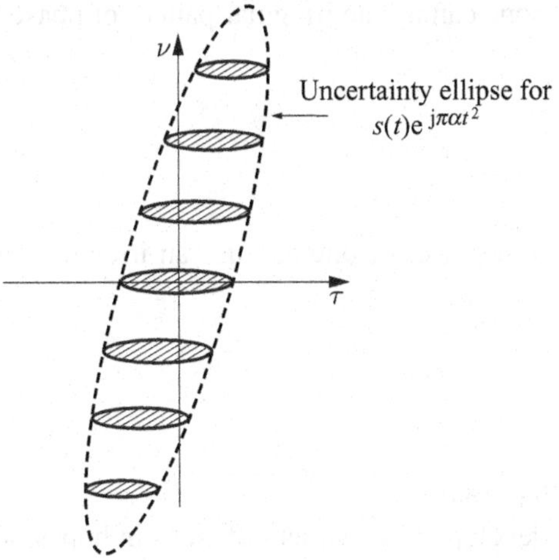

Figure 6.5 Illustrating the formation of doppler grating lobes – chirped pulse

direction that repeat for ν a multiple of $1/T_r$. The ν separation between grating lobes is called the *doppler ambiguity*. Generally, one wishes that both the delay ambiguity and the doppler ambiguity were large. Hence a compromise is always necessary with a uniform pulse train.

Next, for the real pulse $s(t)$, consider the uncertainty ellipse of the chirp pulse $s(t)e^{j\pi\alpha t^2}$ which follows from the quadratic-phase property. This uncertainty ellipse, shown in Figure 6.5, has poor resolution along the line of the major axis. The ambiguity function of the pulse train, however, is relatively narrow in both the τ and the ν directions. Thus a pulse train of chirp pulses can be used to give good resolution in both the τ and ν directions even though the pulse width is very wide in comparison to the width of the ambiguity function in that direction.

6.5 Ambiguity function of a Costas pulse

A uniform pulse train has ambiguities in both the τ and the ν directions. Whenever such ambiguities are undesirable, one must design a waveform with the periodicity suppressed so that the ambiguities will not occur. There are many ways to accomplish this while still maintaining the basic structure of a pulse train. One can use an irregular spacing of the pulses

$$p(t) = \sum_{n=0}^{N-1} s(t - \textstyle\sum_{\ell=1}^{n} T_\ell)$$

where T_ℓ is the ℓth pulse spacing; or one can use an irregular pattern of phase shifts of the pulses

$$p(t) = \sum_{n=0}^{N-1} s(t - nT_r)e^{-j\theta_n}$$

where θ_n is the phase angle of the nth pulse; or one can use an irregular pattern of frequency shifts of the pulses

$$p(t) = \sum_{n=0}^{N-1} s(t - nT_r)e^{-j2\pi\Omega_n t}$$

where Ω_n is the frequency of the nth pulse.

Each of these waveforms can be developed in a variety of ways in hopes of obtaining a satisfactory ambiguity function. We shall only study one such waveform in the remainder of this section: this waveform is based on the method of using an irregular pattern of frequencies. Let

$$p(t) = \sum_{n=0}^{N-1} s(t - nT_r + \tfrac{1}{2}(N-1)T_r)e^{-j2\pi\Omega_n t}$$

where

$$\Omega_n = \frac{\theta_n}{T_r},$$

and the θ_n are a permutation of the integers

$$\{\theta_0, \theta_1, \ldots, \theta_{N-1}\} = \{1, 2, 3, \ldots, N\}.$$

Our goal is to choose the "firing sequence" $\{\theta_0, \theta_1, \ldots, \theta_{N-1}\}$ so that the ambiguity surface has a sharp peak at the origin. For an example of such a waveform, known as a *Costas pulse*, let $s(t) = \text{rect}(t/T_r)$, $N = 4$, and choose $(\theta_0, \theta_1, \theta_2, \theta_3) = (2, 4, 3, 1)$. Figure 6.6 shows the Costas pulse, which can be described as a "frequency hopping" pattern of four "subpulses." The ambiguity function of a Costas pulse, denoted $\chi_p(\tau, \nu)$,

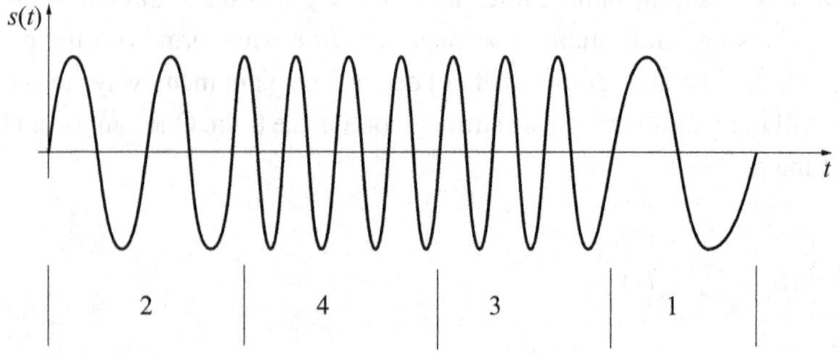

Figure 6.6 Waveform based on $n = 4$ Costas array

can be written

$$\chi_p(\tau, \nu)\mathrm{e}^{-\mathrm{j}\pi(N-1)\nu T_r} = \sum_{n=0}^{N-1}\sum_{m=0}^{N-1}\int_{-\infty}^{\infty} s\left(t+\frac{\tau}{2}-nT_r\right)\mathrm{e}^{-\mathrm{j}2\pi\Omega_n(t+\tau/2)}$$
$$\times s^*\left(t-\frac{\tau}{2}-mT_r\right)\mathrm{e}^{\mathrm{j}2\pi\Omega_m(t-\tau/2)}\mathrm{e}^{-\mathrm{j}2\pi\nu t}\,\mathrm{d}t.$$

Replace t with $t+\frac{m+n}{2}T_r$ and identify $\chi_s(\tau, \nu)$ as the ambiguity function of the subpulse $s(t)$. The same method used in the proof of Theorem 6.4.1 allows us to write

$$\chi_p(\tau, \nu) = \sum_{n=0}^{N-1}\sum_{m=0}^{N-1}\mathrm{e}^{-\mathrm{j}2\pi\frac{m+n}{2}(\nu T_r+\Omega_n-\Omega_m)}\mathrm{e}^{-\mathrm{j}2\pi(\Omega_n+\Omega_m)\frac{\tau}{2}}\chi_s(\tau+(m-n)T_r, \nu+\Omega_n-\Omega_m).$$

There are two kinds of terms in the double sum: those with $m = n$, and those with $m \neq n$. We shall consider the two kinds of terms separately, denoting their subsums as $\chi^{(1)}(\tau, \nu)$ and $\chi^{(2)}(\tau, \nu)$. If $m = n$, the argument of χ_s is the same in every term, so define

$$\chi_p^{(1)}(\tau, \nu) = \chi_s(\tau, \nu)\sum_{n=0}^{N-1}\mathrm{e}^{-\mathrm{j}2\pi n\nu T_r}\mathrm{e}^{-\mathrm{j}2\pi\Omega_n\tau}.$$

Also define

$$\chi_p^{(2)}(\tau, \nu) = \sum_{n=0}^{N-1}\sum_{m\neq n}\mathrm{e}^{-\mathrm{j}2\pi\frac{m+n}{2}(\nu T_r+\Omega_n-\Omega_m)}\mathrm{e}^{-\mathrm{j}2\pi(\Omega_n+\Omega_m)\frac{\tau}{2}}\chi_s(\tau+(m-n)T_r, \nu+\Omega_n-\Omega_m).$$

Then

$$\chi_p(\tau, \nu) = \chi_p^{(1)}(\tau, \nu) + \chi_p^{(2)}(\tau, \nu).$$

Our strategy for understanding $\chi_p(\tau, \nu)$ will be to treat the first term as a desired term and the second term as an undesired term, which will be made small by the choice of permutation. We shall see that the first term will produce the main lobe of the ambiguity function; the second term will produce the sidelobes. We shall refer to the second term as the "self-noise" term.

We first inspect the first term. Although we cannot evaluate the sum analytically for arbitrary τ and ν, we can evaluate it along the τ and ν axes. To evaluate it along the ν axis, set $\tau = 0$,

$$\begin{aligned}\chi_p^{(1)}(0, \nu) &= \left[\chi_s(0, \nu)\sum_{n=0}^{N-1}\mathrm{e}^{-\mathrm{j}2\pi n\nu T_r}\right]\mathrm{e}^{\mathrm{j}\pi(N-1)\nu T_r}\\ &= \chi_s(0, \nu)\,\mathrm{dirc}_N\,\nu T_r\\ &= T\frac{\sin\pi\nu T}{\pi\nu T_r}\frac{\sin N\pi\nu T_r}{\sin\pi\nu T_r}.\end{aligned}$$

Set $T_r = T$, then

$$\chi_p^{(1)}(0, \nu) = NT\,\mathrm{sinc}\,\nu NT,$$

which is the same as the ambiguity function of a unit amplitude pulse of duration NT.

Now compute the function along the τ axis. Set $\nu = 0$,

$$\begin{aligned}
\chi_p^{(1)}(\tau, 0) &= \left[\chi_s(\tau, 0) \sum_{n=0}^{N-1} \mathrm{e}^{-\mathrm{j}2\pi \Omega_n \tau} \right] \mathrm{e}^{\mathrm{j}\pi(N-1)\tau/T} \\
&= \left[(T - |\tau|) \sum_{i=0}^{N-1} \mathrm{e}^{-\mathrm{j}2\pi i \tau/T} \right] \mathrm{e}^{\mathrm{j}\pi(N-1)\tau/T} \qquad |\tau| \leq T \\
&= (T - |\tau|) \frac{\sin N\pi\tau/T}{\sin \pi\tau/T} \qquad |\tau| \leq T \\
&\approx (T - |\tau|) \operatorname{sinc}(\tau N/T).
\end{aligned}$$

This has its first zero crossing at $\tau = T/N$. Thus, in the ν direction, the width of $\chi^{(1)}(\tau, \nu)$ is as if there were a single pulse of amplitude one and duration NT, while in the τ direction, the width of $\chi^{(1)}(\tau, \nu)$ is as if there were a single pulse of amplitude one and duration T/N.

Now we turn to the "self-noise" term $\chi^{(2)}(\tau, \nu)$. The inequality $|\sum_i z_i| \leq \sum_i |z_i|$, which is valid for any set of complex numbers, ensures that this term satisfies

$$|\chi^{(2)}(\tau, \nu)| \leq \sum_{n=0}^{N-1} \sum_{m \neq n} |\chi_s(\tau + (m-n)T, \nu - (\Omega_n - \Omega_m))|.$$

The m, n term of the summation has a peak at $\tau = (n-m)T$ and another at $\nu = (\Omega_n - \Omega_m)$. To keep the self-noise small everywhere, we want to choose the firing sequence $\theta_0, \theta_1, \ldots, \theta_{N-1}$ so that each summand takes its peak at a different place in the τ, ν plane. This means that, for any given pair of integers, (r, s), there can be at most one other pair, (m, n), such that $n - m = s$ and $\Omega_n - \Omega_m = r$. This requirement motivates the following definition.

Definition 6.5.1 *A Costas array, $\boldsymbol{A} = [A_{ij}]$, is an N by N array of zeros and ones such that*

$$\sum_{i=0}^{N-1} A_{ij} = \sum_{j=0}^{N-1} A_{ij} = 1$$

$$\sum_{i=0}^{N-1-r} \sum_{j=0}^{N-1-s} A_{ij} A_{i+r, j+s} \leq 1 \quad \text{if } (r, s) \neq (0, 0).$$

An example of a Costas array is the four-by-four array

$$\boldsymbol{A} = \begin{bmatrix} 0 & 0 & 0 & 1 \\ 1 & 0 & 0 & 0 \\ 0 & 0 & 1 & 0 \\ 0 & 1 & 0 & 0 \end{bmatrix}.$$

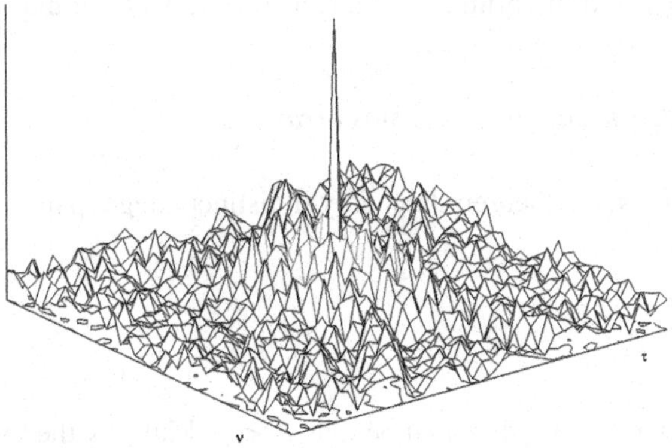

Figure 6.7 Ambiguity function for a Welch–Costas pulse

This array has the firing sequence 2, 4, 3, 1 and leads to the Costas pulse that is shown in Figure 6.6. Figure 6.7 shows the ambiguity function for a larger Costas pulse, this one with ten subpulses and the firing sequence 2, 4, 8, 5, 10, 9, 7, 3, 6, 1. One can form much larger Costas pulses, perhaps with hundreds of subpulses, that have very sharp ambiguity functions with very sharp main lobes, which require the construction of large Costas arrays.

General rules for constructing or classifying all Costas arrays are not known, but several constructions are known for special cases. A simple construction is available when $N + 1$ is a prime. The construction uses modulo-p arithmetic and an integer π between zero and p that has the property that $\pi^{p-1} = 1$ modulo p, and $\pi^n \neq 1$ modulo p for $n < p$. The integer π is said to have the multiplicative order $p - 1$ modulo p. Such an element (of multiplicative order $p - 1$ modulo p) is called a *primitive element* of modulo-p arithmetic. If p is a prime, such a primitive element always exists. For example, 7 is a prime, so there must be a primitive element for modulo-7 arithmetic. One way of finding it is by trial and error. First, try $\pi = 2$ to see if 2 is primitive. Thus $\pi = 2$, $\pi^2 = 4$, and $\pi^3 = 1$. Because $\pi^4 = \pi$, successive powers of π will only repeat this cycle, so $\pi = 2$ is not primitive. Next, try $\pi = 3$ to see if 3 is primitive. Thus $\pi = 3$, $\pi^2 = 2$, $\pi^3 = 6$, $\pi^4 = 4$, $\pi^5 = 5$, and $\pi^6 = 1$. Therefore 3 is a primitive element.

Definition 6.5.2 *Let p be a prime and $N = p - 1$, and let π be a primitive element of the integer arithmetic system modulo p. An N by N* Welch–Costas array *is the N by N array*

$$A_{ij} = \begin{cases} 1 & j = \pi^i \\ 0 & j \neq \pi^i \end{cases}.$$

For example, with $p = 7$ and the primitive element $\pi = 3$, we have the firing sequence 3, 2, 6, 4, 5, 1.

Theorem 6.5.3 *A Welch–Costas array is a Costas array.*

Proof: For any fixed r and s, the theorem fails only if distinct integer pairs (i, j) and (i', j') exist such that

$$\begin{aligned} j &= \pi^i & j' &= \pi^{i'} \\ j + s &= \pi^{i+r} & j' + s &= \pi^{i'+r} \end{aligned}$$

where, without loss of generality, we can assume that $j' < j$. Multiply the upper two equations by π^r, and then eliminate π^{i+r} from each pair of equations to obtain

$$j + s = j\pi^r \quad j' + s = j'\pi^r.$$

Subtracting these two equations gives

$$j - j' = (j - j')\pi^r.$$

Then, either $j - j' = 0$ or $\pi^r = 1$. But any $p - 1$ successive powers of π are distinct and $r < p - 1$, the second condition, cannot be satisfied. Consequently, $j = j'$, and the theorem is proved. □

6.6 The cross-ambiguity function

The ambiguity function measures the similarity of $s(t)$ to time-delayed and frequency-shifted versions of itself. Sometimes, it is necessary to study the similarity of two different pulses, $s_1(t)$ and $s_2(t)$, under the actions of time delay and frequency shift. For this purpose, the cross-ambiguity function is defined.

Definition 6.6.1 *Let $s_1(t)$ and $s_2(t)$ be finite energy pulses. The two-dimensional function*

$$\chi_c(\tau, \nu) = \int_{-\infty}^{\infty} s_1(t + \tau/2)s_2^*(t - \tau/2)\mathrm{e}^{-\mathrm{j}2\pi\nu t}\,\mathrm{d}t$$

is called the cross-ambiguity function of $s_1(t)$ with $s_2(t)$.

The cross-ambiguity function can also be defined in the asymmetric form

$$\chi_c(\tau, \nu) = \int_{-\infty}^{\infty} s_1(t)s_2^*(t - \tau)\mathrm{e}^{-\mathrm{j}2\pi\nu t}\,\mathrm{d}t.$$

The asymmetric form of the cross-ambiguity function has the mathematical structure of a modification of the Fourier transform, given by

$$S_\tau(f) = \int_{-\infty}^{\infty} s(t)g(t - \tau)\mathrm{e}^{-\mathrm{j}2\pi f t}\,\mathrm{d}t,$$

which goes by the name of the *short-time Fourier transform.* Although the cross-ambiguity function and the short-time Fourier transform have the same form, the motivation in the two cases is different and the theory is developed in different directions. The cross-ambiguity function is thought of as "comparing" $s_1(t)$ with $s_2(t)$. The short-time Fourier transform is thought of as weighting $s(t)$ by a sliding "window", $g(t)$, and forming the Fourier transform for each such windowed signal.

Theorem 6.6.2 *The cross-ambiguity function can be written as*

$$\chi_c(\tau, \nu) = \int_{-\infty}^{\infty} S_1(f + \nu/2) S_2^*(f - \nu/2) e^{j2\pi f\tau} \, df.$$

Proof: This proof is similar to the proof of Theorem 6.1.3. □

The cross-ambiguity function $\chi_c(\tau, \nu)$ has other properties that parallel the properties of the ambiguity function. They are proved in the same way.

Property 1 (Antisymmetry)

$$\chi_c(\tau, \nu) = \chi_c^*(-\tau, -\nu)$$

Property 2 (Origin)

$$\chi(0, 0) = \int_{-\infty}^{\infty} s_1(t) s_2^*(t) \, dt \leq \sqrt{E_{p_1} E_{p_2}}$$

where E_{p_1} and E_{p_2} denote the energy of $s_1(t)$ and $s_2(t)$, respectively.

Property 3

$$\int_{-\infty}^{\infty} \int_{-\infty}^{\infty} |\chi_c(\tau, \nu)|^2 \, d\tau \, d\nu = E_{p_1} E_{p_2}.$$

An important instance of the cross-ambiguity function is when $s_1(t)$ is a time-delayed and frequency-shifted version of $s_2(t)$. Suppose that

$$s_2(t) = s(t)$$
$$s_1(t) = s(t - \tau_0) e^{j2\pi \nu_0 t}.$$

Then

$$\begin{aligned}
\chi_c(\tau, \nu) &= \int_{-\infty}^{\infty} s_1(t + \tau/2) s_2^*(t - \tau/2) e^{-j2\pi \nu t} \, dt \\
&= \int_{-\infty}^{\infty} s(t - \tau_0 + \tau/2) s^*(t - \tau/2) e^{-j2\pi(\nu - \nu_0)t} \, dt.
\end{aligned}$$

Let $t = t' + \tau_0/2$.

$$\begin{aligned}\chi_c(\tau, \nu) &= e^{-j2\pi(\nu-\nu_0)\tau_0/2} \int_{-\infty}^{\infty} s\left(t + \frac{\tau - \tau_0}{2}\right) s^*\left(t - \frac{\tau - \tau_0}{2}\right) e^{-j2\pi(\nu-\nu_0)t}\, dt \\ &= e^{-j\pi(\nu-\nu_0)\tau_0} \chi(\tau - \tau_0, \nu - \nu_0).\end{aligned}$$

The term $e^{-j\pi(\nu-\nu_0)\tau_0}$ has no effect on the magnitude. Thus,

$$|\chi_c(\tau, \nu)| = |\chi(\tau - \tau_0, \nu - \nu_0)|.$$

Moreover, whenever $(\nu - \nu_0)\tau_0$ is small, we can make the *radar-imaging approximation*

$$\chi_c(\tau, \nu) \approx \chi(\tau - \tau_0, \nu - \nu_0)$$

in which the phase term is ignored.

For another example, suppose that

$$\begin{aligned}s_2(t) &= s(t) \\ s_1(t) &= s(t - \tau_0)e^{j2\pi\nu_0 t} + s(t - \tau_1)e^{j2\pi\nu_1 t}.\end{aligned}$$

Then

$$\chi_c(\tau, \nu) = e^{-j\pi(\nu-\nu_0)\tau_0} \chi(\tau - \tau_0, \nu - \nu_0) + e^{-j\pi(\nu-\nu_1)\tau_1} \chi(\tau - \tau_1, \nu - \nu_1).$$

Whenever both $(\nu - \nu_0)\tau_0$ and $(\nu - \nu_1)\tau_1$ are small, the radar-imaging approximation gives

$$\chi_c(\tau, \nu) \approx \chi(\tau - \tau_0, \nu - \nu_0) + \chi(\tau - \tau_1, \nu - \nu_1).$$

If the main lobes of $\chi(\tau, \nu)$ are resolved, they will be observed individually in $\chi_c(\tau, \nu)$, and the peaks provide estimates of the coordinates τ_0, ν_0 and τ_1, ν_1.

6.7 The sample cross-ambiguity function

The cross-ambiguity function arises as a natural consequence of the matched filter. Consider the delayed complex baseband pulse $s(t - \tau_0)$ received in additive white noise where τ_0 is a known constant. The received signal is

$$v(t) = s(t - \tau_0) + n(t).$$

The matched filter

$$g(t) = s^*(-t - \tau_0)$$

will maximize the signal-to-noise ratio at time zero. The output of the matched filter at time zero is

$$u(0) = \int_{-\infty}^{\infty} v(t)s^*(t - \tau_0)\,dt.$$

More generally, let the received signal be the pulse with both a known delay, τ_0, and a known frequency offset, ν_0,

$$v(t) = s(t - \tau_0)e^{j2\pi\nu_0 t} + n(t)$$

observed in additive white noise $n(t)$. The matched filter for this case is

$$\begin{aligned} g(t) &= [s(-t - \tau_0)e^{-j2\pi\nu_0 t}]^* \\ &= s^*(-t - \tau_0)e^{j2\pi\nu_0 t} \end{aligned}$$

The output of the matched filter at time zero is

$$\begin{aligned} u(0) &= \int_{-\infty}^{\infty} v(\xi)g(-\xi)\,d\xi \\ &= \int_{-\infty}^{\infty} v(t)s^*(t - \tau_0)e^{-j2\pi\nu_0 t}\,dt. \end{aligned}$$

The matched filter for this case is shown in Figure 6.8. The value $u(0)$ is the value of the cross-ambiguity function at $\tau = \tau_0$ and $\nu = \nu_0$. Thus the matched filter output can be obtained from the computed cross-ambiguity function. To emphasize that the input $v(t)$ is a measured noisy signal, this is sometimes called the *sample cross-ambiguity function.*

Now, suppose that the received signal is,

$$v(t) = s(t - \tau_0)e^{j2\pi\nu_0 t} + n(t),$$

as before, but the values of τ_0 and ν_0 are unknown, and we wish to estimate them. The filter

$$g(t) = s^*(-t - \tau)e^{-j2\pi\nu t}$$

will be the matched filter if $\tau = \tau_0$ and $\nu = \nu_0$. The output of the filter will be the value

$$\chi_c(\tau, \nu) = \int_{-\infty}^{\infty} v(t)s^*(t - \tau)e^{-j2\pi\nu t}\,dt,$$

at the location $\tau = \tau_0$, $\nu = \nu_0$. Consequently, if we compute the sample cross-ambiguity function, we know that no linear functional of $v(t)$ can have a larger signal-to-noise

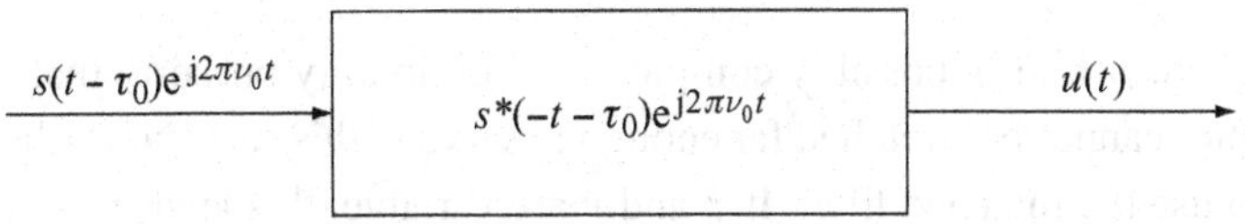

Figure 6.8 Matched filter with time and frequency offset

ratio at $\tau = \tau_0$, $\nu = \nu_0$. We also know that the expected value of $\chi_c(\tau, \nu)$ is

$$\begin{aligned} E[\chi_c(\tau, \nu)] &= \int_{-\infty}^{\infty} E[v(t)]s^*(t-\tau)e^{-j2\pi\nu t}\,dt \\ &= \int_{-\infty}^{\infty} s(t-\tau_0)s^*(t-\tau)e^{j2\pi\nu_0 t}e^{-j2\pi\nu t}\,dt \\ &= e^{-j2\pi(\nu-\nu_0)\tau_0}\chi(\tau-\tau_0, \nu-\nu_0), \end{aligned}$$

which has its maximum magnitude at $\tau = \tau_0$, $\nu = \nu_0$. Consequently, we may estimate (τ_0, ν_0) by computing $\chi_c(\tau, \nu)$ and finding the values of τ and ν at which it achieves its maximum.

The sample cross-ambiguity function is linear in the received signal. Suppose that the received signal consists of the sum of two copies of the known pulse $s(t)$ in the form

$$v(t) = \rho_0 s(t-\tau_0)e^{j2\pi\nu_0 t} + \rho_1 s(t-\tau_1)e^{j2\pi\nu_1 t} + n(t)$$

where ρ_0 and ρ_1 are amplitude parameters. In this case, the optimality property of the matched filter can only be regarded as a suggestion of an estimator because the matched filter was not developed for the sum of two signals. However, it would apply for each signal were the other signal not there. Although there is no assurance of optimality, we are free to process the composite signal $v(t)$ using a cross-ambiguity calculation. One-computes the sample cross-ambiguity function

$$\chi_c(\tau, \nu) = \int_{-\infty}^{\infty} v(t)s^*(t-\tau)e^{-j2\pi\nu t}\,dt$$

because it maximizes the signal-to-noise ratio for each pulse individually. The expected value is

$$E[\chi_c(\tau, \nu)] = \rho_0 e^{-j2\pi(\nu-\nu_0)\tau_0}\chi(\tau-\tau_0, \nu-\nu_0) + \rho_1 e^{-j2\pi(\nu-\nu_1)\tau_1}\chi(\tau-\tau_1, \nu-\nu_1).$$

The magnitude $|E[\chi_c(\tau, \nu)]|$ has a peak at $(\tau, \nu) = (\tau_0, \nu_0)$ and a peak at $(\tau, \nu) = (\tau_1, \nu_1)$, provided the peaks are well resolved. When the peaks are well resolved, it is as if each signal appears in noise uncontaminated by the other signal, and so the matched filter is effectively optimal for each.

In the general case, suppose that $v(t)$ arises as a composite of a continuum of copies of $s(t)$ in the form

$$v(t) = \int_{-\infty}^{\infty}\int_{-\infty}^{\infty} \rho(\tau', \nu')s(t-\tau')e^{j2\pi\nu' t}\,d\tau'\,d\nu' + n(t)$$

for some function, $\rho(\tau', \nu')$, possibly complex. The optimality property that led us to the matched filter cannot be stretched far enough to serve in this case. Nevertheless, we are still free to use the matched filter. If τ and ν are variable, this leads us to compute

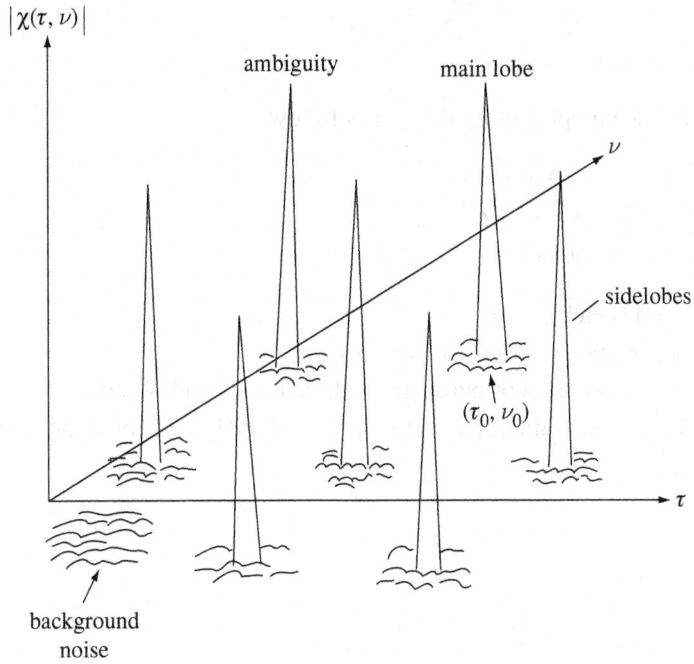

Figure 6.9 Illustrating the sample cross-ambiguity surface

the sample cross-ambiguity function

$$\chi_c(\tau, \nu) = \int_{-\infty}^{\infty} v(t)s^*(t - \tau)e^{-j2\pi\nu t}\,dt.$$

The expected value is

$$E[\chi_c(\tau, \nu)] = \int_{-\infty}^{\infty}\int_{-\infty}^{\infty} \rho(\tau', \nu')e^{-j2\pi(\nu-\nu')\tau'}\chi(\tau - \tau', \nu - \nu')\,d\tau' d\nu'.$$

The radar-imaging approximation, that $(\nu - \nu')\tau'$ is small, is usually satisfied so that the exponential is approximately 1. Then we have the two-dimensional convolution

$$E[\chi_c(\tau, \nu)] = \rho(\tau, \nu) * *\chi(\tau, \nu).$$

Thus, the result of the cross-ambiguity computation has an easily interpreted form. This convenient interpretation provides another partial justification for using the sample cross-ambiguity function. Furthermore, this method of processing is robust and computationally tractable. Moreover, to compute the cross-ambiguity function, there is no need for a prior model of the statistics of $\rho(\tau, \nu)$.

Figure 6.9 shows a sketch of the sample cross-ambiguity function for the case where the reflectivity is a simple impulse, $\rho(\tau, \nu) = \delta(\tau - \tau_0, \nu - \nu_0)$, and the received signal is contaminated by additive noise.

Problems

6.1 Let $s(t)$ be the complex pulse defined as follows:

$$s(t) = \begin{cases} 1 - \mathrm{j} & -T \le t \le 0 \\ 1 + \mathrm{j} & 0 \le t \le T \\ 0 & \text{otherwise.} \end{cases}$$

Compute and sketch $\chi(\tau, \nu)$.

6.2 Let $s(t)$ be a rectangular pulse of width T_1.
Let $S(t)$ be a pulse train of pulse $s(t)$ with PRI $= T_2$ and N pulses.
Let $\mathcal{S}(t)$ be a pulse train of pulse train $S(t)$ with PRI $= T_3$ and R pulse trains.

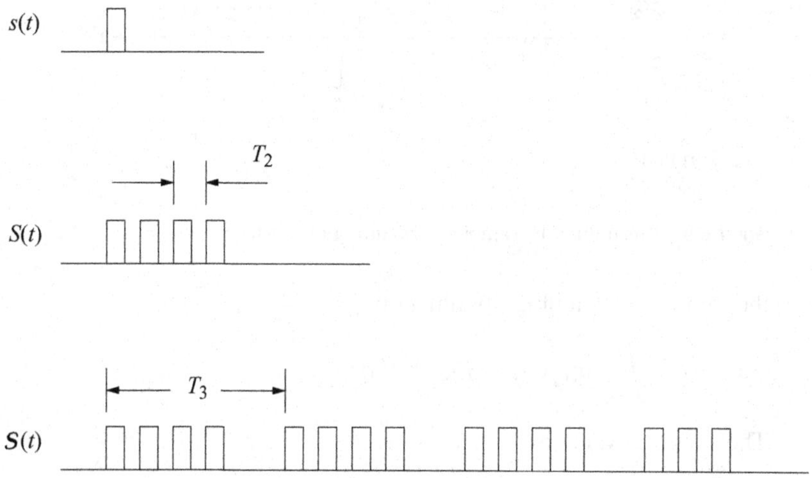

Find $\chi(\tau, \nu)$ for $\mathcal{S}(t)$.

6.3 Compute the ambiguity function of the triangular pulse

$$s(t) = \begin{cases} 1 - |t| & |t| \le 1 \\ 0 & \text{otherwise.} \end{cases}$$

6.4 Derive expressions for the ambiguity functions of $s(t)\cos \pi \alpha t^2$ and $s(t) \sin \pi \alpha t^2$ in terms of the ambiguity function of $s(t)$.

6.5 The *Barker-coded pulse* of length 7, denoted $(+ + + - - + -)$, is sketched as

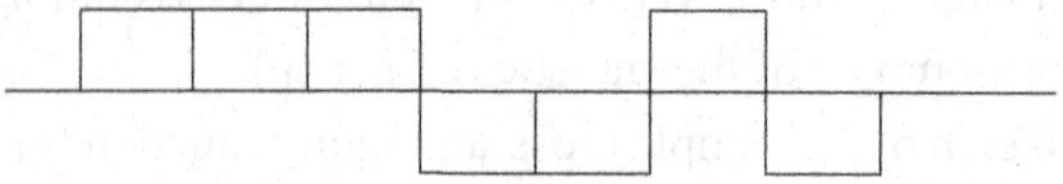

That is, the pulse consists of three counts of $+1$, two counts of -1, one count of $+1$, and one count of -1.

a. Compute the autocorrelation function of this pulse. What is the ratio between the main lobe and the largest sidelobe?

b. Compute and display the ambiguity function of this Barker pulse.

c. Repeat for the Barker-coded pulse of length 13

$$(+, +, +, +, +, -, -, +, +, -, +, -, +)$$

6.6 The *Golay-coded pulse pair* (also called *complementary codes*) of blocklength 4, denoted $((+ + + -), (+ + - +))$, is sketched as

This pair of pulses, denoted $s_1(t)$ and $s_2(t)$, has the property that the sum of their autocorrelation functions has no sidelobes.

a. Compute the two autocorrelation functions of these two pulses. Compute and sketch the sum of the two autocorrelation functions.

b. Compute and display the ambiguity functions of these two pulses. Compute and display the sums of the ambiguity functions.

c. Define the pulse

$$s(t) = s_1(t - T/2) + s_0(t + T/2)$$

for T large. Does $s(t)$ have correlation sidelobes? Does the ambiguity function of $s(t)$ violate the volume property?

d. Repeat for the Golay-coded pulse pair of blocklength 20, given by

$$(+ + + + + - + - - + + + - - + + + - + -)$$

$$(+ - + - - - + + - - + - - + - + + + + +)$$

6.7 A uniform pulse train has N pulses separated in time by a multiple of T_r, and each pulse is stepped in complex frequency by a multiple of Ω. That is,

$$P(t) = \sum_{n=0}^{N-1} p(t - nT_r)e^{-j2\pi n\Omega t}.$$

a. Use the quadratic-phase property of ambiguity functions to find $\chi_P(\tau, \nu)$ in terms of $\chi_p(\tau, \nu)$. How should Ω and T_r be related to make this work out simply?

b. How should $p(t)$ be defined so that $P(t)$ is a simple chirped rectangular pulse? Verify that $\chi_p(\tau, \nu)$, derived in part a, reduces to the ambiguity function of a chirped pulse.

6.8 Derive the ambiguity function of the passband pulse

$$s(t) = s_R(t)\cos 2\pi f_0 t + s_I(t)\sin 2\pi f_0 t$$

in terms of the ambiguity functions and cross-ambiguity function of the modulation components $s_R(t)$ and $s_I(t)$. How does the ambiguity function of the complex baseband pulse

$$s(t) = s_R(t) + js_I(t)$$

relate to the ambiguity function of the passband pulse?

6.9 Let

$$s(t) = \begin{cases} e^{-j2\pi\beta t} & 0 \le t \le T \\ e^{j2\pi\beta t} & -T \le t < 0 \\ 0 & \text{otherwise.} \end{cases}$$

Calculate the ambiguity function of $s(t)$. Is the ambiguity surface symmetric about the τ and ν axes?

6.10 **a.** Prove that

$$\int_{-\infty}^{\infty}\int_{-\infty}^{\infty} |\chi_{12}(\tau, \nu)|^2 \, d\tau \, d\nu = \int_{-\infty}^{\infty} |s_1(t)|^2 \, dt \int_{-\infty}^{\infty} |s_2(t)|^2 \, dt.$$

b. Given the finite-energy signals $s_1(t)$, $s_2(t)$, $s_3(t)$, and $s_4(t)$, prove that the cross-ambiguity functions satisfy

$$\int_{-\infty}^{\infty}\int_{-\infty}^{\infty} \chi_{12}(\tau, \nu)\chi_{34}^*(\tau, \nu)e^{j2\pi(\nu\rho - \mu\tau)} \, d\tau \, d\nu = \chi_{42}(\rho, \mu)\chi_{31}^*(\rho, \mu).$$

6.11 The cross-ambiguity function of $s_1(t)$ and $s_2(t)$ is given by

$$\chi_c(\tau, \nu) = \int_{-\infty}^{\infty} s_1(t + \tau/2)s_2^*(t - \tau/2)e^{-j2\pi\nu t} \, dt.$$

a. Prove that

$$s_1(t) = \frac{1}{s_2^*(0)} \int_{-\infty}^{\infty} \chi_c(t, \nu)e^{j\pi\nu t} \, d\nu.$$

b. Prove that unless $s_1(t) = cs_2(t)$ for all t for some constant c, the cross-ambiguity function is not equal to the ambiguity function $\chi(\tau, \nu)$ of the pulse $p(t)$ for any $p(t)$.

6.12 Prove that if $\chi_1(\tau, \nu)$ and $\chi_2(\tau, \nu)$ are ambiguity functions, then the sum $a\chi_1(\tau, \nu) + b\chi_2(\tau, \nu)$ is an ambiguity function if and only if $\chi_1(\tau, \nu) = c\chi_2(\tau, \nu)$ for some constant c. (In the language of geometry, we say that the space of ambiguity functions contains no "lines" except those through the origin.)
Hint: Use the volume property in the form

$$\int_{-\infty}^{\infty}\int_{-\infty}^{\infty} | \, a\chi_1(\tau, \nu) + b\chi_2(\tau, \nu) \, |^2 \, d\tau \, d\nu = [aE_1 + bE_2]^2.$$

6.13 **a.** Show that if

$$r(t) = s(t - \Delta)e^{-j(\alpha t+\beta)},$$

then

$$|\chi_r(\tau, \nu)| = |\chi_s(\tau, \nu)|.$$

b. Let $p(t)$ be a pulse that is equal to zero for $|t| > T/2$. Define

$$s(t) = p(t) + p(t - 2T)$$
$$r(t) = p(t) - p(t - 2T).$$

Show that $|\chi_r(\tau, \nu)| = |\chi_s(\tau, \nu)|$.

c. Show that the set of all pulses corresponding to a given ambiguity surface can be rather large.

6.14 **a.** Prove the stronger form of the Gabor uncertainty principle given by

$$T_G^2 B_G^2(1 - \rho^2) \geq \frac{1}{(4\pi)^2}.$$

b. Prove that the uncertainty ellipse

$$B_G^2\tau^2 + 2B_G T_G \rho\tau\nu + T_G^2\nu^2 = \frac{1}{4\pi^2}$$

has an area not larger than one with equality if and only if the pulse is gaussian.

6.15 Prove that, if the uncertainty ellipse

$$B_G^2\tau^2 + T_G^2\nu^2 = \frac{1}{4\pi^2}$$

of pulse $s(t)$ has area A, then the uncertainty ellipse of $s(t)e^{j\pi\alpha t^2}$ also has area A.

6.16 Let $a(t)$ be a real, nonnegative function and

$$s(t) = a(t)e^{j\theta(t)}.$$

a. Show that

$$a(t)^2 = \int_{-\infty}^{\infty} \chi(0, \nu)e^{j2\pi\nu t}\, d\nu.$$

b. Show that if $a(t) = 1$,

$$\dot{\theta}(t) = -j\int_{-\infty}^{\infty} \frac{\partial\chi(0, \nu)}{\partial\tau}e^{j2\pi\nu t}\, d\nu.$$

6.17 Find a "firing sequence" for a constant amplitude waveform of duration $16T$ based on a 16 by 16 Costas array. Describe the uncertainty ellipse of this waveform. Describe approximately the relationship between the sidelobes and the mainlobe.

6.18 Let $s'(t) = h(t) * s(t)$. Express $\chi'(\tau, \nu)$ in terms of $\chi(\tau, \nu)$ and $H(f)$.

6.19 Find the Wigner distribution of the pulse $\operatorname{rect}(t/T)$.

6.20 Show that the pair of conditions

$$i) \int_{-\infty}^{\infty} \int_{-\infty}^{\infty} |s(x, y)|^2 \, dx \, dy = C$$

$$ii) \int_{-\infty}^{\infty} \int_{-\infty}^{\infty} s(x, y) e^{-j2\pi(x\xi - y\eta)} \, dx \, dy = s(\xi, \eta)$$

is satisfied by an infinite number of $s(x, y)$. What is the nature of such an $s(x, y)$?

6.21 **a.** Suppose that $s(t)$ has the Fourier transform $S(f)$ and $s'(t)$ has the Fourier transform $S'(f) = S(f)e^{j\pi\beta f^2}$. Prove that, if

$$s(t) \to \chi(\tau, \nu),$$

then

$$s'(t) \to \chi(\tau + \beta\nu, \nu).$$

b. Using part a and the quadratic-phase property, prove that if $\chi(\tau, \nu)$ is an ambiguity function, then for any ϕ,

$$\chi'(\tau, \nu) = \chi(\tau \cos\phi - \nu \sin\phi, \tau \sin\phi + \nu \cos\phi),$$

is also an ambiguity function.

6.22 Let $\Xi_s(f, t)$ denote the Wigner distribution of the pulse $s(t)$. Prove the following:

a. $s(t - t_0)e^{-j2\pi f_0 t}$ has Wigner distribution $\Xi_s(f - f_0, t - t_0)$.
b. $\sqrt{\gamma} s(\gamma t)$ has Wigner distribution $\Xi_s(f/\gamma, \gamma t)$.
c. $s(t)e^{j\pi\alpha t^2}$ has Wigner distribution $\Xi(f - \alpha t, t)$.
d. $s(t)r(t)$ has Wigner distribution $\Xi_s(f, t) *_f \Xi_r(f, t)$.

6.23 **(Apodization)** To reduce the doppler sidelobes of a sample ambiguity function, the definition is modified to the alternative form

$$\chi_c(\tau, \nu) = \int_{-\infty}^{\infty} h(t)s(t)v^*(t - \tau)e^{-j2\pi\nu t} \, dt$$

where $h(t)$ is a function called an *apodizing window*. Let $s(t) = v(t) = \operatorname{rect}(t/T)$, and choose the window

$$h(t) = \cos\pi\frac{t}{T} \quad |t| \leq T/2.$$

a. By introducing $h(t)$, what happens to the first sidelobe?
b. By introducing $h(t)$, what happens to main lobe?
c. By introducing $h(t)$, what happens to the output noise?

6.24 An ambiguity function can be defined for a function of two variables. Let $s(x, y)$ be a bivariate function whose energy $E_p = \int_{-\infty}^{\infty} \int_{-\infty}^{\infty} |s(x, y)|^2 \, dx \, dy$ is finite.

Define

$$\chi(\tau_x, \tau_y, \nu_x, \nu_y) = \int_{-\infty}^{\infty}\int_{-\infty}^{\infty} s\left(x - \frac{\tau_x}{2}, y - \frac{\tau_y}{2}\right) s^*\left(x + \frac{\tau_x}{2}, y + \frac{\tau_y}{2}\right) e^{-j2\pi(\nu_x x + \nu_y y)}\, dx\, dy.$$

What is the ambiguity function of $s(x, y)e^{j\pi\alpha(x^2+y^2)}$ expressed in terms of the ambiguity function of $s(x, y)$?

6.25 The spatial ambiguity function of the two-dimensional pulse $s(x, y)$ is the function

$$\chi(\tau_x, \tau_y, \nu_x, \nu_y) = \int_{-\infty}^{\infty}\int_{-\infty}^{\infty} s\left(x - \frac{\tau_x}{2}, y - \frac{\tau_y}{2}\right) s^*\left(x + \frac{\tau_x}{2}, y + \frac{\tau_y}{2}\right) e^{-j2\pi(\nu_x x + \nu_y y)}\, dx\, dy.$$

a. Is there a quadratic-phase property for the spatial ambiguity function?

b. Suppose that $s(x, y)$ is propagated under Fresnel diffraction. How does the spatial ambiguity function change?

c. Show that the spatial autocorrelation function and the power density spectrum can be recovered from $\chi(\tau_x, \tau_y, \nu_x, \nu_y)$. Describe the "interchange" of these two quantities under Fresnel propagation.

d. Suppose that $s(x, y)$ is passed through an ideal lens. How does the spatial ambiguity function change?

e. How does the spatial ambiguity function change under the cascade of three operations: Fresnel diffraction, followed by an ideal lens, followed by Fresnel diffraction? Does this provide an alternative proof of the lens law?

Notes

The important role that the ambiguity function plays in the design of radar waveforms and radar systems was recognized as early as 1953 by Woodward, but the ambiguity function had been defined earlier by Ville (1948). The basic theorems about the ambiguity function were developed by Siebert (1956, 1958), and also by Lerner (1958), Wilcox (1960), and Price and Hofstetter (1965). Sussman (1962) studied the synthesis of ambiguity waveforms that approximate a desired ambiguity function in the least-squares sense. Klauder (1960) related the ambiguity function to the Wigner distribution, a function that is important within the subject of quantum mechanics. Papoulis (1974) described the role of the ambiguity function in Fourier optics. The inverse problem of determining whether an arbitrary complex function of two variables is an ambiguity function is very difficult, as is the problem of finding a pulse, $s(t)$, that corresponds to a given ambiguity surface. This problem was studied by DeBuda (1970). The algebraic properties of the mapping that takes the function $s(t)$ into its ambiguity function were studied by Auslander and Tolimieri (1985).

The uncertainty principle for waveforms was discussed by Gabor (1946), and by Kay and Silverman (1957). The uncertainty ellipse was named by Helstrom (1960). Additional properties of the ambiguity function were described by Rihaczek (1965) and Grünbaum (1984). The term "chirp" was attributed to Oliver by Klauder, Price, Darlington, and Albersheim (1960). Costas (1984) introduced the Costas array as one way of designing a frequency-hopping pulse with a good ambiguity function. Golomb and Taylor (1984) surveyed known methods for constructing Costas arrays. Efficient digital computation of the ambiguity function of an arbitrary waveform has been studied by Tolimieri and Winograd (1985).

7 Radar imaging systems

A conventional radar consists of a transmitter that illuminates a region of interest, a receiver that collects the signal reflected by objects in that region, and a processor that extracts information of interest from the received signal. A radar processor consists of a preprocessor, a detection and estimation function, and a postprocessor. In the preprocessor, the signal is extracted from the noise, and the entire signal reflected from the same resolution cell is integrated into a single statistic. An *imaging radar* uses the output of the preprocessor to form an image of the observed scene for display. A *detection radar* makes further inferences about the objects in the scene. The detection and estimation function is where individual target elements are recognized, and parameters associated with these target elements are estimated. The postprocessor refines postdetection data by establishing track histories on detected targets.

This chapter is concerned with the preprocessor, which is an essentially linear stage of processing at the front end of the processing chain. The radar preprocessor usually consists of the computation of a sample cross-ambiguity function in some form. Sometimes the computation is in such a highly approximated form that it will not be thought of as the computation of a cross-ambiguity function. The output of the preprocessor can be described in a very compact way, provided that several easily satisfied approximations hold. The output is the two-dimensional convolution of the reflectivity density of the radar scene and the ambiguity function of the transmitted waveform. Thus the ambiguity function plays the role of a two-dimensional pointspread function. This unifying description of the preprocessor output is very powerful, and it makes many aspects of the radar performance clearly evident, especially when rectangular coordinates are appropriate to the situation. An alternative description of radar imaging, which is based on the projection-slice theorem and is more suitable when polar coordinates are appropriate, is given in Chapter 10.

7.1 The received signal

Electromagnetic signals travel through space at a finite speed. In free space, they travel at the free space speed of light c (approximately 3×10^8 meters per second).

Electromagnetic waves consist of propagating vector fields. Because the electric field is a vector, it has an orientation, called the *polarization* of the signal, that is orthogonal to the direction of propagation, but otherwise is arbitrary. Because the polarization is another degree of freedom in the radar signal, it sometimes carries useful information. Unlike radar and optical waves, sonar, seismic, and acoustic waves do not have a polarization. For the most part, we shall not be concerned with the polarization.

Propagation of electromagnetic waves through free space, and the interactions of these waves with reflecting objects and with antennas, are described by using Maxwell's equations. These phenomena are studied in the science of electrodynamics and can be quite complex. In this section, we shall adopt the signal processor's simplified point of view, modeling the signal at the antenna input or at the receiver input to include all pertinent effects, but ignoring the details of the electromagnetic phenomena that give rise to these effects. Thus, for the purposes of this chapter, it suffices to treat an antenna as a linear element, joining free space to the receiver, whose gain is a known function of the angle of arrival and the polarization. The calculation of this function from more elementary considerations will not concern us in this section.

The received signal is presumed to be unknown. If the signal were completely known, there would be no point in receiving it. However, if it were completely unknown, a receiver could not be intelligently designed. This suggests that, although the received signal is not known, it will conform to an appropriate prior parametric model. Common prior models have the form

$$v(t) = s(t, \gamma) + n(t)$$

where the noise-free received signal $s(t, \gamma)$ is dependent on the parameter γ which is the term of interest, and $n(t)$ is additive random noise, commonly gaussian, arising in the receiver electronics. The parameter γ might be rather simple, such as a single real number or a vector of real numbers, or might be as elaborate as a two-dimensional image, or perhaps something even more complicated. For image formation, which is our interest, the parameter γ usually denotes a two-dimensional function $\rho(x, y)$ of the spatial coordinates x and y.

There are two models of reflection by which $s(t, \gamma)$ can contain information about a scene. These are referred to as *specular reflection* and *diffuse reflection*, respectively. In the case of a specular reflector, $s(t, \gamma)$ is modeled as an unknown deterministic function of the parameter γ. In the case of a diffuse reflector, $s(t, \gamma)$ is modeled as an unknown random function of the parameter γ with zero mean and an unknown variance that may be a function of the spatial coordinates x, y. Diffuse reflectors are studied in Chapter 11. In this chapter, we shall study only specular reflectors. The two cases can be combined by modeling a reflector as a gaussian random variable for which the variance is zero for a specular model and for which the mean is zero for the diffuse model.

Typically, specular reflections are due to objects whose significant details are large in comparison to the wavelength of the illumination, while diffuse reflections are due to objects whose significant details are small in comparison to a wavelength. These are sometimes referred to, more simply, as *reflection* and *backscatter*, respectively, though often these terms are not carefully distinguished, and may even be used interchangeably.

If the complex passband signal $s(t)$ is transmitted at one point, then the signal $s'(t)$ received at a distant point at distance R is

$$s'(t) = as(t - R/c).$$

If the complex passband signal $s(t)e^{-j2\pi f_0 t}$ is transmitted at one point, then the complex passband signal received at a distant point at distance R is

$$s'(t)e^{-j2\pi f_0 t} = as(t - R/c)e^{-j2\pi f_0(t-R/c)},$$

or, at complex baseband

$$s'(t) = as(t - R/c)e^{j2\pi f_0 R/c}.$$

This form of the received signal at complex baseband, with the carrier delay explicitly displayed, is the most convenient form with which to deal.

The signal $s'(t)$ differs from $s(t)$ by virtue of an amplitude attenuation, a, and a time delay, R/c. The attenuation a may be a predictable consequence of the antenna gain pattern and the inverse square-law attenuation experienced by propagating radiation. Then it can be offset by amplification in the receiver. In such cases, the received signal is written

$$s'(t) = s(t - R/c)e^{j2\pi f_0 R/c}.$$

The attenuation is not written into the equations in most formulations of the signal processing because it plays no essential role other than establishing the gain needed in the receiver.

A radar system makes use of echoes, so we need to consider the range between the transmitter and the reflector and also the range between the reflector and the receiver. Thus, if a signal, $s(t)$, is transmitted and travels the distance R_1 to a reflector, and the echo travels the distance R_2 to a receiver, then the received signal is

$$s'(t) = a\rho s(t - (R_1 + R_2)/c)e^{j2\pi f_0(R_1+R_2)/c}.$$

In a monostatic radar system, the transmitter and the receiver are together, so $R_1 = R_2 = R$. In monostatic radar, the time delay is $2R/c$, and the amplitude attenuation is $a\rho$ where the constant ρ represents the fraction of the signal incident on the reflector that is actually reflected toward the receiver, and a represents the signal attenuation from all other causes. Again, the attenuation may be predictable. As before, we omit the attenuation from the received signal on the presumption that it is offset by amplification in the receiver. Then the received signal is

$$s'(t) = \rho s(t - 2R/c)e^{j4\pi f_0 R/c}.$$

The parameter ρ is called the *reflectivity*. The parameter ρ is included explicitly because, in most problems of interest, the reflectivity cannot be predicted; it must be treated as an unknown parameter. Indeed, observation of an unknown ρ can be regarded as the usual purpose of a radar. In general, reflectivity is a complex number because there can be a phase shift during the process of reflection. The *radar cross section* of the reflector is defined as $\sigma = |\rho|^2$.

Let us consider reflection in more detail. A physical object has spatial extent, and also has an appearance that depends on the aspect from which it is viewed. Suppose that a sinusoidal signal, such as an electromagnetic wave, is incident on a surface. A portion of this signal is reflected back in the direction of incidence, and a portion is scattered into other directions. We can expect reflection from a real object to be quite complicated. It may be beyond our abilities to completely predict the reflection by even fairly simple objects, although numerical computational methods can often give satisfactory predictions. One can sometimes postulate a parametric model for reflection with parameters that can be filled in by measurement or computation. For our purposes, we will consider only ideal reflectors that, though they may be of complex shape, reflect signals instantly with no internal memory. Thus we will ignore the possibility of certain artifacts that sometimes occur in a radar return signal, such as those due to multiple reflections or electrical currents induced in the reflector. In general, the reflectivity may be dependent on carrier frequency. More generally, the reflectivity varies with time because the aspect, and therefore the scattering interaction with the reflector, will vary as it moves or rotates.

If the spatial details of a reflector are comparable to the resolution of a system, the reflectivity is replaced by the reflectivity density $\rho(x, y)$. Let $s(t)\Delta x \Delta y$ be an incident complex signal on a small cell of area $\Delta x \Delta y$ centered at the point with coordinates x, y. The complex reflected signal from that cell is generally an attenuated and phase-shifted copy of the incident signal. The reflected signal from a small cell at x, y will be proportional to the incident signal by a factor $\rho(x, y)$ called the *reflectivity density*. The reflectivity density may be complex, because it may include a phase shift.

The *radar cross-section density* at x, y is a real, positive function given by

$$\sigma(x, y) = |\rho(x, y)|^2.$$

The integral of $\sigma(x, y)$ over an object or a portion of an object is the radar cross section of that object or a portion of that object.

The reflectivity density and the radar cross-section density also depend on the angle of incidence of the incident radiation, as measured by the angular coordinates ϕ, ψ. To make this dependence explicit, we may write $\rho(x, y, \phi, \psi)$ and $\sigma(x, y, \phi, \psi)$.

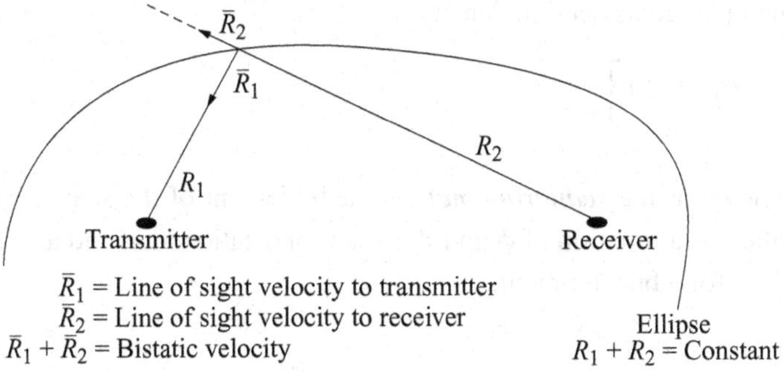

Figure 7.1 Geometry of a bistatic radar

Bistatic reflection

One may also be interested in applications in which the transmitter and the receiver are located in different places. One reason might be so that a receiver near a radar target can be passive and covert, while the transmitter, though active, can be far away. This is called a *bistatic radar* and is illustrated in Figure 7.1. Then one is interested in the signal from direction (ϕ, ψ) that is reflected into direction (ϕ', ψ'). In this way, we can also define the *bistatic* (or bidirectional) *reflectivity density* $(x, y, \phi, \psi, \phi', \psi')$ and the bistatic radar cross-section density $\sigma(x, y, \phi, \psi, \phi', \psi')$.

The following properties of the bistatic cross section may help ease the study of the bistatic case.

- For sufficiently smooth and perfectly conducting bodies, in the limit of small wavelength, the bistatic cross section is approximately the monostatic cross section at the bisector of the bistatic angle between the direction to the transmitter and receiver.
- The bistatic cross section is unchanged if the positions of the transmitter and receiver are interchanged.

Polarization

The reflectivity depends on the polarization of the incident signal. One may consider the vertical and horizontal components of the polarization of the incident signal, and the vertical and horizontal components of the polarization of the reflected signal. Consequently, there are actually four reflectivity density functions. These can be arranged in a matrix of functions,

$$\rho = \begin{bmatrix} \rho_{11}(x, y) & \rho_{12}(x, y) \\ \rho_{21}(x, y) & \rho_{22}(x, y) \end{bmatrix}.$$

Similarly, for the radar cross-section density,

$$\sigma = \begin{bmatrix} \sigma_{11}(x, y) & \sigma_{12}(x, y) \\ \sigma_{21}(x, y) & \sigma_{22}(x, y) \end{bmatrix}.$$

This matrix is known as the *scattering matrix*. Each element of the scattering matrix may also be written as a function of ϕ and ψ for a monostatic radar, and as a function of ϕ, ψ, ϕ', and ψ' for a bistatic radar.

Delay and doppler

A signal $s(t)$ that travels from location $(x'(t), y'(t), z'(t))$ at time t to location $(x(t), y(t), z(t))$ travels the distance

$$R(t) = \sqrt{(x'(t) - x(t))^2 + (y'(t) - y(t))^2 + (z'(t) - z(t))^2}.$$

The complex baseband signal at the second location is the retarded signal

$$s'(t) = s(t - R(t)/c)e^{j2\pi f_0 R(t)/c}.$$

In a bistatic radar system, the transmitter, receiver, and reflector are all at different points. The transmitter is at location $(x'(t), y'(t), z'(t))$, the echo is from an object at location $(x(t), y(t), z(t))$, and the receiver is at location $(x''(t), y''(t), z''(t))$. Then we have two range expressions:

$$R_1(t) = \sqrt{(x'(t) - x(t))^2 + (y'(t) - y(t))^2 + (z'(t) - z(t))^2}$$
$$R_2(t) = \sqrt{(x''(t) - x(t))^2 + (y''(t) - y(t))^2 + (z''(t) - z(t))^2},$$

and the received complex baseband signal is

$$s'(t) = s(t - (R_1(t) + R_2(t))/c)e^{j2\pi f_0 (R_1(t)+R_2(t))/c}.$$

In a monostatic radar system, the transmitter and receiver are at the same place, and

$$s'(t) = s(t - 2R(t)/c)e^{j4\pi f_0 R(t)/c}.$$

The three cases are subsumed by one equation,

$$s'(t) = s(t - \tau(t))e^{j2\pi f_0 \tau(t)},$$

where

$$\tau(t) = R(t)/c,$$

or

$$\tau(t) = (R_1(t) + R_2(t))/c,$$

or

$$\tau(t) = 2R(t)/c.$$

Whenever there is motion in the system, $R(t)$ is not constant, and the retarded signal may become quite involved because of the time-varying delay. In the simplest of such cases, with constant velocity V,

$$R(t) = R + Vt$$

is an adequate approximation. In the case of a single range delay, the received complex baseband signal is

$$s'(t) = s(t - (R + Vt)/c)e^{j2\pi f_0(R+Vt)/c}.$$

The doppler shift f_0V/c and the phase shift f_0R/c in the exponent are the contributions from the complex carrier. The modulating pulse $s(t)$ undergoes a compression or expansion of the time axis (depending on the sign of V), and a range delay. In sonar applications, the dilation of the time axis may be important. In radar applications, the velocity V is very small compared to c so that most properties of $s(t)$ are scarcely changed.

By ignoring the dilation term Vt/c within the pulse, the pulse $s'(t)$ can be expressed more compactly as the complex baseband signal

$$s'(t) = s(t - \tau_0)e^{j2\pi\nu_0 t}e^{j2\pi\theta}$$

where $\tau_0 = R/c$ is a modulation delay, $\nu_0 = f_0V/c$ is a frequency offset, and $\theta = f_0\tau_0$ is a phase offset. The same equation can also be used for the bistatic radar case by setting $\tau_0 = (R_1 + R_2)/c$ and $\nu_0 = f_0(\dot{R}_1 + \dot{R}_2)/c$, and for the monostatic radar case by setting $\tau_0 = 2R/c$ and $\nu_0 = 2V/c$.

The dilation of the time axis might be unnoticeable in the modulation term $s(t)$, but, in the carrier term, it appears as a frequency shift that is quite noticeable. This frequency offset $\nu_0 = f_0V/c$ is known as the *doppler shift* (after the Austrian physicist Christian Doppler, 1803–1853). Typically, in radar systems, V/c is in the range of 10^{-6} to 10^{-7}. Then, for f_0 in the VHF band (above 30×10^6 hertz), the doppler shift is tens of hertz or greater. This is easily measured with a properly designed waveform and suitable equipment.

In general, $R(t)$ is not a straight line. Often $R(t)$ can be adequately approximated by the first several terms of the Taylor series expansion

$$R(t) = R_0 + \dot{R}_0t + \tfrac{1}{2}\ddot{R}_0t^2 + \cdots$$

The acceleration term, and higher-order terms, are sometimes significant, although in most applications the elementary narrowband delay-doppler approximation is entirely adequate. Even when higher order terms are needed, it may suffice to make simple compensation corrections to the received signal and then proceed by using the delay-doppler approximation. In Section 7.2, we shall study the processing of a signal under the delay-doppler approximation. In Section 7.3, we shall study focusing and motion compensation corrections to account for other terms.

7.2 The imaging equation

The ambiguity function was studied in Chapter 6. This function enables a mathematical description of the output of a radar preprocessor in terms of two-dimensional filter theory. Thus, the expected value of the sample cross-ambiguity function is, within good approximations, described as a filtered version of the scene. This expected value is related to the reflectivity density in range-doppler coordinates by a two-dimensional convolution of $\rho(\tau, \nu)$ with the ambiguity function of the imaging waveform $\chi(\tau, \nu)$. This relationship is so useful that we will make it a definition.

Definition 7.2.1 *The radar imaging equation for the pulse $s(t)$ is the two-dimensional convolution*

$$r(\tau, \nu) = \chi(\tau, \nu) ** \rho(\tau, \nu)$$

of the reflectivity density $\rho(\tau, \nu)$ with the ambiguity function $\chi(\tau, \nu)$ of the pulse $s(t)$.

The radar imaging equation is exact because it is a definition. From the radar-imaging approximation, we know that the radar imaging equation, though not an exact description of the result of the cross-ambiguity computation, is usually a close approximation. The radar imaging equation defines the radar image as the two-dimensional scene $\rho(\tau, \nu)$ seen through the two-dimensional filter whose pointspread function is $\chi(\tau, \nu)$. This filtering viewpoint is very powerful because it submerges all the details of the processing into a simple formula that exposes the underlying limit of resolution. This is illustrated in Figure 7.2. If $\chi(\tau, \nu)$ were an impulse, the radar image would equal $\rho(\tau, \nu)$. But $\chi(\tau, \nu)$ is not an impulse, so the convolution will blur the details of $\rho(\tau, \nu)$ by the main lobe of $\chi(\tau, \nu)$. If $\chi(\tau, \nu)$ has ambiguities, multiple copies of $\rho(\tau, \nu)$ will be found in $r(\tau, \nu)$. Based on this formulation as a two-dimensional filter, one can understand methods for the manipulation and processing of radar images by using two-dimensional Fourier transform techniques.

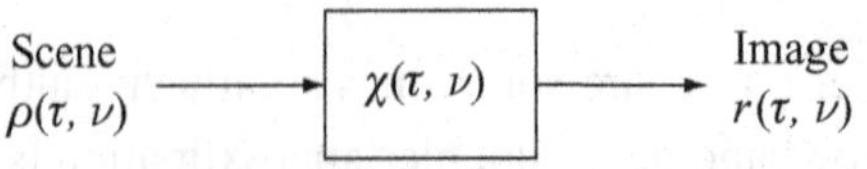

Figure 7.2 A filtering view of imaging

This relationship of the cross-ambiguity function

$$\chi_c(\tau, \nu) = \int_{-\infty}^{\infty} v(t)s^*(t-\tau)e^{-j2\pi\nu t}\,dt$$

to the radar imaging equation will be revisited by setting

$$v(t) = \sum_i \rho_i s(t-\tau_i)e^{j2\pi\nu_i t}.$$

Then, the cross-ambiguity function is

$$\chi_c(\tau, \nu) = \sum_i \rho_i e^{-j\pi(\nu-\nu_i)\tau_i}\chi(\tau-\tau_i, \nu-\nu_i).$$

Whenever the points (τ_i, ν_i) are widely separated in the τ, ν plane compared to the width of the main lobe of $\chi(\tau, \nu)$, the approximation

$$|\chi_c(\tau, \nu)| \approx \sum_i |\rho_i|\,|\chi(\tau-\tau_i, \nu-\nu_i)|$$

is appropriate, and the phase term is eliminated. Even if the points (τ_i, ν_i) are not sparse, the radar approximation

$$\chi_c(\tau, \nu) = \sum_i \rho_i\chi(\tau-\tau_i, \nu-\nu_i)$$

applies whenever the terms $(\nu-\nu_i)\tau_i$ are small.

If instead, there is a continuum of reflectors so that

$$v(t) = \int_{-\infty}^{\infty}\int_{-\infty}^{\infty} \rho(\tau', \nu')s(t-\tau')e^{j2\pi\nu' t}\,d\tau'\,d\nu',$$

then

$$\chi_c(\tau, \nu) = \int_{-\infty}^{\infty}\int_{-\infty}^{\infty} \rho(\tau', \nu')e^{-j\pi(\nu-\nu')\tau'}\chi(\tau-\tau', \nu-\nu')\,d\tau'\,d\nu'.$$

In most high-resolution imaging applications, $(\nu-\nu')$ is a few hertz and τ' is a few milliseconds. Therefore, $(\nu-\nu')\tau'$ is small for the region where the integral is significant, and we can make the radar-imaging approximation

$$\chi_c(\tau, \nu) \approx \int_{-\infty}^{\infty}\int_{-\infty}^{\infty} \rho(\tau', \nu')\chi(\tau-\tau', \nu-\nu')\,d\tau'\,d\nu',$$

which is the imaging equation. Even if the radar-imaging approximation fails far away from the origin, it does not spoil any of the general conclusions about the structure of $\chi_c(\tau, \nu)$, which are largely due to the main lobe of $\chi(\tau, \nu)$. However, the approximation does interfere with any attempt to recover $\rho(\tau, \nu)$ by deconvolution methods.

The imaging equation views the reflectivity density as a two-dimensional signal that is the input to a two-dimensional filter, and the image as the two-dimensional signal that is the output of the filter. The pointspread function of the two-dimensional filter is the ambiguity function $\chi(\tau, \nu)$ of the waveform $s(t)$. If it were possible, one would choose the pointspread function $\chi(\tau, \nu)$ to be a two-dimensional impulse. Then the image, as given by the radar-imaging equation, would be a multiple of the reflectivity density function of the scene, $\rho(\tau, \nu)$. In fact, the resolution in the scene is limited by the resolution of the imaging pointspread function $\chi(\tau, \nu)$. The resolution in the range direction is determined by the τ width of the main lobe of $\chi(\tau, \nu)$; the resolution in the doppler direction is determined by the ν width of the main lobe of $\chi(\tau, \nu)$. The sidelobes and the grating lobes produce various processing artifacts in the image.

The radar-imaging equation makes it easy to see the effect of the major sidelobes that arise in the ambiguity function of a pulse train. Such an ambiguity function has grating lobes in the doppler direction, and delay ambiguities in the delay direction. To study these ambiguities, approximate the grating lobes as impulses. Then, for the ambiguity function of a uniform pulse train,

$$\chi(\tau, \nu) \approx \sum_{i=-I}^{I} \sum_{j=-J}^{J} \delta\left(\tau - iT_r, \nu - j\frac{1}{T_r}\right),$$

which is a two-dimensional array of impulses. The image

$$r(\tau, \nu) = \rho(\tau, \nu) \;**\; \chi(\tau, \nu)$$

then has the form of an array of copies of $\rho(\tau, \nu)$. If $\rho(\tau, \nu)$ is confined to a rectangle, as shown in Figure 7.3, of width T_r in the τ direction and width T_r^{-1} in the ν direction, then there will be no image aliasing and the ambiguities will not be of major importance.

The sampling interval T_r has to be large enough to encompass the delay spread of the scene without folding, and the sampling frequency T_r^{-1} has to be large enough to encompass the doppler spread of the scene without folding. Since making one of these terms larger makes the other smaller, we must compromise in the choice of T_r. This compromise is at the heart of the design of pulse-train waveforms. In some cases, it may

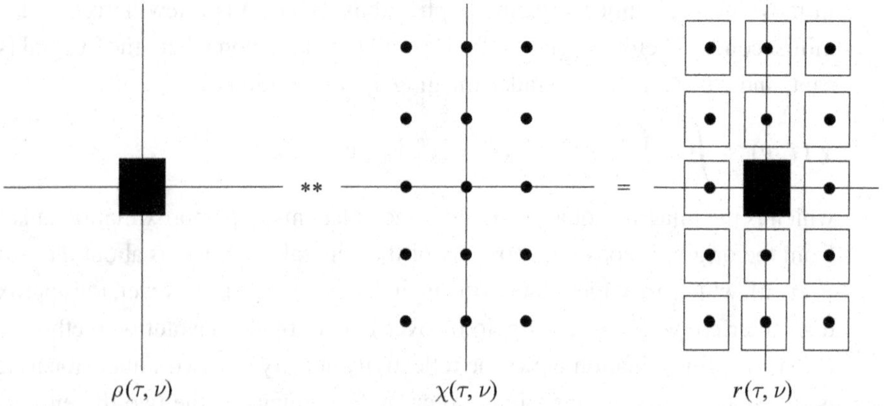

Figure 7.3 Illustrating an image without image aliasing

not be possible to satisfy simultaneously the constraints on both the delay and doppler. Then one must either abandon the use of a pulse waveform or accept the existence of artifacts in the image caused by aliasing.

7.3 Imaging resolution

The quality of a radar image is determined by the ambiguity function of the imaging waveform. The ambiguity function describes ghosts due to major sidelobes, predicts self-clutter due to minor sidelobes, and, most importantly, determines the resolution. The resolution can be understood by considering a point reflector at the center of a scene of interest. Choosing the center of the scene as the origin of the coordinate system, a point reflector at the origin, $\rho(x, y) = \delta(x, y)$, has a complex image equal to the ambiguity function:

$$r(x, y) = \chi(\tau(x, y), \nu(x, y)).$$

The resolution of the image is determined by the width of the main lobe of $r(x, y)$. We can visualize this main lobe in the shape of the uncertainty ellipse, which can be used as one descriptor of resolution. The arguments of the ambiguity function are related to x and y by

$$\tau(x, y) = \frac{2}{c} R(x, y)$$

$$\nu(x, y) = 2\frac{f_0}{c} \dot{R}(x, y).$$

Consequently, imaging performance is also determined by the functions $\tau(x, y)$ and $\nu(x, y)$.

Choose any measure of resolution. When applied to τ or ν, the resolution is related to the resolution in R or $\dot{R}$ by

$$\Delta\tau = \frac{2}{c}\Delta R$$

$$\Delta\nu = 2\frac{f_0}{c}\Delta\dot{R}.$$

In turn, the right side of these equations can be related to the resolution in the x and y coordinate axes as follows

$$\Delta\tau = \frac{2}{c}\left[\frac{\partial R}{\partial x}\Delta x + \frac{\partial R}{\partial y}\Delta y\right]$$

$$\Delta\nu = 2\frac{f_0}{c}\left[\frac{\partial \dot{R}}{\partial x}\Delta x + \frac{\partial \dot{R}}{\partial y}\Delta y\right].$$

For example, a common geometrical configuration, known as a *sidelooking radar*, is shown in Figure 7.4. This origin is at the center of the scene. The velocity vector is

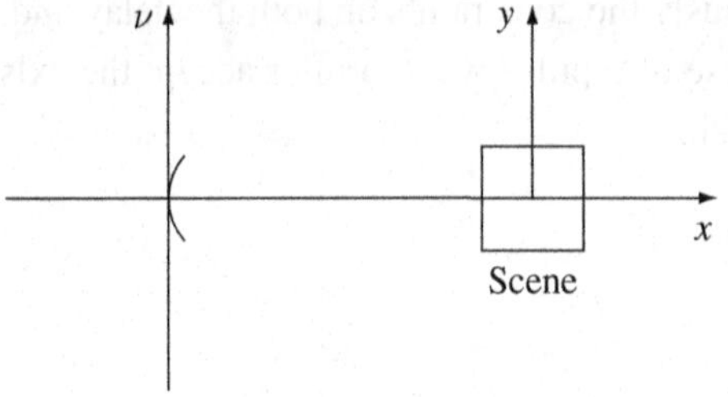

Figure 7.4 Sidelooking geometry

parallel to the y coordinate axis. The radar antenna at time zero is on the x coordinate axis pointing in the direction toward the center of the scene and moves with velocity V. The relationships now take the simple forms

$$\begin{aligned}\tau(x, y) &= \frac{2}{c} R(x, y, x', y') \\ &\approx \frac{2}{c}(x - x'),\end{aligned}$$

and

$$\begin{aligned}\nu(x, y) &= 2\frac{f_0}{c} \dot{R}(x, y, x', y') \\ &= 2\frac{f_0}{c} V \frac{y}{R}.\end{aligned}$$

The expressions for spatial resolution then become

$$\begin{aligned}\Delta x &= \frac{c}{2}\Delta\tau \\ \Delta y &= \frac{R}{V}\frac{c}{2f_0}\Delta\nu = \frac{\lambda R}{2V}\Delta\nu.\end{aligned}$$

The reciprocal of the waveform bandwidth is often used as a measure of the delay resolution, and the reciprocal of the waveform duration is often used as a measure of the doppler resolution. Thus

$$\begin{aligned}\Delta\tau &= \frac{1}{B} \\ \Delta\nu &= \frac{1}{T}\end{aligned}$$

where B is an appropriate measure of the bandwidth of the transmitted waveform, and T is an appropriate measure of the timewidth. For a waveform that is a pulse train, the resolution in cross range can be defined as the first null in the ambiguity function in the ν direction, which is at $1/NT_r$.

The range resolution and cross-range resolution are then expressed as

$$\Delta x = \frac{c}{2B}$$
$$\Delta y = \frac{\lambda R}{2L}$$

where $L = VT$ is the distance that any point on the antenna travels in time T. It is natural to call L the "synthetic-aperture length." For this reason, the real antenna is visualized as sweeping out an interval of length L called the *synthetic aperture*.

7.4 Focusing and motion compensation

In many instances of imaging radar, the delay and doppler alone are not adequate to fully describe the significant effects on carrier phase due to change in delay because of range differences. Then the next term of the Taylor series, which is the second time derivative of the delay, must be included. We shall see, however, that to account for radar motion it is often enough to make this phase correction only for the center of the scene.

The complex baseband signal, received from a point reflector at x, y, is

$$v(t) = s(t - \tau(x, y, t))e^{j2\pi f_0 \tau(x,y,t)}$$

where $\tau(x, y, t)$ is the path delay at time t. In the previous section, we approximated $\tau(x, y, t)$ in the argument of $s(t)$ by a function independent of t,

$$\tau(x, y, t) \approx \tau_o(x, y),$$

and in the exponential term by an expression linear in t,

$$\tau(x, y, t) \approx \tau_o(x, y) + \dot{\tau}_o(x, y)t.$$

We will now look more carefully at these approximations, bringing in higher-order terms as needed, but in the simplest possible way. Specifically, we will find that the next more complicated approximation in the exponential term is

$$\tau(x, y, t) \approx \tau_o(x, y) + \dot{\tau}_o(x, y)t + \tfrac{1}{2}\ddot{\tau}_o(0, 0)t^2.$$

The significance of expressing the quadratic term as independent of x and y is that this term need not be applied separately for each point in the scene. Recall that the matched-filter response of the received signal $v(t)$ from a point reflector at x, y is

$$r(x, y) = \int_{-\infty}^{\infty} v(t)s^*(t - \tau(x, y, t))e^{-j2\pi f_0 \tau(x,y,t)}\, dt$$
$$\approx e^{-j2\pi f_0 \tau_0} \int_{-\infty}^{\infty} \left[v(t)e^{-j\pi \ddot{\tau}_0(0,0)t^2}\right] s^*(t - \tau_0(x, y))e^{-j2\pi \dot{\tau}_0(x,y)t}\, dt.$$

This reduces to the computation of the cross-ambiguity function $\chi(\tau, \nu)$ of $v'(t)$ where $v'(t)$, the "focused received signal," is defined as

$$v'(t) = v(t)e^{-j\pi \ddot{\tau}_0(0,0)t^2}.$$

The quadratic term is described as a *focusing compensation* because it is incorporated into the computation simply by phase shifting the received signal by $\ddot{\tau}_0(0,0)t^2$. This correction alters the phase – up to quadratic terms – to make it appear as if the radar is following a circle centered at the origin of the scene.

To account for irregularities in the radar motion, it is useful to further modify the approximation to the form

$$\tau(x, y, t) = \tau_o'(x, y) + \dot{\tau}_o(x, y)t + \tfrac{1}{2}\ddot{\tau}_o(0, 0)t^2 + \delta\tau(t).$$

Whereas the third term accounts for the difference between a straight line and a circle, the last term accounts for measured deviations in the motion of the moving radar from the reference straight line. That term is referred to as the *motion compensation*.

The focusing and motion compensation terms stabilize the signal from each reflecting element so that it effectively has constant coordinates in the τ, ν plane. The received signal actually is the composite of many echoes from many points, so the true compensation would be different for every point of the scene. However, if the scene is not too large, the compensation is approximately the same for every point in the scene. This is the approximation that $\ddot{\tau}_0$ is independent of x and y.

The validity of this approximation will now be examined with the aid of Figure 7.5, which depicts an imaging radar following a straight line at a constant velocity V, and a depicted scene to be imaged. The range rate and range acceleration are given by

$$\dot{R} = -V \sin\phi$$
$$\ddot{R} = \frac{V^2}{R}\cos^2\phi.$$

The range rate causes a doppler given by

$$\nu = \frac{2f_0}{c}\dot{R},$$

and the range acceleration causes a doppler rate given by

$$\alpha = \frac{2f_0}{c}\ddot{R} = \frac{2V^2}{\lambda R}\cos^2\phi.$$

The doppler rate over time T causes a quadratic-phase change, $\Delta\theta$ (in cycles), given by

$$\begin{aligned}\Delta\theta &= \tfrac{1}{2}\alpha T^2\\ &= \frac{(VT)^2}{\lambda R}\cos^2\phi\end{aligned}$$

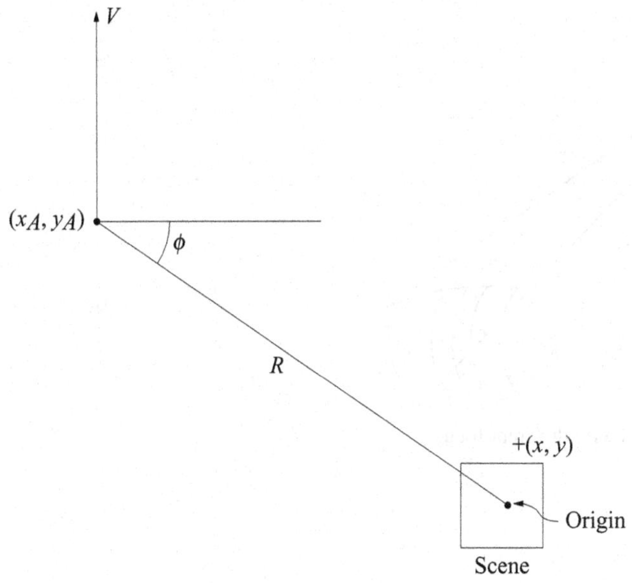

Figure 7.5 Elementary geometry

which is removed by the focusing compensation. For simplicity of exposition, we will consider only the case with $\phi = 0$. Then

$$\Delta\theta = \frac{(VT)^2}{\lambda R}.$$

The cross-range resolution, Δy, given by

$$\Delta y = \frac{\lambda R}{2VT},$$

can be used to eliminate VT in the expression for $\Delta\theta$. Thus

$$\Delta\theta = \frac{\lambda R}{4\Delta y^2}.$$

This is a function of range, so if $\Delta\theta$ is compensated for only one nominal range, then it will be in error for other values of range as given by

$$\delta(\Delta\theta) = \frac{\lambda}{4\Delta y^2}\,\delta R.$$

A design rule of thumb for coherent integration is to keep uncompensated, quadratic-phase deviations below 1/8 cycle.[1] To see how far the doppler-rate correction is valid,

[1] This causes the equivalent of 0.5 dB loss in the signal-to-noise ratio. At 1/4 cycle, the loss would be about 2 dB.

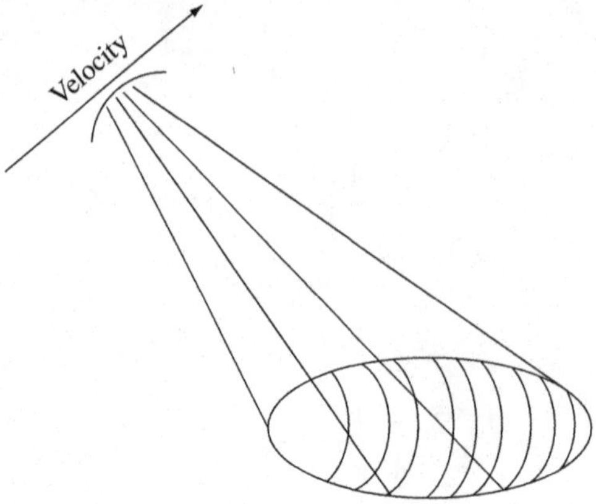

Figure 7.6 Range strips with different focus

solve the equation

$$\frac{1}{8} = \frac{\lambda}{4\Delta y^2} \delta R$$

for δR. This gives

$$\delta R = \frac{1}{2} \frac{(\Delta y)^2}{\lambda}$$

as the half-width of the scene that is satisfactorily imaged. For example, if $\Delta y =$ 10 feet and $\lambda = 0.1$ foot, then the half-width of the scene in the range direction is 500 feet. This does not mean that a larger scene cannot be processed; it only means that the doppler-rate correction must be made separately for each of several subscenes. Thus, if the range spread of the scene is large, it is not possible to focus the entire scene at once. Figure 7.6 shows how a scene can be divided into strips that are individually focused. We can express this by saying there is more than one "doppler-rate bin." Because the focusing takes place in the signal processing, it is straightforward to reprocess the same data many times for different subscenes, as shown in Figure 7.7. Each computation uses a different quadratic-phase correction. Then the appropriate range strip is found in the sample cross-ambiguity function.

Figure 7.8 portrays the doppler-rate compensation in terms of an artificial reference trajectory described by a quadratic. The effect of the phase compensation is to make the received signal appear as if the transmitter/receiver were traversing the quadratic reference trajectory. The compensation in Figure 7.8 is meaningful only while the scene is approximately at broadside so that $\phi = 0°$. The circular trajectory can deviate from the straight line by many thousands of wavelengths even while ϕ is still essentially zero. However, if the coherent processing extends over a long enough time, then other

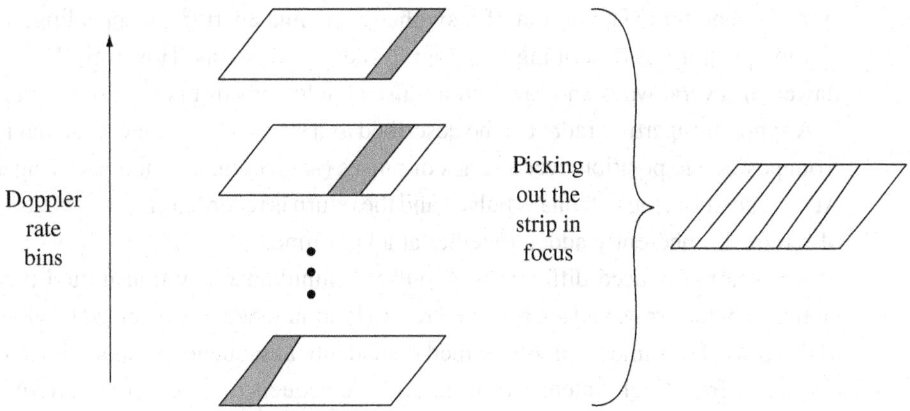

Figure 7.7 Conceptualizing multiple doppler-rate bins

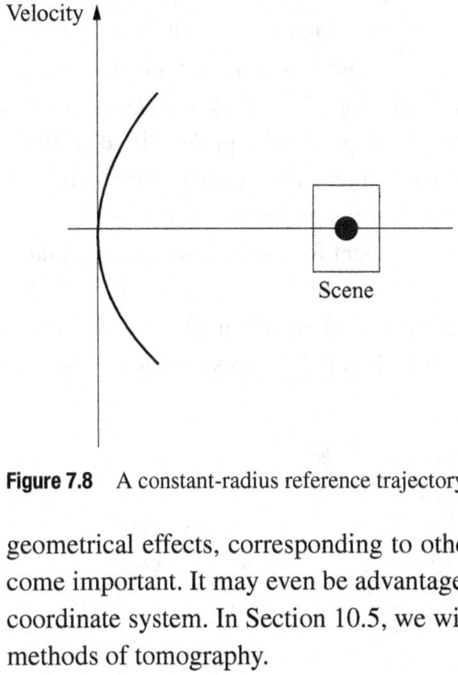

Figure 7.8 A constant-radius reference trajectory

geometrical effects, corresponding to other terms of the Taylor series expansion, become important. It may even be advantageous to reformulate the processing in a polar coordinate system. In Section 10.5, we will discuss processing radar signals using the methods of tomography.

7.5 Structure of typical imaging systems

A high-resolution imaging radar that moves a distance large compared to the size of its physical antenna during the duration of the waveform is called a *synthetic-aperture radar*. This name refers to the notion that by moving the physical antenna, one synthesizes a larger antenna. It is satisfying to describe an imaging radar using the idea of a synthetic aperture. The notion of a synthetic antenna aperture is appealing, and there is a long history of describing an imaging radar in this way. However, the analogy is flawed in several ways and can lead to false conclusions if it is not used with care.

A synthetic aperture radar can be described as a single antenna element that is moved from position to position over a series of pulses (neglecting the motion during a pulse). At each position, the antenna is pulsed and the return is recorded. The recorded sequence of returns is coherently added together at a later time.

A real array is used differently. A pulse is simultaneously transmitted through all elements of the array, and the reflected return is simultaneously received by all elements of the array. The same result is obtained if an identical sequence of pulses is transmitted, one pulse from each antenna in turn, and the sequence of returns at all antennas is recorded and these recordings are coherently added together.

While a synthetic aperture is superficially similar to a real array, the differences are significant. A synthetic array can only receive a return at the same antenna element from which the pulse was transmitted. Any signal that might have been received at the other antenna elements is lost because those antenna elements do not exist. Hence the synthetic array processes less information. The array element is moving while transmitting and receiving. Hence the echo pulse is doppler-shifted in the synthetic array but not in the real array. Because of these differences, one should be wary of pushing the analogies between a synthetic array and a real array too far.

The simplest and most common arrangement for a synthetic-aperture imaging radar places the transmitter and receiver on the same moving platform. This is called a *monostatic imaging radar*. A less common arrangement places the transmitter and receiver at different locations, as shown in Figure 7.9. This is called a *bistatic imaging*

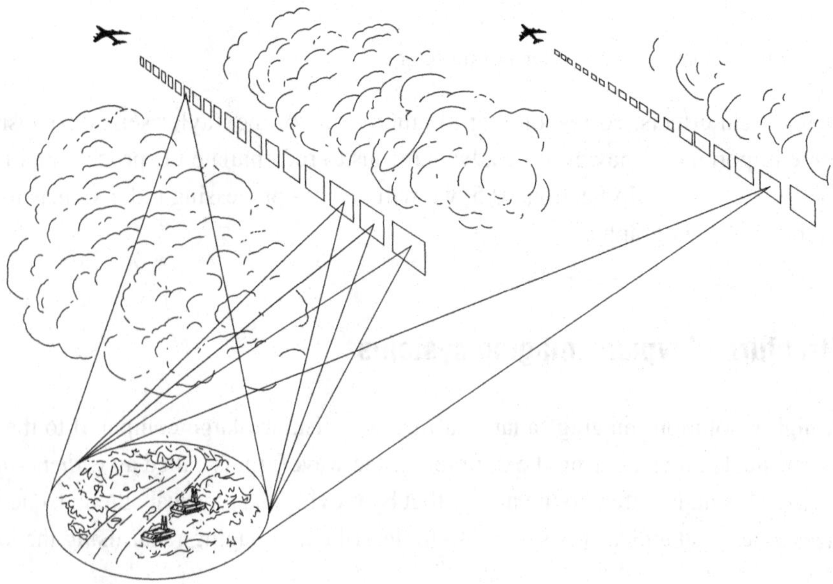

Figure 7.9 Bistatic imaging radar

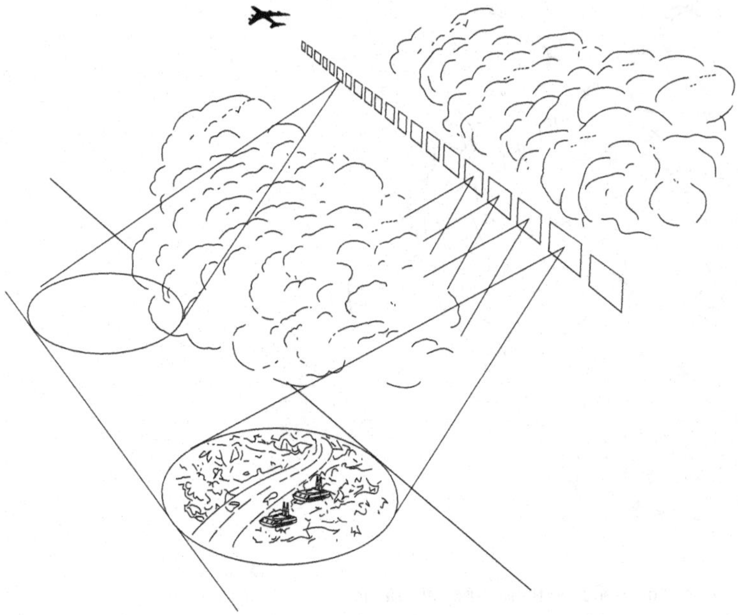

Figure 7.10 Swath-mode synthetic-aperture radar

radar. We shall develop the ideas mainly for monostatic systems. Bistatic systems are a simple generalization and obey the same principles of image formation. The only essential difference is in the coordinate transformation from τ, ν coordinates to rectangular x, y coordinates.

The transmitter or receiver of a synthetic-aperture radar system must be in motion with respect to the target scene. The two most popular arrangements are the *swath-mode imaging radar*, shown in Figure 7.10 and the *spotlight-mode imaging radar*, shown in Figure 7.11. Spotlight-mode imaging suggests a polar coordinate system, which suggests the methods of tomography, as is treated in Section 10.5.

In the simplest case, a swath-mode imaging radar has a straight-line, constant-velocity trajectory. It illuminates and images a continuous strip that is parallel to the trajectory. The resolution is limited by the amount of time a reflecting element is illuminated, and this time duration is limited by the beamwidth of the real antenna. A spotlight-mode imaging radar enhances resolution by keeping the antenna beam on the target area for a longer time by rotating the antenna (or by rotating the radar platform). In the extreme case, a spotlight-mode imaging radar has a constant-speed, circular trajectory centered on a small scene that is to be imaged. A much different geometry, shown in Figure 7.12, occurs in the imaging of distant objects, such as asteroids or planets using the methods of radar astronomy. The basic concept of first imaging in τ, ν coordinates is still valid, but the details of the transformation to x, y coordinates are different because of the nonplanar geometry and the rotation.

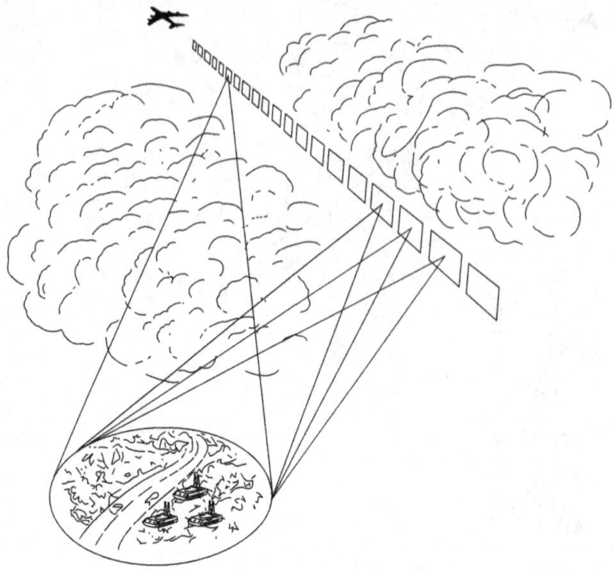

Figure 7.11 Spotlight-mode synthetic-aperture radar

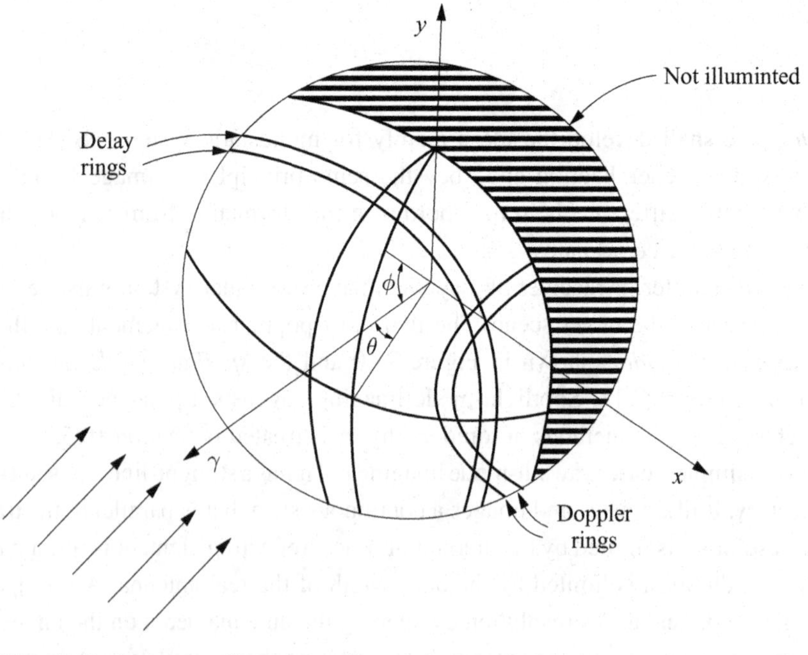

Figure 7.12 Delay-coordinates on a distant sphere

The synthetic-aperture image is created in the τ, ν plane. This leads to two considerations. First, because of changing geometry, objects will not normally have persistent τ, ν coordinates. To obtain good ν resolution, long waveforms are needed and, during the duration of this waveform, ν itself will change. The only exception occurs when the transmitter and receiver are on a circular trajectory about this reflecting element. In general, objects will be out of focus or smeared because their τ and ν coordinates are changing during the exposure. This imposes requirements of focusing and motion compensation as were studied in Section 7.4.

The cross-range resolution for a sidelooking radar was given in Section 7.2 as

$$\Delta y = \frac{\lambda R}{2L}.$$

The duration of the uniform pulse train is approximately NT_r. More precisely, the ambiguity function of the uniform pulse train has its first null at $1/NT_r$, so a common measure of the doppler resolution is $\Delta\nu = 1/NT_r$. Then the *synthetic-aperture length* is $L = VNT_r$, which is the distance[2] that the antenna moves during the duration of the pulse $s(t)$.

It is interesting to compare the cross-range resolution of a synthetic array with the cross-range resolution of a real array of length L. We will choose a real aperture that is uniformly illuminated, so the antenna pattern will be a sinc function. A real aperture of length D that is uniformly illuminated has its first null at

$$\Delta\phi = \frac{\lambda}{D}.$$

The cross-range resolution of a real sidelooking aperture of length L is given by

$$\Delta y = R\Delta\phi$$
$$\Delta y = \frac{\lambda R}{D}.$$

This will be the same as the cross-range resolution of the synthetic aperture if $D = 2L$. A real aperture must be twice as long as a synthetic aperture to obtain the same cross-range resolution. This shows that the notion of a "synthetic aperture" must be used with care. The aperture is not filled simultaneously but in the alternative, sequential manner discussed in Section 5.3. Of course, if the real aperture is an array and each element is used individually to sequentially transmit and receive a pulse, as was described in Chapter 5, the resolution would be the same as the synthetic aperture.

An implicit assumption of our analysis of the imaging radar is that the antenna beam is directed at the scene throughout the duration of the waveform $s(t)$. If VT is large, this may become an issue that cannot be ignored. Figure 7.13 shows an idealized antenna beam illuminating the scene. The width of the beam at range R is $R\Delta\phi$. If VT is smaller than $R\Delta\phi$, then a scene element will be illuminated for the full waveform of duration T, and the antenna beamwidth need not be considered further. If VT is larger than $R\Delta\phi$, then the beam can move off a target element during the waveform duration. A spotlight-mode imaging radar rotates the antenna beam to keep a scene illuminated.

[2] More precisely, $V(N-1)T_r$ is a better expression for the distance, but when N is large, the difference is unimportant.

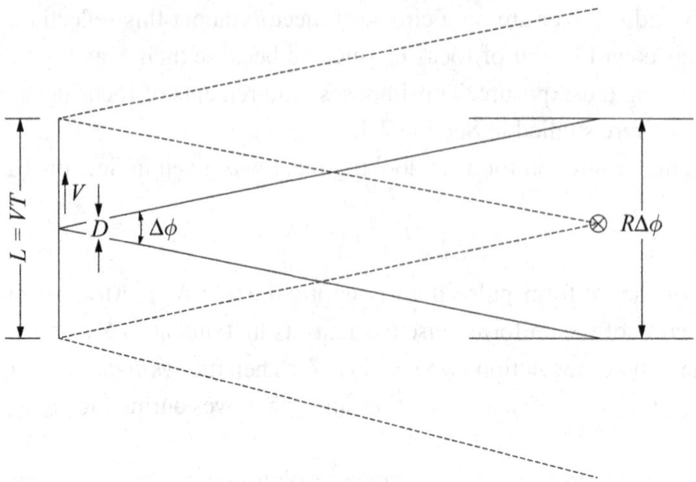

Figure 7.13 Swath-mode limit on illumination time

A swath-mode imaging radar uses a staring antenna beam so that the illuminated region changes with time. The integration time T is then set by

$$VT = R\Delta\phi.$$

Therefore the cross-range resolution of a swath-mode imaging radar is

$$\Delta y = \frac{\lambda R}{2VT}.$$

Because $\Delta\phi = \lambda/D$, and $VT = R\Delta\phi$, this becomes

$$\Delta y = \frac{D}{2}.$$

The cross-range resolution is limited to one-half the physical size of the staring real antenna.

7.6 Computing the cross-ambiguity function

Computation of the sample cross-ambiguity function

$$\chi_c(\tau, \nu) = \int_{-\infty}^{\infty} v(t)s^*(t-\tau)e^{-j2\pi\nu t}\,dt$$

from the received signal $v(t)$ and the transmitted signal $s(t)$ is a central task of radar signal processing. Depending on the circumstances of the application, this might be a massive computation or it might be a trivial computation, as determined by the application and the approximations. Notice that the structure of the computation, when τ is fixed, has the form of a Fourier transform, and when ν is fixed, has the form of a convolution. The structure of a processing implementation, or the choice of waveform, may emphasize one of these details to simplify the computations.

The processing might be performed by analog circuits, either baseband or passband, or by an optical or digital processor. One might choose $s(t)$ cleverly to keep the computations manageable, or one may develop fast computational algorithms that compute $\chi_c(\tau, \nu)$ for an arbitrary $s(t)$.

It is only a slight exaggeration to say that simple radars, including the earliest radars, always use a waveform, $s(t)$, that makes the computation trivial. For a *pulse radar*, we choose $s(t)$ to be a short pulse of duration T such that $\nu T \ll 1$ for all ν of interest. Then $\chi_c(\tau, \nu)$ does not depend on ν for the range of interest, and it is sufficient to do the computations for $\nu = 0$. The computation now takes the simple form

$$\chi_c(\tau, 0) = \int_{-\infty}^{\infty} v(t)s^*(t - \tau)\,dt,$$

which can be computed by passing $v(t)$ through a filter with the impulse response $s^*(-t)$.

Alternatively, for a *doppler radar*, we choose $s(t)$ to be a long pulse of narrow bandwidth B so that $B\tau \ll 1$ for all τ of interest, then $\chi_c(\tau, \nu)$ does not depend on τ for the range of interest, and it is sufficient to do the computation for $\tau = 0$. Then the computation is

$$\chi_c(0, \nu) = \int_{-\infty}^{\infty} v(t)s^*(t)e^{-j2\pi\nu t}\,dt.$$

If $s(t)$ is a rectangular pulse of width T, this takes the simple form

$$\chi_c(0, \nu) = \int_{-T/2}^{T/2} v(t)e^{-j2\pi\nu t}\,dt.$$

In this case, $\chi_c(0, \nu)$ can be sampled by passing $v(t)$ through a bank of passband filters, as can be described by a Fourier transform,

$$\chi_c(0, \nu_k) = \int_{-T/2}^{T/2} v(t)e^{-j2\pi\nu_k t}\,dt,$$

for some discrete set of ν_k.

A waveform that has a resolution in both τ and ν is the pulse train

$$p(t) = \sum_{n=0}^{N-1} s(t - nT_r)$$

where $s(t) = 0$ for $|t| \geq T/2$. Then

$$\chi_c(\tau, \nu) = \int_{-\infty}^{\infty} \sum_{n=0}^{N-1} v(t)s^*(t - nT_r + \tau)e^{-j2\pi\nu t}\,dt.$$

Now move the sum outside the integral and then make a change of variables in each summand so that $t - nT_r + \tau$ becomes replaced by t. This gives

$$\chi_c(\tau, \nu) = e^{j2\pi\nu\tau} \sum_{n=0}^{N-1} e^{-j2\pi\nu n T_r} \int_{-T/2}^{T/2} s^*(t) v(t + nT_r - \tau) e^{-j2\pi\nu t}\, dt.$$

There are now n integrals, each from $-T/2$ to $T/2$. Moreover, if T is chosen so that $\nu T \ll 1$ for all ν of interest, this becomes

$$\chi_c(\tau, \nu) = e^{j2\pi\nu\tau} \sum_{n=0}^{N-1} e^{-j2\pi\nu n T_r} \int_{-T/2}^{T/2} s^*(t) v(t + nT_r - \tau)\, dt.$$

What this amounts to is that each returned pulse is passed through a matched filter to provide an output, say $u_n(\tau)$. Then the sample cross-ambiguity function for each value of τ takes the form of a one-dimensional Fourier transform,

$$\chi_c(\tau, \nu) = e^{j2\pi\nu\tau} \sum_{n=0}^{N-1} u_n(\tau) e^{-j2\pi\nu n T_r}.$$

If T is small in comparison to the pulse repetition interval T_r, there can be considerable computational savings by using a pulse train. Figure 7.14 shows the computation of $\chi_c(\tau, \nu)$ for a pulse train as a signal-processing task. The output of a single matched filter is broken into intervals, each interval corresponding to one of the $u_n(\tau)$ for a pulse train. These time segments are read as rows into a two-dimensional array. The Fourier transform of the array in the column direction then provides the desired sample cross-ambiguity function.

A widely used pulse train is a train of chirped pulses. One reason for this popularity is that the computation of the cross-ambiguity function can be arranged to take the form

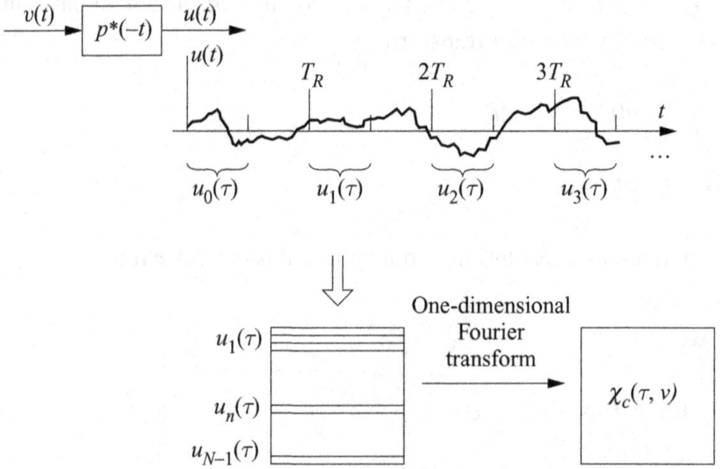

Figure 7.14 Computing the sample cross-ambiguity function

of a two-dimensional Fourier transform. To do this, let $s(t) = e^{j\pi\alpha t^2}$ and write as before

$$\chi_c(\tau, \nu) = e^{j2\pi\nu\tau} \sum_{n=0}^{N-1} e^{-j2\pi\nu nT_r} \int_{-T/2}^{T/2} e^{j\pi\alpha t^2} v(t + nT_r - \tau)\, dt$$
$$= e^{j2\pi\nu\tau} \sum_{n=0}^{N-1} e^{-j2\pi\nu nT_r} u_n(\tau)$$

where

$$u_n(\tau) = \int_{-T/2}^{T/2} e^{j\pi\alpha t^2} v(t + nT_r - \tau)\, dt.$$

For convenience, the time origin for the received signal will be chosen so that τ is zero at the center of the scene. With another change of variables such that $t - \tau$ is replaced by t, this becomes

$$u_n(\tau) = e^{j\pi\alpha\tau^2} \int_{-T/2+\tau}^{T/2+\tau} e^{j2\pi\alpha t\tau} [e^{j\pi\alpha t^2} v(t + nT_r)]\, dt.$$

Define the nth "dechirped" pulse as

$$v_n(t) = e^{j\pi\alpha t^2} v(t + nT_r).$$

The processing now takes the form of a Fourier transform of the dechirped pulse

$$u_n(\tau) = \int_{-T/2+\tau}^{T/2+\tau} e^{j2\pi\alpha t\tau} v_n(t)\, dt.$$

The limits of integration are offset by τ, which disrupts the otherwise clean form of this equation. For this reason, make the approximation that τ is small compared to $T/2$, and set

$$u_n(\tau) \approx \int_{-T/2}^{T/2} e^{j2\pi\alpha t\tau} v_n(t)\, dt.$$

The approximation in the limits of integration means that a part of the signal returned from reflectors near the edge of the scene will be wasted. The image will be attenuated near the edge of the scene in proportion to τ. The computation of the cross-ambiguity function now takes the form

$$\chi_c(\tau, \nu) = \sum_{n=0}^{N-1} e^{j2\pi\nu nT_r} \int_{-T/2}^{T/2} e^{j2\pi\alpha t\tau} v_n(t)\, dt.$$

This computation is in the form of a two-dimensional Fourier transform of the dechirped signal, as shown in Figure 7.15. One axis is a discrete Fourier transform and one axis is a continuous Fourier transform. Depending on the technology used to do the processing, the computation may be approximated in either of two ways. For an optical processor, the computation can be put in the form of a continuous two-dimensional

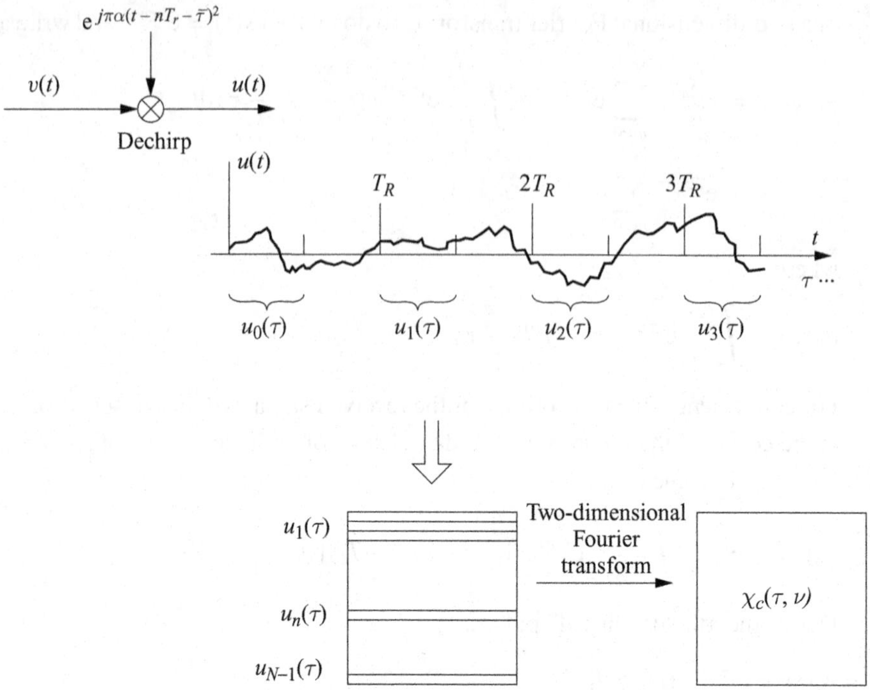

Figure 7.15 Receiver computations for a chirp pulse train

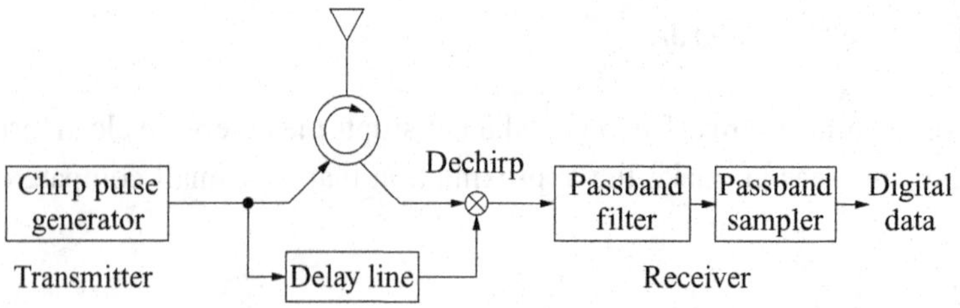

Figure 7.16 Block diagram of a chirp transmitter/receiver

Fourier transform by replacing the discrete Fourier transform by a continuous Fourier transform. For a digital processor, the computation can be put in the form of a discrete two-dimensional Fourier transform by replacing the continuous Fourier transform by a discrete Fourier transform.

Figure 7.16 shows a block diagram of a radar that transmits a chirp pulse and dechirps each pulse of the received signal, a technique sometimes called *stretch* or *frequency compression*. In this version, the transmitted pulse $s(t)$ has a nominal delay to produce the reference chirp pulse

$$s_{\text{ref}}(t) = e^{j\pi\alpha t^2} \qquad |t| \leq T/2,$$

relative to the time origin of the received signal. The echo pulse for a reflector with

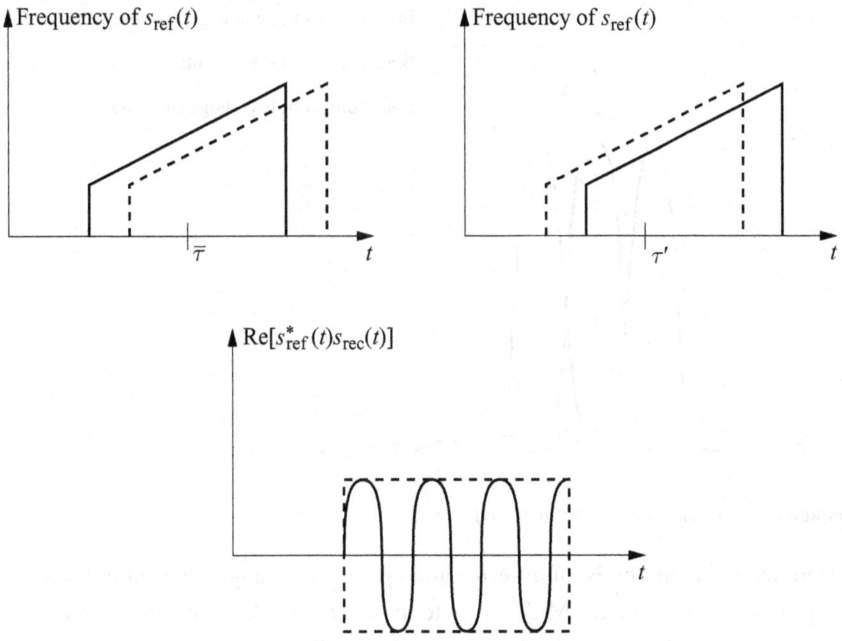

Figure 7.17 Frequency compression

delay τ with respect to the scene center is

$$s_{ref}(t) = e^{j\pi\alpha(t-\tau)^2} \quad |t-\tau| \leq T/2.$$

Then the dechirped signal is

$$s^*_{ref}(t)s_{rec}(t) = e^{j2\pi\alpha\tau t - j\pi\alpha\tau^2}$$

for $|t-\tau| \leq T/2$ and $|t| \leq T/2$. The dechirped signal takes the simple form of a sinusoid at the frequency $\alpha\tau$ with the phase shift $\alpha\tau^2/2$. The frequency of the dechirped signal depends on the range offset. Figure 7.17 illustrates how the process of dechirping changes a delay into a frequency offset that is proportional to range. For a scene with many reflectors, the dechirped signal will be a superposition of many such sinusoids, depicted in Figure 7.18. The bandwidth of the dechirped signals depends on the range spread and on the pulse duration, and can be far smaller than the bandwidth of the chirp pulse. Consequently, if the sampling occurs after dechirping, it can use a much lower sampling rate than if the sampling occurs before dechirping.

7.7 Dual aperture imaging

There are many ways in which multiple antenna apertures can be used to provide additional information for an imaging radar. For example, multiple antennas can be

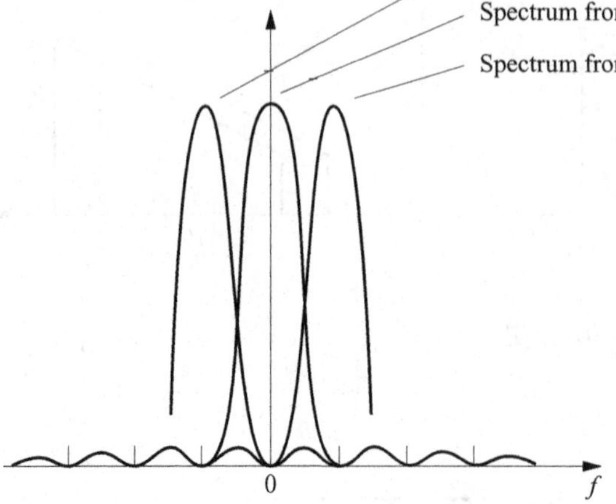

Figure 7.18 Illustrating the compressed signal

used with the methods of interferometry to make angle measurements to reflecting points in the scene. Multiple antennas can also be used to suppress signals or echoes from other directions that interfere with the desired term. Such methods can be placed after the calculation of the sample cross-ambiguity function so the cancellation or interferometric measurement can be different in each range-doppler cell.

One application of interferometry is to annotate a radar image with height measurements. The normal output image of an imaging radar is two dimensional, presenting the image as if it were viewed from directly overhead. The height of objects in the scene is not directly evident, but if an interferometer is incorporated into the imaging radar, the height of individual reflectors can be measured.

The raw signal at the radar antenna is the composite of the echoes from many reflecting elements. Some of these echoes are completely masked by stronger echoes. Therefore it is not possible to perform interferometry directly on the raw signal. Instead, it must be performed after the computation of the sample cross-ambiguity function. Figure 7.19 shows the concept. Two antennas are used, and are separated vertically. The signal is transmitted from one antenna, but the echo is received at both antennas. Each received echo signal will be used to compute a sample cross-ambiguity function. Because the antennas are close together in comparison to the range, both computed images will have nearly the same magnitude. The phase of the signal in each τ, ν cell, however, will be different because of the slight difference in path length from the dominant reflecting element in that cell to each of the two antennas. This phase difference between the cross-ambiguity functions at range-doppler coordinates τ, ν is given by

$$\Delta\theta(\tau, \nu) = 2\pi \frac{d}{\lambda} \sin\phi(\tau, \nu)$$

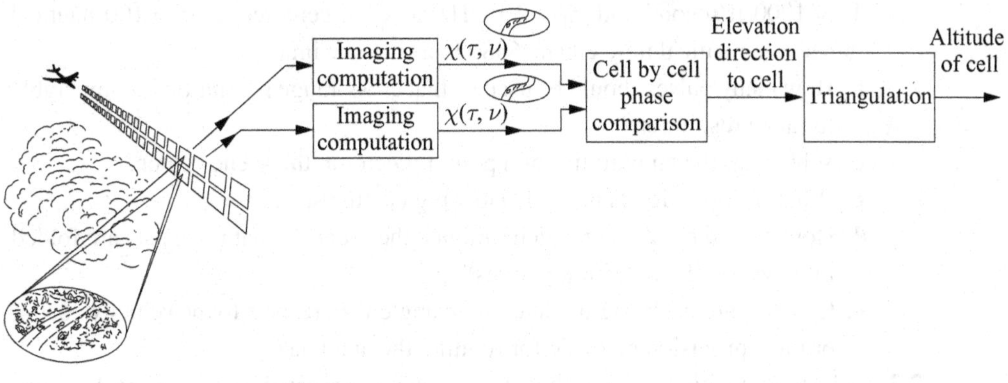

Figure 7.19 Imaging interferometry

where $\phi(\tau, \nu)$ is the declination angle to the dominant reflecting element with range-doppler coordinates τ, ν. Hence at each τ, ν coordinate, $\phi(\tau, \nu)$ can be easily computed from the phase difference $\Delta\theta(\tau, \nu)$ and, because range is known from the measurement of τ, the height can be easily computed by elementary trigonometry for the reflector with coordinates τ, ν.

Alternatively, if two antennas are placed side by side, then the azimuth angle to each reflecting point can be computed by the same methods of interferometry. It may seem that this is unnecessary because the azimuth angle could be computed instead from the doppler and the flight geometry. This is true, but an accurate determination of azimuth requires that the radar velocity vector be precisely known and the reflector be stationary. In contrast, an interferometric measurement gives an independent measurement of azimuth with respect to the antenna baseline with no need to know the velocity vector of the radar and reflector. The interferometric measurement gives an estimate of azimuth even for moving reflectors. Indeed, by comparing this estimate of azimuth to the estimate based on doppler, one can detect motion of the reflector.

One can also use two antennas to form a null to reject interference from another reflector that is spatially separated but has the same τ, ν coordinates. The rotating sphere that was shown in Figure 7.12 is an example of this situation. Each hemisphere has a matching set of τ, ν coordinates. With a single antenna, the radar image will consist of the superposition of both hemispheres. With two antennas, the relative phase can be used to separate the images of the two hemispheres.

Problems

7.1 A waveform for a sidelooking imaging radar uses a uniform pulse train with rectangular chirped pulses of duration $T = 1$ microsecond and chirp rate $\alpha = 100$ MHz/microsecond. The pulse spacing is $T_r = 1$ millisecond. With

$V = 1000$ ft/second and $f_0 = 10$ GHz, a scene centered at $R = 100$ nautical miles (one nautical mile equals 6080 feet) is to be imaged.

a. How many pulses should be used so that cross-range resolution is comparable to range resolution?

b. Where are the ambiguities of a point reflector at the scene center?

c. What $\ddot{R}$ correction is needed? How big can the scene be?

d. How must the radar antenna illuminate the scene so that no ghosts are folded into the scene via the ambiguities?

e. Can the antenna be rigidly fixed at an angle with respect to the velocity vector, or must provision be made for rotating the antenna?

7.2 A bistatic imaging radar with its transmitter and receiver separated by 150 nautical miles (one nautical mile equals 6080 feet) images a 1 NM by 1 NM scene that is centered at a point 100 NM from the transmitter and 100 NM from the receiver. Sketch the situation if the 1 NM by 1 NM scene is aligned with the local τ, ν coordinate system.

Does one nanosecond of τ resolution provide better or poorer spatial resolution than for the case of a monostatic imaging radar?

7.3 **a.** Suppose that an imaging radar uses a real antenna of length d that is rigidly fixed so that its beam is perpendicular to the velocity vector. If the aperture of length d is uniformly illuminated, what is the real antenna pattern?

b. Approximate the beam by an ideal beam of width $\Delta\phi = \lambda/d$. That is,

$$A(\phi) = \begin{cases} 1 & \text{if } |\phi| \leq \frac{1}{2}(\lambda/d) \\ 0 & \text{otherwise.} \end{cases}$$

Within this approximation, and in terms of d, find a limit on the cross-range resolution of a synthetic-aperture imaging radar.

7.4 A real aperture has half the resolution of a synthetic aperture in cross range. Suppose that a stationary real aperture consists of an array of elements with spacing of $L_r = VT_r$ where V is the velocity of the synthetic aperture and T_r is the pulse spacing of the synthetic-aperture waveform.

Describe how the real aperture can be used so that the resolution of the real aperture becomes the same as the resolution of a synthetic aperture of the same length.

7.5 A synthetic-aperture imaging radar is used to image a scene at a nominal angle of 45° from the radar velocity vector. Does the image placement depend on knowledge of the magnitude of the velocity vector? If so, how does the image change because of a 1% error in knowledge of the velocity magnitude?

7.6 An optical beam is focused by using a quadratic-phase lens. A radar beam is focused by using a parabolic antenna. Can these two statements be reconciled? How?

7.7 Consider a coherent pulse train that consists of n identical, equispaced pulses.

a. What is the ambiguity surface for this pulse train?

b. Give an expression for the sum of the individual ambiguity surfaces of the individual pulses.

c. Compare these two cases.

d. A proposed imaging radar computes the sample cross-ambiguity function for each individual echo pulse, then adds the magnitudes. Is this a reasonable proposal? Why?

7.8 An aircraft moving along the y axis at a velocity of 1000 feet/second is crossing the x axis at $t = 0$. The aircraft carries a sidelooking, synthetic-aperture imaging radar that uses a carrier frequency of 10 GHz. Along the x axis is a perfectly straight highway on which two cars are moving at location $x = 100$ miles and $y = \pm 10$ feet (and within the beam): one moving with $\dot{x} = 100$ feet/second, and one moving with $\dot{x} = -100$ feet/second. Describe the position and appearance of these cars in the radar image if the waveform duration is 0.1 second and the waveform bandwidth is 10 MHz.

7.9 Show that if a reflecting object viewed by a synthetic-aperture radar has a velocity in the range direction, it shifts the position of the imaged object in the cross-range direction, and if it has a velocity in the cross-range direction, it blurs the image of the object in the range direction.

7.10 An imaging radar at a carrier frequency of 10 GHz and a pulse repetition interval of 1 millisecond is carried by an aircraft with a velocity of 1000 feet/second. Describe the locus of ground points corresponding to each doppler grating lobe.

Describe an interferometric imaging radar that will reject large discrete objects from entering the image through a grating lobe (by determining in which grating lobe the signal entered). Will this work for small distributed objects in the grating lobe? What advantage might this have over a scheme that uses a sharp antenna beam to illuminate only objects in the main doppler lobe?

7.11 A helicopter has a rotor with three uniformly spaced blades, each 10 feet long, rotating at 60 rpm. Each blade has a reflectivity density uniformly distributed along its length. Each blade has a radar cross section of one square foot, and the body of the helicopter has a radar cross section of 10 square feet.

Describe the general appearance of the sample cross-ambiguity function if the helicopter is illuminated by a radar with a 10 GHz carrier and a pulse train waveform consisting of 100 pulses of pulse width 1 microsecond spaced 100 microseconds apart. An understandable sketch will suffice.

7.12 A common waveform for a synthetic-aperture radar is a uniformly spaced pulse train of chirp pulses. If the real antenna is a sidelooking antenna in uniform motion at velocity V, and the pulse interval is T_r, what is the angle between grating lobes? By spacing two antenna elements, design a two-element phased array for the real antenna that will have a real antenna null exactly on the first grating lobe. How must T_r be controlled as a function of V to make this work? What happens to the second and third grating lobes?

7.13 Explain how an interferometric technique can be used to erase moving vehicles from a synthetic-aperture radar image. Explain how this technique can be used to place the images of moving vehicles at their proper place in the scene. Select parameters for a sidelooking synthetic-aperture radar with an aircraft velocity of 300 meters per second and a range of 100 kilometers. What baseline distance should be used for the interferometer?

7.14 Show that the doppler effect moves a spectroscopic line at wavelength λ_0 to a spectroscopic line at wavelength λ given by

$$\lambda = \lambda_0 \frac{1 + (V/c)\cos\phi}{\sqrt{1 + (V/c)^2}}.$$

The denominator is significant only when the velocity V is large enough to produce relativistic effects. Then there can be a "doppler" effect even if the motion is orthogonal to the line of sight.

Notes

The early history of imaging radar is presented in the articles by Sherwin, Ruina, and Rawcliff (1962) and Brown and Porcello (1969). This early work was dominated by the need to process massive amounts of data at a time when digital computers did not yet exist. The processing was made possible by the technique of dechirping of chirp pulses which transformed the processing problem into the form of a two-dimensional Fourier transform that is suitable for optical processing. Later work, as by Brown and Fredericks (1969), studied various motion effects. The use of range-doppler sorting in radar astronomy to separate radar returns from different features of a planet so as to form an image was suggested by Green (1962). He described the magnitude image as a convolution of the radar cross section with the square of the ambiguity function, and he realized that his ideas were essentially those of a synthetic-aperture radar applied to the radar astronomy problem. Tagfors and Campbell (1973) gave an early survey of radar astronomy. A radar astronomy system would be seriously flawed because of doppler ambiguity were it not for the use of interferometry to resolve this ambiguity, so it is not surprising that the incorporation of interferometry into synthetic-aperture radar first occurred in radar astronomy. It was proposed by Manasse (1959) and demonstrated by Campbell (1971).

Synthetic-aperture radar was originally conceived as a way to image an essentially two-dimensional surface. Graham (1974) described the incorporation of interferometry into a synthetic-aperture radar as a way of annotating the image with altitude. Further work on interferometric synthetic-aperture radar was reported by Zebker and

Goldstein (1986), and by Hirasawa and Kobayashi (1986). A landmark paper by Munson, O'Brien, and Jenkins (1983) presented the then novel idea that, under the right conditions, synthetic aperture radar can be given a tomographic formulation. The use of a tomographic formulation of imaging radar suggests the use of a three-dimensional projection-slice theorem with the constraint that the image lies on a surface, not necessarily a plane. This method of treating three-dimensional objects motivated by the topic of imaging the moon and planets, was studied by Webb and Munson (1995), and by Webb, Munson, and Stacy (1998). A general formulation of a theory of radar imaging of three-dimensional objects was presented by Jakowatz and Thompson (1995).

8 Diffraction imaging systems

Because of their very small wavelengths (usually less than a nanometer), X-rays can be used to obtain valuable information about the structure of very small objects. The diffraction of X-rays by a crystal can provide information about the lattice structure of the crystal, the shape of the molecules making up the crystal, and the irregularities in the lattice structure.

The molecules of a crystal are arranged in a regular lattice with a spacing between them that is comparable to the wavelength of X-rays. An incident X-ray propagates as an electromagnetic wave and is diffracted by the crystal. The diffracted wave forms sharp grating lobes in the far-field diffraction region because of the periodic structure of the crystal. The distribution of these grating lobes in solid angle depends on the lattice structure of the crystal, while the relative amplitudes of the grating lobes depend on the electron structure of the molecules making up the crystal. The width of the grating lobes depends on the irregularities in the crystal structure.

We shall begin this chapter with a discussion of the three-dimensional Fourier transform. Then the diffraction of a plane wave by a three-dimensional crystal will be described by using the three-dimensional Fourier transform.

8.1 The three-dimensional Fourier transform

A function, $s(x, y, z)$, possibly complex, of three variables, is called a *three-dimensional function*, or a *three-dimensional signal*. In many applications, $s(x, y, z)$ will be a real-valued function; indeed, $s(x, y, z)$ often will be a nonnegative, real-valued function that can be visualized as the density of a diffuse cloud. The density of the cloud at point x, y, z is $s(x, y, z)$.

Given the three-dimensional function $s(x, y, z)$, possibly complex, whose energy

$$E_p = \int_{-\infty}^{\infty} \int_{-\infty}^{\infty} \int_{-\infty}^{\infty} |s(x, y, z)|^2 \, dx \, dy \, dz$$

is finite, the three-dimensional Fourier transform is defined as

$$S(f_x, f_y, f_z) = \int_{-\infty}^{\infty}\int_{-\infty}^{\infty}\int_{-\infty}^{\infty} s(x, y, z)e^{-j2\pi(f_x x + f_y y + f_z z)}\,dx\,dy\,dz,$$

and is denoted by a triply shafted arrow

$$s(x, y, z) \Leftrightarrow S(f_x, f_y, f_z).$$

It is apparent that the inverse three-dimensional Fourier transform is given by

$$s(x, y, z) = \int_{-\infty}^{\infty}\int_{-\infty}^{\infty}\int_{-\infty}^{\infty} S(f_x, f_y, f_z)e^{j2\pi(f_x x + f_y y + f_z z)}\,df_x\,df_y\,df_z.$$

This is an immediate generalization of the inverse one-dimensional Fourier transform.

Other properties of the three-dimensional Fourier transform can be obtained easily as generalizations of familiar properties of the one-dimensional Fourier transform. Some of the basic properties are as follows:

1. **Linearity:** For any constants a and b, possibly complex,

$$as_1(x, y, z) + bs_2(x, y, z) \Leftrightarrow aS_1(f_x, f_y, f_z) + bS_2(f_x, f_y, f_z).$$

2. **Sign Reversal:**

$$\begin{aligned} s(-x, y, z) &\Leftrightarrow S(-f_x, f_y, f_z) \\ s(x, -y, z) &\Leftrightarrow S(f_x, -f_y, f_z) \\ s(x, y, -z) &\Leftrightarrow S(f_x, f_y, -f_z). \end{aligned}$$

3. **Conjugation:**

$$s^*(x, y, z) \Leftrightarrow S^*(-f_x, -f_y, -f_z).$$

If $s(x, y, z)$ is real, then $S^*(-f_x, -f_y, -f_z) = S(f_x, f_y, f_z)$.

4. **Scaling:** For any real nonzero constants a, b, and c,

$$s(ax, by, cz) \Leftrightarrow \frac{1}{|abc|}S\left(\frac{f_x}{a}, \frac{f_y}{b}, \frac{f_z}{c}\right).$$

5. **Origin Translation:** For any real constants a, b, and c,

$$s(x - a, y - b, z - c) \Leftrightarrow S(f_x, f_y, f_z)e^{-j2\pi(af_x + bf_y + cf_z)}.$$

6. **Modulation:** For any real constants a, b, and c,

$$s(x, y, z)e^{j2\pi(ax + by + cz)} \Leftrightarrow S(f_x - a, f_y - b, f_z - c).$$

7. **Convolution:**

$$g(x, y, z) *** h(x, y, z) \Leftrightarrow G(f_x, f_y, f_z)H(f_x, f_y, f_z)$$

where $***$ denotes a three-dimensional convolution, given by

$$g(x, y, z) *** h(x, y, z) = \int_{-\infty}^{\infty}\int_{-\infty}^{\infty}\int_{-\infty}^{\infty} g(\xi, \eta, \zeta)h(x-\xi, y-\eta, z-\zeta)\,\mathrm{d}\xi\,\mathrm{d}\eta\,\mathrm{d}\zeta.$$

8. Product:

$$g(x, y, z)h(x, y, z) \Leftrightarrow G(f_x, f_y, f_z) *** H(f_x, f_y, f_z).$$

9. Parseval's Formula:

$$\int_{-\infty}^{\infty}\int_{-\infty}^{\infty}\int_{-\infty}^{\infty} g(x, y, z)h^*(x, y, z)\,\mathrm{d}x\,\mathrm{d}y\,\mathrm{d}z = \int_{-\infty}^{\infty}\int_{-\infty}^{\infty}\int_{-\infty}^{\infty} G(f_x, f_y, f_z)H^*(f_x, f_y, f_z)\,\mathrm{d}f_x\,\mathrm{d}f_y\,\mathrm{d}f_z.$$

10. Energy Relation:

$$\int_{-\infty}^{\infty}\int_{-\infty}^{\infty}\int_{-\infty}^{\infty} |s(x, y, z)|^2\,\mathrm{d}x\,\mathrm{d}y\,\mathrm{d}z = \int_{-\infty}^{\infty}\int_{-\infty}^{\infty}\int_{-\infty}^{\infty} |S(f_x, f_y, f_z)|^2\,\mathrm{d}f_x\,\mathrm{d}f_y\,\mathrm{d}f_z.$$

11. Coordinate Transformation: If $s(x, y, z) \Leftrightarrow S(f_x, f_y, f_z)$ and $\boldsymbol{A}$ is any three-by-three invertible matrix, then

$$s\left(\boldsymbol{A}^T\begin{bmatrix}x\\y\\z\end{bmatrix}\right) \Leftrightarrow \frac{1}{\det \boldsymbol{A}} S\left(\boldsymbol{A}^{-1}\begin{bmatrix}f_x\\f_y\\f_z\end{bmatrix}\right).$$

8.2 Transforms of some useful functions

A list of three-dimensional Fourier transforms of various useful functions is given in Table 8.1. Some of these will be derived in this section.

The *three-dimensional rectangle function* is defined as

$$\mathrm{rect}(x, y, z) = \mathrm{rect}(x)\mathrm{rect}(y)\mathrm{rect}(z).$$

The Fourier transform of the three-dimensional rectangle function is easily derived by a separation of variables. It is

$$S(f_x, f_y, f_z) = \mathrm{sinc}(f_x, f_y, f_z)$$

where the *three-dimensional sinc function* is defined as

$$\mathrm{sinc}(f_x, f_y, f_z) = \mathrm{sinc}(f_x)\mathrm{sinc}(f_y)\mathrm{sinc}(f_z).$$

Table 8.1 *A table of three-dimensional Fourier transform pairs*

$s(x, y, z)$	$S(f_x, f_y, f_z)$
$\text{rect}(x, y, z)$	$\text{sinc}(f_x, f_y, f_z)$
$\text{sinc}(x, y, z)$	$\text{rect}(f_x, f_y, f_z)$
$\delta(x, y, z)$	1
1	$\delta(f_x, f_y, f_z)$
$\text{cyln}(x, y, z)$	$\text{jinc}(f_x, f_y)\text{rect}(f_z)$
$\text{sphr}(x, y, z)$	$\text{tinc}\sqrt{f_x^2 + f_y^2 + f_z^2}$
$\text{comb}_N(x, y, z)$	$\text{dirc}_N(f_x, f_y, f_z)$
$\text{comb}(x, y, z)$	$\text{comb}(f_x, f_y, f_z)$
$\text{comb}(x)$	$\text{comb}(f_x)\delta(f_y, f_z)$
$\text{helx}(x, y, z)$	$\sum_n J_n(2\pi f)\text{e}^{-\text{j}n\psi}\delta(n - 2\pi f_z)$

The scaled rectangle function

$$s(x, y, z) = \text{rect}\left(\frac{x}{a}, \frac{y}{b}, \frac{z}{c}\right)$$

has the three-dimensional Fourier transform

$$S(f_x, f_y, f_z) = abc\ \text{sinc}(af_x, bf_y, cf_z),$$

which immediately follows from the scaling property of the three-dimensional Fourier transform.

The *three-dimensional impulse function* is defined by

$$\delta(x, y, z) = \delta(x)\delta(y)\delta(z).$$

The separation of variables immediately yields

$$\delta(x, y, z) \Leftrightarrow 1$$

as a three-dimensional Fourier transform pair.

The three-dimensional gaussian pulse is defined by

$$\text{e}^{-\pi(x^2+y^2+z^2)} = \text{e}^{-\pi x^2}\text{e}^{-\pi y^2}\text{e}^{-\pi z^2}.$$

The separation of variables immediately leads to the Fourier transform pair

$$\text{e}^{-\pi(x^2+y^2+z^2)} \Leftrightarrow \text{e}^{-\pi(f_x^2+f_y^2+f_z^2)}.$$

The *three-dimensional comb function* is defined by

$$\text{comb}(x, y, z) = \text{comb}(x)\text{comb}(y)\text{comb}(z).$$

The comb function can be visualized as an infinitely small, infinitely dense grain at each point of the integer lattice $\mathbf{Z}^3$. The three-dimensional comb function satisfies the Fourier transform

$$\text{comb}(x, y, z) \Leftrightarrow \text{comb}(f_x, f_y, f_z).$$

We also have the Fourier transform pair

$$\text{comb}_N(x, y, z) \Leftrightarrow \text{dirc}_N(f_x, f_y, f_z)$$

where the *three-dimensional finite comb function* is given by

$$\text{comb}_N(x, y, z) = \text{comb}_N(x)\text{comb}_N(y)\text{comb}_N(z),$$

and the *three-dimensional dirichlet function* is given by

$$\text{dirc}_N(x, y, z) = \text{dirc}_N(x)\text{dirc}_N(y)\text{dirc}_N(z).$$

The *three-dimensional cylinder function* is defined in terms of the two-dimensional circle function $\text{circ}(x, y)$ and the one-dimensional rectangle function $\text{rect}(z)$ by

$$\text{cyln}(x, y, z) = \text{circ}(x, y)\text{rect}(z).$$

The three-dimensional Fourier transform of $\text{cyln}(x, y, z)$ is

$$S(f_x, f_y, f_z) = \text{jinc}\left(\sqrt{f_x^2 + f_y^2}\right)\text{sinc}(f_z),$$

which is easily derived by a separation of variables.

The *three-dimensional sphere function* is defined by

$$\text{sphr}(x, y, z) = \begin{cases} 1 & \text{if } \sqrt{x^2 + y^2 + z^2} \le 1/2 \\ 0 & \text{otherwise.} \end{cases}$$

The sphere is defined to have a unit diameter rather than a unit radius so that it is consistent with the rectangle function in one dimension, which has a unit diameter. Because $\text{sphr}(x, y, z)$ is spherically symmetric, its Fourier transform must also be spherically symmetric; it can only depend on $f_x^2 + f_y^2 + f_z^2$. The three-dimensional Fourier transform of $\text{sphr}(x, y, z)$, evaluated at $f_x = 0$, $f_y = 0$, $f_z = f$, is

$$S(0, 0, f) = \iiint\limits_{\sqrt{x^2+y^2+z^2} \le 1/2} e^{-j2\pi f z}\, dx\, dy\, dz.$$

Expressed in spherical coordinates, this integral is

$$S(0, 0, f) = \int_0^{1/2}\int_0^{\pi}\int_0^{2\pi} e^{-j2\pi f r \cos\phi} r^2 \sin\phi\, d\psi\, d\phi\, dr.$$

The triple integral is evaluated by elementary methods to give

$$S(0,0,f) = \frac{\pi}{2}\frac{\sin \pi f - \pi f \cos \pi f}{(\pi f)^3}.$$

The *tinc function* or *tinc pulse* is defined as

$$\text{tinc}(t) = \frac{\sin \pi t - \pi t \cos \pi t}{2\pi^2 t^3}.$$

Because the sphere function has the Fourier transform $S(0,0,f) = \text{tinc}(f)$, we conclude that

$$\text{sphr}(x,y,z) \Leftrightarrow \text{tinc}\sqrt{f_x^2 + f_y^2 + f_z^2}$$

is a three-dimensional Fourier transform pair. The tinc pulse should be contrasted with the sinc pulse and the jinc pulse. The tinc function is compared to the jinc function and the sinc function in Figure 8.1. The value of the tinc pulse at the origin is $\text{tinc}(0) = \pi/6$, which is the volume of a sphere of diameter one. For large t, the zeros of $\text{tinc}(t)$ approach the half odd integers. The sidelobes of $|\text{tinc}(t)|$ decay with t as t^{-2}. The expression

$$\text{tinc}(x,y,z) = \text{tinc}\sqrt{x^2 + y^2 + z^2}$$

defines the *three-dimensional tinc function.*

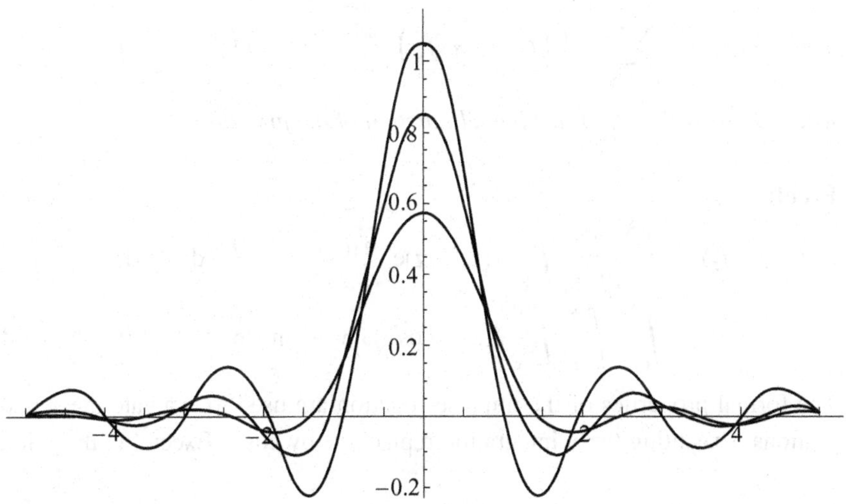

Figure 8.1 Comparison of the sinc, jinc, and tinc functions

The next example is the helix.[1] The helix does not have a close analog in lower dimensions. An infinite helical curve is given parametrically by

$$x = a\cos\psi$$
$$y = a\sin\psi$$
$$z = b\psi$$

where ψ ranges from $-\infty$ to ∞. A finite helical curve with T turns is obtained if ψ ranges from 0 to $2\pi T$.

An ideal infinite helix, then, can be defined as an infinitely thin, infinitely dense "solid" lying along a helical curve. This can be rewritten precisely in terms of a two-dimensional impulse function, provided care is taken in interpreting the formal properties of the impulse function. Define the *helix function* as

$$\begin{aligned}\text{helx}(x, y, z) &= \delta(x - \cos z)\delta(y - \sin z)\\ &= \delta(x - \cos z,\, y - \sin z).\end{aligned}$$

The helix function has radius 1 and period 1. A helix of radius a and period b is given by

$$\text{helx}\left(\frac{x}{a}, \frac{y}{a}, \frac{z}{b}\right) = \delta\left(\frac{x}{a} - \cos\frac{z}{b}\right)\delta\left(\frac{y}{a} - \sin\frac{z}{b}\right).$$

Because the helix is periodic in the z direction, we can expect that the Fourier transform will be discrete – that is, a Fourier series – in the f_z variable.

Theorem 8.2.1 *The three-dimensional Fourier transform of the helix function is given by*

$$S(f_x, f_y, f_z) = \sum_{n=-\infty}^{\infty} J_n\left(2\pi\sqrt{f_x^2 + f_y^2}\right) e^{jn\tan^{-1}(f_x/f_y)}\delta\left(f_z - \frac{n}{2\pi}\right)$$

where $J_n(a)$ is the nth-order Bessel function of the first kind.

Proof:

$$\begin{aligned}S(f_x, f_y, f_z) &= \int_{-\infty}^{\infty}\int_{-\infty}^{\infty}\int_{-\infty}^{\infty} s(x, y, z)e^{-j2\pi(f_x x + f_y y + f_z z)}\,dx\,dy\,dz\\ &= \int_{-\infty}^{\infty}\int_{-\infty}^{\infty}\int_{-\infty}^{\infty} \delta(x - \cos z)\delta(y - \sin z)e^{-j2\pi(f_x x + f_y y + f_z z)}\,dx\,dy\,dz.\end{aligned}$$

The formal properties of the impulse function are used to evaluate the x and y integrations. Executing the x integration replaces x by $\cos z$. Executing the y integration

[1] Using prior knowledge of the constituent molecular fragments, Watson and Crick constructed their famous model of the DNA molecule after they (and others) observed that the diffraction pattern of DNA displayed the characteristics of a helical structure.

replaces y by $\sin z$. Therefore

$$S(f_x, f_y, f_z) = \int_{-\infty}^{\infty} e^{-j2\pi(f_x \cos z + f_y \sin z + f_z z)} dz.$$

Now let

$$f_x = f \sin \gamma$$
$$f_y = f \cos \gamma.$$

Then

$$S(f_x, f_y, f_z) = \int_{-\infty}^{\infty} e^{-j2\pi(f \sin(z+\gamma) + f_z z)}\, dz.$$

To evaluate the integral, we will make use of the identity[2]

$$e^{-ja \sin t} = \sum_{n=-\infty}^{\infty} J_n(a) e^{jnt}$$

with $a = 2\pi f$ and $t = z + \gamma$. Then

$$\begin{aligned} S(f_x, f_y, f_z) &= \sum_{n=-\infty}^{\infty} J_n(2\pi f) \int_{-\infty}^{\infty} e^{jn(z+\gamma) - j2\pi f_z z}\, dz \\ &= \sum_{n=-\infty}^{\infty} J_n(2\pi f) e^{jn\gamma} \int_{-\infty}^{\infty} e^{-j2\pi\left(f_z - \frac{n}{2\pi} z\right)}\, dz \\ &= \sum_{n=-\infty}^{\infty} J_n(2\pi f) e^{jn\gamma}\, \delta\left(f_z - \frac{n}{2\pi}\right), \end{aligned}$$

as was to be proved. □

The expression in Theorem 8.2.1 can be decomposed into an infinite set of two-dimensional functions,

$$S_n(f_x, f_y) = J_n\left(2\pi\sqrt{f_x^2 + f_y^2}\right) e^{jn \tan^{-1}(f_x/f_y)},$$

in the planes at $f_z = n/2\pi$. These planes are sometimes called *layers* or, when viewed in cross section, *layer lines*. The layer $S_0(f_x, f_y)$ is the Fourier transform of an infinite, hollow, cylindrical surface, satisfying the heuristic view that an infinite cylinder is a

[2] To derive this identity, recall that any periodic $s(t)$ satisfies the Fourier series relationship

$$s(t) = \sum_{n=-\infty}^{\infty} S_n e^{jnt} \quad S_n = \frac{1}{2\pi}\int_{-\pi}^{\pi} s(t) e^{-jnt}\, dt.$$

In our case, $s(t) = e^{-ja \sin t}$, and

$$S_n = \frac{1}{2\pi}\int_{-\pi}^{\pi} e^{-ja \sin t} e^{-jnt}\, dt,$$

which is the definition of $J_n(a)$.

simple approximation to an infinite helix. By filtering out all other layers of the three-dimensional Fourier transform, the helix is converted to an infinite cylinder.

8.3 Diffraction by three-dimensional objects

The diffraction pattern of a three-dimensional object is a consequence of the scattering of waves. Consider a plane wave and a three-dimensional function, $s(x, y, z)$, which describes the density of scatterers. An important example is the scattering of X-rays by electrons. The real, nonnegative function $s(x, y, z)$ may be thought of as a density of electrons in a molecule, so an estimate of $s(x, y, z)$ amounts to an estimate of the shape of the molecule. The function $s(x, y, z)$ need not be centered in any way at the origin because the origin is only a point of reference, though, of course, it is natural to choose the origin so that $s(x, y, z)$ is concentrated near it.

The idealized scattering problem that we shall study is as follows. A plane wave passes through a partially transparent object, $s(x, y, z)$. Within an infinitesimal volume at x, y, z is a scatterer, $s(x, y, z)\, dx\, dy\, dz$. When the scatterer is illuminated with the plane wave, it omnidirectionally and coherently reradiates, or scatters, a spherical wavelet with the same phase and amplitude in all directions. As the incident wave sweeps through the transparent object $s(x, y, z)$, each infinitesimal volume reradiates some of the incident energy. We shall assume that the fraction of energy reradiated is negligible, an assumption known as the *Born approximation*. The Born approximation means that the incident wave maintains its amplitude as it sweeps through the scattering object. This assumption allows us to model the scattered radiation from two identical infinitesimal cells, at different places within the scatterer, as equal except for phase. The Born approximation also allows us to ignore multiple scattering in which scattered energy is rescattered.

In the next section, we shall consider the scattering object to be a periodic crystal; in this section, $s(x, y, z)$ is an arbitrary function of finite energy. Because we consider only the scattered wave in the far field, the Fraunhofer approximation is implicit in this discussion of diffraction.

Consider a single point (x, y, z) within the scatterer $s(x, y, z)$. First, as shown in Figure 8.2, we will restrict the situation to two dimensions and find the difference in the length of two paths to a common point in the far field. The point in the far field is in the x, z plane at angle ϕ from the z axis. One path to this point is the line originating at the point $(0, 0, 0)$; the other path is the line originating at the point $(x, 0, z)$. The Fraunhofer approximation amounts to the approximation that these two paths are parallel. If ϕ is zero, then both propagation paths are parallel to the z axis. The path from the origin to

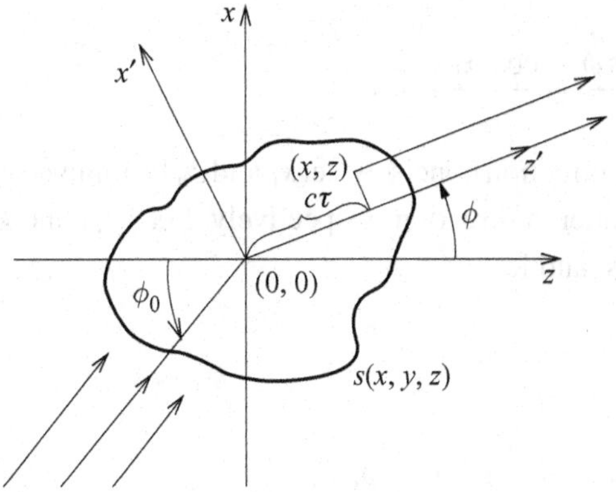

Figure 8.2 Illustrating the phase delay

a point in the far field has length $c\tau$, and the path from the point $(x, 0, z)$ to the same point has length $c\tau - z$, so the path difference is z. This means that the phase difference is $2\pi f_0 z/c$. If ϕ is not zero, rotate the coordinate system to a new coordinate system, $(x', 0, z')$, with the z' axis in the direction of propagation. Then

$$z' = x \sin\phi + z\cos\phi,$$

and the path difference with respect to a scatterer at the origin is z'. Hence the phase difference after scattering due to the path difference is

$$\Delta\theta = \frac{2\pi f_0}{c}(x \sin\phi + z\cos\phi).$$

There will also be a contribution due to the path difference in the incident wave prior to scattering. The incident angle ϕ_0 is defined so that this contribution has a negative sign:

$$\Delta\theta = \frac{-2\pi f_0}{c}(x \sin\phi_0 + z\cos\phi_0).$$

The contribution to the signal in the far field, at angle ϕ from this scattering cell, is

$$dv = s(x, y, z)\mathrm{e}^{-\mathrm{j}2\pi(f_0/c)[(x\sin\phi + z\cos\phi - x\sin\phi_0 - z\cos\phi_0)]}\,\mathrm{d}x\,\mathrm{d}y\,\mathrm{d}z.$$

The total signal in the far field, then, is the superposition of all these contributions

$$v(\phi) = \int_{-\infty}^{\infty}\int_{-\infty}^{\infty}\int_{-\infty}^{\infty} s(x, y, z)\mathrm{e}^{-\mathrm{j}(2\pi/\lambda)[x(\sin\phi - \sin\phi_0) + z(\cos\phi - \cos\phi_0)]}\,\mathrm{d}x\,\mathrm{d}y\,\mathrm{d}z.$$

This integral is recognized to have the form of a Fourier transform. Therefore, the diffraction pattern is

$$v(\phi) = S\left(\frac{\sin\phi - \sin\phi_0}{\lambda}, 0, \frac{\cos\phi - \cos\phi_0}{\lambda}\right).$$

In general, to specify an arbitrary direction concisely, let $\boldsymbol{s}_i$ and $\boldsymbol{s}_o$ be unit vectors along the incident direction and the scattered direction, respectively. Let $\boldsymbol{i}$, $\boldsymbol{j}$, and $\boldsymbol{k}$ be unit vectors along the x, y, and z axes, and let

$$\boldsymbol{x} = x\boldsymbol{i} + y\boldsymbol{j} + z\boldsymbol{k}.$$

Then

$$v(\phi, \psi) = \int_{-\infty}^{\infty}\int_{-\infty}^{\infty}\int_{-\infty}^{\infty} s(x, y, z)e^{-j(2\pi/\lambda)\boldsymbol{x}\cdot(\boldsymbol{s}_i - \boldsymbol{s}_o)}dx\,dy\,dz$$

where

$$\begin{aligned}\boldsymbol{x}\cdot(\boldsymbol{s}_i - \boldsymbol{s}_o) = {} & x(\cos\psi\sin\phi - \cos\psi_0\sin\phi_0) + y(\sin\psi\sin\phi - \sin\psi_0\sin\phi_0)\\ & + z(\cos\phi - \cos\phi_0).\end{aligned}$$

Therefore, with

$$\begin{aligned}f_x &= \frac{1}{\lambda}\boldsymbol{i}\cdot(\boldsymbol{s}_i - \boldsymbol{s}_o)\\ f_y &= \frac{1}{\lambda}\boldsymbol{j}\cdot(\boldsymbol{s}_i - \boldsymbol{s}_o)\\ f_z &= \frac{1}{\lambda}\boldsymbol{k}\cdot(\boldsymbol{s}_i - \boldsymbol{s}_o),\end{aligned}$$

the received signal is expressed neatly as

$$\begin{aligned}v(\phi, \psi) &= S(f_x, f_y, f_z)\\ &= S\left(\frac{\cos\psi\sin\phi - \cos\psi_0\sin\phi_0}{\lambda}, \frac{\sin\psi\sin\phi - \sin\psi_0\sin\phi_0}{\lambda}, \frac{\cos\phi - \cos\phi_0}{\lambda}\right).\end{aligned}$$

This is an important and simplifying result. It says that the ideal X-ray diffraction pattern can be expressed simply as the three-dimensional Fourier transform of the scattering density. Computing a diffraction pattern then reduces to the task of computing a three-dimensional Fourier transform.

The *scattering function* is defined as a function of spherical angle as

$$A(\phi, \psi) = \frac{1}{\lambda^3}S\left(\frac{\cos\psi\sin\phi - \cos\psi_0\sin\phi_0}{\lambda}, \frac{\sin\psi\sin\phi - \sin\psi_0\sin\phi_0}{\lambda}, \frac{\cos\phi - \cos\phi_0}{\lambda}\right).$$

The scattering function is defined so that the scaling property of the Fourier transform makes $A(\phi, \psi)$ a function of d/λ where d is a scaling parameter. Thus, under scaling by d,

$$A(\phi, \psi) = \left(\frac{d}{\lambda}\right)^3 S\left(\frac{\cos\psi\sin\phi - \cos\psi_0\sin\phi_0}{\lambda/d}, \frac{\sin\psi\sin\phi - \sin\psi_0\sin\phi_0}{\lambda/d}, \frac{\cos\phi - \cos\phi_0}{\lambda/d}\right).$$

Finally,

$$v(\phi, \psi) = \lambda^3 A(\phi, \psi)$$

is the scattered signal.

For example, under the Born approximation, the scattering function of a sphere of diameter d is

$$A(\phi, \psi) = \left(\frac{d}{\lambda}\right)^3 \operatorname{tinc}\left(\frac{\cos\psi \sin\phi - \cos\psi_0 \sin\phi_0}{\lambda/d}, \frac{\sin\psi \sin\phi - \sin\psi_0 \sin\phi_0}{\lambda/d}, \frac{\cos\phi - \cos\phi_0}{\lambda/d}\right).$$

This expression describes the scattering from a spherical ball of unit diameter; all points in the interior of the sphere are scatterers. Because of symmetry, there is no loss of generality in choosing $\psi_0 = \phi_0 = 0$. Then

$$A(\phi, \psi) = \left(\frac{d}{\lambda}\right)^3 \operatorname{tinc}\left(\frac{\sqrt{2 - 2\cos\phi}}{\lambda/d}\right).$$

Then

$$v(\phi, \psi) = d^3 \operatorname{tinc}\left(\frac{\sqrt{2 - 2\cos\phi}}{\lambda/d}\right)$$

describes the signal in the far field at angle ϕ, ψ.

8.4 Observation from diffraction data

An X-ray sensor can only measure the magnitude $|v(\phi, \psi)|$ of the diffraction pattern,[3] but not the phase. This is because the wavelength of an X-ray may be on the order of a tenth of a nanometer or less, and the many and various spatial factors that affect the phase of the measurement cannot be controlled to an accuracy of a fraction of a wavelength.

If a measurement of $S(f_x, f_y, f_z)$ for all values of f_x, f_y, f_z could be obtained, then estimating the scattering object $s(x, y, z)$ would reduce to the task of computing the inverse three-dimensional Fourier transform. However, this is not so straightforward because only the absolute value $|v(\phi, \psi)|$ can be measured, and this gives $|S(f_x, f_y, f_z)|$ only for some values of f_x, f_y, f_z. Then the task is to estimate $s(x, y, z)$ when given only the magnitude $|S(f_x, f_y, f_z)|$ of its Fourier transform. Moreover, the magnitude is available only for some values of f_x, f_y, f_z. Equivalently, the measurement can be regarded as a measurement of the square $|S(f_x, f_y, f_z)|^2$, which is the Fourier transform of the autocorrelation $s(x, y, z) *** s(-x, -y, -z)$ (sometimes called the *Patterson map*). The general signal-processing task of recovering the nonnegative real function $s(x, y, z)$ from $|S(f_x, f_y, f_z)|$ is the task of phase retrieval. The two-dimensional phase-retrieval problem will be discussed in Chapter 9, in Sections 9.4 and 9.5. As described

[3] The diffraction pattern is called the *visibility function* in the field of X-ray crystallography.

in these sections, various constraints, including the fact that $s(x, y, z)$ is known to be real and nonnegative, may nearly compensate for the fact that the phase of $S(f_x, f_y, f_z)$ is not known.

The magnitude of $S(f_x, f_y, f_z)$ can be measured only for those values of f_x, f_y, f_z of the form

$$f_x = (\cos\psi \sin\phi - \cos\psi_0 \sin\phi_0)/\lambda$$
$$f_y = (\sin\psi \sin\phi - \sin\psi_0 \sin\phi_0)/\lambda$$
$$f_z = (\cos\phi - \cos\phi_0)/\lambda.$$

To determine which values of f_x, f_y, and f_z can be so obtained by an appropriate choice of ϕ, ψ, ϕ_0, and ψ_0, first let $\phi_0 = \psi_0 = 0$. Then the equations become

$$\lambda f_x = \cos\psi \ \sin\phi$$
$$\lambda f_y = \sin\psi \ \sin\phi$$
$$\lambda f_z + 1 = \cos\phi.$$

Thus

$$\lambda^2 f_x^2 + \lambda^2 f_y^2 + (\lambda f_z + 1)^2 = 1.$$

This is the equation of a sphere, called an *Ewald sphere*, with the radius $1/\lambda$ passing through the origin and with its center at $f_z = -\lambda^{-1}$. All points (f_x, f_y, f_z) on the surface of this sphere in frequency space can be observed by the choice of ϕ and ψ. Moreover, the center of this sphere can be relocated by the choice of ϕ_0 and ψ_0 to any point at distance λ^{-1} from the origin, thereby forming another Ewald sphere. The union of the surfaces of all Ewald spheres forms the set of all points in the interior of a sphere of radius $2\lambda^{-1}$. Thus, by the choice of ϕ_0, ψ_0, ϕ, and ψ, any frequency point (f_x, f_y, f_z) within a sphere of radius $2\lambda^{-1}$ can be observed. We conclude that the observable data are $|S(f_x, f_y, f_z)|$ for all f_x, f_y, f_z satisfying $f_x^2 + f_y^2 + f_z^2 \leq 2\lambda^{-1}$. We think of this as an *observable data sphere* in frequency space. The size of the observable data sphere is limited by the choice of λ. To measure large spatial frequencies (corresponding to small details in $s(x, y, z)$), a small value of λ is needed.

Of course, it is usually impractical to observe the entire data sphere because this requires full angular coverage of both the illumination and the scattering. This means that the estimate of $s(x, y, z)$ must be computed with only partial Fourier data. Moreover, only a finite number of samples can be observed in practice. The sampling theorem or, in this case, the dual of the sampling theorem can be used to determine a sufficient set of samples of $|S(f_x, f_y, f_z)|$.

The sampling theorem in three dimensions states that if the support of $S(f_x, f_y, f_z)$ lies inside the unit cube centered at the origin, then the sampling operation

$$s'(x, y, z) = \text{comb}(x, y, z) s(x, y, z)$$

can be inverted by the expression

$$S(f_x, f_y, f_z) = \text{rect}(f_x, f_y, f_z)S'(f_x, f_y, f_z),$$

which leads to the Nyquist–Shannon interpolation formula in three dimensions:

$$s(x, y, z) = \text{sinc}(x, y, z) *** s'(x, y, z).$$

We need the dual of this theorem because, in the diffraction problem, sampling takes place in Fourier space. The dual of the sampling theorem states that if the support of $s(x, y, z)$ lies inside a unit cube, then the sampling operation on the spectrum

$$S'(f_x, f_y, f_z) = \text{comb}(f_x, f_y, f_z)S(f_x, f_y, f_z)$$

can be inverted by

$$s(x, y, z) = \text{rect}(x, y, z)s'(x, y, z),$$

or

$$S(f_x, f_y, f_z) = \text{sinc}(f_x, f_y, f_z) *** S'(f_x, f_y, f_z).$$

The general case in which the support of $s(x, y, z)$ is contained in a cube of side L can now be treated by using the scaling property. Then it is sufficient to know $S(f_x, f_y, f_z)$ for values of (f_x, f_y, f_z) given by the points of the cubic lattice with spacing L^{-1}. These three-dimensional Nyquist samples are the points with coordinates $(i_x/L, i_y/L, i_z/L)$ where i_x, i_y, and i_z are integers. The values of this set of samples completely determine $s(x, y, z)$.

However, our problem is more difficult. The observations are samples of $|S(f_x, f_y, f_z)|$, not of $S(f_x, f_y, f_z)$. But sampling $|S(f_x, f_y, f_z)|$ is equivalent to sampling $|S(f_x, f_y, f_z)|^2$, which is the Fourier transform of the autocorrelation $s(x, y, z) *** s(-x, -y, -z)$. If the support of the autocorrelation is contained in a cube of side $2L$, then it is sufficient to know $|S(f_x, f_y, f_z)|$ for $(f_x, f_y, f_z) = (i_x/2L, i_y/2L, i_z/2L)$. Thus the loss of phase not only replaces $s(x, y, z)$ with its autocorrelation, but unless side information is used, it also creates the need for a higher sampling density.

8.5 X-ray diffraction by arrays

A crystal is a three-dimensional array of molecular elements, as shown in Figure 8.3. An elementary crystal consists of an array of identical molecules, which is the case we will consider. For the more familiar examples, the molecules are simple, but even very large molecules[4] such as proteins can often be made to form crystals. By forming

[4] It is a matter for chemistry to decide whether a particular molecule can be made to form a crystal.

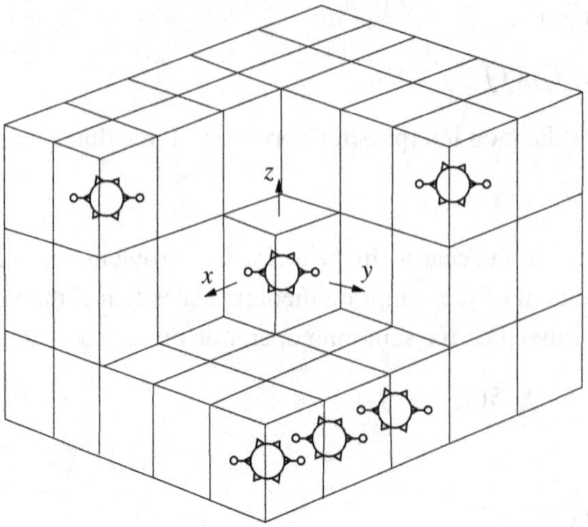

Figure 8.3 A crystal as a periodic array

a crystal and using it as a three-dimensional diffraction grating, one has a method for probing the crystal structure or forming an image of the molecule. The wavelength of the incident radiation must be comparable to the crystal spacing, making X-rays a natural choice to use for most crystallography because a wavelength somewhat smaller than the crystal spacing is appropriate to produce grating lobes.

From a mathematical point of view, the formation of crystallographic diffraction patterns is an exercise in Fourier transform theory – in this case, the theory of the three-dimensional Fourier transform. Indeed, this section may be viewed as a mathematical extension of the theory of antenna arrays, studied in Section 5.3, from two dimensions into three dimensions.

Let the three-by-three matrix $\boldsymbol{M}$ specify a three-dimensional lattice. The lattice is the set of points given by

$$\boldsymbol{x} = \boldsymbol{M}\boldsymbol{i},$$

where $\boldsymbol{i}$ is any three-dimensional vector of integers. A common lattice is a rectangular lattice; in this case, $\boldsymbol{M}$ is a diagonal matrix.

An infinitely large crystal is defined as a three-dimensional convolution of $s(x, y, z)$ with a three-dimensional array of impulses on the lattice points

$$\begin{aligned} c(x, y, z) &= s(x, y, z) *** \sum_{i_x} \sum_{i_y} \sum_{i_z} \delta(\boldsymbol{x} - \boldsymbol{M}\boldsymbol{i}) \\ &= \sum_{i_x} \sum_{i_y} \sum_{i_z} s(\boldsymbol{x} - \boldsymbol{M}\boldsymbol{i}) \end{aligned}$$

where $s(\boldsymbol{x} - \boldsymbol{M}\boldsymbol{i})$ denotes the translation of $s(x, y, z)$ to the lattice point $\boldsymbol{M}\boldsymbol{i}$.

A finite crystal is obtained if the lattice points are restricted to a finite region of space. A finite crystal on a rectangular lattice can be written

$$c(x, y, z) = \sum_{i_x=0}^{N_x-1} \sum_{i_y=0}^{N_y-1} \sum_{i_z=0}^{N_z-1} s(x - d_x i_x, y - d_y i_y, z - d_z i_z).$$

Figure 8.3 shows an embodiment of this equation in which a complex organic molecule, $s(x, y, z)$, forms the crystal $c(x, y, z)$.

The simplest (and first) task of crystallography is to determine the crystal's lattice structure by estimating the matrix $\boldsymbol{M}$ from the diffraction data. A harder (and more recent) task is to estimate the structure of the individual molecule, that is, to form an image of $s(x, y, z)$. Other tasks include the study of thermal fluctuations or imperfections in the crystal by observation of their effects on the diffraction pattern. These topics will be briefly treated in Sections 8.8 and 8.9.

Whereas the single molecule $s(x, y, z)$ is physically too small to scatter observable waves, the array $c(x, y, z)$ can be very large if N_x, N_y, and N_z are very large, and the scattered waves can be easily observed. We need only to find an equation for the scattered wave in terms of $s(x, y, z)$. Then we can invert that equation to find $s(x, y, z)$ as a function of the scattered wave. Any implementable solution to this inverse problem gives an imaging algorithm.

We shall describe the diffraction of a wavefront by a finite rectangular crystal. Figure 8.4 shows a two-dimensional cross section of the crystal, specifically the cross section lying in the x, z plane. As compared to the time at which it passes the origin, a point on a wave moving along the z axis reaches the position with coordinate z with the time delay $\tau = z/c$. To derive the diffraction pattern directly, we would start with the simple geometry of Figure 8.4. We would first calculate the signal scattered in the

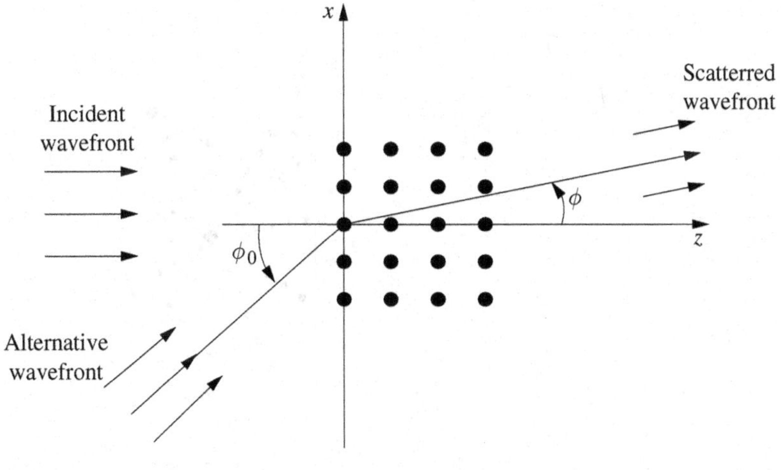

Figure 8.4 A two-dimensional crystal

x, z plane at the angle ϕ from the z axis by a plane, monochromatic incident wave that moves along the z axis. Then we would examine the scattering at an arbitrary direction.

However, this analysis is not different from the analysis in Section 8.3 in which we found $v(\phi, \psi)$ in terms of $S(f_x, f_y, f_z)$ for scattering from an object $s(x, y, z)$, so there is no need to repeat it. Thus, for the crystal, $v(\phi, \psi)$ can be described in terms of $C(f_x, f_x, f_z)$. The diffraction pattern for the complete three-dimensional, rectangular crystal can now be written in terms of the element transform $S(f_x, f_y, f_z)$ multiplied by a dirichlet function on each axis:

$$v(\phi, \psi) = C\left(\frac{\cos\psi\sin\phi - \cos\psi_0\sin\phi_0}{\lambda}, \frac{\sin\psi\sin\phi - \sin\psi_0\sin\phi_0}{\lambda}, \frac{\cos\psi - \cos\psi_0}{\lambda}\right)$$

where

$$C(f_x, f_y, f_z) = S(f_x, f_y, f_z)\mathrm{dirc}_{N_x}(d_x f_x)\mathrm{dirc}_{N_y}(d_y f_y)\mathrm{dirc}_{N_z}(d_z f_z).$$

This signal $v(\phi, \psi)$ will be large only when all three dirichlet functions are large. In effect, the grating lobes sample $S(f_x, f_y, f_z)$. At other values of ϕ and ψ, $v(\phi, \psi)$ will be negligibly small. This set of amplitude-modulated grating lobes is called the *Laue scattering pattern* of the crystal; a typical Laue scattering pattern is shown in Figure 8.5.

Grating lobes of the dirichlet function are found if and only if the following three equations, known as the *Bragg–Laue equations* for a rectangular crystal, are satisfied:

$$(\boldsymbol{s}_i - \boldsymbol{s}_0)\cdot\boldsymbol{i} = \ell_x\lambda/d_x$$
$$(\boldsymbol{s}_i - \boldsymbol{s}_0)\cdot\boldsymbol{j} = \ell_y\lambda/d_y$$
$$(\boldsymbol{s}_i - \boldsymbol{s}_0)\cdot\boldsymbol{k} = \ell_z\lambda/d_z$$

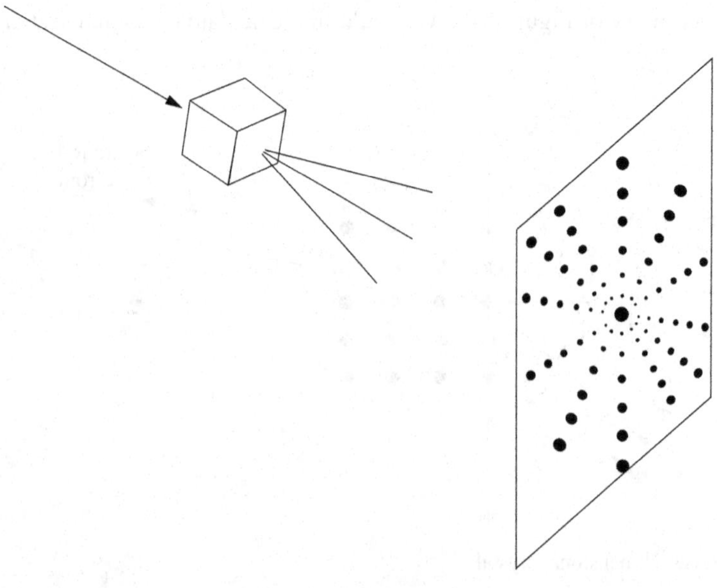

Figure 8.5 Illustrating formation of a Laue scattering pattern

where ℓ_x, ℓ_y, and ℓ_z are integers. If the crystal lattice is not rectangular, then the Bragg–Laue equations take a more general form. Measuring the angles of the grating lobes determines the terms on the left sides of the Bragg–Laue equations. From these equations, d_x, d_y, and d_z can then be computed.

The left side of each of the Bragg–Laue equations cannot be larger than two because all terms are unit vectors, and a dot product of two unit vectors has a magnitude not larger than one. This gives the maximum values for the integers ℓ_x, ℓ_y, and ℓ_z. The grating lobes can be observed for all integers satisfying

$$\ell_x \leq 2d_x/\lambda$$
$$\ell_y \leq 2d_y/\lambda$$
$$\ell_z \leq 2d_z/\lambda.$$

To ensure that grating lobes are formed, λ must be chosen smaller than twice the crystal spacing. Otherwise, no positive integer will satisfy these inequalities. To illustrate this point, the Bragg–Laue equations are written as

$$\cos\psi \sin\phi = \ell_x\lambda/d_x + \cos\psi_0 \sin\phi_0$$
$$\sin\psi \sin\phi = \ell_y\lambda/d_y + \sin\psi_0 \sin\phi_0$$
$$\cos\phi = \ell_z\lambda/d_z + \cos\phi_0.$$

The Bragg–Laue equations specify a relationship between the angles of incidence and the angles of reflection. In general, both sets of angles can be varied to observe the grating lobes. A solution exists only if the right side of each equation has a magnitude not larger than one. The equations may fail to have a solution in ψ and ϕ for a specified triple of integers (ℓ_x, ℓ_y, ℓ_z).

If the solution exists, it has the form

$$\psi = \tan^{-1}\frac{(\ell_x\lambda/d_x) + \cos\psi_0 \sin\phi_0}{(\ell_y\lambda/d_y) + \sin\psi_0 \sin\phi_0}$$

$$\phi = \tan^{-1}\frac{[[\quad]^2 + [\quad]^2]^{\frac{1}{2}}}{(\ell_z\lambda/d_z) + \cos\psi_0}$$

where the open brackets contain the terms in the numerator and denominator of the equation for ψ. The width of a grating lobe on each axis is inversely proportional to N_x, N_y, and N_z. These are extremely large integers for a typical crystal. Therefore, from a mathematical point of view, the grating lobes are exquisitely thin. However, from a physical point of view, the spacings in a crystal are not exact and, because of thermal noise, are fluctuating in time. This causes the actual grating lobes to be much wider than predicted from the idealized array. Furthermore, because the X-ray is not monochromatic, the incident beam will contain a range of wavelengths, and each wavelength will have a grating lobe at a slightly different angle.

8.6 Diffraction imaging

The more difficult task of computational crystallography is the general inverse problem of forming an estimated image of the arbitrary scattering function $s(x, y, z)$ from the signal scattered from a crystal. Measuring the scattered signal along the grating lobes gives $S(\ell_x/d_x, \ell_y/d_y, \ell_z/d_z)$, the Fourier coefficients of $s(x, y, z)$. In principle, the Fourier coefficients can be measured and inverted to find $s(x, y, z)$. In practice, however, only the magnitude of a Fourier coefficient, not the phase, is measured. Although phase retrieval from Fourier magnitude is possible in principle, it is difficult in practice. Moreover, often only partial data are available because of the limited viewing angle or undersampling. This means that the inversion usually must use some prior knowledge such as knowledge of the atomic species making up the molecule.

A useful way to find the inverse Fourier transform in simple cases is by pattern recognition. One recognizes from past experience some characteristic details of the magnitude of a Fourier transform. This requires familiarity with common cases. For example, one should be familiar with the magnitude of a Fourier transform of the simple helix.

Theorem 8.6.1 *The scattering function of the ideal infinite helix, illuminated along the longitudinal axis, is*

$$A(\phi, \psi) = \frac{a^2 b}{\lambda^3} \sum_{n=-\infty}^{\infty} J_n\left(2\pi \frac{a}{\lambda} \sin \phi\right) e^{jn(\psi+\pi/2)} \delta\left(\frac{\cos \phi - 1}{\lambda} + \frac{n}{2\pi b}\right).$$

Proof: This formula follows easily from the relationship

$$A(\phi, \psi) = \frac{1}{\lambda^3} S\left(\frac{\cos \psi \sin \phi}{\lambda}, \frac{\sin \psi \sin \phi}{\lambda}, \frac{\cos \phi - 1}{\lambda}\right)$$

by substitution of the trigonometric terms into the Fourier transform of the helix. □

The scattering function of Theorem 8.6.1 is zero everywhere except on a finite set of values of ϕ, given by

$$\cos \phi_n = 1 - \frac{n}{2\pi} \frac{\lambda}{b},$$

which has solutions (not larger than 90°) only for $n = 0, \ldots, \lfloor 2\pi b/\lambda \rfloor$. Thus, except for a finite number of values of ϕ, the function $A(\phi, \psi)$ is zero. For each ϕ_n satisfying the above equation, the scattering function is

$$A(\phi_n, \psi) = \frac{a^2 b}{\lambda^3} J_n\left(2\pi \frac{a}{\lambda} \sin \phi_n\right)$$

and

$$|A(\phi_n, \psi)| = \frac{a^2 b}{\lambda^3} \left| J_n \left(2\pi \frac{a}{\lambda} \sin \phi_n \right) \right|.$$

The function can be visualized as a finite set of coaxial cones, the nth cone at angle ϕ_n having an amplitude determined by its angle ϕ_n in terms of the nth-order Bessel function.

For a long helix of finite length, the Fourier transform of the infinite helix is convolved with a sinc function along the z axis. Now the layers of the Fourier transform become broadened and partially overlap. This means that each of the coaxial cones forming $|A(\phi_n, \psi)|$ is diffuse in ϕ. If the helix is sufficiently long, the broadening is negligible.

Finally, for an N_x by N_y by N_z array of finite-length helices, the function $A(\phi, \psi)$ is sampled by multiplying it by a three-dimensional dirichlet function. This means that the values of $|A(\phi_n, \psi)|$ can be observed only at the grating lobes. If these samples have the right Bessel function amplitudes, it may be reasonable to hypothesize that $s(x, y, z)$ is a helix, even when the Bragg–Laue samples are sparse and with not nearly enough samples to do the inversion. Additional data could be collected by illuminating the helix from directions other than the longitudinal axis.

We can calculate a library of Fourier transforms of various shapes, including variations of a helix, that may be used in this way – at least to form an initial estimate of the inverse Fourier transform. In general, pattern-matching techniques of this kind are useful only to find the coarsest details of an image.

8.7 Model formation from diffraction data

The task of image formation from diffraction data is to compute $s(x, y, z)$ from values of $|S(f_x, f_y, f_z)|$ sampled within the observable data sphere. For small molecules, the number of atoms of each atomic species contained in the molecule is usually known, and the electron density is approximately the same function $a(x, y, z)$ for all atoms of the same atomic species, provided the atoms are not too close to each other. If the molecule $s(x, y, z)$ consists of a moderate number of known atoms, it may be sufficient to determine only the coordinates of each atom. Then it may be possible to form an image by working with a parametric model of the form

$$s(x, y, z) = \sum_{\ell=1}^{n} a_\ell(x - x_\ell, y - y_\ell, z - z_\ell)$$

where $a_\ell(x, y, z)$ is the electron density of the ℓth atom indexed in order by weight. The coordinates (x_ℓ, y_ℓ, z_ℓ) for $\ell = 1, \ldots, n$ are the unknowns. There are $3n$ unknowns in the parametric model. If the number of atoms is not too large, there may be far more than $3n$ samples of $|S(f_x, f_y, f_z)|$. This means that one can expect, in principle, that $s(x, y, z)$ can be computed from the magnitude data by computing the unknown parameters. However, one needs a systematic way of working through the data.

The autocorrelation function

$$\phi(x, y, z) = s(x, y, z) *** s(-x, -y, -z)$$

has the Fourier transform

$$\Phi(f_x, f_y, f_z) = |S(f_x, f_y, f_z)|^2,$$

which is straightforward to compute from the observed data $|S(f_x, f_y, f_z)|$. We will assume here that the Bragg–Laue samples of $|S(f_x, f_y, f_z)|$ are spaced closely enough to be Nyquist samples for $\Phi(f_x, f_y, f_z)$, so it is straightforward to compute $\phi(x, y, z)$ from $|S(f_x, f_y, f_z)|$. We are concerned only with the computation of $s(x, y, z)$ from $\phi(x, y, z)$.

The autocorrelation function can be written in the form

$$\phi(x, y, z) = \sum_{\ell=1}^{n} \sum_{\ell'=1}^{n} a_\ell(x - x_\ell, y - y_\ell, z - z_\ell) *** a_{\ell'}(-x - x_{\ell'}, -y - y_{\ell'}, -z - z_{\ell'}).$$

This is a kind of parametric model: the n coordinate points (x_ℓ, y_ℓ, z_ℓ) for $\ell = 1, \ldots, n$ are the unknowns. There are n^2 terms here, so it is reasonable to consider an expression in this form if n is small, say up to a few hundred, but may be unreasonable if n is more than one thousand.

If the $a_\ell(x, y, z)$ were unit impulses, then $\phi(x, y, z)$ would consist of a sum of impulses of the form

$$\phi(x, y, z) = \sum_{\ell=1}^{n} \sum_{\ell'=1}^{n} \delta(x - (x_\ell - x_{\ell'}), y - (y_\ell - y_{\ell'}), z - (z_\ell - z_{\ell'})).$$

The computational task would be to determine the individual terms in this expression. From this information, one could determine all atomic vector separations and infer the molecular structure. This technique will hold as long as the $a_\ell(x, y, z)$ are narrow compared to the atomic spacing. Even if the $a_\ell(x, y, z)$ have widths comparable to the atomic spacing, it may be possible to force useful inferences by proceeding along these lines.

Often, there are only a small number of atomic species and only one copy, or perhaps only a few copies, of the largest atom. The largest atom has an electron density, $a_1(x, y, z)$, that may be known. If the large atoms are relatively sparse within the molecule, then the largest peaks in $\phi(x, y, z)$ that are not at the origin will correspond to pairwise differences in the positions of these large atoms, or to the pairwise difference in the position of a large atom and a smaller atom.

Because the origin of the coordinate system is arbitrary, we can choose it such that $x_1 = y_1 = z_1 = 0$. Then the autocorrelation function is written

$$\begin{aligned}\phi(x, y, z) &= a_1(x, y, z) *** a_1(x, y, z)\\ &+ \sum_{\ell=2}^{n} a_\ell(x - x_\ell, y - y_\ell, z - z_\ell) *** a_1(x, y, z)\\ &+ a_1(x, y, z) *** \sum_{\ell'=2}^{n} a_{\ell'}(-x - x_{\ell'}, -y - y_{\ell'}, -z - z_{\ell'})\\ &+ \sum_{\ell=2}^{n}\sum_{\ell'=2}^{n} a_\ell(x - x_\ell, y - y_\ell, z - z_\ell) *** a_{\ell'}(-x - x_{\ell'}, y - y_{\ell'}, z - z_{\ell'}).\end{aligned}$$

The first term is not useful. The last term will be small if the other a_ℓ are small in comparison with a_1. The two middle terms, if the atoms are not too close, will have peaks at the locations of individual atoms and at the mirror images of these locations. In this way, one may obtain initial estimates $(\widehat{x}_\ell, \widehat{y}_\ell, \widehat{z}_\ell)$ for $\ell = 2, \ldots, n$. The estimated molecule then is

$$\widehat{s}(x, y, z) = \sum_{\ell=1}^{n} a_\ell(x - \widehat{x}_\ell, y - \widehat{y}_\ell, z - \widehat{z}_\ell).$$

This estimate can be refined by various techniques that use the difference between $|\widehat{S}(f_x, f_y, f_z)|$ and $|S(f_x, f_y, f_z)|$ as an error term and set up an iteration.

Such methods of parametric model fitting may require some manual intervention in the computations, but a skilled analyst can eventually reconstruct a small molecule in this way. Indeed, such parametric methods are responsible for some early successes in diffraction imaging of moderate-sized molecules.

There are two formulas, the *Sayre formula* and the *Karle–Hauptman formula*, that are useful in devising computational estimation procedures along these lines. These formulas can be used to state constraints in the frequency domain that may be applied in some computational methods of image formation.

The Sayre formula interrelates the complex Fourier coefficients for the case where n is not too large and the n atoms are identical and individually resolvable. The Sayre formula states that, under these conditions,

$$S(f_x, f_y, f_z)B(f_x, f_y, f_z) = S(f_x, f_y, f_z) *** S(f_x, f_y, f_z)$$

where the function $B(f_x, f_y, f_z)$ depends only on $a(x, y, z)$ and is independent of the number of atoms n. The function $B(f_x, f_y, f_z)$ can be computed from the case with $n = 1$. Thus

$$A(f_x, f_y, f_z)B(f_x, f_y, f_z) = A(f_x, f_y, f_z) *** A(f_x, f_y, f_z),$$

so that

$$B(f_x, f_y, f_z) = \frac{A(f_x, f_y, f_z) *** A(f_x, f_y, f_z)}{A(f_x, f_y, f_z)}.$$

Whenever $a(x, y, z)$ is known, $A(f_x, f_y, f_z)$ is also known, so $B(f_x, f_y, f_z)$ can be computed.

The proof of the following proposition derives the Sayre formula, which explicitly refers to nonoverlapping, identical atoms. The condition is satisfied if each atom is confined to a circular support and these supports are nonintersecting.

Proposition 8.7.1 *The Fourier transform of a molecule, $s(x, y, z)$, composed of n nonoverlapping identical atoms, $a(x, y, z)$, satisfies the Sayre formula:*

$$\begin{aligned} &S(f_x, f_y, f_z)[A(f_x, f_y, f_z) *** A(f_x, f_y, f_z)] \\ &\quad = A(f_x, f_y, f_z)[S(f_x, f_y, f_z) *** S(f_x, f_y, f_z)]. \end{aligned}$$

Proof: In the space domain, the formula to be proved is

$$s(x, y, z) *** [a(x, y, z)]^2 = a(x, y, z) *** [s(x, y, z)]^2,$$

which we can write more concisely as

$$s(\boldsymbol{x}) *** a(\boldsymbol{x})^2 = a(\boldsymbol{x}) *** s(\boldsymbol{x})^2$$

with the notation $\boldsymbol{x} = (x, y, z)$. We will start with the term on the right and develop it as follows:

$$\begin{aligned} a(\boldsymbol{x}) *** s(\boldsymbol{x})^2 &= \int_{-\infty}^{\infty}\int_{-\infty}^{\infty}\int_{-\infty}^{\infty} a(\boldsymbol{x} - \boldsymbol{\xi}) \sum_{\ell=1}^{n} a(\boldsymbol{\xi} - \boldsymbol{x}_\ell) \sum_{\ell'=1}^{n} a(\boldsymbol{\xi} - \boldsymbol{x}_{\ell'})\, \mathrm{d}\boldsymbol{\xi} \\ &= \sum_{\ell=1}^{n}\sum_{\ell'=1}^{n} \int_{-\infty}^{\infty}\int_{-\infty}^{\infty}\int_{-\infty}^{\infty} a(\boldsymbol{x} - \boldsymbol{\xi}) a(\boldsymbol{\xi} - \boldsymbol{x}_\ell) a(\boldsymbol{\xi} - \boldsymbol{x}_{\ell'})\, \mathrm{d}\boldsymbol{\xi} \end{aligned}$$

where $\mathrm{d}\boldsymbol{\xi} = \mathrm{d}\xi_x\, \mathrm{d}\xi_y\, \mathrm{d}\xi_z$. The assumption of nonoverlapping atoms means that $a(\boldsymbol{\xi} - \boldsymbol{x}_\ell)a(\boldsymbol{\xi} - \boldsymbol{x}_{\ell'}) = 0$ unless $\ell = \ell'$. Hence the equation collapses to

$$\begin{aligned} a(\boldsymbol{x}) *** s(\boldsymbol{x})^2 &= \sum_{\ell=1}^{n} \int_{-\infty}^{\infty}\int_{-\infty}^{\infty}\int_{-\infty}^{\infty} a(\boldsymbol{x} - \boldsymbol{\xi}) a^2(\boldsymbol{\xi} - \boldsymbol{x}_\ell)\, \mathrm{d}\boldsymbol{\xi} \\ &= \sum_{\ell=1}^{n} \int_{-\infty}^{\infty}\int_{-\infty}^{\infty}\int_{-\infty}^{\infty} a(\boldsymbol{x} - \boldsymbol{x}_\ell - \boldsymbol{\xi}) a^2(\boldsymbol{\xi})\, \mathrm{d}\boldsymbol{\xi} \\ &= \int_{-\infty}^{\infty}\int_{-\infty}^{\infty}\int_{-\infty}^{\infty} a^2(\boldsymbol{\xi}) \sum_{\ell=1}^{n} a(\boldsymbol{x} - \boldsymbol{x}_\ell - \boldsymbol{\xi})\, \mathrm{d}\boldsymbol{\xi} \\ &= s(\boldsymbol{x}) *** a^2(\boldsymbol{x}), \end{aligned}$$

as was to be proved. □

The next general formula to be proved is the Karle–Hauptman formula. This is a generalization of the simple statement that if $s(t)$ is any real and nonnegative function, then its Fourier transform $S(f)$ satisfies

$$\det \begin{bmatrix} S(0) & S(\Delta) \\ S(-\Delta) & S(0) \end{bmatrix} \geq 0.$$

To verify this elementary inequality, write the determinant as

$$S(0)^2 - S(\Delta)S(-\Delta) = \int_{-\infty}^{\infty} s(t)\,dt \int_{-\infty}^{\infty} s(t')\,dt' - \int_{-\infty}^{\infty} s(t)e^{-j2\pi\Delta t}\,dt \int_{-\infty}^{\infty} s(t')e^{j2\pi\Delta t'}\,dt',$$

which is real because it is invariant under conjugation. But then

$$\begin{aligned} \int_{-\infty}^{\infty}\int_{-\infty}^{\infty} s(t)s(t')e^{-j2\pi\Delta(t-t')}\,dt\,dt' &= \int_{-\infty}^{\infty}\int_{-\infty}^{\infty} s(t)s(t')\cos 2\pi\Delta(t-t')\,dt\,dt' \\ &\leq \int_{-\infty}^{\infty}\int_{-\infty}^{\infty} s(t)s(t')\,dt\,dt', \end{aligned}$$

and the assertion follows. This simple two by two case clearly holds as well for the three-dimensional Fourier transform.

A general three-dimensional Karle–Hauptman matrix is defined as

$$\boldsymbol{M} = \begin{bmatrix} S(\mathbf{0}) & S(\boldsymbol{f}_1) & S(\boldsymbol{f}_2) & \dots & S(\boldsymbol{f}_n) \\ S^*(\boldsymbol{f}_1) & S(\mathbf{0}) & S(\boldsymbol{f}_1) & \dots & S(\boldsymbol{f}_{n-1}) \\ S^*(\boldsymbol{f}_2) & S^*(\boldsymbol{f}_1) & S(\mathbf{0}) & \dots & S(\boldsymbol{f}_{n-2}) \\ \vdots & \vdots & \vdots & \vdots & \vdots \\ S^*(\boldsymbol{f}_n) & S^*(\boldsymbol{f}_{n-1}) & S^*(\boldsymbol{f}_{n-2}) & \dots & S(\mathbf{0}) \end{bmatrix}$$

where $\boldsymbol{f}_\ell = (f_{x\ell}, f_{y\ell}, f_{z\ell})$. Define $\boldsymbol{f}_{-\ell} = -\boldsymbol{f}_\ell$. Then the ℓk element, denoted $m_{\ell k}$, of matrix $\boldsymbol{M}$ is $S(\boldsymbol{f}_{\ell-k})$. The determinant of $\boldsymbol{M}$ is real because $\boldsymbol{M}^* = \boldsymbol{M}^T$ implies that

$$\begin{aligned} (\det \boldsymbol{M})^* &= \det \boldsymbol{M}^* = \det \boldsymbol{M}^T \\ &= \det \boldsymbol{M}. \end{aligned}$$

The following proposition states that, moreover, the determinant of $\boldsymbol{M}$ is nonnegative.

Proposition 8.7.2 *A Karle–Hauptman matrix, $\boldsymbol{M}$, for the nonnegative function $s(x, y)$ satisfies*

$$\det \boldsymbol{M} \geq 0.$$

Proof: We shall prove only the one-dimensional version of this theorem. Both the two-dimensional version and the three-dimensional version are proved in the same way. The one-dimensional version will be simply obtained as the dual of the statement that a covariance matrix is positive-semidefinite. To recall this statement, let $X_1, \dots, X_n$ be random variables with zero mean, and let $\boldsymbol{\Sigma}$, with elements $\mathrm{E}[X_i X_j]$, denote the n by n covariance matrix. A covariance matrix is always positive-semidefinite because

$$\boldsymbol{a}\boldsymbol{\Sigma}\boldsymbol{a}^T = \sum_i \sum_j a_i a_j \mathrm{E}[X_i X_j] = \mathrm{E}\left[(\Sigma_i a_i X_i)^2\right] \geq 0.$$

We shall apply this to a stationary random process, $X(t)$, by choosing any set of time points, $t_1, \ldots, t_n$, and setting $X_i = X(t_i)$. Then

$$\mathrm{E}[X_i X_j] = \mathrm{E}[X(t_i)X(t_j)] = \phi(t_i - t_j)$$

where $\phi(\tau) = \mathrm{E}[X(t)X(t+\tau)]$ is the correlation function of $X(t)$. Any nonnegative real function, $R(f)$, that has an inverse Fourier transform can be regarded as the power density spectrum of some stationary random process $X(t)$, and the inverse Fourier transform of $R(f)$ is the autocorrelation function $\phi(\tau)$ of this random process.

To proceed with the proof, recall that the spectrum $S(f)$ has an inverse Fourier transform, $s(x)$, that is nonnegative, while any correlation function $\phi(\tau)$ has a Fourier transform, $R(f)$, that is nonnegative. By changing notation, the elements of the Karle–Hauptman matrix $\boldsymbol{M}$ are samples of $\phi(\tau)$ and $\boldsymbol{M} = \boldsymbol{\Sigma}$. Because $\boldsymbol{\Sigma}$ is positive-semidefinite, we can conclude that $\boldsymbol{M}$ is positive-semidefinite. This means that all eigenvalues are nonnegative, so the determinant, as the product of the eigenvalues, is nonnegative. Now this is the statement of the Karle–Hauptman formula except that the time and frequency variables have been interchanged. This change does not affect the conclusion. □

8.8 Diffraction from fiber arrays

Some molecules cannot be made to form a regular crystal, hence the standard techniques of diffraction imaging cannot be used to form images for these molecules. However, even a disordered collection of molecules will diffract X-rays, and the diffraction pattern does depend in some way on the shape of the individual molecule. When there are a large number of molecules, individually disordered, the expected diffraction pattern is the statistical average of the individual diffraction patterns. Even though the information in this diffraction pattern of a disordered array of molecules is much weaker than the information in the diffraction pattern for an ordered array, there still may be useful information.

The nature of the disorder can take various forms, and each possibility requires individual analysis. One very disordered case that we might consider consists of identical molecules strewn randomly at the points of a three-dimensional poisson process, which is a random set of locations in three dimensions. In addition, each molecule has a random orientation that is described by a rotation matrix that is uniformly distributed and independent of the rotation matrix defining the orientation of any other molecule. A variation of this is a powder in which the individual grains, though themselves regular crystals, are completely disordered. Each grain is randomly oriented in angle, so each has a diffraction pattern that is randomly oriented. The diffraction pattern of the powder is a composite of the diffraction patterns of the individual grains. Evidentally, the diffraction pattern has the form of a series of concentric shells because the only spatial dependence that is not averaged out is the radial dependence.

A simpler case of importance in practice is a regular array of molecules, but with the orientation of each molecule independent and uniformly distributed over all rotations in three-dimensional space. We shall discuss an even simpler case – a random rotation about the z axis of each molecule in an otherwise regular array. This is called a *fiber array*.

Let $s(x, y, z)$ be any function, and let

$$s_\psi(x, y, z) = s(x \cos\psi - y \sin\psi, x \sin\psi + y \cos\psi, z).$$

This function has the Fourier transform

$$S_\psi(f_x \cos\psi - f_y \sin\psi, f_x \sin\psi + f_y \cos\psi, f_z).$$

Define the fiber array as

$$p(x, y, z) = \sum_{\ell_x=0}^{N-1}\sum_{\ell_y=0}^{N-1}\sum_{\ell_z=0}^{N-1} s_{\psi_{\ell_x\ell_y\ell_z}}(x - \ell_x, y - \ell_y, z - \ell_z)$$

where the $\psi_{\ell_x\ell_y\ell_z}$ are independent random variables that are uniform on the interval $[0, 2\pi)$. With this model, each molecule is at a lattice point indexed by the integers ℓ_x, ℓ_y, and ℓ_z, but has a random rotation about the z axis, denoted by $\psi_{\ell_x\ell_y\ell_z}$. The scattering function is given in terms of the Fourier transform

$$P(f_x, f_y, f_z) = \sum_{\ell_x=0}^{N-1}\sum_{\ell_y=0}^{N-1}\sum_{\ell_z=0}^{N-1} S_{\psi_{\ell_x\ell_y\ell_z}}(f_x, f_y, f_z)e^{-j2\pi(\ell_x f_x + \ell_y f_y + \ell_z f_z)}.$$

The ensemble average of this random function $P(f_x, f_y, f_z)$ is the expectation

$$\mathrm{E}[P(f_x, f_y, f_z)] = \sum_{\ell_x=0}^{N-1}\sum_{\ell_y=0}^{N-1}\sum_{\ell_z=0}^{N-1} \mathrm{E}[S_{\psi_{\ell_x\ell_y\ell_z}}(f_x, f_y, f_z)]e^{-j2\pi(\ell_x f_x + \ell_y f_y + \ell_z f_z)}.$$

Because the expectation term is the same for every value of ℓ_x, ℓ_y, and ℓ_z, it can be moved outside the summation signs. Then the sums can be executed (and the residual phase term ignored) to write

$$\mathrm{E}[P(f_x, f_y, f_z)] = \mathrm{E}[S_\psi(f_x, f_y, f_z)]\mathrm{dirc}_N(f_x, f_y, f_z)$$

where

$$\mathrm{E}[S_\psi(f_x, f_y, f_z)] = \frac{1}{2\pi}\int_0^{2\pi} S(f_x \cos\psi - f_y \sin\psi, f_x \sin\psi + f_y \cos\psi, f_z)\,d\psi.$$

A similar analysis applied to the squared magnitude of $P(f_x, f_y, f_z)$ gives

$$\mathrm{E}[|P(f_x, f_y, f_z)|^2] = \mathrm{E}[|S_\psi(f_x, f_y, f_z)|^2]\mathrm{dirc}_N(f_x, f_y, f_z)$$

for the expected intensity.

As an example, we will consider an array of helices, each helix randomly rotated around the z axis by a uniformly distributed angle. A finite helix of length L has the Fourier transform

$$S(f_x, f_y, f_z) = \sum_{n=-\infty}^{\infty} J_n\left(2\pi\sqrt{f_x^2 + f_y^2}\right) \mathrm{e}^{-\mathrm{j}n\tan^{-1}(f_x/f_y)}\mathrm{sinc}(L(f_z - n/2\pi)).$$

The complex exponential in $S_\psi(f_x, f_y, f_z)$ can be written

$$\mathrm{e}^{-\mathrm{j}n\tan^{-1}\left(\frac{f_x\cos\psi - f_y\sin\psi}{f_x\sin\psi + f_y\cos\psi}\right)} = \mathrm{e}^{-\mathrm{j}n\tan^{-1}(f_x/f_y)}\mathrm{e}^{\mathrm{j}n\psi}.$$

The expectation operator applied to this expression will lead to the term $\mathrm{E}[\mathrm{e}^{\mathrm{j}n\psi}]$, which will be zero except for the term with $n = 0$. Consequently,

$$\mathrm{E}[S_\psi(f_x, f_y, f_z)] = \mathrm{sinc}(Lf_z)J_0\left(2\pi\sqrt{f_x^2 + f_y^2}\right),$$

and

$$\mathrm{E}[P(f_x, f_y, f_z)] = \mathrm{sinc}(Lf_z)J_0\left(2\pi\sqrt{f_x^2 + f_y^2}\right)\mathrm{dirc}_N(f_x, f_y, f_z).$$

This is the same as the diffraction pattern for an array of cylindrical shells. By allowing each helix in the array to have an arbitrary angle about the z axis, the details of the helix have been ensemble-averaged into a cylindrical shell. This is the ensemble mean. The ensemble variance can be evaluated in a similar way. It is, approximately,

$$\mathrm{E}[|P(f_x, f_y, f_z)|^2] = \sum_{n=-\infty}^{\infty} J_n^2(2\pi f)\mathrm{sinc}^2(L(f_z - n/2\pi))\mathrm{dirc}_N(f_x, f_y, f_z)$$

where $f^2 = f_x^2 + f_y^2$. This is an approximation because the interaction between layers has been ignored as negligible.

8.9 Diffraction from excited arrays

The individual molecules in a crystal array always have thermal energy, which causes them to randomly vibrate. This vibration leads to time-varying irregularities in the crystal spacing that affects the diffraction pattern. Moreover, even if there were no thermal vibration, the molecules would not be placed perfectly at the lattice points; there would be imperfections in the spacing and the orientation. Because the diffraction pattern is a Fourier transform, these imperfections can be studied as the effect of irregularities on the Fourier transform of an array.

The irregularities in the positions of the elements in an array manifest themselves primarily as phase errors in the terms making up the dirichlet functions. The effect of phase errors on the Fourier transform, which holds for irregularities in the spacings of any array, is studied in general in Chapter 15. In this section, we shall refer to that general theory to describe the effect that the vibration and imperfections of a crystal have on the diffraction pattern. This is important because observation of this effect in the diffraction pattern provides an indirect means of observing the vibration energy and measuring the amount of irregularity in the crystal.

Proposition 15.3.1 of Chapter 15 describes the expected Gabor bandwidth of a one-dimensional pulse $s(t)$ contaminated by phase noise as given by $s(t)e^{j\theta(t)}$. This expression for Gabor bandwidth is

$$E[B_G^2(v)] = B_G^2(s) + \int_{-\infty}^{\infty} f^2\Phi(f)\,df$$

where $B_G^2(s)$ is the Gabor bandwidth of the pulse $s(t)$, and $\Phi(f)$ is the power density spectrum of the phase noise. If the first term on the right is small, it may be adequate to use the approximation

$$\int_{-\infty}^{\infty} f^2\Phi(f)\,df \approx E[B_G^2(v)]$$

as a simple way to estimate the bandwidth of the phase noise. How does this broadening of the Gabor bandwidth relate to crystal diffraction? In the case of a crystal, the time variable t is replaced by the three space variables x, y, and z. Because there are three space variables, the situation is somewhat more elaborate but not really more complicated. It is enough to consider only one space variable. The Gabor bandwidth B_G^2 is determined by the width of a one-dimensional grating lobe, which is negligibly small if the number of cells of the crystal is large. Replacing the simple variable time by the three space variables does not change the nature of this conclusion. Thus, by observing the Gabor bandwidth of the main lobe of the diffraction pattern, one can estimate the bandwidth of the spatial phase noise. The variance of position errors in the elements of the crystal can be estimated by describing the effect of position error on the phase error, then estimating the three-dimensional bandwidth of the phase error from the Gabor bandwidth of the main lobe of the diffraction pattern.

There is more to say if the array also exhibits thermal vibration. This is because there are two consequences of thermal vibration of an array: vibration affects both the space structure and the time structure of the grating lobes. The vibration will affect the space structure because the irregularities in the space structure of the diffracted wave cause the grating lobes to be widened. The vibration will affect the time structure because then the spatial irregularities are time varying, so the structure of the grating lobes will display time fluctuations. By measuring the shape of the lobes as a function of time, it is possible to make inferences about the thermal fluctuations of the crystal and the strength of its bonds.

The diffraction pattern of a crystal is usually regarded as a space function, but the diffraction pattern can also be made to be a function of time by taking diffraction patterns at successive time intervals $[(k-1)\Delta T, k\Delta T]$ each of length ΔT. Then the diffraction pattern becomes $s(x, y, z, k\Delta T)$. The peak of this function in space will vary with k, as will the width, and this dispersion is an indicator of thermal motion of the elements. The variance of the main lobe then gives a measure of the second derivative of the correlation function at the origin.

Problems

8.1 **(Coordinate Transformation)** Prove the three-dimensional Fourier transform relationship

$$s\left(\boldsymbol{A}^T\begin{bmatrix} x \\ y \\ z \end{bmatrix}\right) \Leftrightarrow \frac{1}{|\boldsymbol{A}|} S\left(\boldsymbol{A}^{-1}\begin{bmatrix} f_x \\ f_y \\ f_z \end{bmatrix}\right)$$

where $|\boldsymbol{A}|$ is the determinant of $\boldsymbol{A}$. State and prove the coordinate rotation property in three dimensions.

8.2 Find the three-dimensional Fourier transform of $\mathrm{rect}(x, y, z)$ and $\mathrm{cyln}(x, y, z)$.

8.3 A three-dimensional function $s(x, y, z)$ is to be estimated from its Fourier transform $S(f_x, f_y, f_z)$ measured only at those points within the sphere described by

$$f_x^2 + f_y^2 + f_z^2 \leq k^2.$$

a. Suppose that $s(x, y, z)$ is computed by simply setting $S(f_x, f_y, f_z) = 0$ at all f_x, f_y, f_z outside this sphere and computing the three-dimensional inverse Fourier transform of this cropped data set. How is $s(x, y, z)$ degraded?

b. Can one make any better inferences about $S(f_x, f_y, f_z)$ outside this sphere?

8.4 Show that $\mathrm{tinc}(0) = \pi/6$, first by using l'Hôpital's rule, and then by using a Taylor series expansion.

8.5 **a.** Given a general nonrectangular lattice,

$$v = \boldsymbol{M}\boldsymbol{i},$$

defined by matrix $\boldsymbol{M}$, find the diffraction pattern for a crystal defined by this lattice and the scattering function $s(x, y, z)$.

b. State the general form of the Bragg–Laue equations for a nonrectangular lattice.

8.6 **a.** The autocorrelation of a circle function is the two-dimensional *Chinese hat function*

$$\text{chat}(x, y) = \text{circ}(x, y) ** \text{circ}(x, y).$$

Find the Fourier transform of $\text{cyln}(x, y, z) *** \text{cyln}(x, y, z)$.

b. Find the general features of the autocorrelation function of a hexagonal molecule, that is, find the autocorrelation function of a nonoverlapping hexagonal arrangement of six cylinder functions in the x, y plane.

c. Find the intensity of the scattering function of a hexagonal arrangement of six nonoverlapping circle (or cylinder) functions in the x, y plane. Give a "signature" or "template" for recognizing a hexagonal component in a molecule from its scattering intensity.

8.7 A three-dimensional generalization of the two-dimensional hat function $\text{chat}(x, y)$, called the *three-dimensional hat function*, is defined as

$$\text{that}(x, y, z) = \text{sphr}(x, y, z) *** \text{sphr}(x, y, z).$$

Give a closed form expression for $\text{that}(x, y, z)$. What is the three-dimensional Fourier transform of $\text{that}(x, y, z)$?

8.8 Describe the function

$$s(x, y, z) = \text{circ}(x, y)\text{comb}(z)$$

and find its Fourier transform.

8.9 **a.** Prove the Nyquist–Shannon sampling theorem for three-dimensional sampling on a cubic lattice.

b. Let $s(x, y, z)$ be a function whose Fourier transform satisfies $S(f_x, f_y, f_x) = 0$ if $f_x^2 + f_y^2 + f_z^2 > \frac{1}{2}$. Derive a tinc-interpolation formula for samples of $s(x, y, z)$ on a cubic lattice. Why is this better than three-dimensional sinc interpolation?

c. The *grocer's lattice* in three dimensions is familiar from the commonplace packing of spheres. Define this lattice as

$$\begin{bmatrix} x \\ y \\ z \end{bmatrix} = \boldsymbol{M} \begin{bmatrix} i \\ i' \\ i'' \end{bmatrix}$$

for an appropriate matrix $\boldsymbol{M}$. Use this lattice to define a sampling pattern for a function, $s(x, y, z)$, with the Fourier transform satisfying $S(f_x, f_y, f_z) = 0$ for $f_x^2 + f_y^2 + f_z^2 > \frac{1}{2}$. By how much is the sampling density improved compared to cubic sampling? Give an interpolation formula.

d. Do grocers use the grocer's lattice to save space or for another reason?

8.10 Wall paint is colored by microscopic particles suspended within a transparent liquid that dries to a solid. Suppose that light is scattered (in the Born approximation) from all atoms in the interior of each particle and that the sum of the volumes of all scattering particles is fixed. Use the formula for the three-dimensional Fourier transform of a sphere to explain how the diameter of these particles should be chosen to maximize the hiding power of the paint (at wavelength λ). How should the diameter be chosen to maximize the hiding power if the scattering is only from the surface of each particle?

8.11 Derive the identity

$$e^{-ja\sin t} = \sum_{n=-\infty}^{\infty} J_n(a)e^{-jnt}.$$

8.12 Show that the three-dimensional Fourier transform of the ideal finite helix is

$$\text{helx}(x, y, z)\text{rect}\left(\frac{z}{L}\right)$$

$$\Leftrightarrow \sum_{n=-\infty}^{\infty} J_n\left(2\pi\sqrt{f_x^2 + f_y^2}\right) e^{-jn\tan^{-1}(f_x/f_y)} L\,\text{sinc}\left(Lf_z - L\frac{n}{2\pi}\right).$$

8.13 Consider the ideal infinite helix with $b = \lambda/2$ and $a = \lambda$. Sketch the scattering function $A(\phi, \psi)$ for the helix illuminated along the longitudinal axis. Sketch the scattering function for the helix illuminated normal to the longitudinal axis. How does the scattering function change if the infinite helix is truncated to a helix of length 100λ?

8.14 For fixed x and y, expand $\text{helx}(x, y, z)$ in a Fourier series to give

$$\text{helx}(x, y, z) = \sum_{n=-\infty}^{\infty} s_n(x, y)e^{jnz}.$$

Compute the Fourier transform of $\text{helx}(x, y, z)$, starting from this expansion. What is $s_n(x, y)$ and what is its two-dimensional Fourier transform?

8.15 **a.** Show that

$$\text{sinc}(z) \Leftrightarrow \delta(f_x, f_y)\text{rect}(f_z).$$

b. Show that

$$\delta(x, y)\text{rect}(z) *** \text{helx}(x, y, z) = \text{ring}(x, y, z)$$

where $\text{ring}(x, y, z)$ is the infinite cylindrical surface $\text{ring}(x, y)$.

c. Use the convolution theorem to find the Fourier transform of $\text{cyln}(x, y, z)$.

8.16 How does the Fourier transform of $\text{helx}(x, y, z)$ change if the helix is rotated around the z axis by 90°? How does the Fourier transform of $\text{helx}(x, y, z)$ change if the helix is translated along the z axis by $1/4$?

8.17 The discontinuous helix is a set of discrete points uniformly spaced along the helix. It can be defined as

$$\mathrm{dhlx}(x, y, z) = \mathrm{helx}(x, y, z)\mathrm{comb}\left(\frac{z}{a}\right)$$

where a is a parameter that determines the spacing. Find the Fourier transform of the discontinuous helix. What happens if $a = 1$?

8.18 The Sayre formula can be concisely written as

$$S(f) = \frac{1}{B(f)} S(f) * S(f)$$

$$B(f) = \frac{1}{A(f)} A(f) * A(f).$$

Suppose that $S(f)$ and $A(f)$ can be approximated by the sum of a finite number of impulses

$$S(f) = \sum_k S_k \delta(f - f_k).$$

Show that the Sayre formula reduces to

$$S_k = \frac{1}{B_k} \sum_h S_h S_{k-h}.$$

8.19 **(Structure Invariants)** Let $S(f_x, f_y, f_z) = |S(f_x, f_y, f_z)|e^{j\phi(f_x, f_y, f_z)}$ be the Fourier transform of $s(x, y, z)$. Let $S_0(f_x, f_y, f_z) = |S_0(f_x, f_y, f_z)|e^{j\phi_0(f_x, f_y, f_z)}$ be the Fourier transform of $s(x - x_0, y - y_0, z - z_0)$.

a. Prove that

$$\phi_0(f_x, f_y, f_z) + \phi_0(f'_x, f'_y, f'_z) + \phi_0(f''_x, f''_y, f''_z)$$

does not depend on (x_0, y_0, z_0), provided

$$(f_x, f_y, f_z) + (f'_x, f'_y, f'_z) + (f''_x, f''_y, f''_z) = (0, 0, 0).$$

b. Generalize this statement to a sum of N terms.

c. Explain how the structure invariants might be useful in an attempt to experimentally recover phase.

Notes

X-ray diffraction was introduced into crystallography by Laue (1912) to demonstrate the wave properties of X-rays. It was quickly realized that X-rays provide a way to probe the structure of crystallized molecules. The two Braggs, father and son (1929, 1942), extensively studied the problem of reconstruction of the scattering distribution from the scattered waves. This led to the widespread realization that the shapes of molecules could be deduced from the X-ray diffraction pattern.

Patterson (1935) noticed that the autocorrelation function of the scattering function could be recovered as the inverse Fourier transform of the diffraction intensity, and that, for sparse molecules, this autocorrelation function provides all pairwise interatomic distances. Sayre (1952) made the observation that the Nyquist spacing for the autocorrelation function is half the Nyquist spacing for the original function.

The classic paper of Cochran, Crick, and Vand (1952) discussed the Fourier transform of the helix as an important signature in studying many complex molecules such as polymers. Watson and Crick (1953) used this observation to great advantage in deducing the double helix structure of DNA. Harker and Kasher (1948), Karle and Hauptman (1950), and Sayre (1952) introduced various direct constraints on the Fourier transform phase in terms of the Fourier transform magnitude for special cases. Hauptman and Karle (1953) developed direct methods for recovering the scattering function of small molecules based on the structure invariants and a least-squares formulation. Such direct methods use prior knowledge of the number and kinds of atoms composing the molecule. Bricogne (1984) proposed the use of statistical methods to solve the phase-retrieval problem for large molecules.

9 Construction and reconstruction of images

Image formation is the task of constructing an image of a scene when given a set of noisy data that is dependent on that scene. Possibly some prior information about the scene is also given. Image formation also includes the task of refining a prior image when given additional fragmentary or degraded information about that image. Then the task may be called image restoration.

In the most fundamental problem of image restoration, one is given an image of a two-dimensional scene, but the detail of the image, in some way, is limited. For example, the image of the scene may be blurred or poorly resolved in various directions. Sophisticated signal-processing techniques, called *deconvolution* or *deblurring*, can enhance such an image. When the blurring function is not known but must be inferred from the image itself, these techniques are called *blind deconvolution* or *blind deblurring*. Problems of deconvolution are well known to be prone to computational instability, and great care is needed in the implementation of deconvolution algorithms.

Another task of image construction is estimating an image from partial knowledge of some of the properties of the image. An important instance of this task is estimating an image from the magnitude of its two-dimensional Fourier transform. This process is known as *phase retrieval* because the phase of the Fourier transform is implicit in the estimated image. In effect, by recovering the image, the phase of the Fourier transform is inferred from the magnitude of the Fourier transform.

Other topics of image estimation are deferred to later chapters. Of these, one important topic deals with the problem of estimating an image from a set of its projections. This topic, known as *tomography*, is important enough to be given its own chapter, which follows this one. Another important topic, discussed in Chapter 11, is the study of likelihood methods. The maximum-likelihood principle is the basis of probabilistic methods to determine the image that best accounts for a set of fragmentary data related to the image. Such methods use raw data that may be of a form quite different from the image, though dependent in some way on that image.

9.1 Deconvolution and deblurring

The one-dimensional signal $s(t)$ is dispersed in a known way when passed through the known, linear, time-invariant filter $h(t)$. Because the filter $h(t)$ is linear and time invariant, this behavior is described by the one-dimensional convolution

$$r(t) = \int_{-\infty}^{\infty} h(t - \xi)s(\xi)\,d\xi.$$

In general, a filter need not be time invariant. Then it is written

$$r(t) = \int_{-\infty}^{\infty} h(t|\xi)s(\xi)\,d\xi$$

where $h(t|\xi)$ is the time-varying impulse response of the filter.

Likewise, the two-dimensional signal $s(x, y)$ passed through the linear, space-invariant filter $h(x, y)$ is described by the two-dimensional convolution

$$r(x, y) = \int_{-\infty}^{\infty}\int_{-\infty}^{\infty} h(x - \xi, y - \eta)s(\xi, \eta)\,d\xi\,d\eta.$$

The effect of an unwanted two-dimensional filter is called *blurring*. In general, a two-dimensional filter need not be space invariant. The effect is still called blurring and is written

$$r(x, y) = \int_{-\infty}^{\infty}\int_{-\infty}^{\infty} h(x, y|\xi, \eta)s(\xi, \eta)\,d\xi\,d\eta$$

where $h(x, y|\xi, \eta)$ is the space-varying pointspread function of the filter.

The inverse operation – the task of removing the effect of the convolution or blurring – in whole or in part – is called *deconvolution* or *deblurring*. Deblurring is the task of solving an integral equation of the form above[1] for the unknown function $s(x, y)$ when the function $r(x, y)$ and the pointspread function $h(x, y|\xi, \eta)$ are both given, and the integral is over some set known to be the domain of $s(x, y)$. The deblurring problem reduces to the deconvolution problem whenever the pointspread function depends only on the spatial difference between the points (x, y) and (ξ, η), and not on each point separately. Thus two-dimensional deconvolution then is the estimation of the function $s(x, y)$ that satisfies the equation

$$r(x, y) = \int_{-\infty}^{\infty}\int_{-\infty}^{\infty} h(x - \xi, y - \eta)s(\xi, \eta)\,d\xi\,d\eta$$

when $r(x, y)$ and $h(x, y)$ are both known. The term "deconvolution" is natural because

[1] In the situation in which s is unknown, these are all examples of a Fredholm integral equation of the first kind. Many inverse problems can be seen as the task of solving a Fredholm integral equation.

the given convolution

$$r(x, y) = h(x, y) ** s(x, y)$$

is to be undone.

The deconvolution problem is notorious for the instability exhibited when a numerical solution is attempted. Small changes in $h(\xi|t)$ or in $h(t)$ can result in large changes in the solution. Seemingly insignificant choices in an implementation for producing numerical solutions can also have a large effect on the result. A problem that has this behavior is called an *ill-posed problem.* In general, a problem is ill-posed if it is not well-posed, and a problem is *well-posed* if it has a unique solution that is continuous in the data. Inverse problems that are formulated on a continuous space are almost always ill-posed. Problems on finite-dimensional parameter spaces are usually well-posed, though often *ill-conditioned,* meaning that the solution varies rapidly, though continuously, with changes in the data. Ill-posed behavior is well studied and has been addressed by using various methods for regularizing the problem. The process of *regularization* of the problem is the process of restricting the class of acceptable solutions based on heuristic judgment. The purpose of regularization is to replace an ill-posed problem, such as a problem on a continuous parameter space, by a well-posed problem on a finite-dimensional space, as may be obtained by pixelization. For example, the solution may be expected to satisfy certain smoothness conditions, and these conditions become a part of the problem.

The *inverse filter*, if it exists, for the filter $h(x, y)$ is the filter $h^{-1}(x, y)$ such that

$$h^{-1}(x, y) ** h(x, y) = \delta(x, y).$$

Then the convolution

$$r(x, y) = h(x, y) ** s(x, y)$$

can be undone by

$$s(x, y) = h^{-1}(x, y) ** r(x, y).$$

This simple statement may seem to settle the matter, but, in fact, the formula is unsatisfactory for many reasons. One difficulty is that $h^{-1}(x, y)$ does not exist as a proper function with finite energy. The inverse filter stated in the Fourier transform domain is

$$H^{-1}(f_x, f_y) = \frac{1}{H(f_x, f_y)}.$$

Clearly, even though $H(f_x, f_y)$ has finite energy, $H^{-1}(f_x, f_y)$ will have infinite energy, which means that $h^{-1}(x, y)$ has infinite energy. Another difficulty is that $r(x, y)$ may not be precisely known, and the simple deconvolution procedure can amplify errors. Thus if only the noisy signal

$$v(x, y) = h(x, y) ** s(x, y) + n(x, y)$$

is known, then one may compute

$$h^{-1}(x, y) ** v(x, y) = s(x, y) + h^{-1}(x, y) ** n(x, y).$$

Although the blur has been removed, the noise term, in general, is made worse. The inverse filter may even change noise of finite power into noise of infinite power, which clearly is unacceptable.

The Wiener filter

A better approach to deconvolution is to find a filter, $g(x, y)$, that is a compromise between removing the blur and controlling the noise. The filter output forms the estimate of $s(x, y)$ given by

$$\begin{aligned}\widehat{s}(x, y) &= g(x, y) ** v(x, y) \\ &= g(x, y) ** [h(x, y) ** s(x, y) + n(x, y)].\end{aligned}$$

One widely accepted criterion for choosing the filter $g(x, y)$ is to minimize the combined mean-square error of the signal and the noise. The two-dimensional filter that minimizes the mean-square error is known as the two-dimensional *Wiener filter*. To construct such a filter, the unknown signal $s(x, y)$ is modeled as a stationary random process with zero mean and power density spectrum $\Phi_s(f_x, f_y)$, and the noise $n(x, y)$ is modeled as stationary noise with zero mean and power density spectrum $\Phi_n(f_x, f_y)$.

Proposition 9.1.1 *The Wiener filter is given by*

$$G(f_x, f_y) = \frac{H^*(f_x, f_y)\Phi_s(f_x, f_y)}{|H(f_x, f_y)|^2\Phi_s(f_x, f_y) + \Phi_n(f_x, f_y)}.$$

Proof: We shall treat only the case of real signals. The Wiener filter for real signals is derived by starting with the error signal, which is given by

$$\Delta(x, y) = s(x, y) - g(x, y) ** v(x, y)$$

where

$$v(x, y) = h(x, y) ** s(x, y) + n(x, y).$$

The error signal $\Delta(x, y)$ is a real, stationary random process with zero mean and variance

$$\sigma^2 = \mathrm{E}[s(x, y)^2] - 2\mathrm{E}[s(x, y)[g(x, y) ** v(x, y)]] + \mathrm{E}[[g(x, y) ** v(x, y)]^2].$$

Only terms of σ^2 that are quadratic in $s(x, y)$ or $n(x, y)$ have a nonzero expectation. Terms that are linear in $s(x, y)$ or $n(x, y)$ have a zero expectation and need not be considered further. Our next step is to transform the expression for σ^2 into the frequency domain. This step is simplified by using the general rules (see Problem 9.8)

$$\mathrm{E}[s(x,y)[b(x,y) ** s(x,y)]] = \int_{-\infty}^{\infty}\int_{-\infty}^{\infty} B(f_x,f_y)\Phi_s(f_x,f_y)\,\mathrm{d}f_x\,\mathrm{d}f_y$$

and

$$\mathrm{E}[b(x,y) ** s(x,y)]^2 = \int_{-\infty}^{\infty}\int_{-\infty}^{\infty} \left|B(f_x,f_y)\right|^2 \Phi_s(f_x,f_y)\,\mathrm{d}f_x\,\mathrm{d}f_y,$$

which are easily verified for any pointspread function $b(x, y)$ and stationary random process $s(x, y)$. With the aid of these rules, we have

$$\begin{aligned}\sigma^2 = \phi_s(0,0) &- 2\int_{-\infty}^{\infty}\int_{-\infty}^{\infty} G(f_x,f_y)H(f_x,f_y)\Phi_s(f_x,f_y)\,\mathrm{d}f_x\,\mathrm{d}f_y \\ &+ \int_{-\infty}^{\infty}\int_{-\infty}^{\infty} |G(f_x,f_y)|^2|H(f_x,f_y)|^2\Phi_s(f_x,f_y)\,\mathrm{d}f_x\,\mathrm{d}f_y \\ &+ \int_{-\infty}^{\infty}\int_{-\infty}^{\infty} |G(f_x,f_y)|^2\Phi_n(f_x,f_y)\,\mathrm{d}f_x\,\mathrm{d}f_y.\end{aligned}$$

The calculus of variations now can be used to choose $G(f_x, f_y)$. To see how this works, we will minimize the simpler expression

$$\sigma^2 = 2\int_{-\infty}^{\infty} G(f)A(f)\,\mathrm{d}f + \int_{-\infty}^{\infty} |G(f)|^2 B(f)\,\mathrm{d}f$$

over the choice of $G(f)$ with the understanding that all functions in the time domain are real-valued. The method of the calculus of variations replaces $G(f)$ with $G(f) + \epsilon\eta(f)$ and sets the derivative with respect to ϵ equal to zero. Thus

$$\frac{\partial}{\partial\epsilon}\left[2\int_{-\infty}^{\infty} [G(f)+\epsilon\eta(f)]A(f)\,\mathrm{d}f + \int_{-\infty}^{\infty} |G(f)+\epsilon\eta(f)|^2\, B(f)\,\mathrm{d}f\right]_{\epsilon=0} = 0.$$

This leads to

$$2\int_{-\infty}^{\infty} \eta(f)A(f)\,\mathrm{d}f + \int_{-\infty}^{\infty} G^*(f)\eta(f)B(f)\,\mathrm{d}f + \int_{-\infty}^{\infty} G(f)\eta^*(f)B(f)\,\mathrm{d}f = 0.$$

By Parseval's formula, the third term is real, so it may be replaced by its conjugate to write

$$2\int_{-\infty}^{\infty} \left[A(f) + G^*(f)B(f)\right]\eta(f)\,\mathrm{d}f = 0.$$

Because this holds for arbitrary $\eta(f)$, the bracketed term must be zero. Otherwise, for each f, $\eta(f)$ could be chosen to have the same sign as the bracketed term, thereby violating the equality. Therefore $A(f) + G^*(f)B(f) = 0$. For our problem, this corresponding expression is

$$-H(f_x,f_y)\Phi_s(f_x,f_y) + G^*(f_x,f_y)\left[|H(f_x,f_y)|^2\Phi_s(f_x,f_y) + \Phi_n(f_x,f_y)\right] = 0$$

from which the Wiener filter follows. □

If the noise is white, then the noise power density spectrum is constant, conventionally written as $\Phi_n(f_x, f_y) = N_0/2$. Therefore,

$$G(f_x, f_y) = \frac{H^*(f_x, f_y)\Phi_s(f_x, f_y)}{|H(f_x, f_y)|^2\Phi_s(f_x, f_y) + N_0/2}$$

is the Wiener filter for white noise.

The clean algorithm

The Wiener filter makes use of the power density spectrum of the noise process, but does not use any other known properties of the image. The Wiener filter cannot create a new spectral component in the image that is not already in the support of $H(f_x, f_y)$. It can only amplify or attenuate existing components. If $H(f_x, f_y)$ is zero at some f_x, f_y, then the reconstruction will also be zero at that frequency.

Alternative processing methods are available that incorporate other known properties of the image. Techniques that use prior information to reconstruct frequencies not in the support of $H(f_x, f_y)$ are referred to as *superresolution* methods. These methods are always nonlinear. We shall discuss two important instances – specifically, the case in which the image is known to be real and nonnegative is discussed in the next section. The case in which the image is known to have limited support is discussed in this section. In an extreme case, an image may be known to consist of a sparse collection of impulses. An image known to have limited support is sometimes called a *mostly-black image*, because outside of its support the image will be all black (or all white).

Whenever the image $s(x, y)$ is known to consist of a sparse set of positive impulse-like pulses, or can be so modeled, a method of deconvolution known as the *clean algorithm* may be appropriate. The clean algorithm is intuitively satisfying, but it is not developed from a fundamental principle. Suppose that $s(x, y)$ has the form

$$s(x, y) = \sum_{\ell=1}^{L} a_\ell \delta(x - x_\ell, y - y_\ell)$$

where the amplitudes a_ℓ are positive and unknown, and the locations (x_ℓ, y_ℓ) are also unknown. The number of terms L is unknown as well. The noisy blurred image is given by

$$v(x, y) = \int_{-\infty}^{\infty}\int_{-\infty}^{\infty} h(x - \xi, y - \eta)s(\xi, \eta)\,d\xi\,d\eta + n(x, y)$$

where $h(x, y)$ is the known pointspread function of the system. Because presumably it has undesirable sidelobes, $h(x, y)$ is sometimes called the *dirty pointspread function*. Often the dirty pointspread function has remote sidelobes that create artifacts elsewhere in the image. The task is to estimate $s(x, y)$ – or at least an improved version of $s(x, y)$ – when given the noisy blurred image $v(x, y)$. The clean algorithm operates iteratively in an obvious way by estimating the terms $a_\ell \delta(x - x_\ell, y - y_\ell)$, one by one starting with the largest, and removing each term from $v(x, y)$ as it is estimated.

With the initialization $\widehat{s}(x, y) = 0$, the clean algorithm consists of the following steps.

Step 1

$$a = \max_{x,y} v(x, y)$$
$$(\widehat{x}, \widehat{y}) = \operatorname{argmax}_{x,y} v(x, y).$$

Step 2

$$v(x, y) \leftarrow v(x, y) - ah(x - \widehat{x}, y - \widehat{y})$$
$$\widehat{s}(x, y) \leftarrow \widehat{s}(x, y) + a\widehat{h}(x - \widehat{x}, y - \widehat{y}).$$

Step 3 Return to Step 1 or halt.

The iterations are halted when they no longer improve the image. The function $\widehat{h}(x, y)$ is called the *clean pointspread function.* It is typically a two-dimensional gaussian pulse and is used to broaden the replacement impulses so that the image is more attractive. The amplitude and the width of the gaussian pulse are typically chosen to match the amplitude and width of the dirty pointspread function. This also prevents the image from displaying an artificially good resolution.

9.2 Deconvolution of nonnegative images

In many problems of signal processing, such as image formation, one is asked to find a function in the space of real functions satisfying certain properties. In some problems, the correct formulation of this task is to find the function that minimizes the euclidean distance between that function and some other function, usually originating as a set of measured data. However, if the function must lie in the space of *nonnegative* real functions, the problem changes considerably. The euclidean distance is not the appropriate discrepancy measure on the space of nonnegative real functions.

Let

$$v(x, y) = h(x, y) ** s(x, y) + n(x, y)$$

be a blurred, two-dimensional nonnegative signal in the presence of noise. To obtain a good estimate, $\widehat{s}(x, y)$, of the image, we must remove as much of the blur as possible without undue amplification of the noise. The Wiener filter, described in the previous section, is a real-valued inverse filter that is chosen to satisfy a mean-square-error criterion. However, the Wiener filter can produce negative values at its output so it is not fully suitable for problems with the prior condition that the image is nonnegative. Of course, one may proceed naively by again minimizing euclidean distance as by using the Wiener filter and then simply replacing negative values of the resulting function by zero. This procedure can sometimes give acceptable results, but it cannot be formally justified. The procedure is a forced solution and does not satisfy any optimality criterion, and it might give poor results. We shall describe an alternative procedure.

Any relationship between the blurred input signal and an estimated deblurred signal that satisfies the nonnegativity constraint must, of necessity, be nonlinear. Therefore this relationship cannot take the form of a linear filter. In this section, we shall describe the *Richardson–Lucy algorithm*, which is an algorithm for computing a nonnegative signal or image estimate in the presence of noise. A satisfying intuitive development will be given in this section as an aid in understanding how the algorithm works, though in this section it is not developed from first principles. In Section 12.3, we shall give a more formal development.

The motivation for the Richardson–Lucy algorithm is the Bayes formula of probability theory. Recall that a bivariate probability distribution can be decomposed into marginals and conditionals in two ways. These are $P_{jk} = p_j Q_{k|j}$ and $P_{jk} = q_k P_{j|k}$ where the marginals $\boldsymbol{p} = [p_j]$ and $\boldsymbol{q} = [q_k]$ are given by $p_j = \sum_k P_{jk}$ and $q_k = \sum_j P_{jk}$, and the conditionals $\boldsymbol{Q} = [Q_{k|j}]$ and $\boldsymbol{P} = [P_{j|k}]$ are given by $Q_{k|j} = P_{jk}/p_j$ and $P_{j|k} = P_{jk}/q_k$. Then the Bayes formula is

$$P_{j|k} = \frac{p_j Q_{k|j}}{\sum_i p_i Q_{k|i}}.$$

An easy problem is to compute the marginals and conditionals from the joint distribution $\boldsymbol{P} = [P_{jk}]$. A difficult problem is to compute a $\boldsymbol{P} = [P_{jk}]$ that has a specific marginal $\boldsymbol{q} = [q_k]$ and a specific conditional $\boldsymbol{Q} = [Q_{k|j}]$. This latter task has the nature of an inverse problem. The required solution need not even exist. Indeed, any given probability vector $\boldsymbol{q}$ of length K, and any given K by J probability transition matrix $\boldsymbol{Q}$, need not be compatible in the sense that a $[P_{jk}]$ need not exist with this marginal $\boldsymbol{q}$ and conditional $\boldsymbol{Q}$. This might be the case, for example, in a physical application in which $\boldsymbol{q}$ and $\boldsymbol{Q}$ are each measured, but each with some error.

From the Bayes formula, one can write

$$\begin{aligned} p_j &= \sum_k q_k P_{j|k} \\ &= p_j \sum_k \frac{Q_{k|j}}{\sum_i p_i Q_{k|i}} q_k. \end{aligned}$$

This suggests that, if the probability $\boldsymbol{q}$ and the conditional probability $\boldsymbol{Q}$ are both known, then the iteration

$$p_j^{(r+1)} = p_j^{(r)} \sum_k \frac{Q_{k|j}}{\sum_i p_i^{(r)} Q_{k|i}} q_k$$

might be a way to iteratively compute the $\boldsymbol{p}$ for which $q_k = \sum_j p_j Q_{k|j}$. Clearly, the iteration produces a sequence of probability vectors because for each r, $p_j^{(r+1)}$ is nonnegative and the $p_j^{(r+1)}$ sum to one over j. It is natural in this iteration to choose the initial value $\boldsymbol{p}^{(0)}$ to be the uniform distribution.

If the iteration reaches a limit point, then this limit point is clearly a probability vector and may be the desired $\boldsymbol{p}$. A fixed point to which the iteration converges may exist even if the expression $q_k = \sum_j p_j Q_{k|j}$ does not have a probability vector $\boldsymbol{p}$ as a solution, as may occur when $\boldsymbol{q}$ and $\boldsymbol{Q}$ are contaminated by measurement errors. In this case, the fixed point may be regarded as a best fit to the stated conditions.

One can mimic this discussion to develop a similar iterative algorithm for the general problem of deblurring a nonnegative function. Let a known nonnegative blurred function $v(t)$ be given as

$$v(\xi) = \int_{-\infty}^{\infty} h(\xi\,|\,t)s(t)\,dt,$$

where, for each t, $h(\xi\,|\,t)$ is a known nonnegative function of ξ whose integral $\int_{-\infty}^{\infty} h(\xi\,|\,t)\,d\xi$ does not depend on t, and $s(t)$ is an unknown nonnegative function to be computed from $v(\xi)$ and $h(\xi\,|\,t)$. By rescaling $s(t)$, we can ensure that the integral of $v(\xi)$ is one. Because it does not depend on t, the integral of $h(\xi\,|\,t)$ can be chosen to be one as well by rescaling and imposing the constraint that $s(t)$ integrates to one.

The special condition that characterizes this deblurring problem is that $s(t)$ is required to be real and nonnegative. Define

$$g(t\,|\,\xi) = \frac{h(\xi\,|\,t)s(t)}{\int_{-\infty}^{\infty} h(\xi\,|\,t)s(t)\,dt}.$$

The denominator on the right is $v(\xi)$, so this gives

$$v(\xi)g(t\,|\,\xi) = h(\xi\,|\,t)s(t),$$

and

$$\int_{-\infty}^{\infty} v(\xi)g(t\,|\,\xi)\,d\xi = \int_{-\infty}^{\infty} h(\xi\,|\,t)s(t)\,d\xi = s(t).$$

Therefore the unknown $s(t)$ satisfies

$$s(t) = s(t)\int_{-\infty}^{\infty} v(\xi)\frac{h(\xi\,|\,t)}{\int_{-\infty}^{\infty} h(\xi\,|\,t)s(t)\,dt}\,d\xi.$$

With this equality as a motivation, the Richardson–Lucy algorithm is now stated as the iteration

$$s^{(r+1)}(t) = s^{(r)}(t)\int_{-\infty}^{\infty} v(\xi)\frac{h(\xi\,|\,t)}{\int_{-\infty}^{\infty} h(\xi\,|\,t)s^{(r)}(t)\,dt}\,d\xi.$$

At iteration r, the old estimate $s^{(r)}(t)$ is known and the new estimate $s^{(r+1)}(t)$ is computed. It is clear that if $s^{(r)}(t)$ is nonnegative, then $s^{(r+1)}(t)$ is also nonnegative. It can be easily verified that the integral of $s^{(r)}(t)$ is independent of r. If $s^{(r)}(t)$ converges to a fixed point, then we may test the fixed point to see that it is the desired solution to the integral equation. If the fixed point does not satisfy the original integral equation, then one might suppose that the integral equation does not have a solution satisfying the nonnegativity constraint. In this case, the fixed point might be regarded as the best possible fit to the specified conditions. Indeed in many applications the algorithm has been found empirically to converge to a fixed point that provides a satisfactory solution. In Chapter 11, we shall describe a more formal procedure for developing this algorithm, and we will prove a formal statement regarding convergence, even in the presence of noise.

The Richardson–Lucy algorithm applies not only to nonnegative one-dimensional functions, but also to more general cases. Therefore, we can immediately write

$$s^{(r+1)}(x, y) = s^{(r)}(x, y) \int_{-\infty}^{\infty} \int_{-\infty}^{\infty} v(\xi, n) \frac{h(\xi, \eta | x, y)}{\int_{-\infty}^{\infty} \int_{-\infty}^{\infty} h(\xi, \eta | x, y) s^{(r)}(x, y) \, dx \, dy} \, d\xi \, d\eta$$

as the two-dimensional Richardson–Lucy algorithm. This algorithm is suitable for the deconvolution of nonnegative images.

9.3 Blind image deconvolution

It is common experience that a projected optical image can be focused even if the projected image and the blurring function are not previously known. This is an everyday instance of a kind of blind image enhancement similar to the blind deconvolution studied in this section. This familiar example of focusing an optical image illustrates the topic of this chapter, but in this instance, the "deconvolution" takes place within the optical field prior to converting the observed image into an intensity image in the detector, so it is not really the same as the problem we will consider. The term *blind deconvolution* is used for the simpler problem of removing the effect of an unknown linear filter from an image after the image is formed.

In this section, we shall derive algorithms for blind deconvolution that remove an unknown space-invariant blur from a blurred image. Blind deconvolution is a more difficult version of the deconvolution problem.

A one-dimensional convolution has the form

$$v(t) = \int_{-\infty}^{\infty} h(t - \xi) s(\xi) \, d\xi.$$

The task of blind deconvolution in one dimension is to compute $s(t)$ from $v(t)$ even though $h(t)$ is unknown. In the frequency domain, this becomes

$$V(f) = H(f)S(f).$$

Then the task is to find $H(f)$ and $S(f)$ when given $H(f)S(f)$. This is not a well-formulated problem, and one should not expect that it has a satisfactory solution as given because the problem may have many solutions. Moreover, noise has not been considered. A more realistic form of the problem, including noise, is written

$$v(t) = h(t) * s(t) + n(t).$$

The noise $n(t)$ is unknown, but its mean and correlation function are usually presumed known.

Similarly, a two-dimensional convolution has the form

$$v(x, y) = \int_{-\infty}^{\infty} \int_{-\infty}^{\infty} h(x - \xi, y - \eta) s(\xi, \eta) \, d\xi \, d\eta.$$

The task of blind deconvolution in two dimensions is to compute $s(x, y)$ from $v(x, y)$ even though $h(x, y)$ is unknown. To account for noise, the problem is written

$$v(x, y) = h(x, y) ** s(x, y) + n(x, y).$$

The task then is to estimate the signal $s(x, y)$, and perhaps $h(x, y)$, from the noisy observation $v(x, y)$. Again, the noise $n(x, y)$ is unknown, but its mean and two-dimensional correlation function are usually known.

Conditions for blind image deconvolution

Perhaps surprisingly, the problem of blind deconvolution of two-dimensional signals and the problem of blind deconvolution of one-dimensional signals are quite different. Simple side conditions can suffice to make the blind deconvolution problem in two dimensions solvable. One important side condition is a Nyquist condition on $S(f_x, f_y)$ and $H(f_x, f_y)$ so that $s(x, y)$ and $h(x, y)$ may be sampled. This condition is important because the discrete version of two-dimensional convolution is highly constrained, while the continuous version is not. Another common side condition is a nonnegativity constraint on the image $s(x, y)$, on the pointspread function $h(x, y)$, or on both. There may also be a boundedness constraint on the domains of $s(x, y)$ and $h(x, y)$, such as the constraint that $s(x, y)$ can be nonzero only if $(x, y) \in \Omega_1$, for some region of support Ω_1, and $h(x, y)$ can be nonzero only if $(x, y) \in \Omega_2$, for some region of support Ω_2.

In any case, the notion of a solution must be relaxed to allow for certain kinds of ambiguity in blind deconvolution. The reason is that, for any real numbers, a, b, and c,

$$h(x, y) ** s(x, y) = ch(x - a, y - b) ** c^{-1} s(x - a, y - b),$$

so we cannot hope to recover a, b, and c without additional information. Thus we will be content to recover $h(x, y)$ and $s(x, y)$ up to amplitude and translation. It is also clear that an algorithm cannot determine which of the two functions, $h(x, y)$ and $s(x, y)$, is the image and which is the filter. Other considerations must resolve this ambiguity.

Another, more egregious, ambiguity arises if either $h(x, y)$ or $s(x, y)$ can be decomposed as the convolution of two terms

$$h(x, y) = h_1(x, y) ** h_2(x, y),$$

or

$$s(x, y) = s_1(x, y) ** s_2(x, y)$$

possibly with some approximation error. If both are true,

$$h(x, y) ** s(x, y) = h_1(x, y) ** h_2(x, y) ** s_1(x, y) ** s_2(x, y).$$

Now there are $2^4 - 2$ ways to partition the four terms on the right into two groups, excluding the two uninteresting partitions in which all terms are in the same group. In general, if $h(x, y)$ is an m-fold convolution and $s(x, y)$ is an n-fold convolution, there will be $2^{m+n} - 2$ nontrivial ways to decompose the convolution $h(x, y) ** s(x, y)$, and each decomposition, called an *intrinsic ambiguity*, corresponds to another solution.

The existence of meaningful blind deconvolution algorithms requires that most $h(x, y)$ and $s(x, y)$ cannot be decomposed into convolutions of more than one term. Because measurements are always noisy and computations are inexact, this statement must be relaxed to allow for approximations in the decomposition. Thus we will need to argue that $s(x, y)$ for which $s(x, y) \approx s_1(x, y) ** s_2(x, y)$ are rare. Before we can make this statement, we shall need to prepare appropriate preconditions. Otherwise, we could always choose a factorization in the frequency domain, such as

$$S(f_x, f_y) = \sqrt{S(f_x, f_y)}\sqrt{S(f_x, f_y)} = S_1(f_x, f_y)S_2(f_x, f_y)$$

so that $s(x, y) = s_1(x, y) ** s_2(x, y)$. Thus, there is too much freedom in the continuous form of this problem. Fortunately, the discrete form, which is actually the preferred form for processing, does not have this freedom. Therefore, rather than dealing with this topic further in the continuous form, we will restrict the problem by a kind of regularization. Specifically, we will assume that the appropriate Nyquist conditions are satisfied and the problem has been discretized accordingly.

In the discrete case, the image is the array $\boldsymbol{s} = [s_{ij}]$ and the blur is the array $\boldsymbol{h} = [h_{ij}]$. The blurred image is the discrete convolution $\boldsymbol{v} = \boldsymbol{h} ** \boldsymbol{s}$ defined by

$$v_{ij} = \sum_k \sum_\ell h_{i-k, j-\ell} s_{k\ell}.$$

Our goal is to show that the arrays $\boldsymbol{h}$ and $\boldsymbol{s}$, in general, cannot be decomposed as $\boldsymbol{h}_1 ** \boldsymbol{h}_2$ and $\boldsymbol{s}_1 ** \boldsymbol{s}_2$.

Consider any polynomial[2] $p(x, y) = \sum_i \sum_j p_{ij} x^i y^j$. Then $\boldsymbol{p} = \boldsymbol{p}_1 \;**\; \boldsymbol{p}_2$ if and only if $p(x, y) = p_1(x, y)p_2(x, y)$. To show that it is unusual for an array $\boldsymbol{p}$ to decompose as $\boldsymbol{p} = \boldsymbol{p}_1 \;**\; \boldsymbol{p}_2$, we refer to the fact that it is rare for a bivariate polynomial to factor. This is a well-known statement of mathematics. We will make the statement plausible by reasoning in euclidean space. We will argue from the fact that just as a finite number of curves cannot fill $\boldsymbol{R}^2$, and a finite number of surfaces cannot fill $\boldsymbol{R}^3$, so too a finite number of lower-dimensional manifolds cannot fill $\boldsymbol{R}^n$.

First, consider a two by two array corresponding to the bivariate polynomial

$$\begin{aligned} p(x, y) &= \sum_{i=0}^{1} \sum_{j=0}^{1} p_{ij} x^i y^j \\ &= p_{11}xy + p_{10} + p_{01}y + p_{00} \\ &= p_{11}(xy + ax + by + c). \end{aligned}$$

Because the leading coefficient p_{11} can be factored out, it is enough to study only the factorizability of the polynomial

$$p(x, y) = xy + ax + by + c.$$

Such a polynomial, with the coefficient of the highest-degree term equal to zero, is called a *monic polynomial*. The set of coefficients (a, b, c) can be considered as a point in three-dimensional euclidean space $\boldsymbol{R}^3$. If $p(x)$ factors nontrivially, then it factors as

$$\begin{aligned} p(x, y) &= (x + A)(y + B) \\ &= xy + Bx + Ay + AB. \end{aligned}$$

Thus, if $p(x, y)$ factors, the point $(a, b, c) \in \boldsymbol{R}^3$ is describable in terms of two parameters (A, B), so it lies on a two-dimensional surface in $\boldsymbol{R}^3$. If the point (a, b, c) is not on this surface, the polynomial $p(x, y)$ does not factor. Because an arbitrary point almost surely does not lie on this surface, an arbitrary point in $\boldsymbol{R}^3$ corresponds to a polynomial that almost surely does not factor. Moreover, an arbitrary point (a, b, c) almost surely is contained in a ball such that every point in this ball corresponds to a polynomial that does not factor. The radius of this ball varies from point to point and indicates a tolerance to noise or computational error.

In the same way, consider a real monic bivariate polynomial $p(x, y)$ of second degree in x and second degree in y. This polynomial has eight free coefficients so it may be regarded as a point in $\boldsymbol{R}^8$. Suppose that $p(x, y) = p_1(x, y)p_2(x, y)$ and that this is a nontrivial factorization into monic factors. Then either the x degree or the y degree of $p_1(x, y)$ is one. Suppose, for example, that both are equal to one. Then the factorization

[2] There is a potentially confusing transition of notation here. Because $s_{ij} = s(i\Delta x, j\Delta y)$, the polynomial $p(x, y)$ is not replacing the function $s(x, y)$.

has the form

$$p(x, y) = (xy + Ax + By + C)(xy + Dx + Ey + F)$$

which has six free coefficients. Hence the set of $p(x, y)$ that factor in this way forms a six-dimensional manifold in $\boldsymbol{R}^8$. Alternatively, if $s_1(x, y)$ has degree one in x and degree two in y, then

$$p(x, y) = (xy^2 + Axy + Bx + Cy^2 + Dy + E)(x + F).$$

Because there are six free coefficients, this corresponds to another six-dimensional manifold in $\boldsymbol{R}^8$. Other factorizations correspond to other manifolds. Thus in $\boldsymbol{R}^8$ only the points on one of a finite set of six-dimensional (or smaller) manifolds correspond to factorable polynomials. These manifolds do not fill $\boldsymbol{R}^8$. Other points, not on one of these manifolds, correspond to polynomials that do not factor. Indeed each of these other points has a ball around it that does not intersect one of the stated manifolds, so each of these points has some noise tolerance. The amount of noise tolerance varies from point to point.

For large n, this argument becomes even more compelling. An n by n real monic bivariate polynomial has $(n + 1)^2 - 1$ real coefficients. The array of these coefficients can be regarded as a point in $\boldsymbol{R}^{n^2+2n}$. An n by n bivariate polynomial might factor, for example, as the product of two $n/2$ by $n/2$ bivariate polynomials if n is even. This replaces $n^2 + 2n$ coefficients with $2(n/2 - 1)^2 - 2 = n^2/2 + 2n$ coefficients. The set of $n/2$ by $n/2$ bivariate polynomials is a manifold in $\boldsymbol{R}^{n^2+2n}$ of dimension $n^2/2 + 2n$. Other factorizations define other manifolds of other dimensions not larger than $n^2 + n$, of which there are only a finite number and so they do not fill $\boldsymbol{R}^{n^2+2n}$.

An algorithm for blind image deconvolution

In general, the observed signal is contaminated by noise. Then

$$v(x, y) = h(x, y) \;**\; s(x, y) + n(x, y).$$

The task, then, is to find $h(x, y)$ and $s(x, y)$ so that $h(x, y) \;**\; s(x, y)$ is the best fit to $v(x, y)$ in the sense of euclidian distance. We shall describe here a rather straightforward procedure for blind deconvolution, constructed heuristically as a simple iteration. This somewhat naive procedure illustrates the nature of an iterative solution for such a problem. In Section 11.4, we shall set up a formal procedure and derive a better algorithm.

Given the noise-free expression

$$v(x, y) = h(x, y) \;**\; s(x, y)$$

with $v(x, y)$ known, the goal is to find the two nonnegative unknown functions $h(x, y)$ and $s(x, y)$ whose Fourier transforms have bounded domains. In the Fourier transform domain, the equivalent equation is

$$V(f_x, f_y) = H(f_x, f_y)S(f_x, f_y),$$

so the task is to factor $V(f_x, f_y)$. Because the problem is formulated as a product, a factorization does exist. Therefore the only formal difficulty is that it might not be unique. We have seen that this is only a minor concern if the problem is appropriately discretized. It is reasonable to presume that if an iterative algorithm converges to a fixed point, then that fixed point is usually the desired solution.

The nature of the problem suggests the following iterative algorithm, as appropriately regularized. The iteration proceeds in the transform domain by choosing some initial nonnegative function for $h(x, y)$, and then alternating the following two steps with appropriate Fourier transforms and inverse Fourier transforms computed as needed.

Step 1

a.

$$\widehat{S}(f_x, f_y) = \begin{cases} \frac{V(f_x, f_y)}{H(f_x, f_y)} & \text{if } (f_x, f_y) \in \Omega_2 \\ 0 & \text{otherwise.} \end{cases}$$

b.

$$s(x, y) = \begin{cases} \widehat{s}(x, y) & \text{if } \widehat{s}(x, y) \geq 0 \\ 0 & \text{otherwise.} \end{cases}$$

Step 2

a.

$$\widehat{H}(f_x, f_y) = \begin{cases} \frac{V(f_x, f_y)}{S(f_x, f_y)} & \text{if } (f_x, f_y) \in \Omega_1 \\ 0 & \text{otherwise.} \end{cases}$$

b.

$$h(x, y) = \begin{cases} \widehat{h}(x, y) & \text{if } \widehat{h}(x, y) \geq 0 \\ 0 & \text{otherwise.} \end{cases}$$

If $s(x, y)$ and $h(x, y)$ are initialized as real functions, and Ω_1 and Ω_2 are symmetric about the origin, then $s(x, y)$ and $h(x, y)$ remain real. The iterations may be halted if and when the squared error

$$d^2(v, h ** s) = \int_{-\infty}^{\infty} \int_{-\infty}^{\infty} [v(x, y) - h(x, y) ** s(x, y)]^2 \, dx \, dy$$

falls below some desired value. Otherwise, the iterations are halted when an iteration counter exceeds some specified value.

9.4 Phase retrieval

The one-dimensional Fourier transform $S(f)$ of the one-dimensional signal $s(t)$ uniquely determines $s(t)$, and the two-dimensional Fourier transform $S(f_x, f_y)$ of the two-dimensional signal $s(x, y)$ uniquely determines $s(x, y)$. This follows from the existence of the inverse Fourier transform. If, however, some part of $S(f_x, f_y)$ is not known, then the signal $s(x, y)$ is not uniquely determined unless that incomplete knowledge of the Fourier transform can be compensated for in some way by side information. For example, in many instances of image formation, the magnitude of the two-dimensional Fourier transform can be measured, but the phase cannot. In many such applications, prior side information may be known and that side information may completely or partially compensate for the missing phase of the Fourier transform. Any computational algorithm that uses side information to recover the signal when the phase of the Fourier transform is unknown is called a *phase-retrieval algorithm* because recovering the signal is equivalent to recovering the missing phase of the Fourier transform. Phase-retrieval algorithms, either in two dimensions or in three dimensions, are important in crystallography, optics, microscopy, and many other fields.

Conditions for phase retrieval

Suppose that only the magnitude $|S(f_x, f_y)|$ of a two-dimensional Fourier transform,

$$S(f_x, f_y) = \int_{-\infty}^{\infty}\int_{-\infty}^{\infty} s(x, y)\mathrm{e}^{-\mathrm{j}2\pi(f_x x+f_y y)}\,\mathrm{d}x\,\mathrm{d}y,$$

is given. The problem of phase retrieval is to recover $s(x, y)$, given that $s(x, y)$ satisfies known side conditions. The function $s(x, y)$ can be written

$$s(x, y) = \int_{-\infty}^{\infty} |S(f_x, f_y)|\mathrm{e}^{\mathrm{j}\theta(f_x, f_y)}\mathrm{e}^{\mathrm{j}2\pi(f_x x+f_y y)}\,\mathrm{d}f_x\,\mathrm{d}f_y$$

where $\theta(f_x, f_y)$ is the unknown phase function. Certainly, the unknown phase function may be chosen in an infinite number of ways which means that an infinite number of $s(x, y)$ are associated with each $|S(f_x, f_y)|$. However, relatively benign side conditions on $s(x, y)$ usually suffice to make the choice essentially unique. Only those phase functions $\theta(f_x, f_y)$ that give a function $s(x, y)$ consistent with the side conditions can be used. Often, for example, it is known that $s(x, y)$ is real and nonnegative. This section will argue that the side information is nearly equivalent to the missing phase information, at least for the discretized version of the problem. However, this side information is not completely equivalent to the missing phase information because some information is irretrievably lost. For example, translation of the real image $s(x, y)$ gives another real image $s(x - a, y - b)$, which corresponds to a simple multiplication of $S(f_x, f_y)$ by a complex exponential to give $S(f_x, f_y)e^{-j2\pi(f_x a + f_y b)}$, and this complex exponential is lost by the absolute value operation. Thus the real image can be recovered from $|S(f_x, f_y)|$ only up to translation. In addition, the mirror image of $s(x, y)$ is $s(-x, -y)$, which has the Fourier transform $S^*(f_x, f_y)$. This conjugation is lost when taking the absolute value of the Fourier transform, which means that $s(x, y)$ cannot be distinguished from $s(-x, -y)$. Thus we can hope to solve the phase-retrieval problem only up to translation and mirror image. Fortunately, these ambiguities are usually regarded as acceptable.

A more egregious failure occurs if $s(x, y)$ can be decomposed as a convolution because, if $s(x, y) = s_1(x, y) ** s_2(x, y)$, then $|S(f_x, f_y)| = |S_1(f_x, f_y)||S_2(f_x, f_y)|$. But this is also the magnitude of the Fourier transform of the function $s_1(x, y) ** s_2(-x, -y)$. Consequently, the magnitude data cannot distinguish between $s_1(x, y) ** s_2(x, y)$ and $s_1(x, y) ** s_2(-x, -y)$. These are very different, and this is an ambiguity that is unacceptable. Even more seriously, if $s(x, y)$ can be decomposed as an N-fold convolution,

$$s(x, y) = s_1(x, y) ** s_2(x, y) ** \cdots ** s_N(x, y),$$

then there are 2^N solutions, called *intrinsic ambiguities*, because any number of the $s_\ell(x, y)$ can be replaced by $s_\ell(-x, -y)$, and most such replacements give an incorrect image. Fortunately, this kind of ambiguity is not a serious problem because it is rare for a bivariate function to decompose as a convolution of two bivariate functions.

The problem of phase retrieval can be restated in the space domain by noting that $|S(f_x, f_y)|^2$ can be immediately computed from $|S(f_x, f_y)|$. The Fourier transform relationship

$$s(x, y) ** s(-x, -y) \Leftrightarrow |S(f_x, f_y)|^2$$

means that the autocorrelation function

$$\phi(x, y) = s(x, y) ** s(-x, -y)$$

can be easily computed as the inverse Fourier transform of $|S(f_x, f_y)|^2$. Thus the problem of computing $s(x, y)$ from $|S(f_x, f_y)|$ is equivalent to the problem of computing $s(x, y)$ from its two-dimensional autocorrelation function. In this sense, phase retrieval is a variant of the problem of blind deconvolution.

The behavior of the two-dimensional Fourier transform with regard to phase retrieval is very different from the corresponding behavior of the one-dimensional Fourier transform. In the one-dimensional case, the function $s(t)$ is not very well identified by the magnitude of its Fourier transform $|S(f)|$. This striking contrast is ultimately related to the fact that every polynomial in one variable can be factored (in the complex field), whereas most polynomials in two variables cannot be factored.

To demonstrate the nonuniqueness of phase retrieval in one dimension, we note that restricting the function $s(t)$ to zero outside of a specified interval does not adequately compensate for the missing phase because there are multiple functions $s(t)$ on a given finite support with the spectrum magnitude $|S(f)|$. The phase-retrieval problem is equivalent to the problem of finding $s(t)$ when given its autocorrelation function $\phi(t)$, because if $|S(f)|$ is known, then $|S(f)|^2$ is also known, and so, by the convolution theorem, $s(t) * s(-t)$ is known as well.

If $s(t)$ has a finite width, the sampling theorem (applied in the frequency domain) says that $S(f)$ and $|S(f)|^2$ can be recovered from their samples, provided the samples are sufficiently close in frequency to avoid time-domain aliasing. Moreover, $S(f)$ goes to zero as f goes to infinity because the integral of $|S(f)|^2$ is finite. Only a finite number of samples of $S(f)$ are needed for reconstructing $s(t)$ to within a given precision.

The z transform of the vector $s = (s_0, \ldots, s_{n-1})$ is defined as the polynomial

$$\widetilde{s}(z) = \sum_{i=0}^{n-1} s_i z^i$$

and the factorization of monovariate polynomials is straightforward. The fundamental theorem of algebra says that a polynomial of degree $n-1$ has $n-1$ zeros in the complex plane, some possibly repeated. Thus we have the factorization

$$\widetilde{s}(z) = \widetilde{s}_0 \prod_{i=0}^{n-1} \left(1 - \frac{z}{a_i}\right)$$

and

$$\widetilde{s}(z)\widetilde{s}^*(z^*) = |\widetilde{s}_0|^2 \prod_{i=0}^{n-1} \left(1 - \frac{z}{a_i}\right)\left(1 - \frac{z}{a_i^*}\right),$$

from which $|S_k|^2$ is obtained by setting $z = \mathrm{e}^{-\mathrm{j}2\pi k/n}$, where

$$S_k = \widetilde{s}(\mathrm{e}^{-\mathrm{j}2\pi k/n}).$$

But if a_i is replaced by a_i^*, $\widetilde{s}(z)$ changes, but the product $\widetilde{s}(z)\widetilde{s}^*(z^*)$ does not change. Hence $|S_k|^2$ does not change. Because there are n such a_i, there are 2^n possibilities for the polynomial $s(z)$ that are consistent with the $|S_k|^2$. Consequently, there are 2^n different signal vectors consistent with $|\widetilde{S}_k|$. Thus, there are multiple solutions to the one-dimensional phase-retrieval problem. Of course, in some cases, some of these 2^n signal vectors can be rather similar because two zeros of $\widetilde{s}(z)$ are close together. In the typical case they will be quite different.

The discussion cannot be generalized to two dimensions because there is no fundamental theorem of algebra for bivariate polynomials. Though one may introduce a polynomial in two variables, $\widetilde{s}(z, w)$, as in the one-dimensional case, this polynomial will almost never factor. Indeed, because the phase retrieval problem is equivalent to solving the autocorrelation

$$\phi(x, y) = s(x, y) ** s(-x, -y)$$

for $s(x, y)$, this problem is a variant of blind demodulation and our discussion of that topic can be adapted to this topic. In fact, a surprisingly small amount of side information suffices to allow the reconstruction of a discrete two-dimensional image.

Algorithms for phase retrieval

In this section, we shall study nonprobabilistic phase-retrieval algorithms. In Section 11.7, we shall study a larger problem in which the data consist of noisy measurements. Then a probabilistic formulation is used to determine a maximum-likelihood two-dimensional image directly from the data. In that formulation, the phase-retrieval task becomes an implicit consequence of the larger problem and is not explicitly identified as a subtask.

In this section, we shall study the *Fienup algorithm*. The Fienup algorithm, shown in Figure 9.1, is a useful method for recovering $s(x, y)$ from $|S(f_x, f_y)|$, though it is not a formal algorithm because it contains heuristics. Any arbitrary image, $s_0(x, y)$, is chosen to initialize an estimate, $\widehat{s}(x, y)$. At each iteration, a new estimate, $\widehat{s}(x, y)$, is computed as follows. A Fourier transform, $S'(f_x, f_y)$, is formed with the magnitude $|S(f_x, f_y)|$ and phase given by the phase of the Fourier transform of $\widehat{s}(x, y)$. The inverse Fourier

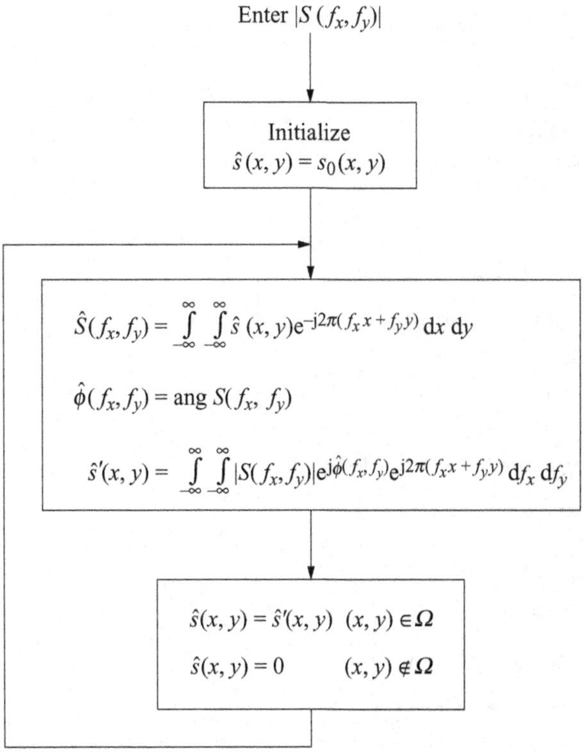

Figure 9.1 The Fienup algorithm

transform of $S'(f_x, f_y)$, denoted $s'(x, y)$, becomes the new estimate $\widehat{s}(x, y)$ wherever it meets the constraints. At each iteration, let Ω be the set of (x, y) for which $s'(x, y)$ satisfies the image constraints. For example, Ω may be the set consisting of those points that are in the prior support of $s(x, y)$ and on which $s'(x, y)$ is nonnegative. The new estimate $\widehat{s}(x, y)$ is given by

$$\widehat{s}(x, y) = \begin{cases} s'(x, y) & (x, y) \in \Omega \\ 0 & \text{otherwise.} \end{cases}$$

Although the Fienup algorithm is widely used and has been quite successful, it is an ad hoc algorithm with little theoretical motivation; no general proof of convergence is known. Indeed, there is no evident reason even to suppose that the first iteration improves the image; this is only an empirical observation. Furthermore, the algorithm is flawed by a stagnation problem: the iterated image may reach a point where it no longer changes noticeably with subsequent iterations even though various error measures may be converging to values strictly larger than zero. It then requires intervention by the user to force the algorithm away from a point of stagnation. Thus there is an art in using the algorithm.

Many variations of the Fienup algorithm are used. The alternative version, shown in Figure 9.2, employs a different update rule in order to speed convergence.

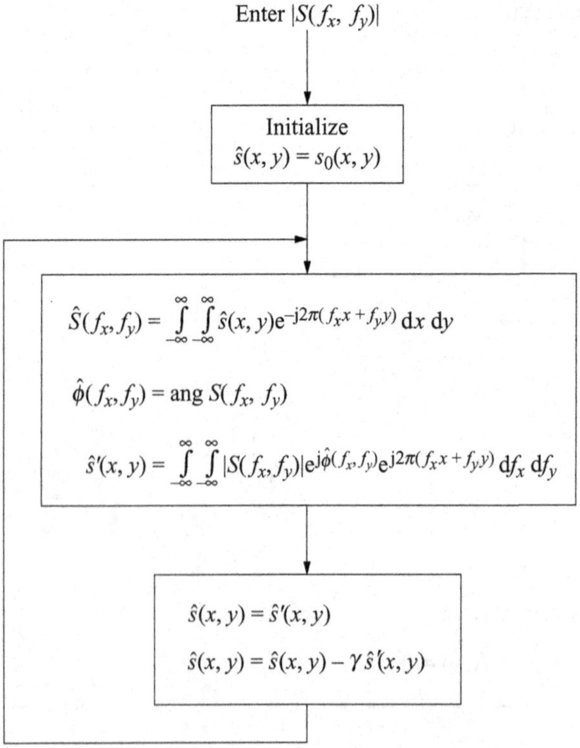

Figure 9.2 Alternative form of Fienup algorithm

The euclidean distance of the estimate from the true image is given by

$$d(\widehat{s}, s) = \int_{-\infty}^{\infty} |s(x, y) - \widehat{s}(x, y)|^2 \, dx \, dy$$
$$= \int_{-\infty}^{\infty} |S(f_x, f_y) - \widehat{S}(f_x, f_y)|^2 \, df_x \, df_y.$$

Because $s(x, y)$ is not known, $d(\widehat{s}, s)$ cannot be computed. Instead, the empirical distance

$$\widehat{d}(\widehat{s}, s) = \int_{-\infty}^{\infty} (|S(f_x, f_y)| - |\widehat{S}(f_x, f_y)|)^2 \, df_x \, df_y$$

can be computed as a means of monitoring the iterations. The goal of the problem is to find an image, $\widehat{s}(x, y)$, for which $\widehat{d}(\widehat{s}, s)$ is zero. A *stagnation point* of the algorithm is a stable point of the algorithm at which $\widehat{d}(\widehat{s}, s)$ is nonzero.

9.5 Optical imaging from point events

Modern physics tells us that electromagnetic signals can exhibit either the behavior of waves or particles. Very weak signals must be treated as a stream of particles called *photons*. The detection of weak incident light can be thought of as the intermittent detection of photons, each of which is detected by means of a photoelectron conversion. To form an image,[3] a large number of events must be collected. If the number of photons is very large, then, by the law of large numbers, the number of photons collected in each pixel is nearly equal to the intensity of the image in that pixel, so it is sufficient to accumulate photoconversion events in each pixel with a photodetector array or a photographic film. Thus, if the light intensity is high or the exposure time is long, the number of photons in each cell of the photodetector will be very large, and the quantized nature of light can be ignored. However, this requires that the situation be stable for the duration of the collection process. If either the propagating medium or the imaging sensor were changing or moving during the time it takes to collect these photoconversion events, then the conventional optical image would be blurred and possibly rendered useless. Because of the disturbances, the conventional ideas of optical image formation will be unsatisfactory.

In this section, we shall develop algorithms that compute images directly from a large file of detected photoelectric conversions and their measured parameters. In this way, the "camera" becomes a computational algorithm that processes a data file of measurements on individual photoconversions. Today it is practical to record individually as many as

[3] There are other examples of imaging from multiple point events. The technique known as positron-emission tomography (PET) is important in medical imaging. This topic is discussed in Section 10.8.

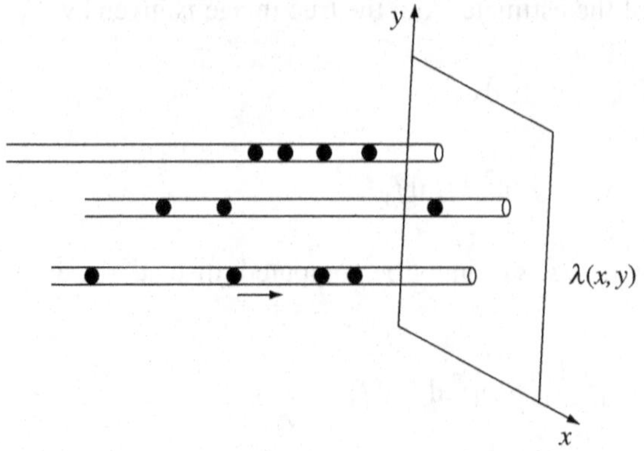

Figure 9.3 Illustrating spatially distributed poisson arrivals

one million photoconversions and their parameters and to process this record to form an image of an object.

Let

$$u(x, y) = h(x, y) \ast\ast \ \rho(x, y)$$

be the optical signal arising from an object, $\rho(x, y)$, as received at an array of photodetectors. We shall suppose that the incident light is very weak, so the signal must be regarded as a stream of individual photons. The stream of photons at an infinitesimal cell of size $\mathrm{d}x\,\mathrm{d}y$ at location x, y is a poisson process with an infinitesimal arrival rate, $\lambda(x, y)\,\mathrm{d}x\,\mathrm{d}y$, proportional to the intensity in that cell $u(x, y)\,\mathrm{d}x\,\mathrm{d}y$.

Figure 9.3 shows a way of visualizing[4] photons arriving at each cell as individual poisson streams. This illustration is not meant to be understood literally as a physical model of photon travel; it is only meant to clarify the meaning of $\lambda(x, y)$.

By disregarding the x, y coordinate, the full array of infinitesimal poisson streams can be pooled to form a single stream. Collectively, they form a poisson process with the arrival rate

$$\lambda = \int_{-\infty}^{\infty}\int_{-\infty}^{\infty} \lambda(x, y)\,\mathrm{d}x\,\mathrm{d}y.$$

This is the overall arrival rate of photoconversions at the photodetector array without regard for the x, y coordinate of the photoconversion. The ratio of $\lambda(x, y)$ to λ is the fraction expected in an infinitesimal cell at coordinate x, y, so this ratio can be interpreted as a probability density function. Given that a photon has arrived, the position of arrival on the x, y plane is a random variable with the probability density

[4] This model is adequate for our mathematical needs, though physically it is not really proper to visualize photons as having well-defined paths.

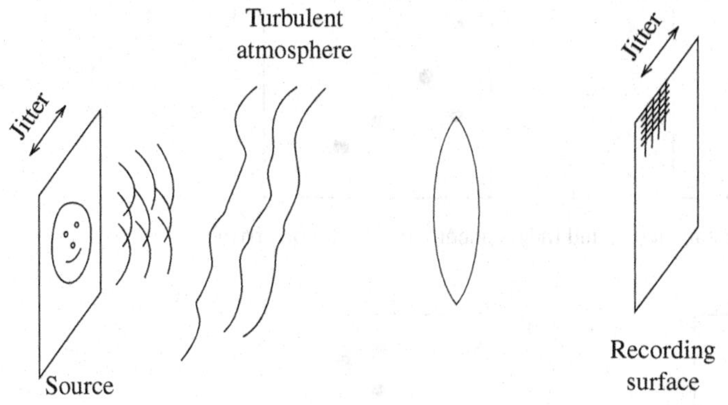

Figure 9.4 Some time-varying impairments

function

$$p(x, y) = \frac{\lambda(x, y)}{\lambda}.$$

In turn, this can be related to the intensity

$$p(x, y) = \frac{u(x, y)}{u}$$

where

$$u = \int_{-\infty}^{\infty} \int_{-\infty}^{\infty} u(x, y)\, dx\, dy.$$

The distribution of photons is described by the propagation of electromagnetic waves, as shown in Figure 9.4. Two impairments to the image-forming process are also shown. These impairments are time-varying changes in the propagation medium, such as atmospheric turbulence and random jitter of the recording surface. In the absence of impairments, the history of photon arrivals should be a list of measured parameters of the form

$$\begin{array}{ccc} x_1 & y_1 & t_1 \\ x_2 & y_2 & t_2 \\ x_3 & y_3 & t_3 \\ & \vdots & \\ & \vdots & \end{array}$$

In the presence of impairments, a photon arriving at time t is displaced from its correct position in the image by $\Delta x(t)$ and $\Delta y(t)$. Because of impairments, the observed history

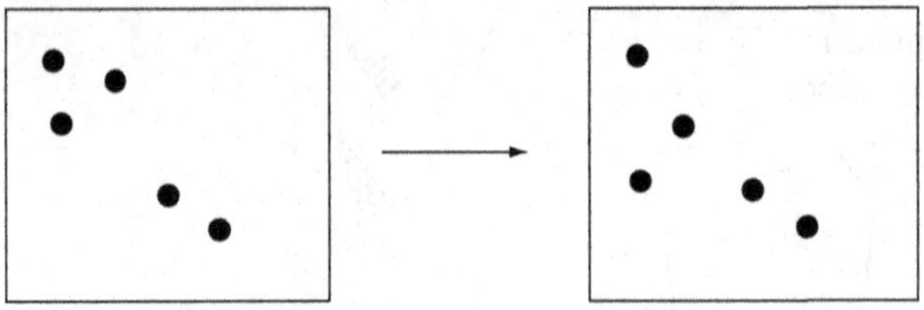

Figure 9.5 Illustrating actual and independently offset photon arrivals

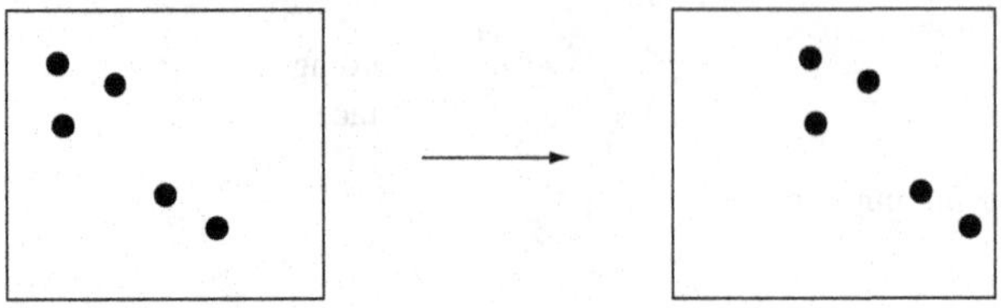

Figure 9.6 Illustrating actual and collectively offset photon arrivals

is the list

$$\begin{array}{lll} x_1 + \Delta x(t_1) & y_1 + \Delta y(t_1) & t_1 \\ x_2 + \Delta x(t_2) & y_2 + \Delta y(t_2) & t_2 \\ x_3 + \Delta x(t_3) & y_3 + \Delta y(t_3) & t_3 \\ & \vdots & \\ & \vdots & \end{array}$$

Figure 9.5 shows the actual and independently offset positions of the first five photons for the case where $\Delta x(t)$ and $\Delta y(t)$ change significantly from photon to photon. Figure 9.6 shows the same five photons for the case where $\Delta x(t)$ and $\Delta y(t)$ remain constant over the interval during which those five photons arrive. Whereas Figure 9.5 suggests that the image is irretrievably lost from the data, Figure 9.6 suggests a situation where useful information remains in the data, at least over short intervals of time. Although the absolute position is not correct, the position difference is correct. This suggests that the data can be preprocessed to remove the effect of the unknown translation, such as by subtracting the position of the first photoconversion from the remaining four. This process, known as *position differencing*, reduces the data set to the form

$$\begin{array}{lll} x_2 - x_1 & y_2 - y_1 & t_2 \\ x_3 - x_1 & y_3 - y_1 & t_3 \\ x_4 - x_1 & y_4 - y_1 & t_4 \\ x_5 - x_1 & y_5 - y_1 & t_5. \end{array}$$

The probability density function of a sum is given by the convolution of the probability density functions of the summands; the probability density function of the difference

$(u, v) = (x - x', y - y')$ is given by a convolution together with a sign reversal. Thus

$$p_D(u, v) = p(x, y) ** p(-x, -y).$$

This is the autocorrelation function of $p(x, y)$ and is proportional to the autocorrelation function of the image $u(x, y)$.

The imaging procedure is as follows. Given x, y, and t for a set of nN detected photoconversions, partition the data into N batches of n consecutive photoconversions. The batch size n is chosen small enough so that $\Delta x(t)$ and $\Delta y(t)$ can be adequately approximated as constant within each batch. Within each batch, subtract the position of the first event from the positions of the remaining events. Then the task is to estimate the image from the sequence of position differences.

We will explain this further for the case in which n equals two. In this case, partition the $2N$ data points into N pairs of data points, compute pairwise position differences, and form a histogram[5] of the N position differences given by $v = N(x, y)/N$. By the law of large numbers, $N(x, y)/N$ converges to

$$p_D(x, y) = p(x, y) ** p(-x, -y),$$

which is proportional to the autocorrelation function of the image. For large N, the two-dimensional Fourier transform of $N(x, y)$ is proportional to the magnitude squared of the Fourier transform of the image

$$P_D(f_x, f_y) = |P(f_x, f_y)|^2,$$

and so

$$|P(f_x, f_y)| = \sqrt{P_D(f_x, f_y)}.$$

Image recovery from $N(x, y)$, for large enough N, is essentially the task of phase retrieval, as described in Section 9.4. The Fienup algorithm is one possible algorithm to recover the image from the magnitude of its Fourier transform.

9.6 Coded aperture imaging

An electromagnetic wave in or near the optical band can be focused into an image of the wave's source by the use of a lens. The higher-energy photons of X-rays, however, interact with all species of atoms and so will not penetrate any material used to make a lens. Because a transparent material is not available at these frequencies, a conventional lens cannot be constructed for X-rays or gamma rays. An alternative method of focusing

[5] On the continuous x, y plane, $N(x, y)$ is a scatter diagram of impulses. To speak of a histogram, the plane must be discretized into a grid of pixels (as it always is in practice), and N should be much larger than the number of pixels.

uses reflection lenses at high angles of incidence. This requires extreme precision in the construction of the reflecting surface. Not only is it expensive but, for wavelengths approaching 0.1 nanometer, impractical. Another alternative for forming an image is to use a Fresnel zone plate. For X-rays, the circular zones would have a width of the order of a nanometer or less. Clearly, it is not practical to construct a Fresnel zone plate at X-ray or gamma-ray frequencies.

Coded aperture image formation is yet a third alternative based on the geometrical optics approximation used to describe the pinhole camera. If an aperture containing a simple pinhole is placed in the path of the radiation, each ray, or incident photon, follows a straight line through the pinhole and onto a screen behind the pinhole. The density distribution forming an image on the screen is an inverted copy of the radiation intensity distribution emitted by the source. The disadvantage of the pinhole camera used for X-ray imaging is the same disadvantage as for the pinhole camera used for optical imaging: it has very poor sensitivity because only the area of the pinhole aperture captures radiation. This means that a long exposure time may be necessary in order to collect enough light, and this is not acceptable in many applications. An alternative is to use multiple pinholes, which collectively comprise the aperture.

An array of multiple pinholes is shown in Figure 9.7. The arrangement of pinholes must be designed to facilitate the purpose of image formation. The active area of an array of N pinholes is N times the area of a single pinhole, so the aperture captures N times as much energy. Coded apertures for X-ray or gamma-ray telescopes may use an N of more than 30 000, thereby increasing the amount of captured light energy by this factor.

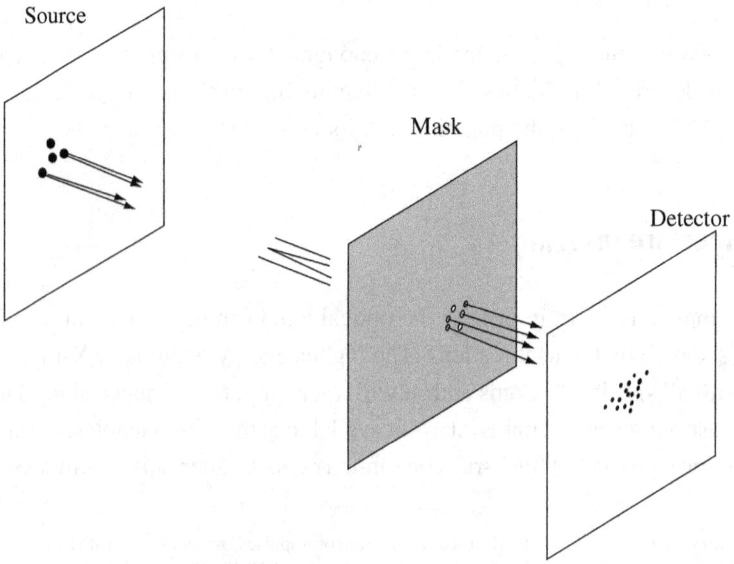

Figure 9.7 A coded aperture

The wavelength of an X-ray or a gamma-ray is so small that the geometrical optics approximation[6] is quite adequate for scales of distance larger than molecular scales. Thus we model a pinhole camera at X-ray frequencies very accurately as a pinhole through which X-ray photons pass. In the geometrical optics approximation, a non-coherent source of intensity density, $\sigma(x, y)$ at distance d to the left of a single pinhole at the origin produces an image at distance d' to the right of the pinhole, given by

$$r(x, y) = \sigma\left(-\frac{x}{M}, -\frac{y}{M}\right)$$

where M is the ratio of the distances. To suppress the image magnification and inversion from the equations, let $s(x, y) = \sigma(-x/M, -y/M)$. A pinhole at the origin then produces the image $s(x, y)$, while a pinhole at position (ξ, η) produces the image

$$r(x, y) = s(x - \xi, y - \eta).$$

This has the form of a convolution of $s(x, y)$ with an impulse at the location (ξ, η)

$$r(x, y) = \delta(x - \xi, y - \eta) ** s(x, y).$$

An arrangement of L pinholes is given by

$$g(x, y) = \sum_{\ell=1}^{L} \delta(x - \xi_\ell, y - \eta_\ell),$$

where (ξ_ℓ, η_ℓ) is the location of the ℓth pinhole. This arrangement of pinholes produces the image

$$r(x, y) = \sum_{\ell=1}^{L} s(x - \xi_\ell, y - \eta_\ell).$$

In the presence of additive sensor noise, the recorded image is the noisy convolution

$$v(x, y) = g(x, y) ** s(x, y) + n(x, y).$$

The object $s(x, y)$ cannot be recognized in the recorded image $v(x, y)$ because the many pinholes cause the recorded image to consist of many overlapping copies of $s(x, y)$. The source $s(x, y)$ must be estimated by processing $v(x, y)$ using some form of deconvolution. The quality of the estimated image depends on the pinhole pattern $g(x, y)$. The pinhole pattern must be chosen to enable a satisfactory deconvolution. It is clear that it is desirable for $|G(f_x, f_y)|^2$ to be nonzero for all (f_x, f_y). Otherwise, $G(f_x, f_y)S(f_x, f_y)$ would be zero and $S(f_x, f_y)$ could not be recovered at that frequency. Even better, if the noise is white, $|G(f_x, f_y)|$ should be more or less constant so that the noise is not unduly amplified at any frequency.

One standard method of deconvolution in noise is to use a Wiener filter, as described in Section 9.1, and, in principle, this method is suitable here. However, the Wiener filter

[6] The absence of diffraction implies the absence of interference between the holes of the aperture.

is computationally expensive and the usual implementation would require the raw data to be converted to digital form.

The Wiener filter is based on the criterion of minimizing mean-squared error. It does not attempt to directly control the largest sidelobe. An alternative method that is used in coded-aperture imaging is to use a pair of two-dimensional, binary pinhole arrays $h(x, y)$ and $g(x, y)$ in cascade, such that $h(x, y) \ast\ast\ g(x, y)$ approximates an impulse, and the largest sidelobe is small. An instance of this method is to choose $h(x, y) = g(x, y)$ and to choose $g(x, y)$ to be a binary pattern whose autocorrelation function $\phi(x, y) = g(x, y) \ast\ast\ g(-x, -y)$ has small sidelobes and so approximates an impulse. We will consider the case in which the "pinholes" are constrained to be squares selected from a checkerboard grid.

Definition 9.6.1 *An m by m' coded aperture is an arrangement of black squares on an m by m' uniform array of unit squares.*

The squares that are not black can be regarded as white squares. The black squares are the pinholes against a white background. A binary indicator function, $g(x, y)$, defined on the coded aperture is equal to one on the black squares and equal to zero on the white squares. The task is to design the coded aperture so that $g(x, y)$ has a satisfactory autocorrelation function. One successful design approach is based on the notion of a cyclic difference set.

Definition 9.6.2 *A cyclic difference set modulo for an odd integer n is a set of $(n-1)/2$ integers smaller than n such that every integer from 1 to $n-1$ occurs as a cyclic difference the same number of times.*

For example, the set $\{1, 2, 3, 5, 6, 9, 11\}$ is a cyclic difference set modulo 15. To check this, notice that the difference 1 occurs three times (as $2-1$, $3-2$, and $6-5$); the difference 2 occurs three times (as $3-1$, $5-3$, and $11-9$); the difference 3 occurs three times (as $5-2$, $6-3$, and $9-6$), the difference 4 occurs three times (as $5-1$, $6-2$, and $9-5$), the difference 5 occurs three times (as $6-1$, $11-6$, and $1-11$); and so forth.

Any cyclic difference set modulo n for which n is the product of two coprime factors can be used to construct a coded aperture as follows. Let $n = mm'$ where m and m' are coprime. Simply number the squares of an m by m' array by the integers running down the extended diagonal. Because m and m' are coprime, the extended diagonal will pass through every cell of the array, and every cell will be assigned a unique integer from 1 to n. Those squares labeled with elements of the cyclic difference set are the holes of the coded aperture. Let $g_{ii'} = 1$ if (i, i') is an element of the coded aperture. Otherwise,

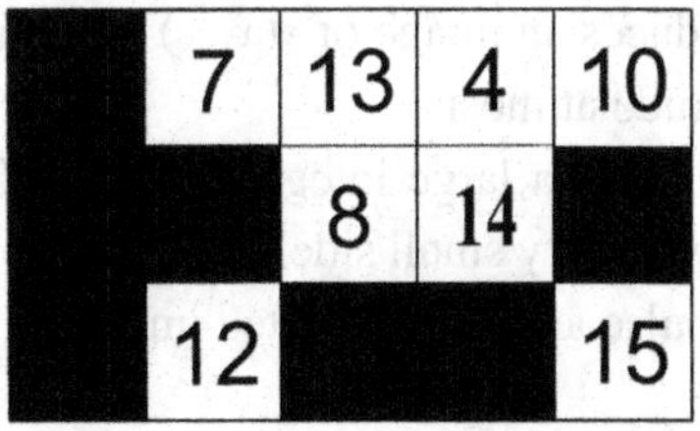

Figure 9.8 A difference-set coded aperture

let $g_{ii'} = 0$. The correlation function is

$$\phi_{rr'} = \sum_{i=0}^{m-1} \sum_{i'=0}^{m'-1} g_{ii'} g_{i+r,i'+r'}$$

where $g_{i+r,i'+r'} = 0$ if either $i + r$ or $i' + r'$ is larger than one. Alternatively the cyclic correlation function

$$\phi^{\circ}_{rr'} = \sum_{i=0}^{m-1} \sum_{i'=0}^{m'-1} g_{ii'} g_{((i+r)),((i'+r'))}$$

where $((i + r))$ denotes $i + r$ modulo m and $((i' + r'))$ denotes $i' + r'$ modulo m'. This is equivalent to a linear correlation with the array $\boldsymbol{g}$ periodically repeated in two dimensions. Then the product $g_{ii'} g_{((i+r)),((i'+r'))}$ is always defined, and

$$\phi^{\circ}_{rr'} = \begin{cases} n & \text{if } r = 0 \text{ (mod } m) \text{ and } r' = 0 \text{ (mod } m') \\ d & \text{otherwise} \end{cases}$$

where d is the number of times that each difference occurs in the difference set. The acyclic correlation function then satisfies

$$\phi_{rr'} = n \text{ if } (r, r') = (0, 0)$$
$$\phi_{rr'} \leq d \text{ otherwise.}$$

An example of a three by five coded aperture with $n = 7$, shown in Figure 9.8, is based on the cyclic difference set $\{1, 2, 3, 5, 6, 9, 11\}$. Starting in the upper left corner, the integers are written into a three by five array down the extended diagonal. The squares indexed by the elements of the cyclic difference set are shaded and become the holes of the coded aperture.

The correlation function cannot be larger than 3 if $(r', r'') \neq (0, 0)$ because the underlying cyclic difference set has all cyclic differences equal to 3. Therefore we know that the function

$$\phi(x, y) = g(x, y) ** g(-x, -y)$$

is equal to 7 at the origin and has no sidelobe larger than 3. This means that the cascade of pinhole arrays $g(x, y)$ and $g(-x, -y)$ produces an image of $s(x, y)$ amplified by 7, but marred by several ghost images of amplitude at most 3.

By constructing a cyclic difference set modulo a large integer n, say 30 000, one can construct a correlation function $\phi(x, y)$ with very small sidelobes compared to the mainlobe. Then $\phi(x, y)$ approximates an impulse and the estimated image

$$\begin{aligned}\widehat{s}(x, y) &= g(-x, -y) ** [g(x, y) ** s(x, y)]\\ &= \phi(x, y) ** s(x, y)\end{aligned}$$

will be acceptable.

Problems

9.1 By discussion of the following two examples, show that knowledge of $|S(f)|$ may not be enough information to determine $s(t)$ uniquely (up to translation and sign reversal) even with the side information that $s(t)$ is known to be real and nonnegative.

a. Sketch $s(t)$ and $|S(f)|$ if

$$s(t) = (1 \pm \epsilon \cos 2\pi f_0 t)\mathrm{sinc}^2\left(\frac{t}{T_1}\right).$$

b. Sketch both versions of $s(t)$ and $|S(f)|$ if

$$s(t) = r(t) * r(t/2)$$

where $r(t) = \mathrm{rect}(t) \pm \frac{1}{2}\mathrm{rect}(t-1)$.

9.2 Let

$$s(x, y) = \mathrm{rect}(x/a)\mathrm{rect}(y/b)$$

with the Fourier transform $S(f_x, f_y)$. Let

$$s'(x, y) = \mathrm{rect}(x/a)\mathrm{rect}(y/b)\cos(2\pi x/A)$$

with the Fourier transform $S'(f_x, f_y)$. Show that, in some cases, when given $|S(f_x, f_y)|$ and the initial estimate $s(x, y) = s'(x, y)$, the Fienup iteration does not change the estimate $s(x, y)$. From this observation, give an explanation of the empirical fact that the Fienup algorithm sometimes produces false strips as artifacts on a computed image.

9.3 Given that $s(t)$ has a contiguous support, can the support of $s(t)$ be determined from the support of $s(t) * s(-t)$? Given that $s(x, y)$ has contiguous support, can the support of $s(x, y)$ be determined from the support of $s(x, y) ** s(-x, -y)$?

(A support is contiguous if any two points of the support can be connected by a curve lying in the interior of the support.)

9.4 The arrival times of a stream of photons form a poisson random process.

a. What is the probability density function on the waiting time, starting at time zero, for the arrival of the first photon?

b. What is the probability density function on the waiting time, starting at time zero, for the arrival of the second photon?

c. What is the probability density function on the time interval between pairs of photons?

9.5 **a.** Given that

$$h(x, y) \;**\; s(x, y) = \mathrm{e}^{-\pi(x^2+y^2)}$$

what are $h(x, y)$ and $s(x, y)$? Is there a unique answer to this question?

b. Suppose that the side conditions

$$S(f_x, f_y) = 0 \quad \text{if} \quad f_x^2 + f_y^2 \geq 10$$
$$H(f_x, f_y) = 0 \quad \text{if} \quad f_x^2 + f_y^2 \geq 10$$

are given. Does the problem now have a solution? Does the problem have an approximate solution? What criteria should be used to define an approximate solution?

c. Numerically find an (approximate) solution.

9.6 **a.** In the set of monovariate polynomials with real coefficients, the irreducible polynomials have a degree not larger than two. What is the fewest number of irreducible factors (with real coefficients) that a real-valued polynomial of degree $2m$ can have? Give bounds on the number of ways in which a real-valued, monovariate polynomial can be factored into two real-valued, monovariate polynomials.

b. Given that a real-valued, monovariate polynomial of degree $2m$ has m irreducible factors, is it possible, in general, to perturb each coefficient by an amount less than ϵ so that more than m irreducible factors occur?

9.7 Derive the *multichannel Wiener filter* to solve the problem in which $s(x, y)$ is observed through two filters

$$v_1(x, y) = h_1(x, y) \;**\; s(x, y) + n_1(x, y)$$
$$v_2(x, y) = h_2(x, y) \;**\; s(x, y) + n_2(x, y)$$

where $n_1(x, y)$ and $n_2(x, y)$ are independent noise processes with the power density spectra $\Phi_1(f_x, f_y)$ and $\Phi_2(f_x, f_y)$, respectively, and the signal process has the power density spectrum $\Phi_s(f_x, f_y)$.

9.8 If $s(x, y)$ is a covariance-stationary random process with the two-dimensional correlation function $\phi(x, y)$, and $h(x, y)$ is any pointspread function, prove that

a.

$$\mathrm{E}[s(x, y)[h(x, y) ** s(x, y)]] = \int_{-\infty}^{\infty}\int_{-\infty}^{\infty} h(\xi, \eta)\phi(\xi, \eta)\, \mathrm{d}\xi\, \mathrm{d}\eta$$
$$= \int_{-\infty}^{\infty}\int_{-\infty}^{\infty} H(f_x, f_y)\Phi(f_x, f_y)\, \mathrm{d}f_x\, \mathrm{d}f_y.$$

b.

$$\mathrm{E}[h(x, y) ** s(x, y)]^2$$
$$= \int_{-\infty}^{\infty}\int_{-\infty}^{\infty}\left[\int_{-\infty}^{\infty}\int_{-\infty}^{\infty} h(\xi, \eta)h(\xi - u, \eta - v)\, \mathrm{d}\xi\, \mathrm{d}\eta\right]\phi(u, v)\, du\, dv$$
$$= \int_{-\infty}^{\infty}\int_{-\infty}^{\infty} \left|H(f_x, f_y)\right|^2 \Phi(f_x, f_y)\, \mathrm{d}f_x\, \mathrm{d}f_y.$$

9.9 The deconvolution problem is usually complicated by the fact that the blurred image is truncated by the finite support of the recording frame. Therefore only

$$r(x, y) = \mathrm{rect}\left(\frac{x}{L_1}, \frac{y}{L_2}\right)[h(x, y) ** s(x, y)]$$

is known, rather than $h(x, y) ** s(x, y)$. What then is the known signal in the frequency domain? How does this change the problem of deconvolution?

9.10 A given image, $s(x, y, \lambda)$, is a function of wavelength (color), denoted by λ, as well as position (x, y). Under what conditions can the image be represented by three "color samples" $s(x, y, \lambda_1)$, $s(x, y, \lambda_2)$, and $s(x, y, \lambda_3)$? How does this change if the filtered image $s'(x, y, \lambda)$ is sampled, where $s'(x, y, \lambda) = s(x, y, \lambda) * g(\lambda)$? How does this change if three different filters are used, one for each sample? If two such systems use different filters, can the reconstructions agree?

9.11 A discrete array $s = [s_{ij}]$ consists of 512 by 512 randomly chosen eight-bit numbers. How many such arrays are there? What can be said about the fraction of these arrays that can be decomposed as the convolution of two smaller arrays (within the precision of the fixed-point numbers)?

Notes

Practical methods of phase retrieval were first developed in the field of X-ray crystallography, as by Hauptman and Karle (1953). Early methods assumed that the image is composed of a finite set of discrete points and set up a finite system of equations. O'Neill and Walter (1963) recognized that mathematical side conditions may compensate for the missing phase. Methods for the more general task of imaging a continuously distributed object emerged later. The Fienup algorithm (1978) is a modification of the Gerchberg–Saxton algorithm (1972). Bruck and Sodin (1979), and later Bates (1982, 1984), argued in favor of the essential uniqueness of phase retrieval. By introducing the notions of *zero sheets*, Izraelevitz (1987) proposed a noniterative, computational algorithm that explicitly computes the image from the zeros of a two-dimensional polynomial. This was also discussed by Lane, Fright, and Bates (1987). A more recent approach is to craft a likelihood function based on a probabilistic formulation of the problem, as will be discussed in Chapter 11, thereby suppressing explicit reference to phase retrieval. The image is obtained directly by maximizing the likelihood; thus the phase of the Fourier transform never needs to be considered.

Wiener (1949) introduced the study of optimal filtering via the least-squares error criteria. The solution to various formulations of the problem are all called Wiener filters, including problems with and without noise, and problems with and without constraints on the filter, such as constraints on its support. Deconvolution algorithms were developed independently in many fields, including seismic processing (1979) and radio astronomy (1974). The Richardson (1972)–Lucy (1974) iterative method of deblurring was developed to recover nonnegative images of objects from noisy blurred images. This method was independently rediscovered by Shepp and Vardi (1982) in the context of medical imaging. The problem of blind deconvolution arose independently in many fields and a large literature now exists. Lane and Bates (1987) used ideas similar to those used in the study of phase retrieval to better understand blind deconvolution, arguing from the nature of zero sheets in four-dimensional space that blind deconvolution is essentially unique. Ghiglia, Romero, and Mastin (1996) demonstrated that zero sheet methods are sensitive to noise. Ayers and Dainty (1982) studied iterative methods for blind convolution. Nayar and Nakagawa (1994) studied the relationship between focusing and shape estimation. The clean algorithm was introduced by Högbom (1974) for applications in radio astronomy. The performance of the clean algorithm, and its relationship to least-squares, was studied by Schwarz (1978, 1979).

The statistics of speckle was studied by Goodman (1976), and by Tue, Chin, and Goodman (1982). Speckle in ultrasound systems was studied by Bucharest (1978), who established that a slight change in perspective would give independent random speckle, thereby making possible the smoothing of speckle by averaging. In astronomy, the term "speckle" is used to describe the appearance of short-exposure images through a turbulent atmosphere. Labeyrie (1970) pointed out that multiple, short-exposure images collectively retain diffraction-limited information and that suitable phase-retrieval algorithms can recover the diffraction-limited image. Synge (1928) had earlier introduced the method of near-field scanning optical microscopy which obtains superresolution by using observations at multiple positions through a subwavelength aperture.

The method of coded aperture image formation was introduced by Dicke (1968) and Ables (1968) by replacing the simple single pinhole of a pinhole camera by many pinholes in an irregular pattern. Methods of coding the pinhole pattern to facilitate image recovery were discussed by Fenimore and Cannon (1978), and Skinner (1988). Pasedach and Haase (1981) discussed using a pair of coded apertures for applications in which coherence can be maintained.

10 Tomography

Suppose that one is given several images of a two-dimensional (or a multidimensional) object, but that the detail of each of these images is limited in some way. For example, the images may be projections of a multidimensional object onto a lower-dimensional space. By using sophisticated signal-processing techniques, many such limited images of a common object can be combined to construct a single enhanced image.

Techniques for combining multiple one-dimensional projections into a single two-dimensional image are known collectively as *tomography* (Greek *toma*: a cut + *graphy*). The term may also be used to describe techniques for combining several poor images into a single improved image. This is different from the practice of enhancing a single image by signal-processing techniques, although, of course, the two tasks are closely related.

The most widespread form of tomography, known as *projection tomography*, reconstructs an image from its projections. Projection tomography has a simple mathematical structure. The most familiar instance of projection tomography uses X-rays as the source of illumination and X-ray absorption as the observed phenomenon. In this case, the way the X-ray illumination is used is quite different from the case of molecular imaging where the observation in the far field is based on scattering of the illuminating X-rays. In projection tomography, the observation is based on attenuation of the illuminating rays in the geometrical-optics approximation.

Many kinds of data may be considered for tomography such as: absorption, emission, scattering, or diffusion. Absorption refers to the attenuation of a signal, such as an X-ray, while passing through an object. The negative logarithm of the attenuation, at an infinitesimal region at (x, y, z) is denoted $s(x, y, z)\,dx\,dy\,dz$. Emission refers to the generation of a signal, such as a stream of radioactive particles, from within an object; the emission from an infinitesimal region at (x, y, z) is denoted $s(x, y, z)\,dx\,dy\,dz$. Scattering refers to the dispersal of a signal from the interior of an object into specified directions; the scattering from an infinitesimal region at (x, y, z) is again denoted $s(x, y, z)\,dx\,dy\,dz$. In each of these situations, we wish to estimate $s(x, y, z)$ by observing the projection data.

10.1 Projection tomography

The reconstruction of images from projections is called *projection tomography* (or *transmission tomography*). Projection tomography is familiar to us through important applications in the field of medical imaging. Within a designated region of the human body, an unknown function, $s(x, y, z)$, is associated with a phenomenon of interest. The three-dimensional function $s(x, y, z)$ is to be estimated, in part, based on an observation of its projections.

In many applications, it is adequate to form a single two-dimensional section of $s(x, y, z)$ in the z direction. To turn a three-dimensional problem into a collection of two-dimensional problems, it is common practice to sample $s(x, y, z)$ in the z direction by defining the sections $s_k(x, y) = s(x, y, k\Delta z)$. Each of these sections $s_k(x, y)$ can be treated individually as a two-dimensional function $s(x, y)$. For this reason, we need only consider the problem of forming a two-dimensional image. One can also develop methods to form a three-dimensional image directly. Techniques in three dimensions are much more burdensome computationally if the data consist of three-dimensional projections. The use of two-dimensional projections confined to two-dimensional sections reduces the computational burden.

Mathematically, the two-dimensional form of the tomography problem is as follows. Given the projections

$$p_\theta(t) = \int_{-\infty}^{\infty} s(t\cos\theta - r\sin\theta, t\sin\theta + r\cos\theta)\, dr$$

for some, or all, values of the parameter θ, find $s(x, y)$. The projections, defined in Section 3.4, are described in more detail as follows. A line at angle θ, as shown in Figure 10.1, is described as the set of (x, y) that satisfy the equation[1]

$$t = x\cos\theta + y\sin\theta.$$

The projection of $s(x, y)$ along this line is the integral of $s(x, y)$ along this line. As the value of t changes the line moves in the x, y plane parallel to the first line, as shown in Figure 10.2, and the projection, $p_\theta(t)$, becomes a function of the variable t. Finally the value of angle θ can change, as shown in Figure 10.3, to describe a change in direction of the bundle of parallel rays. Computation of $s(x, y)$ from the set of all projections at all θ is known as *parallel-beam* projection tomography.

[1] For fixed t, this line can be written

$$x = t\cos\theta - r\sin\theta$$
$$y = t\sin\theta + r\cos\theta$$

where r is a parameter that varies along the line.

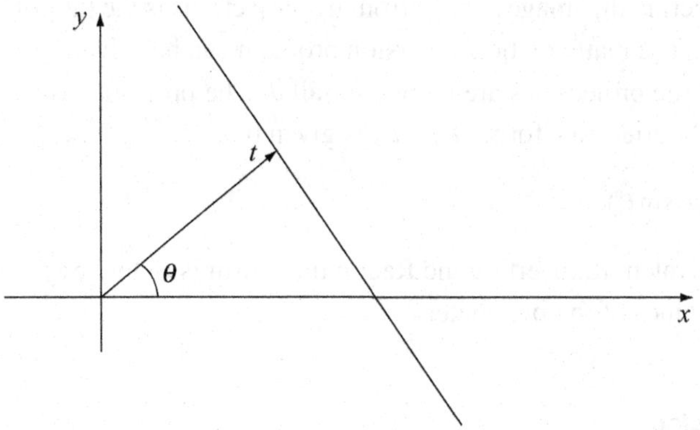

Figure 10.1 A ray at angle θ

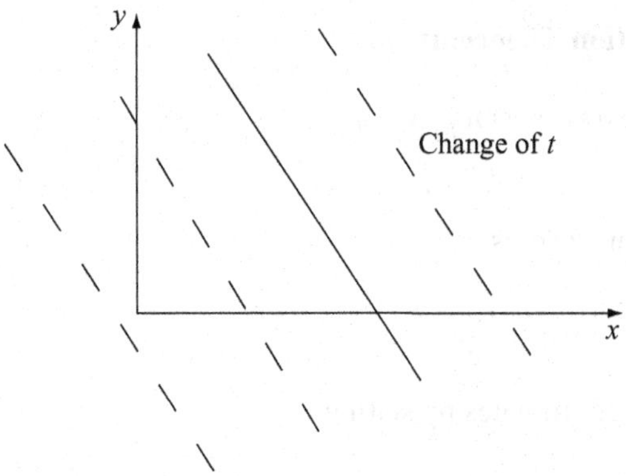

Figure 10.2 More rays at angle θ

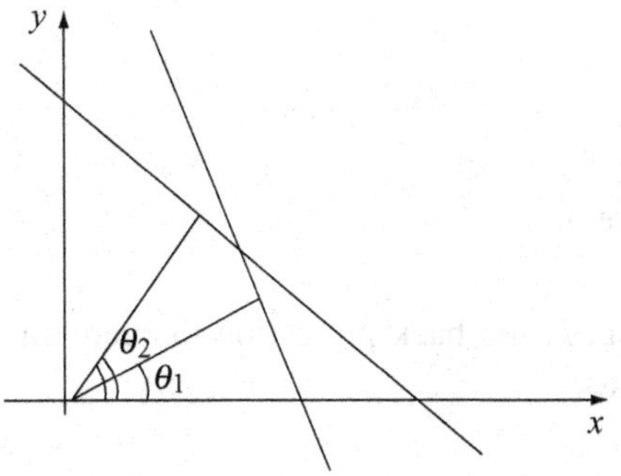

Figure 10.3 A change in angle

When $p_\theta(t)$ is known for all θ from 0 to π, the set of projections constitutes the *Radon transform* of $s(x, y)$, denoted by

$$p(t, \theta) = \int_{-\infty}^{\infty} s(t \cos\theta - r \sin\theta, t \sin\theta + r \cos\theta)\, dr.$$

The task of reconstructing the image $s(x, y)$ from its projections is the task of inverting the Radon transform. The mathematical inversion problem can be solved exactly when there is no noise and the projections are known for all θ. The projection-slice theorem says that $P_\theta(f)$, the Fourier transform of $p_\theta(t)$, is given by

$$P_\theta(f) = S(f\cos\theta, f\sin\theta).$$

Therefore a task equivalent to inverting the Radon transform is to find $S(f_x, f_y)$ when given $P_\theta(f)$ for all values of the parameter θ.

Parallel beam inversion

The next theorem gives an inverse for the Radon transform.

Theorem 10.1.1 (Back-Projection Theorem)

$$s(x, y) = \int_0^{\pi}\int_{-\infty}^{\infty} P_\theta(f)\mathrm{e}^{\mathrm{j}2\pi f(x\cos\theta + y\sin\theta)}|f|\,\mathrm{d}f\,\mathrm{d}\theta.$$

Proof: The inverse Fourier transform is

$$s(x, y) = \int_{-\infty}^{\infty}\int_{-\infty}^{\infty} S(f_x, f_y)\mathrm{e}^{\mathrm{j}2\pi(f_x x + f_y y)}\,\mathrm{d}f_x\,\mathrm{d}f_y.$$

Change the integration to polar coordinates by setting

$$f_x = f\cos\theta$$
$$f_y = f\sin\theta.$$

Then

$$s(x, y) = \int_0^{\pi}\int_{-\infty}^{\infty} S(f\cos\theta, f\sin\theta)\mathrm{e}^{\mathrm{j}2\pi f(x\cos\theta + y\sin\theta)}|f|\,\mathrm{d}f\,\mathrm{d}\theta,$$

which completes the proof of the theorem. □

By setting $x = r\cos\phi$ and $y = r\sin\phi$, the back-projection theorem can also be expressed in polar coordinates as follows.

Corollary 10.1.2

$$s(r\cos\phi, r\sin\phi) = \int_0^{\pi}\left[\int_{-\infty}^{\infty} P_\theta(f)|f|\mathrm{e}^{\mathrm{j}2\pi f r\cos(\theta-\phi)}\,\mathrm{d}f\right]\mathrm{d}\theta.$$

Proof: The proof follows immediately from the change of variables to polar coordinates. □

Methods of reconstruction based on the back-projection theorem are referred to as *back projection*. The structure of the back-projection operation may be easier to see if it is decomposed into two steps as follows:

$$g_\theta(t) = \int_{-\infty}^{\infty} |f| P_\theta(f) e^{j2\pi f t}\, dt$$

$$s(x, y) = \int_0^{\pi} g_\theta(x\cos\theta + y\sin\theta)\, d\theta,$$

as is illustrated in the flow chart of Figure 10.4.

The first integral is in the form of an inverse Fourier transform of the modified Fourier transform slice $|f|P_\theta(f)$. Consequently, one may suppose that the result of the inverse Fourier transform can be described formally as the convolution of the projection $p_\theta(t)$ with the function $h(t)$ where $h(t)$ is the inverse Fourier transform of $|f|$. Although the inverse Fourier transform of $|f|$ does not exist as a proper function, the inverse Fourier transform of $|f|$ does exist as a generalized function that behaves the right way under formal manipulations. Specifically, the second derivative of $\frac{1}{2}|f|$ is the impulse function $\delta(f)$, so $|f|$, formally, has the inverse Fourier transform $-2(2\pi t)^{-2}$. Thus, with $h(t) = -\frac{1}{2}(\pi t)^{-2}$, the back-projection operation can be written as

$$g_\theta(t) = h(t) * p_\theta(t).$$

The term "back projection" refers to the fact that $p_\theta(t)$ is somehow spread along the line by the function $h(t)$. Then

$$s(x, y) = \int_0^{\pi} g_\theta(x\cos\theta + y\sin\theta)\, d\theta.$$

This formulation of back projection has the drawback that $h(t)$ is a generalized function and has a singularity. The use of a generalized function can be avoided whenever $S(f_x, f_y) = 0$ for $\sqrt{f_x^2 + f_y^2} > B$ for some constant B, either naturally or because these high frequencies have been suppressed by filtering. Then it is enough to replace $|f|$ by any filter for which $H(f) = |f|$ for $|f| \le B$, and otherwise $H(f)$ is an arbitrary function of finite energy. We may refer to any $h(t)$ with such an $H(f)$ as a *ramp filter*. The use of a ramp filter will also ensure that the two-dimensional Fourier transform falls off with frequency sufficiently quickly to have an inverse Fourier transform $h(t)$. Then we may replace the $|f|$ in the back-projection formula by $H(f)$ because $|f|P_\theta(f) = H(f)P_\theta(f)$ and $H(f)P_\theta(f) \leftrightarrow h(t) * p_\theta(t)$. The function $h(t)$ is called the *kernel* of the reconstruction.

We may choose to filter either the signal $s(x, y)$ or its projections $p_\theta(t)$. The following theorem gives a condition for which these two methods are equivalent. Let $h(t)$ be a one-dimensional filter with Fourier transform $H(f)$, and let $h(x, y)$ be the two-dimensional

$p(t, \theta)$
for all $\theta \in [0, \pi]$

Fourier transform
$$P(f, \theta) = \int_{-\infty}^{\infty} p(t, \theta) e^{-j2\pi ft} \, dt$$
for all $\theta \in [0, \pi]$

$G(f, \theta) = |f| P(f, \theta)$

Inverse Fourier transform
$$g(t, \theta) = \int_{-\infty}^{\infty} G(f, \theta) e^{j2\pi ft} \, dt$$

$$s(x, y) = \int_{-\infty}^{\infty} g(x \cos \theta + y \sin \theta, \theta) \, d\theta$$

$s(x, y)$

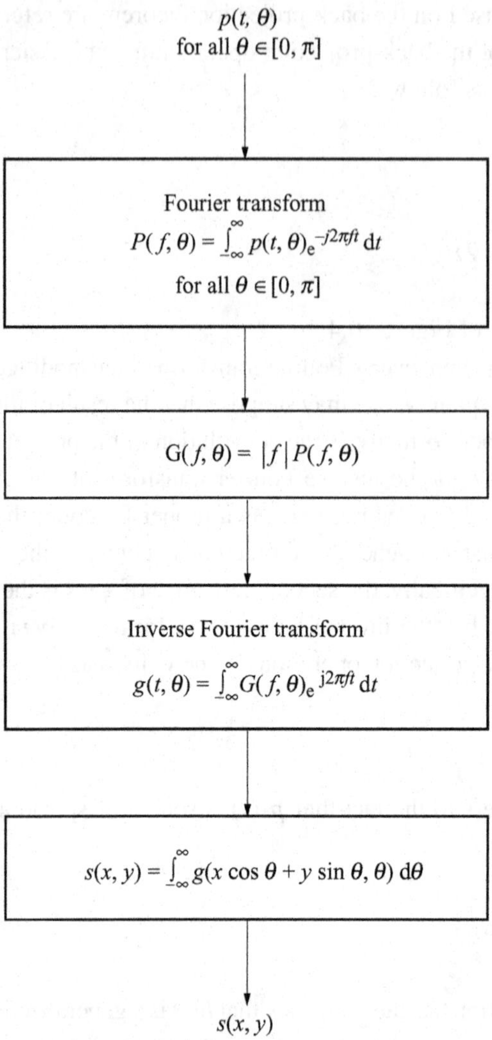

Figure 10.4   Computational description of back-projection theorem

circularly symmetric filter whose Fourier transform

$$H(f_x, f_y) = H\left(\sqrt{f_x^2 + f_y^2}\right)$$

is defined in terms of $H(f)$.

Theorem 10.1.3 *Let $p_\theta(t)$ be the projection at angle θ of $s(x, y)$. If $g_\theta(t) = h(t) * p_\theta(t)$ and $g(x, y) = h(x, y) * * s(x, y)$, then $g_\theta(t)$ is the projection at angle θ of $g(x, y)$.*

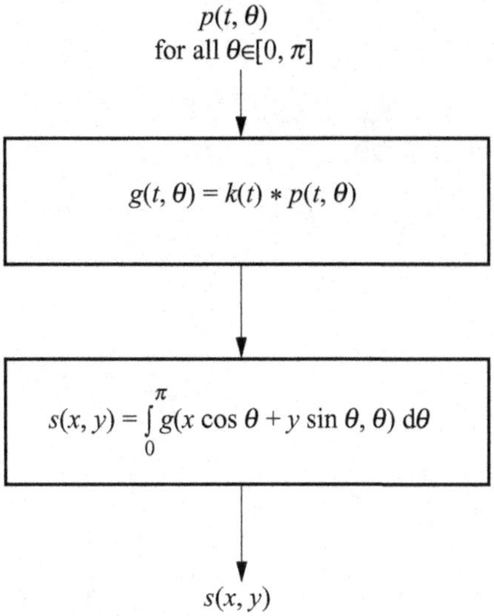

Figure 10.5 Filtered back projection

Proof: The projection of $g(x, y)$ at angle θ will have the Fourier transform

$$\begin{aligned} G(f\cos\theta, f\sin\theta) &= H(f\cos\theta, f\sin\theta)S(f\cos\theta, f\sin\theta) \\ &= H(f)P_\theta(f), \end{aligned}$$

which completes the proof of the theorem. □

With the aid of Corollary 10.1.2, we can now give

$$g(r\cos\theta, r\sin\theta) = \int_0^\pi \left[\int_{-\infty}^{\infty} H(f)P_\theta(f)e^{j2\pi f r\cos(\theta-\phi)}\,df\right] d\theta$$

as an alternative form of back projection. This alternative form, shown in Figure 10.5, is known as *filtered back projection.*

The filter $H(f)$ can be chosen so that all frequencies of interest in $s(x, y)$ are passed. In that case, the filter will have no effect on $s(x, y)$, but may be of considerable benefit in the computations. In this case, let

$$h(t) = \int_{-\infty}^{\infty} H(f)e^{j2\pi f t}\,dt.$$

Define

$$g_\theta(t) = h(t) * p_\theta(t).$$

Then

$$s(x, y) = \int_0^\pi g_\theta(x\cos\theta + y\sin\theta)\,d\theta.$$

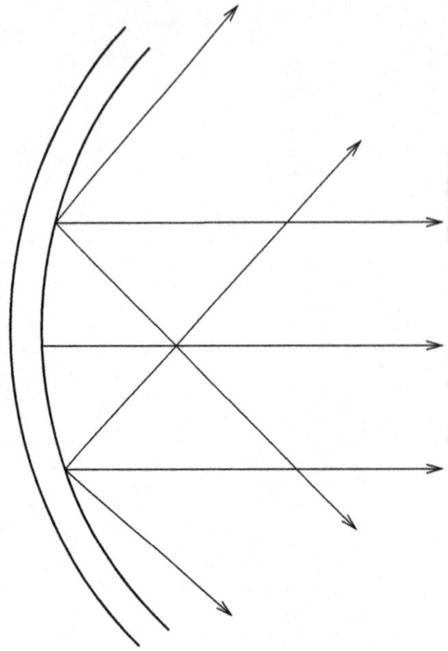

Figure 10.6 Finding parallel beams in fan beams

We now have the back-projection computation broken into a two-step process given by these two equations. First every projection is passed through a filter with the impulse response $h(t)$, then an integral is evaluated.

Fan-beam inversion

A parallel-beam tomography system collects, at each angle θ, a set of projections along each of a large set of parallel rays. This approach uses a large number of sources and detectors because each ray uses one X-ray source and one X-ray detector. Usually, an array of sources and a corresponding array of detectors are provided at only one angle, and measurements are taken sequentially at many angles by stepping the arrays in angle. One way to simplify this further is by mechanically stepping a single source–detector pair, but then the data collection time is much greater. Since data are usually collected sequentially in angle for many angles, it would be undesirable to extend the collection time by also sequentially measuring each ray at each angle.

An alternative to parallel-beam tomography is fan-beam tomography. Fan-beam tomography uses a single wide-beam X-ray source at each angle θ, as shown in Figure 10.6. At the radiation source, a fan beam of illumination is generated, the multiple rays from one source are detected by multiple sensors distributed in angle. Of course, an equivalent way to form a fan beam is to use multiple X-ray sources and only one sensor, but this would take more time to collect the data because the sources cannot be active at the same time.

In either case, the projection data consist of groups of rays in the form of fans, rather than of groups of parallel rays. However, inspection of Figure 10.6 makes it clear that the data in the rays of each fan can be regrouped into groups of parallel rays by choosing one ray from each fan, provided the sampling intervals are consistent. In this way fan-beam data can be sorted into parallel-beam data, then processed by any suitable algorithm, such as back projection, for parallel-beam data. It is apparent, however, that instead of restructuring the data, it should be possible to restructure the algorithm to accept the fan-beam data directly.

Data limitations

The back-projection algorithm provides a mathematical solution for projection tomography that is theoretically correct, but there are many practical considerations that arise in applications. Real data sets are degraded or limited in various ways. For example, projection data, in practice, is always sampled data. Data are collected at only a finite number of θ, and at each θ, data are sensed at only a finite number of t. This means, in effect, that data is collected on a two-dimensional sampling grid in polar coordinates. In turn, the computations of back projection are implemented on a digital computer and can be described as digital inversion of sampled data in polar coordinates. The sampling theorem does not have a completely satisfactory version in polar coordinates. Moreover, the back-projection procedure is formally exact only if $p_\theta(t)$ is known for all θ. In practice, $p_\theta(t)$ will be known only for a finite number of θ, usually on a polar sampling grid that approximates the requirements of the sampling theorem. Possibly there will be some larger angular gaps in the data. When $p_\theta(t)$ is known only for partial data, then one needs to compensate for the missing data, or suffer processing artifacts in the image. The back-projection theorem, based on the Radon transform, and Fourier reconstruction both require data with full angular coverage. If data are not available for some angles, then the algorithms may be modified to account for the missing data.

Another consideration is noise in the projections. A simple noise model is additive noise in which the measured data are

$$v_\theta(t) = p_\theta(t) + n_\theta(t)$$

where, for each θ, the additive term $n_\theta(t)$ is a noise process whose mean and variance are known. One may choose to build some sort of filtering into the computations to suppress the effect of noise. It is straightforward to incorporate a circularly symmetric noise filter, $H\left(\sqrt{f_x^2 + f_y^2}\right)$, into filtered back-projection. In other models of noise, the mean and variance of $n_\theta(t)$ may depend on $p_\theta(t)$, in which case the processing may be more complicated.

In some situations, prior data may be available. If one has a partial prior knowledge of $s(x, y)$, such as the region of the x, y plane where $s(x, y)$ can be nonzero, one may try to exploit this prior knowledge to improve the processing. Prior data about $s(x, y)$

may allow one to reduce the set of θ for which projections are needed. For an extreme but artificial example, if it is known that $s(x, y)$ factors as

$$s(x, y) = s'(x)s''(y),$$

we may readily expect that a projection along the x axis and a projection along the y axis should suffice to define $s(x, y)$.

10.2 Fourier and algebraic reconstruction

Filtered back projection has complexity proportional to n^3 and so can be computationally unattractive in its elementary form. Moreover, it may be necessary to form images from partial data or from noisy data. For these reasons, alternative computational methods of image formation are desirable.

Fourier representation

An alternative to back-projection image formation, which views the tomography processing problem in terms of the polar coordinates of the Radon transform, is Fourier reconstruction, which views the problem in terms of rectangular coordinates. The projection-slice theorem can be used to create slices of the two-dimensional Fourier transform $S(f_x, f_y)$ from the projections $p_\theta(t)$. However, these slices of $S(f_x, f_y)$ are in polar coordinates. It is computationally awkward to transform these samples to obtain the image $s(x, y)$ because the usual fast algorithms for digitally computing a two-dimensional Fourier transform are formulated for data that are sampled on a rectangular grid. To use these fast algorithms, the data samples must be converted from a polar grid, as shown in Figure 10.7, to a rectangular grid. This requires a two-dimensional

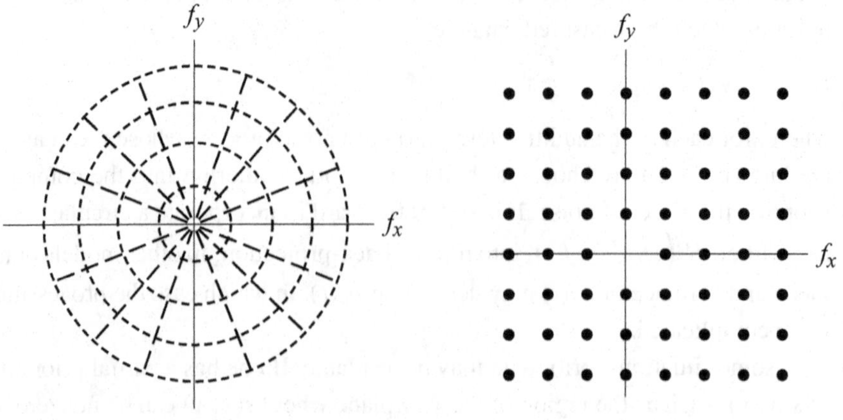

Figure 10.7 A polar grid and a rectangular grid

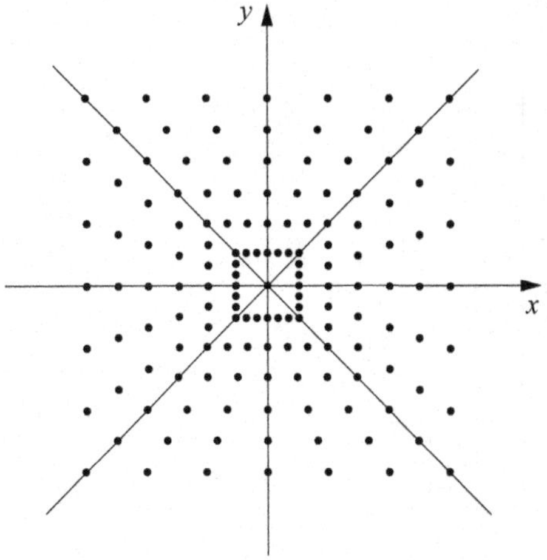

Figure 10.8 A concentric-squares grid

interpolation of the polar samples into samples on the rectangular grid. For example, one may choose to use a simple two-dimensional interpolation scheme in which each sample on the rectangular grid is computed as the weighted average of the four nearest samples on the polar grid. The interpolation error directly affects the quality of the image. Indeed, the interpolation error is usually the dominant source of computational noise in an image because interpolation is the most demanding part of the computation. Furthermore, because the polar samples are further apart at high frequencies, interpolation errors will be more severe at high frequencies unless the interpolation procedure varies across the f_x, f_y plane.

An alternative to the conventional discrete polar grid is the *concentric-squares grid*, shown in Figure 10.8, which uses samples on each slice that are spaced uniformly in $\tan\theta$ (or in $\cot\theta$) rather than in θ. In addition, the radial sample spacing varies with the angle in such a way that the sample points line up horizontally and vertically. The reason for using the consecutive-squares grid is so that a one-dimensional interpolation can be used to compute the samples on a two-dimensional grid.

Algebraic reconstruction

Fourier reconstruction works best if the Fourier sampling grid satisfies the Nyquist criterion and the entire Fourier sampling grid is filled with measured data. If some Fourier data samples are not measured, as is often the case, they might be replaced by zeros, but artifacts will occur in the image.

An alternative approach to tomography is *algebraic reconstruction*. Algebraic reconstruction is a computationally intensive approach that can be used for sparse data sets, or to account for unusual prior constraints on the support of the image. The approach is to set up a large system of linear equations in the unknown pixels, denoted s_{ij}, and

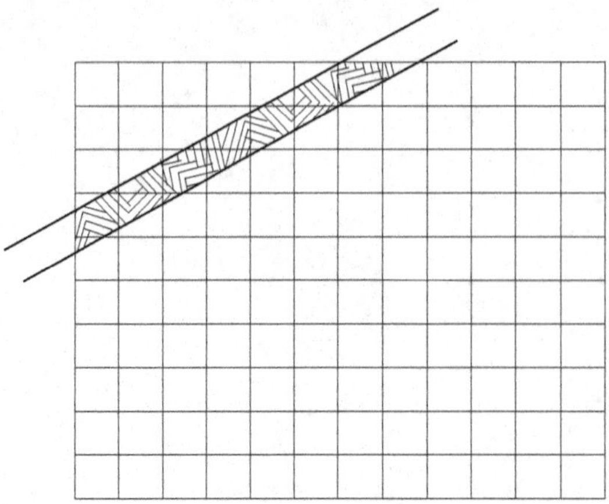

Figure 10.9 Rays overlapping pixels

the known rays, then to solve for the pixel values using the methods of linear algebra. Consider the image to be an n by n grid of pixels with value s_{ij} in the ij pixel. Regard a ray to have a width comparable to the pixel size, but oriented at angle θ. A ray overlaps each pixel by a different amount, and does not overlap most pixels at all. Let $\gamma_{ij\ell\theta}$ denote the overlap area of the ℓth ray at angle θ with the ijth pixel as shown in Figure 10.9. Then projection $p_{\ell\theta}$ is related to the pixel values by $p_{\ell\theta} = \sum_{ij} \gamma_{ij\ell\theta} s_{ij}$. There is one such linear equation for each ray. The number of linear equations is equal to the number of rays, which must be comparable to the number of unknown pixels and may be very large. For a 512 by 512 image, there are 262 144 pixels, so the number of linear equations must be of a similar number. If the I pixels are reindexed by a single index i, say by a raster scan, and regarded as a single one-dimensional vector, and the projections are reindexed by a single index ℓ, then the system of linear equations becomes

$$p_\ell = \sum_i \gamma_{i\ell} s_i.$$

Now the task of image formation is seen as the task of solving the matrix-vector equation of the form $\boldsymbol{p} = \boldsymbol{\Gamma s}$ where $\boldsymbol{\Gamma}$ is a large sparse matrix. In this formulation, the projection data set is simply the point $\boldsymbol{p}$, the image is the point $\boldsymbol{s}$, and $\boldsymbol{s} = \boldsymbol{\Gamma}^{-1}\boldsymbol{p}$. Of course, if $\boldsymbol{\Gamma}$ has dimension 262 144, or even a much smaller dimension, conventional techniques for solving linear systems of equations cannot be used. However, attractive iterative projection methods, suitable for sparse matrices, are available. To explain such methods, consider first only two equations in two unknowns

$$\gamma_{11}s_1 + \gamma_{12}s_2 = p_1$$
$$\gamma_{21}s_1 + \gamma_{22}s_2 = p_2.$$

Each of these two equations describes a line in the s_1, s_2 plane as shown in Figure 10.10. One way to find the intersection of the two lines is to start at any point and project

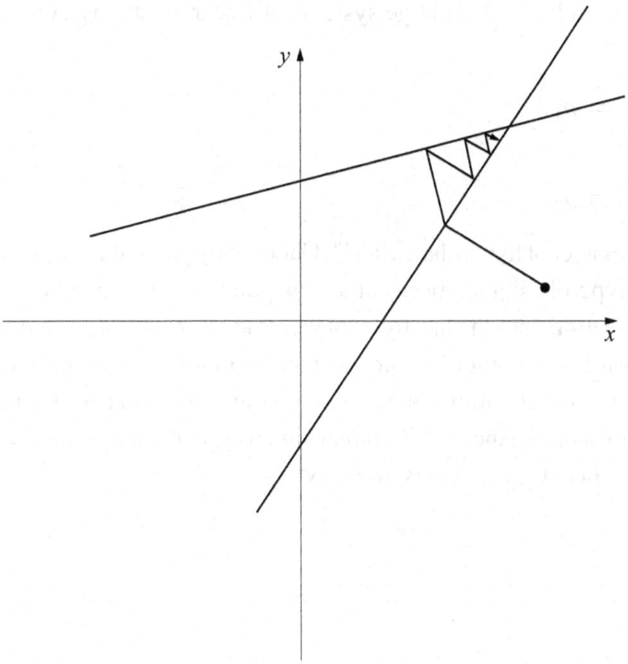

Figure 10.10 Projection operations in the plane

that point onto one of the lines. Then project the new point onto the other line. By repeating this process of successive projections, as shown in Figure 10.10, the point gradually moves to the point of intersection.

To derive this recursion, we pose a simple problem in the familiar x, y plane. What point on the line $x\cos\theta + y\sin\theta = c$ is closest to a given point x_0, y_0? It is simple to compute

$$x = c\cos\theta + (x_0\sin\theta - y_0\cos\theta)\sin\theta$$
$$y = c\sin\theta - (x_0\sin\theta - y_0\cos\theta)\cos\theta$$

as the answer to this question. Similarly, the closest point on the line $Ax + By = c$ to the point x_0, y_0 is given by

$$x = \frac{Ac}{A^2+B^2} + (x_0B - y_0A)\frac{B}{A^2+B^2}$$
$$y = \frac{Bc}{A^2+B^2} - (x_0B - y_0A)\frac{A}{A^2+B^2}.$$

The iterative projection method for two linear equations in the plane is now evident. Simply apply the projection alternately to these two lines, first to one line, then to the other.

The same process can be used on large systems of linear equations. Any system of m linear equations

$$\gamma_{11}s_1 + \cdots + \gamma_{1n}s_n = p_1$$

$$\cdots$$

$$\gamma_{m1}s_1 + \cdots + \gamma_{mn}s_n = p_m$$

in n unknowns defines a set of hyperplanes in $\boldsymbol{R}^n$. Choose any point of $\boldsymbol{R}^n$ and project that point onto the first hyperplane, then project the new point onto the second hyperplane, and so on. After projecting onto the last hyperplane, start over. The projected point will gradually move towards the point of common intersection if the system of equations is invertible. To derive the iteration, one uses the simple fact that in n-dimensional euclidian space $\boldsymbol{R}^n$, the point of the $(n-1)$-dimensional hyperplane $\sum_i a_i x_i = c$ closest to a given point $\boldsymbol{x}^\circ = (x_1^\circ, x_2^\circ, \ldots, x_n^\circ)$ is given by

$$x_i = \lambda a_i + x_i^\circ$$

where

$$\lambda = \frac{c - \sum_i a_i x_i^\circ}{\sum_i a_i^2}.$$

An alternative method is to form groups of r constraint equations, which correspond to $(n-r)$-dimensional hyperplanes. By using this group of r constraint equations at once, each step is a projection onto an $(n-r)$-dimensional hyperspace. In this variation, there will be fewer projection steps, but each projection step will be more complicated.

The method of successive projections can be applied to the problem of image formation because this problem can be regarded as the task of inverting the system of linear equations given by $\boldsymbol{p} = \boldsymbol{\Gamma s}$. The image $s(x, y)$ is then obtained from $\boldsymbol{s}$ by the usual interpolation formula

$$s(x, y) = \sum_i \sum_j s_{ij} h(x - i\Delta x, y - j\Delta y)$$

where $h(x, y)$ is an appropriate interpolation function.

Modal-based imaging

Although algebraic reconstruction is computationally intensive, it can easily tolerate missing data and can accommodate prior information about the support of the image simply by setting certain pixel values s_{ij} to zero. The method, in fact, is more general. The image need not be represented in terms of pixels. The method still can be applied even if the image is represented in terms of any basis of functions as

$$s(x, y) = \sum_i a_i b_i(x, y)$$

where the basis $\{b_i(x, y)\}$ is a set of functions spanning the space of possible images, which space may be restricted by prior constraints. This is called *model-based imaging*. For example, if $s(x, y)$ is known to be an image of a section of a human brain, then the basis $\{b_i(x, y)\}$ must be adequate to represent such sections as a linear combination of basis functions, but other kinds of images need not be representable in terms of that basis. Such a basis for a given class of images might be found by a training process involving a large number of instances of such images. Then a set of projection data is expressed in terms of that learned basis. This again, in principle, can be stated as the inversion of a large system of linear equations, and solved by successive projections.

10.3 Merging of multiple images

Deconvolution, studied in Section 9.1, enhances the quality of a single image by suppressing the blur of a pointspread function. Tomography, studied in Section 10.1, reconstructs an image from a collection of its projections. Midway between these two topics is the topic of combining several noisy images, each of which has been blurred by a different pointspread function. Let $h_\ell(x, y)$ for $\ell = 1, \ldots, L$ be a set of L pointspread functions, and let

$$p_\ell(x, y) = h_\ell(x, y) \quad ** \quad s(x, y) \quad \ell = 1, \ldots, L.$$

We will mention two special cases. In the special case for which $L = 1$, this problem reduces to the problem of deconvolution. In another special case for which

$$h_\ell(x, y) = \delta(t \cos\theta_\ell - r \sin\theta_\ell) \quad \ell = 1, \ldots, L$$

this reduces to the problem of tomography. Thus, a projection is a special form of filtered image using the filter

$$h(x, y) = \delta(t \cos\theta - r \sin\theta).$$

Consequently, the task of merging multiple images can be seen either as a generalization of deconvolution or as a generalization of tomography.

The problem of estimating $s(x, y)$ is applicable, for example, to the case where several images of a common scene are available, formed, perhaps, at different times or from different perspectives. It may be that each image has one direction along which resolution is good and another direction along which resolution is poor. It is possible to merge these images to obtain an improved image that combines the best features of each of the several original images.

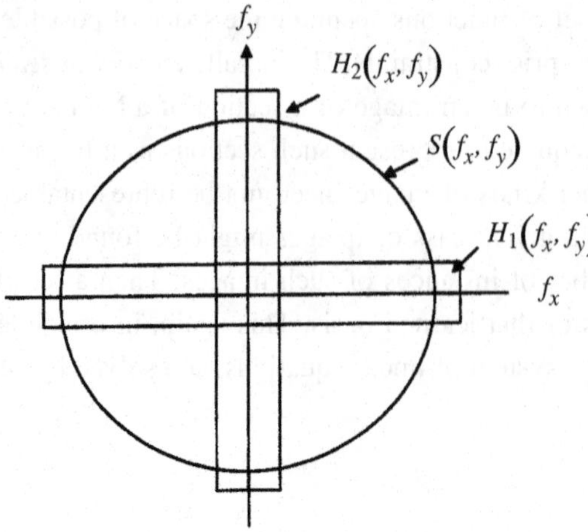

Figure 10.11 Illustrating the problem of image merging

For example, we may have two filtered images, $p_1(x, y)$ and $p_2(x, y)$, of a common scene, $s(x, y)$, given by

$$p_1(x, y) = h_1(x, y) \;**\; s(x, y)$$
$$p_2(x, y) = h_2(x, y) \;**\; s(x, y)$$

where $h_1(x, y)$ is the pointspread function of the first filter and $h_2(x, y)$ is the pointspread function of the second filter.

A more general task is to estimate $s(x, y)$ from noisy versions of the filtered images, as given by

$$v_1(x, y) = h_1(x, y) ** s(x, y) \; + \; n_1(x, y)$$
$$v_2(x, y) = h_2(x, y) ** s(x, y) \; + \; n_2(x, y).$$

The situation is easy to describe in the two-dimensional Fourier plane. The signal $s(x, y)$ may have actual frequency content within some large circle, as shown in Figure 10.11. Each low-resolution image may provide spectral components only within one of two strips, which together cover more of frequency space than either individually, but not all of the pertinent frequency space. The corners of the circle in the figure are not covered by either strip. Hence the original signal $s(x, y)$ cannot be fully reconstructed from only $p_1(x, y)$ and $p_2(x, y)$. Other information is necessary. This motivates the large question of when it is possible to recover $s(x, y)$ if appropriate prior side conditions are known, a question that we do not address here.

In the absence of noise, the two blurred images in the frequency domain are

$$P_1(f_x, f_y) = H_1(f_x, f_y)S(f_x, f_y)$$
$$P_2(f_x, f_y) = H_2(f_x, f_y)S(f_x, f_y).$$

Figure 10.11 illustrates how $P_1(f_x, f_y)$ and $P_2(f_x, f_y)$ may each contain part of the two-dimensional spectrum $S(f_x, f_y)$; other parts of $S(f_x, f_y)$ may not appear in either $P_1(f_x, f_y)$ or $P_2(f_x, f_y)$.

The task is to estimate $S(f_x, f_y)$ from the combination of $P_1(f_x, f_y)$ and $P_2(f_x, f_y)$. Insight into the problem can be obtained by simply writing down a reasonable deterministic inversion formula in the absence of noise. One obvious estimate is

$$\widehat{S}(f_x, f_y) = G_1(f_x, f_y)P_1(f_x, f_y) + G_2(f_x, f_y)P_2(f_x, f_y)$$

where

$$G_1(f_x, f_y) = \frac{H_1^*(f_x, f_y)}{|H_1(f_x, f_y)|^2 + |H_2(f_x, f_y)|^2}$$

$$G_2(f_x, f_y) = \frac{H_2^*(f_x, f_y)}{|H_1(f_x, f_y)|^2 + |H_2(f_x, f_y)|^2}.$$

This gives the correct inverse at each point (f_x, f_y) for which either $H_1(f_x, f_y)$ or $H_2(f_x, f_y)$ is nonzero. If both terms are zero at any (f_x, f_y), then $S(f_x, f_y)$ cannot be determined without additional information.

In the presence of noise, the inversion formula can be used just as it is written. However, at frequencies for which $H_1(f_x, f_y)$ and $H_2(f_x, f_y)$ are both near zero, noise will be amplified. To reduce noise amplification, one can imitate the form of the Wiener filter to write

$$\widehat{S}(f_x, f_y) = \frac{H_1^*(f_x, f_y)P_1(f_x, f_y) + H_2^*(f_x, f_y)P_2(f_x, f_y)}{|H_1(f_x, f_y)|^2 + |H_2(f_x, f_y)|^2 + C}$$

where C is an appropriate positive constant that protects against undue noise enhancement when $H_1(f_x, f_y)$ and $H_2(f_x, f_y)$ are both small.

10.4 Diffraction tomography

The important methods of projection tomography are based on the assumption that geometrical optics is an adequate model of wave propagation. This holds in many common situations in which the propagation medium affects the illuminating signal only by integrated attenuation along rays. This model holds in a variety of applications, but there are also applications in which it does not hold. Then a more complete description of interaction must be used, and other forms of tomography called *diffraction tomography* or *diffusion tomography* are appropriate. These are studied in this and the next section.

The need for diffraction tomography might arise in certain medical applications for which X-rays are not suitable for a variety of possible reasons. Then one may use radiation of a much longer wavelength for which the geometrical optics approximation is not valid. In such cases, diffraction cannot be ignored. Other applications for which diffraction may be significant could involve seismic waves, ultrasound, or microwaves. In such cases – those in which observed details are comparable to or smaller than the wavelength – diffraction will need to be considered. The distinction between diffraction tomography and projection tomography is that a deconvolution of the diffraction pointspread function must somehow be embedded in the back-projection algorithm. The first step toward understanding such modifications to the theory is the appropriate reformulation of the projection-slice theorem.

The projection-slice theorem of projection tomography equates each projection to a slice of the Fourier transform. If the projection is perturbed by a small amount of diffraction, then evidentally the slice must also be perturbed by a small amount in some way. This much is rather obvious, but it is not obvious how the slice will be perturbed. Is the slice perturbed by bending it slightly? Is it perturbed by broadening it slightly? Our task in this section is to determine the appropriate generalization of the projection-slice theorem for diffusion tomography.

First recall the analysis of the projection process without diffraction but with attenuation that was given in Section 1.4. We will review that analysis before revising it in this section to describe diffraction. It is convenient in this section to choose the direction of wave propagation as the z direction, thereby allowing the wave amplitude to be written as a function of both x and y at each z. The attenuation of a ray traversing a small interval of length Δz centered at z_1 is $s(z_1)\Delta z$ where $s(z)$ is the attenuation density at z. The relationship between the input intensity I and the output intensity I' of the ray passing through this small interval is

$$I' = I[1 - s(z_1)\Delta z].$$

More generally, the attenuation depends on x and y. Suppose that a wave is passing through a thin attenuating screen in the x, y plane with intensity attenuation $s(x, y, z_1)\Delta z_1$. Then

$$I'(x, y) = I(x, y)[1 - s(x, y, z_1)\Delta z].$$

The analysis of the nondiffracting attenuation proceeds by approximating the bracketed term by an exponential. Then, after traversing many such attenuating screens, one corresponding to each interval of length Δz, the full attenuation is

$$I_{out}(x, y) = I_{in}(x, y)\mathrm{e}^{-\sum_i s(x,y,z_i)\Delta z}.$$

In the limit as Δz goes to zero, the projection in the z direction is

$$\begin{aligned} p(x, y) &= \log \frac{I_{in}(x, y)}{I_{out}(x, y)} \\ &= \int_{-\infty}^{\infty} s(x, y, z)\,\mathrm{d}z. \end{aligned}$$

This equation is the basis of projection tomography.

We shall modify this analysis to describe the case of diffraction. The first change is that the input illumination signal is now an amplitude signal $a_{in}(x, y)$, rather than an

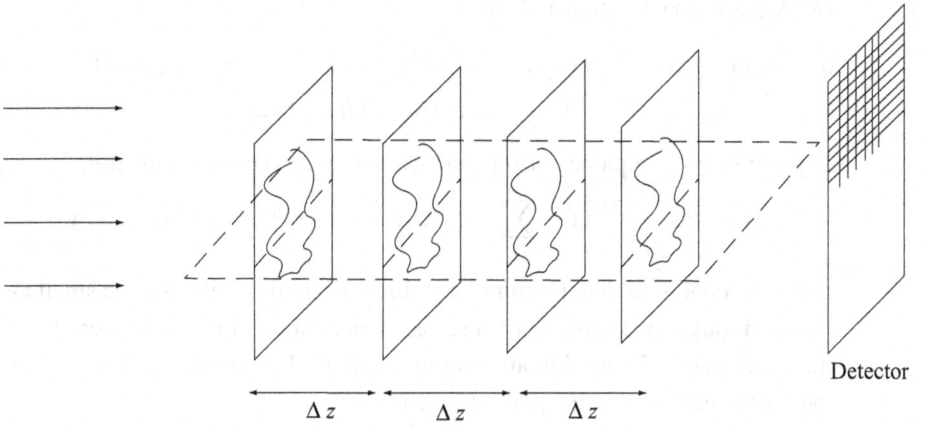

Figure 10.12 Multiple attenuating screens

intensity, and the medium weakly attenuates amplitude as described by an attenuation function $s(x, y, z)$. We require the input illumination signal to be a constant a_{in} independent of x and y. Otherwise there would be diffraction of the illumination term. We will regard the medium $s(x, y, z)$ to be approximated by a sequence of thin screens as shown in Figure 10.12. The screen at z_i has amplitude attenuation $s(x, y, z_i)\Delta z$ where Δz is the separation between screens. First consider only a single screen at position z_1. The wave reaching this screen is phase-delayed by $2\pi z_1/\lambda$. After the wave passes through this thin attenuating screen at position z_1, it is diffracted. The signal reaching the x, y plane at position d is

$$a_{out}(x, y) = a_{in}\mathrm{e}^{\mathrm{j}2\pi z_1/\lambda}[1 - s(x, y, z_1)\Delta z] ** h_{d-z_1}(x, y)$$

where $h_d(x, y)$ is the appropriate diffracting pointspread function of free space for distance d. The complex exponential on the left of the right side accounts for the propagation phase delay prior to the screen. Because a uniform plane wave does not diffract, the diffraction does not change the first term other than by a phase shift by $(d - z_1)/\lambda$. Then, by distributing the term $h_{d-z_1}(x, y)$, this expression can be written as

$$\begin{aligned} a_{out}(x, y) &= a_{in}\mathrm{e}^{\mathrm{j}2\pi z_1/\lambda}[\mathrm{e}^{\mathrm{j}2\pi(d-z_1)/\lambda} - s(x, y, z_1)\Delta z ** h_{d-z_1}(x, y)] \\ &= a_{in}\mathrm{e}^{\mathrm{j}2\pi d/\lambda}[1 - \mathrm{e}^{-\mathrm{j}2\pi(d-z_1)/\lambda}s(x, y, z_1)\Delta z ** h_{d-z_1}(x, y)]. \end{aligned}$$

If there are two phase screens, one screen at z_1 and a later screen at z_2, then the illumination launches a diffracting wave from each screen. Thus, there is one diffracting term from each phase screen. Provided the (first-order) Born approximation applies, the same illuminating plane wave is incident on each screen and the responses of the two screens do not interact. Thus,

$$a_{out}(x, y) \approx a_{in} e^{j2\pi d/\lambda}[1 - e^{-j2\pi(d-z_1)/\lambda} s(x, y, z_1)\Delta z ** h_{d-z_1}(x, y)$$
$$-e^{-j2\pi(d-z_2)/\lambda} s(x, y, z_2)\Delta z ** h_{d-z_2}(x, y)].$$

In general, for a sequence of screens, we will write the approximation

$$a_{out}(x, y) \approx a_{in} e^{j2\pi d/\lambda}(1 - \sum_i s(x, y, z_i)\Delta z ** e^{-j2\pi(d-z_i)/\lambda} h_{d-z_i}(x, y))$$

under the condition that the cumulative loss of signal continues to be small as the wave passes through the sequence of screens. Hence, in the limit as Δz goes to zero, under the assumption of a weakly attenuating medium that admits the Born approximation, the diffracted signal in the plane $z = d$ is

$$\frac{a_{out}(x, y)}{a_{in}} = e^{j2\pi d/\lambda}[1 - \int_{-\infty}^{\infty} s(x, y, z) ** e^{-j2\pi(d-z)/\lambda} h(x, y, d-z)\,dz]$$

with $h(x, y, z) = h_z(x, y)$. The first term is of no interest and will be dropped. The double asterisk in the second term refers to a two-dimensional convolution in the x, y plane, while the integration is a convolution in the z direction. Therefore, the projection in the x, y plane at any z is defined as the attenuated and diffracted signal

$$p(x, y, z) = e^{j2\pi z/\lambda}[s(x, y, z) *** e^{-j2\pi z/\lambda} h(x, y, z)].$$

A simple consequence of the three-dimensional convolution theorem is that this becomes

$$P(f_x, f_y, f_z + \lambda^{-1}) = S(f_x, f_y, f_z)H(f_x, f_y, f_z + \lambda^{-1})$$

or

$$P(f_x, f_y, f_z) = S(f_x, f_y, f_z - \lambda^{-1})H(f_x, f_y, f_z)$$

in the frequency domain. The misleading symmetry of this three-dimensional convolution formula almost makes it seem that the direction of the wave propagation effects the equation only by the offset by λ^{-1}. However, this is quite incorrect, because the diffraction pointspread function $h(x, y, z)$ has a very different dependence on z than on the other two variables. Moreover, the formula is not exact; approximations were made in the derivation.

To verify that this equation has the right behavior in the absence of diffraction, recall that the pointspread function in the absence of diffraction is given by the geometrical-optics approximation

$$h(x, y, z) = e^{j2\pi z/\lambda}\delta(z).$$

Therefore

$$p(x, y) = \int_{-\infty}^{\infty} s(x, y, z)\,dz.$$

Thus, in the absence of diffraction, the statement reduces to the amplitude form of weak projection tomography in three dimensions, as it must.

Next, to express our formulation of diffraction tomography in a form parallel to that of projection tomography in two dimensions, we shall suppress the y coordinate from the situation and from the notation. This amounts to assuming that $s(x, y, z)$ is independent of y. Then the diffracted projection $p(x)$ at distance d when the angle θ is equal to zero is the convolution

$$p(x) = e^{j2\pi d/\lambda} \int_{-\infty}^{\infty} \int_{-\infty}^{\infty} s(\xi, \eta) e^{-j2\pi(d-\eta)} h(x - \xi, d - \eta)\, d\xi\, d\eta.$$

We are now ready to give a version of the projection-slice theorem for a diffracted projection. The following theorem gives the formula for the case where θ equals 0. The general case is obtained afterward by a coordinate rotation and stated as a corollary.

Theorem 10.4.1 (Diffracting Projection-Slice Theorem) *The Fourier transform of a diffracted projection along the x axis is*

$$P(f) = S\left(f, \sqrt{\lambda^{-2} - f^2} + \lambda^{-1}\right) e^{j2\pi d\sqrt{\lambda^{-2} - f_x^2 - f_y^2}}.$$

Proof: The two-dimensional Huygens–Fresnel pointspread function $h(x, y, d)$ has the two-dimensional Fourier transform

$$H_d(f_x, f_y) = e^{j2\pi d\sqrt{\lambda^{-2} - f_x^2 - f_y^2}}.$$

This holds with d replaced by an arbitrary z. Define the three-dimensional function $h(x, y, z)$ as $h_z(x, y)$. This function has the three-dimensional Fourier transform

$$\begin{aligned} H(f_x, f_y, f_z) &= \int_{-\infty}^{\infty} e^{j2\pi z\sqrt{\lambda^{-2} - f_x^2 - f_y^2}} e^{-j2\pi f_z z}\, dz \\ &= \delta\left(f_z - \sqrt{\lambda^{-2} - f_x^2 - f_y^2}\right). \end{aligned}$$

Therefore

$$P(f_x, f_y, f_z) = S(f_x, f_y, f_z - \lambda^{-1})\delta\left(f_z - \sqrt{\lambda^{-2} - f_x^2 - f_y^2}\right).$$

The two-dimensional projection $p(x, y)$ in the plane at $z = d$ has the two-dimensional Fourier transform $P(f_x, f_y)$ given by

$$\begin{aligned} P(f_x, f_y) &= P(f_x, f_y, d) \\ &= \left.\int_{-\infty}^{\infty} P(f_x, f_y, f_z) e^{j2\pi f_z z}\, df_z \right|_{z=d} \\ &= \int_{-\infty}^{\infty} S(f_x, f_y, f_z - \lambda^{-1})\delta\left(f_z - \sqrt{\lambda^{-2} - f_x^2 - f_y^2}\right) e^{j2\pi f_z z}\, df_z \\ &= S\left(f_x, f_y, \sqrt{\lambda^{-2} - f_x^2 - f_y^2} - \lambda^{-1}\right) e^{j2\pi d\sqrt{\lambda^{-2} - f_x^2 - f_y^2}}. \end{aligned}$$

To restrict the treatment to a projection in two dimensions, assuming that $s(x, y, z)$ is independent of y, set $f_x = f$, and drop f_y from the notation. Then

$$P(f) = S\left(f, \sqrt{\lambda^{-2} - f^2} - \lambda^{-1}\right) e^{j2\pi d\sqrt{\lambda^{-2}-f^2}}.$$

We can now conclude that the projection $p(x)$ has a Fourier transform $P(f)$ equal to $S\left(f, \sqrt{\lambda^{-2} - f^2} - \lambda^{-1}\right)$ multiplied by a phase term. □

The Fourier transform $P(f)$ is a slice of $S(f_x, f_z)$ along the curve $f_z = \sqrt{\lambda^{-2} - f_x^2} - \lambda^{-1}$, which can be rewritten in the standard form of a circle as $(\lambda f_z + 1)^2 + \lambda^2 f_x^2 = 1$. Notice that for small λ, the expression $\lambda^{-1}\sqrt{1 - \lambda^2 f^2} - \lambda^{-1}$ is approximately λf^2. By neglecting this term, $|P(f)|$ becomes $|S(f, 0)|$. Thus, within this approximation, the diffracted projection-slice theorem reduces to the situation without diffraction.

Corollary 10.4.2 *The Fourier transform of a diffracted projection $p_\theta(t)$ at angle θ is*

$$P_\theta(f)$$
$$= S\left(f \cos\theta - \left(\sqrt{\lambda^{-2} - f^2} - \lambda^{-1}\right)\sin\theta,\ f \sin\theta + \left(\sqrt{\lambda^{-2} - f^2} - \lambda^{-1}\right)\cos\theta\right).$$

Proof: This is an immediate consequence of the rotation property of the Fourier transform. Simply rotate the coordinate system by angle θ. □

The theorem says that because of diffraction, the "slice" of $S(f_x, f_z)$ now takes place on the circle

$$\lambda^2 f_x^2 + (\lambda f_z - 1)^2 = 1 \quad f_z \geq 0.$$

This is the two-dimensional version of an Ewald sphere. Because f_x must be positive, this is the equation of a half circle in the f_x, f_z plane, as shown in Figure 10.13. The half-circle extends to a point $\sqrt{2}/\lambda$ from the origin. By varying the angle of the projection, as described in the corollary, this semicircular slice can be reoriented. In this way, all values of the two-dimensional Fourier transform for which the point (f_x, f_z) is interior to the circle

$$f_x^2 + f_z^2 = 2\lambda^{-2}$$

can be measured. If $S(f_x, f_y)$ has support inside this circle, then $s(x, y)$, in principle, can be recovered as the inverse two-dimensional Fourier transform of $S(f_x, f_y)$, which can be recovered slice by slice.

If $S(f_x, f_y)$ has support that extends outside of the observable circle in the frequency plane, then $S(f_x, f_y)$ can be measured only for (f_x, f_y) inside the circle. If $s(x, y)$ is estimated as the inverse Fourier transform of the observable portion of $S(f_x, f_y)$, then what is actually computed is $s(x, y)$ blurred by a jinc function.

10.5 Diffusion tomography

A wave can be scattered, rescattered, then rescattered again. The extreme situation in which the signal passes through an infinity of infinitesimal scattering events is called *diffusion*. For example, electromagnetic radiation at a wavelength of about 850 nanometers interacts with human tissue by diffusion. A sharp beam at this wavelength entering into one side of a thickness of such a diffusing material will emerge as a diffuse glow at the other side. Other wavelengths in and around the optical band are heavily attenuated by such tissue and have not been found suitable for imaging. A diffused wavefront can be detected and a collection of such diffuse projections can be processed to obtain an image of the scattering medium, although with great difficulty. Inversion of the diffusion process to form an image of the diffusion medium is called *diffusion tomography*. Hence diffusion tomography, though subject to severe sensitivity and resolution problems, is important in some medical applications.

The theory of diffusion tomography is much more difficult to formulate than is the theory of diffraction tomography. The diffusion mechanism can be motivated by reference to an arrangement of multiple screens as shown in Figure 10.12. In the case of diffusion, however, the screens are not attenuating screens; they are scattering screens. Each screen multiplies the incoming wave by a complex function and scatters it in both the forward and backward directions. Eventually a propagating wave will be incident on both sides of each screen, and will be scattered with random phase shifts in both directions. To analyze this situation in the limit as the number of screens goes to infinity and the spacing goes to zero is beyond our goals.

A different model of diffusion can be set up in terms of rays (or photons) that are randomly bent upon passing through a screen. The collection of these rays forms an intensity field that is measured by a sensor array. Again, the analysis of this model of diffusion is quite difficult. The ray paths describe a random walk analogous to the particle walks under brownian motion. The particle density under brownian motion obeys a diffusion equation.

We will not attempt to analyze the fundamentals of the diffusion mechanism. Instead, we will be content to simply define diffusion as the response to the diffusion equation

$$\nabla^2 s(x, y, z) - k^2 s(x, y, z) = 0$$

where the constant k in this context is called the *diffusion coefficient*. The value of k depends on the diffusion medium. To solve the diffusion equation, we notice that except

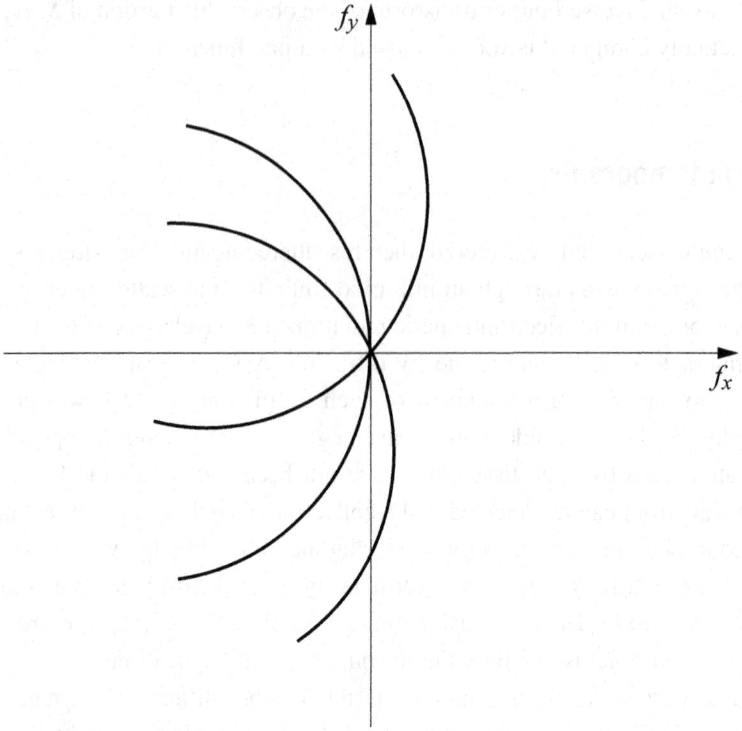

Figure 10.13 Diffracted slices

for the sign of k^2, it is the same as the wave equation given by

$$\nabla^2 s(x, y, z) + k^2 s(x, y, z) = 0,$$

which describes diffraction, where, in this equation, the constant k is the wave number. The wave equation corresponds to the Huygens–Fresnel pointspread function of free space

$$h_d(x, y) = \frac{1}{2\pi}\left[\frac{-\mathrm{j}dk}{(d^2 + x^2 + y^2)} + \frac{d}{(d^2 + x^2 + y^2)^{3/2}}\right] \mathrm{e}^{\mathrm{j}k\sqrt{d^2+x^2+y^2}}.$$

This suggests that the pointspread function for the diffusion equation can be obtained simply by replacing k by jk. Thus, the pointspread function for diffusion is

$$h_d(x, y) = \frac{1}{2\pi}\left[\frac{dk}{(d^2 + x^2 + y^2)} + \frac{d}{(d^2 + x^2 + y^2)^{3/2}}\right] \mathrm{e}^{-k\sqrt{d^2+x^2+y^2}}.$$

The diffusion pointspread function is a real, decaying exponential with decay constant $1/k$. Now, by using the two-dimensional Fourier transform pair

$$\left[\frac{b/2\pi}{(b^2 + x^2 + y^2)^{1/2}} + \frac{ab}{b^2 + x^2 + y^2}\right] \mathrm{e}^{-2\pi a\sqrt{b^2+x^2+y^2}} \Leftrightarrow \mathrm{e}^{-2\pi b\sqrt{a^2+f_x^2+f_y^2}}$$

we see that

$$H_d(f_x, f_y) = \mathrm{e}^{-d\sqrt{k^2+(2\pi)^2(f_x^2+f_y^2)}}.$$

This is the two-dimensional Fourier transform of the wavefront in that x, y plane for which $z = d$. To extend it to a three-dimensional Fourier transform, replace d by z and define the function

$$H(f_x, f_y, z) = \mathrm{e}^{-z\sqrt{k^2+(2\pi)^2(f_x^2+f_y^2)}}.$$

Now take the Fourier transform with respect to z, recalling the one-dimensional Fourier transform pair

$$\mathrm{e}^{-at}, t \geq 0 \leftrightarrow \frac{1}{a - \mathrm{j}2\pi f}.$$

Then

$$H(f_x, f_y, f_z) = \frac{1}{\sqrt{k^2 + (2\pi)^2(f_x^2 + f_y^2)} - \mathrm{j}2\pi f_z}.$$

Moreover,

$$P(f_x, f_y, f_z) = S(f_x, f_y, f_z)H(f_x, f_y, f_z).$$

The two-dimensional Fourier transform, denoted $P_d(f_x, f_y)$ or simply $P(f_x, f_y)$, of the projection in the plane at $z = d$ is

$$\begin{aligned} P(f_x, f_y) &= \int_{-\infty}^{\infty} P(f_x, f_y, f_z)\mathrm{e}^{\mathrm{j}2\pi f_z z}\,\mathrm{d}f_z \Big|_{z=d} \\ &= \int_{-\infty}^{\infty} S(f_x, f_y, f_z)\frac{1}{\sqrt{k^2 + (2\pi)^2(f_x^2 + f_y^2)} - \mathrm{j}2\pi f_z}\mathrm{e}^{\mathrm{j}2\pi f_z d}\,\mathrm{d}f_z. \end{aligned}$$

As was done for diffraction tomography, the formulation of diffusion tomography can be put in a form parallel to that of two-dimensional projection tomography in the x, z plane by eliminating the y coordinate. To do so, we suppose that $s(x, y, z)$ is independent of y. Then we suppress f_y from the situation and also from the notation to write

$$P(f) = \int_{-\infty}^{\infty} S(f, f_z)\frac{1}{\sqrt{k^2 + 4\pi^2 f^2} - \mathrm{j}2\pi f_z}\mathrm{e}^{\mathrm{j}2\pi f_z d}\,\mathrm{d}f_z$$

as the Fourier transform of $p(x)$, the diffused projection onto the x axis at $z = d$. The projection, then, at any angle θ has Fourier transform

$$P_\theta(f) = \int_{-\infty}^{\infty} S(f\cos\theta - f_z\sin\theta, f\sin\theta + f_z\cos\theta)\frac{1}{\sqrt{k^2 + 4\pi^2 f^2} - \mathrm{j}2\pi f_z}\,\mathrm{d}f_z$$

as a consequence of the rotation property of the Fourier transform.

Evanescent wave tomography

Closely related to the mathematics of diffusion tomography is near-field imaging or evanescent-wave optical imaging used to image small features – features smaller than the optical wavelength – by wave fields containing strong evanescent components. The methods of evanescent-wave optical imaging are important to near-field microscopy and are related to modern methods of lithography that operate beyond the diffraction limit.

To generate an evanescent wave, the object is illuminated by light passed through a subwavelength aperture. In recent years, the role of the subwavelength aperture is played by the tip of a tapered optical fiber. The light is passed through a fiber whose tip is tapered to less than a wavelength, and this leads to an evanescent wave leaving the tip. Because the evanescent wave decays quickly, the tip must be within a few wavelengths of the object being viewed, and only the region of the object near the tip is illuminated. The tip can be scanned in the x, y plane to view the entire object.

There is a strong analogy between diffusing waves and evanescent waves. The quickest way to see the analogy is to notice that a homogeneous evanescent wave in the z direction does satisfy the wave equation, but with an imaginary wave number k instead of a real k. Replacing k by jk in the wave equation effectively changes the sign of k and becomes the diffusion equation. Indeed, in the plane wave expansion formalism, the defusing waves then may be seen to consist of nothing but evanescent values whereas the usual diffracting waves generally consist of both evanescent and propagating components.

10.6 Coherent and noncoherent radar tomography

The image formation procedure for an imaging radar was formulated in Chapter 7 in the range-doppler rectangular coordinate system in terms of the sample cross-ambiguity function, which is an implementation of the matched filter. If the waveform is appropriate, as is a chirp pulse train, the image formation procedure for an imaging radar can be formulated using the methods of tomography. This formulation clarifies the behavior of an imaging radar by exploring it from another point of view. It may be the superior point of view when the synthetic aperture is long enough so that a polar (r, θ) coordinate system fits the situation better than a rectangular (x, y) coordinate system.

We shall adopt a tomographic formulation which says that every pulse of the waveform is transmitted and received at its own viewing angle, as shown in Figure 10.14. We can process each pulse individually to obtain one aspect angle on the scene, and then process the batch of these projections to form a tomographic image. An individual

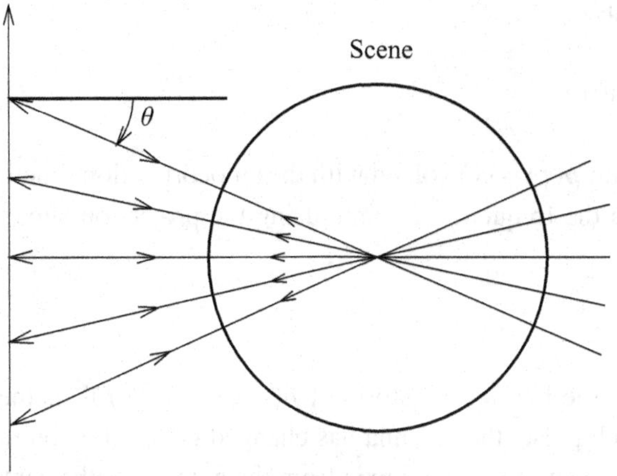

Figure 10.14 Tomographic depiction of synthetic-aperture radar

pulse is very short, so we can choose to ignore the doppler shift on the individual pulse. Thus, although the radar antenna is physically moving during each pulse to obtain different viewing angles, this approximation regards the antenna as stationary during each pulse.

To fully process each pulse by computing a sample cross-ambiguity function on each pulse would give for each pulse

$$\chi_s(\tau, \nu) = \chi^{(p)}(\tau, \nu) \;**\; \rho(\tau, \nu)$$

by the radar imaging approximation, where $\chi^{(p)}(\tau, \nu)$ is the ambiguity function of a single pulse and τ, ν is in a coordinate system appropriate for that pulse. Because the pulse is very short, $\chi^{(p)}(\tau, \nu)$ is very wide in the ν direction, perhaps even wider than the illuminated scene. Therefore $\chi^{(p)}(\tau, \nu)$ can be approximated as independent of ν. Thus

$$\chi^{(p)}(\tau, \nu) = \phi(\tau).$$

The image computed for one pulse then is

$$\chi_s(\tau, \nu) = \phi(\tau) \;**\; \rho(\tau, \nu)$$

where $\phi(\tau)$ is the autocorrelation function of the pulse. Because $\phi(\tau)$ does not depend on ν, the double convolution becomes a single convolution

$$\chi_s(\tau, \nu) = \phi(\tau) * \int_{-\infty}^{\infty} \rho(\tau, \nu)\, d\nu.$$

The integral is a projection along the ν direction. The left side is independent of ν and we denote it by $p'(\tau)$. Thus

$$p'(\tau) = \phi(\tau) * \int_{-\infty}^{\infty} \rho(\tau, \nu)\, d\nu.$$

This says that the projection $p(\tau)$ is convolved with the autocorrelation function of the illuminating pulse $s(t)$. In the frequency domain, using the projection-slice theorem, this becomes

$$P'(f) = |S(f)|^2 R(f, 0)$$

where $R(f, 0)$ is a slice of the Fourier transform of $\rho(\tau, \nu)$ and $|S(f)|^2$ is the Fourier transform of $\phi(\tau)$. For each pulse, the antenna has changed perspective on the scene, and the local τ, ν coordinate system has a τ axis from the antenna to the center of the scene. We can regard it as a polar coordinate system with coordinates r and θ. In the r, θ polar coordinate system, centered at the center of a scene, the return from reflectors at range r after matched filtering is

$$p_\theta(r) = \phi(r) * \int_{-\infty}^{\infty} \rho(r, r')\, dr'.$$

Thus the pulse transmitted at angle θ provides a projection at angle θ. These projections can be used to form an image by the methods of tomography.

In the special case where the pulse is a chirp pulse, $e^{j\pi t^2}$ with Fourier transform $(1/\sqrt{2})(1 + j)e^{-j\pi f^2}$, the power spectrum is $|S(f)|^2 = \sqrt{2}$. More realistically, the chirp pulse has a finite duration T. Thus $s(t) = e^{j\pi t^2}\text{rect}(t/T)$ and

$$|S(f)|^2 = \left| \frac{1+j}{\sqrt{2}} e^{-j\pi f^2} * \text{sinc}(fT) \right|^2.$$

To better understand the processing for the chirp pulse it may be helpful to develop the processing equations directly. The radar return seen by a pulse $s(t)$ at angle θ is

$$v_\theta(t) = \int_{-\infty}^{\infty} p_\theta(r) s(t - 2(R + r)/c)\, dr$$

where

$$p_\theta(r) = \int_{-\infty}^{\infty} \rho(r, r')\, dr'.$$

If the chirp pulse has a duration T that is long in comparison with the dispersion in delay across the scene, we can be a little casual with regard to the end points of the pulse $s(t)$ without causing significant error. Thus the received pulse is

$$v_\theta(t) = \int_{-\infty}^{\infty} p_\theta(r) e^{-j[2\pi f_0(t-2(R+r)/c)+\pi\alpha(t-2(R+r)/c)^2]}\, dr.$$

Define the dechirped pulse as

$$\begin{aligned} c_\theta(t) &= v_\theta(t)e^{j[2\pi f_0(t-2R/c)+\pi\alpha(t-2R/c)^2]} \\ &= \int_{-\infty}^{\infty} p_\theta(r)e^{j[4\pi f_0 r/c+\pi\alpha(2rt/c-4Rr/c^2+2r^2/c^2)]}\, dr. \end{aligned}$$

This has the appearance of a Fourier transform except for the phase term that is quadratic in r. One could compensate, either exactly or approximately, for the quadratic phase terms, but we will simply drop this term under the assumption that the scene size is such that $2\alpha r^2/c^2$ is a negligible angle. Then

$$c_\theta(t) = \int_{-\infty}^{\infty} p_\theta(r)e^{j2\pi(2f_0+\alpha(t-\tau_o))r/c}\, dr$$

where $\tau_o = 2R/c$. Consequently, we now recognize $c_\theta(t)$ as a Fourier transform. Specifically,

$$c_\theta(t) = P_\theta(2f_0 + \alpha(t-\tau_o)/c),$$

and $c_\theta(t)$ is the Fourier transform of a projection. This means that a pulse transmitted at angle θ will generate a slice at angle θ of the Fourier transform. A collection of such pulses for various θ gives a sequence of slices. These can be processed by the methods of tomography.

10.7 Emission tomography from magnetic excitation

Emission tomography depends on radiation that is emitted by an object of interest. This requires that the object be infused with a suitable source of energy to fuel the radiation. We shall discuss here the important instance of emission tomography known as *magnetic-resonance imaging* in which a magnetic field is used to excite nuclei in the chosen object. A strong bias magnetic field is overlaid with a time-varying and space-varying magnetic field which causes the nuclei to precess and radiate electromagnetic waves in the radio frequencies. By designing the spatial distribution of the magnetic fields appropriately, the sensed electromagnetic waves have the form of projections, from which the object can be imaged by the methods of tomography.

Magnetic-resonance imaging is an important method of forming an image of the density of a selected species of atomic nuclei, that is, of isolated protons. It is also possible to target other species of nuclei to form images of the density of those nuclei, though it is not common to use other than hydrogen nuclei.

Hydrogen is an element that occurs in very high concentrations in biological organisms. The nucleus of each hydrogen atom is an isolated proton. The density of hydrogen atoms $\rho(x, y, z)$, which can be regarded as the density of free protons in the organism, provides an image of the organism. The soft tissue of the organism tends to stand out because soft tissue is dominated by water molecules, and these contain hydrogen atoms. The density of isolated protons (hydrogen nuclei) $\rho(x, y, z)$ can be imaged with excellent resolution and so can give a meaningful image even for very small details. A high-resolution image of the hydrogen density $\rho(x, y, z)$ can be regarded as a high-resolution image of some associated details of an organism under study. This imaging modality only uses the magnetic moment and the angular momentum of the nuclei.

Because atomic nuclei with an odd number of nucleons have a spin, they act like small magnets that align themselves with the direction of an ambient magnetic field. Magnetic-resonance imaging is a technique that uses magnetic fields of appropriate frequency to cause protons to align and precess. The radiation from the precessing protons is then detected and processed to form an estimate of $\rho(x, y, z)$. Magnetic-resonance images are obtained by placing a subject inside the bore of a powerful magnet that aligns all the target nuclei in a common direction. A second, oscillating magnetic field is then applied to the subject, which causes the hydrogen nuclei to move out of alignment with the first, thereby generating radiated signals that can be detected and mathematically transformed into images of the body.

Magnetic-resonance imaging uses the fact that an atomic nucleus has a quantum-mechanical spin and the spin axis precesses about an external magnetic field B. The frequency of precession, known as the *Larmor frequency*, is given by $f = \gamma B$ where γ is a constant known as the *gyromagnetic ratio*. The gyromagnetic ratio of a hydrogen nucleus is 42.58 megahertz per tesla. The external magnetic field can vary with position and is usually written in the form $B_0 + B(x, y, z)$ where B_0 is a constant and $B(x, y, z)$ is a spatially-varying magnetic field. Then

$$f(x, y, z) = f_0 + \gamma B(x, y, z)$$

where $f_0 = \gamma B_0$ plays the role of a carrier frequency for the radiated signal. A strong bias field B_0 is usually used, perhaps in the range of one to four tesla, corresponding to a Larmor frequency of the order of 60 megahertz. However, some systems now in use have fields in excess of ten telsa, corresponding to Larmor frequencies in excess of 200 megahertz.

After the nucleus is made to precess at the Larmor frequency, it gives up its energy by radiating radio frequency signals at that frequency. There are two relaxation mechanisms for the processing nucleus, with relative strengths depending on the environment of that nucleus. These are the longitudinal relaxation, known as the T_1 relaxation, and the transverse relaxation, known as the T_2 relaxation. By a differential image comparison between the image made from the T_1 signal and the image made from the T_2 signal, an image of certain structure variations within a subject can be obtained.

Magnetic-resonance image formation is based on the projection-slice theorem. The projections are measured one by one; each projection requires a different magnetic gradient. By controlling the direction of the space-varying magnetic gradient, the direction of the projection is controlled. The spatial magnetic structure is designed so that the sequence of received projections can be processed to form an image.

To describe the processing, suppose that we are able to specify a chosen spatial aperture, $A(x, y, z)$, such that we are able to selectively excite only those protons in the chosen spatial aperture and to excite those protons at the position (x, y, z) to radiate at the carrier frequency f_0 at the complex amplitude $e^{-j2\pi(f_x x+f_y y)}$. If the density of the protons is $\rho(x, y, z)$ then the passband signal radiated from these protons in an infinitesimal cell at (x, y, z) is

$$\widetilde{v}(t)\,dx\,dy\,dz = \rho(x, y, z)e^{-j2\pi(f_x x+f_y y)}e^{-j2\pi f_0 t}\,dx\,dy\,dz.$$

Then the complex baseband signal $v(t)$ satisfies

$$v(t)e^{-j2\pi f_0 t} = \int_{-\infty}^{\infty}\int_{-\infty}^{\infty}\int_{-\infty}^{\infty} \rho(x, y, z)A(x, y, z)e^{-j2\pi(f_x x+f_y y)}e^{-j2\pi f_0 t}\,dx\,dy\,dz$$

so that

$$v(t) = \int_{-\infty}^{\infty}\int_{-\infty}^{\infty}\int_{-\infty}^{\infty} \rho(x, y, z)A(x, y, z)e^{-j2\pi(f_x x+f_y y)}\,dx\,dy\,dz.$$

In this form, the sensor data are three-dimensional, and the formulation is very general. To reduce the computations, the problem can be reduced to two dimensions. If the chosen aperture $A(x, y, z)$ is thin in the z direction, centered at z_0, and such that $s(x, y, z)$ is approximately independent of z over the thin aperture and uniformly covers the region of x, y of interest, this can be regarded as the two-dimensional expression

$$S(f_x, f_y) = \int_{-\infty}^{\infty}\int_{-\infty}^{\infty} s(x, y)e^{-j2\pi(f_x x+f_y y)}\,dx\,dy.$$

By repeating the measurement for many values of f_x, f_y, the entire Fourier plane may be sampled to produce $S(f_x, f_y)$. At this point, the task of computing $s(x, y)$ reduces to the task of computing an inverse two-dimensional Fourier transform, and the topic now resembles many other topics of image formation that we have studied.

For the purpose of this brief introduction, it only remains to discuss methods of magnetic excitation. The excitation field at position (x, y, z) is the magnetic field $B_0 + B(x, y, z)$. The term B_0 is responsible for the carrier frequency and can be set aside. The term $B(x, y, z)$ creates a frequency $f(x, y, z) = \gamma B(x, y, z)$. By choosing $-\gamma B(x, y, z) = ax + by$, we have a frequency with the spatial distribution given by $f(x, y, z) = ax + by$. Then

$$S(f_x, f_y) = \int_{-\infty}^{\infty}\int_{-\infty}^{\infty} s(x, y)e^{-j2\pi(ax+by)t}\,dx\,dy$$

which has the desired form.

10.8 Emission tomography from decay events

In Section 9.5, we studied imaging in weak light as one instance of imaging from multiple point events. In this section, we shall discuss an instance of emission tomography based on the detection of multiple point events. These examples are important modalities in medical imaging. Emission tomography depends on radiation that is emitted by an object of interest. This will require the object to be infused with a suitable source of radiation, as by tagging an appropriate kind of biomolecule with a radionuclide. One may use either a radionucliotide that emits a single photon or one that emits a positron. The first case is called *photon-emission tomography*; the second is called *positron-emission tomography*. In contrast to X-ray tomography, which images structure for medical applications, positron-emission tomography images function because it images a specific kind of tissue that attracts the chosen radionuclide. Positron emission has proved to be attractive because each positron quickly reacts with an electron to create a pair of photons. The fact that such photons occur in pairs provides considerably more information from each decay event and a great simplification in the implementation.

Positron-emission tomography images the distribution of positron sources indirectly by imaging the density of photon-pair production created by positron decay. This equivalence is valid because the photon pair is created quite close to the positron emission. Positron density is $\rho(x, y, z)$, which means that $\rho(x, y, z)\,dx\,dy\,dz$ is (proportional to) the probability density that the infinitesimal volume $dx\,dy\,dz$, centered at (x, y, z), will emit a positron in a unit time interval. Each positron is almost immediately annihilated by an electron, producing two photons that are required to travel in opposite directions by the laws of conservation of energy and momentum. The two photons are detected by an array of photodetectors surrounding the region, as shown in Figure 10.15. The pair of photons is parameterized by the line of travel, which is the line connecting the two detectors sensing the photoelectrons, and by the differential time of arrival.

The computational task of photon or positron emission tomography is to form an estimate of the emission density $\rho(x, y)$ when given a large file of detected photons or photon pairs, described by their position of arrival or by both their line of travel and the differential time of arrival for each pair. The source is an inhomogeneous poisson process whose density function $\lambda(x, y)$ is proportional to the concentration of radionuclide $\rho(x, y)$. The detected data are different for photon emission and positron emission events. For photon emission events, the detected data have the form of the x, y coordinates at a focal-plane photodetector array. Let $p(x, y|u, v)$ be the probability that a photoconversion occurs at x, y given that a photon is emitted at u, v. At the photodetector, the density of arrival events

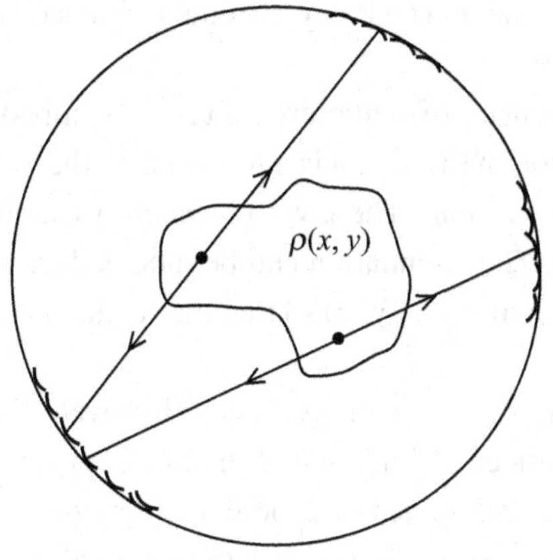

Figure 10.15 Positron-emission tomography

at (x, y) is

$$\mu(x, y) = \int_{-\infty}^{\infty}\int_{-\infty}^{\infty} p(x, y \mid u, v)\lambda(u, v)\, du\, dv + n(x, y)$$

where $n(x, y)$ is an independent noise process. The maximum-likelihood image is then

$$\widehat{\lambda}(x, y) = \text{argmax}_{\lambda(x,y)}\left[\int_{-\infty}^{\infty}\int_{-\infty}^{\infty} p(x, y|u, v)\lambda(u, v)\, du\, dv + n(x, y)\right].$$

This apparently does not have an analytic solution. It can be solved numerically by using an iterated algorithm of the kind described in Chapter 11.

10.9 Coherence tomography

Optical-coherence tomography is an imaging modality based on coherence-gated echoes that is used to form high-resolution images of optical scatterers near the surface of a partially transparent object. An important example is the imaging of very small blood vessels within a few millimeters of the skin's surface. A resolution of several microns is practical.

Optical-coherence tomography has the unusual property that it uses an optical beam with poor spatial coherence because good spatial coherence would actually degrade the resolution. Optical-coherence tomography uses a low-coherence optical beam to illuminate the interior of a partially transparent object. The beam penetrates the object and is reflected from structural details within the object and emerges as a scattered beam from it. Meanwhile, a reference beam has been formed and preserved by splitting the optical beam into an illumination beam and a reference beam, then propagating the reference beam and reflecting it from a mirror. The delay τ in the reference beam can be varied as desired by moving the mirror.

A simple modal of the medium is a density of scatterers, $\rho(x, y, z)$, embedded in a homogeneous attenuating medium. Although the signal is attenuated by the medium, it may be possible to regard the scattering as weak. Then, with respect to scattering, the Born approximation may be acceptable. The attenuation can be modeled explicitly or it can be accommodated by combining it with $\rho(x, y, z)$ so that the scatterers appear to be weaker with depth.

The optical features in most applications are spatially fixed. However, if some of the scattering features are moving, the reflected signal will contain a doppler shift that can be detected as such. In this way, the techniques of coherence tomography can be augmented to observe motion. Blood flow in biological tissue is a good example of such an application.

The incident optical beam at complex baseband at $z = 0$ is

$$c(x, y, t) = \mathrm{e}^{-\mathrm{j}\theta(t)} A(x, y)$$

where $\theta(t)$ is a gaussian random process with zero mean and correlation function $\phi(\tau)$, and $A(x, y)$ describes the beam spatially. To get good resolution in x and y, a spatially narrow beam is used. To observe a range of x and y, the beam is spatially scanned.

In the geometrical-optics approximation, diffraction is neglected. The reference beam at complex baseband is

$$c'(x, y, t) = \mathrm{e}^{\mathrm{j}\theta(t-\tau)}$$

and the reflected beam is

$$v(x, y, t) = \int_0^\infty A(x, y)\rho(x, y, z)\mathrm{e}^{-\mathrm{j}\theta(t-z)}\,\mathrm{d}z.$$

The reflected beam is processed by multiplying it by the reference beam and integrating. Thus

$$r(x, y) = \int_0^T c'(x, y, t)v^*(x, y, t)\,\mathrm{d}t.$$

The expectation is

$$\mathrm{E}[r(x, y)] = \int_0^T \mathrm{E}[c'(x, y, t)v^*(x, y, t)]\,\mathrm{d}t.$$

The sample cross-correlation is

$$\mathrm{E}[c'(x, y, t)v^*(x, y, t)] = \int_0^\infty A(x, y)\rho(x, y, z)\mathrm{E}[\mathrm{e}^{\mathrm{j}\theta(t-z)}\mathrm{e}^{-\mathrm{j}\theta(t-\tau)}]\,\mathrm{d}z.$$

The expectation can be evaluated by using Corollary 15.2.2 to give

$$\begin{aligned} \mathrm{E}[c'(x, y, t)v^*(x, y, t)] &= \int_0^{\infty} A(x, y)\rho(x, y, z)\mathrm{e}^{-\mathrm{E}[\theta(t-z)-\theta(t-\tau)]^2/2}\,\mathrm{d}z \\ &= A(x, y)\int_0^{\infty} \rho(x, y, z)\mathrm{e}^{-\sigma_\theta^2(1-\phi(z-\tau))}\,\mathrm{d}z. \end{aligned}$$

Let

$$h(z) = \mathrm{e}^{\sigma_\theta^2(1-\phi(z))}.$$

Then

$$\frac{1}{T}\mathrm{E}[r(x, y)] = A(x, y)[h(z) * \rho(x, y, z)].$$

If a single spatial sensor is used, as is the normal method, this is integrated across the beam. Then

$$\frac{1}{T}\mathrm{E}[r(x, y)] = h(z) * \rho(z)$$

where

$$\rho(z) = \int_{-\infty}^{\infty}\int_{-\infty}^{\infty} A(x, y)\rho(x, y, z)\,\mathrm{d}x\,\mathrm{d}y.$$

Problems

10.1 Give two different functions $s(x, y)$ and $s'(x, y)$ that have identical projections on the x axis and identical projections on the y axis. Are the f_x and f_y slices of $S(f_x, f_y)$ and $S'(f_x, f_y)$ the same?

10.2 Prove the following generalization of the projection-slice theorem. Let

$$p_\theta(t) = \int_{-\infty}^{\infty} s(t\cos\theta - r\sin\theta, t\sin\theta + r\cos\theta)\,\mathrm{d}r.$$

Then for all $w(t)$,

$$\int_{-\infty}^{\infty} p_\theta(t)w(t)\,\mathrm{d}t = \int_{-\infty}^{\infty}\int_{-\infty}^{\infty} s(x, y)w(x\cos\theta + y\sin\theta)\,\mathrm{d}x\,\mathrm{d}y$$

provided the integrals exist.

10.3 Suppose that $s(x, y)$ separates

$$s(x, y) = s_1(x)s_2(y).$$

Prove that $s(x, y)$ is uniquely determined by two of its projections, given the side information that it so separates.

10.4 A two-dimensional image of a three-dimensional scene can be constructed either as a slice or as a projection. Explain this statement. Under what circumstances is a set of projections sufficient to recover the three-dimensional scene?

10.5 Prove that

$$\int_0^{\pi} \delta(x\cos\theta + y\sin\theta)\, d\theta = \frac{1}{\sqrt{x^2+y^2}}.$$

10.6 Prove that

$$\int_0^{2\pi}\int_0^{\infty} f(r\cos\theta, r\sin\theta) r\, dr\, d\theta = \int_0^{\pi}\int_{-\infty}^{\infty} f(r\cos\theta, r\sin\theta)|r|\, dr\, d\theta.$$

10.7 Let

$$s(x, y) = \text{circ}(x-a, y)\text{circ}(x+a, y).$$

a. Describe the slices of the Fourier transform of $s(x, y)$.
b. Describe the slices of the Fourier transform of $s(x, y) ** s(x, y)$.

10.8 **a.** In n-dimensional euclidean space $\boldsymbol{R}^n$, what point of the $(n-1)$-dimensional hyperplane $\sum_i a_i x_i = c$ is closest to the point $\boldsymbol{x}' = (x_1', x_2', \ldots, x_n')$?
b. In n-dimensional euclidean space $\boldsymbol{R}^n$, what point of the $(n-2)$-dimensional hyperspace $\sum_i a_{i1}x_i = c_1$ and $\sum_i a_{i2}x_i = c_2$ is closest to the point $\boldsymbol{x}' = (x_1', \ldots, x_n')$?
c. In n-dimensional euclidean space $\boldsymbol{R}^n$, what point of the $(n-r)$-dimensional manifold $\sum_i x_{i\ell} = c_\ell$ for $\ell = 1, \ldots, r$ is closest to the point $(x_1', \ldots, x_n')$?
d. Give an alternating projection algorithm for inverting an n by n system of linear equations that projects onto the intersection of r hyperplanes in each step. Must r divide n? What can be said about complexity?

10.9 Consider the two by two matrix

$$\boldsymbol{S} = \begin{bmatrix} s_{11} & s_{12} \\ s_{21} & s_{22} \end{bmatrix}.$$

Given the "row projections" $p_j^R = \sum_i s_{ij}$ and the "column projections" $p_i^C = \sum_j s_{ij}$, can the elements of the matrix be recovered? How does this generalize to a three by three matrix? What if diagonal projections are included?

10.10 **(Three-Dimensional Projection-Slice Theorem)** Let $s(x, y, z)$ have finite energy and Fourier transform $S(f_x, f_y, f_z)$. Let $\boldsymbol{A}$ be a unitary matrix defining a coordinate rotation:

$$\begin{bmatrix} t \\ r \\ q \end{bmatrix} = \boldsymbol{A} \begin{bmatrix} x \\ y \\ z \end{bmatrix}.$$

The *projection* of $s(x, y, z)$ onto the t, r plane is defined as

$$p_{\mathbf{A}}(t, r) = \int_{-\infty}^{\infty} s\left(\mathbf{A}^T \begin{bmatrix} t \\ r \\ q \end{bmatrix}\right) \mathrm{d}q$$

where the integrand denotes that function of t, r, and q obtained by substituting for x, y, and z in $s(x, y, z)$ as indicated. Prove that

$$P_{\mathbf{A}}(f_t, f_r) = S(A_{11} f_t + A_{21} f_r, A_{12} f_t + A_{22} f_r, A_{13} f_t + A_{23} f_r)$$

where $P_{\mathbf{A}}(f_t, f_r)$ is the two-dimensional Fourier transform of $p_{\mathbf{A}}(t, r)$.

10.11 Let $s(x, y, z)$ have finite energy and Fourier transform $S(f_x, f_y, f_z)$. As in Problem 10.10, let $\mathbf{A}$ be a unitary matrix defining a coordinate rotation:

$$\begin{bmatrix} t \\ r \\ q \end{bmatrix} = \mathbf{A} \begin{bmatrix} x \\ y \\ z \end{bmatrix}.$$

The *conjection* of $s(x, y, z)$ onto the t axis is defined as

$$p_{\mathbf{A}}(t) = \int_{-\infty}^{\infty} \int_{-\infty}^{\infty} s\left(\mathbf{A}^T \begin{bmatrix} t \\ r \\ q \end{bmatrix}\right) \mathrm{d}r\, \mathrm{d}q.$$

Prove that

$$P_{\mathbf{A}}(f) = S(A_{11} f, A_{12} f, A_{13} f)$$

where $P_{\mathbf{A}}(f)$ is the Fourier transform of $p_{\mathbf{A}}(t)$.

10.12 A three-dimensional object, $s(x, y, z)$, is reduced to a set of two-dimensional sections by integrating over intervals in the z direction. Thus

$$s_k(x, y) = \int_{(k-\frac{1}{2})\Delta z}^{(k+\frac{1}{2})\Delta z} s(x, y, z)\, \mathrm{d}z.$$

a. Show that this is equivalent to convolution in the z direction with a rectangle function, followed by sampling in z.

b. Describe tomographic images formed from projections of these sections.

c. Describe how to combine the images of the sections using Nyquist–Shannon interpolation. How does the result relate to $s(x, y, z)$?

d. Can the distortion be completely removed by deconvolution?

e. Repeat this analysis for the case in which the sections are defined by

$$s_k(x, y) = \int_{-\infty}^{\infty} \mathrm{e}^{-(z-k\Delta z)^2/2\Delta z^2} s(x, y, z)\, \mathrm{d}z.$$

10.13 Show that the complexity of filtered back-projection tomography is proportional to N^3. What part of the computations can be done while the data are still being collected? What are the computations that can be done only after the full set of data is collected?

10.14 Suppose that the projections of a projection tomography system are contaminated by noise so that the actual measurements are

$$v_\theta(t) = p_\theta(t) + n_\theta(t),$$

where $n_\theta(t)$ is independent noise of power density spectrum $N(f)$ for each θ. If the image is reconstructed using filtered back projection, how does the ramp filter affect the noise? How does the filtered noise affect the image?

10.15 **(Helical Scan)** A three-dimensional function $s(x, y, z)$ has Nyquist slices $s(x, y, k\Delta z)$ satisfying the Nyquist condition in the z direction. A helical-scan sensor takes projections

$$p_\theta(t) = \int_{-\infty}^{\infty} (t\cos\theta - r\sin\theta, t\sin\theta + r\cos\theta, \theta/\Delta z)$$

a. Is this set of projections sufficient to recover $s(x, y, z)$?

b. Describe an interpolation algorithm.

10.16 A fan beam for projection tomography takes data at M uniformly-spaced angles indexed by m. At each angle a fan of N uniformly spread beams, indexed by n, measures projections.

a. What must be the relationship in the angular spacing $\Delta\phi$ of the fan origins and the spacing $\Delta\gamma$ of the beams within each fan so that the fan-beam data can be re-sorted into parallel beams?

b. Specify the jth parallel beam at the ith angle in terms of the nth fan beam at the mth angle.

10.17 Let $s(x_1, \ldots, x_n)$ denote an n-dimensional function of finite energy, and let $S(f_1, f_2, \ldots, f_n)$ denote its Fourier transform. Prove the n-dimensional generalization of the projection-slice theorem stated as follows. Let $\boldsymbol{x}'$ denote a set of coordinates, given by $\boldsymbol{x} = \boldsymbol{x}'\boldsymbol{A}$, where $\boldsymbol{A}$ is an orthogonal matrix, and let $\boldsymbol{f}' = (f_1', \ldots, f_n')$ denote a set of frequency coordinates given by $\boldsymbol{f} = \boldsymbol{f}'\boldsymbol{A}$. Then a *projection* onto an $(n-1)$-dimensional hyperplane is defined as

$$p_{x_i'}(x_1', x_2', \ldots, x_{i-1}', x_{i+1}', \ldots, x_n') = \int_{-\infty}^{\infty} s(\boldsymbol{x}\boldsymbol{A})\, dx_i'.$$

The *slice* of the Fourier transform at f_i' is $S(f_1', f_2', \ldots, f_{i-1}', 0, f_{i+1}', \ldots, f_n')$. Then the projection along x_i' has an $(n-1)$-dimensional Fourier transform equal to the slice of $S(\boldsymbol{f}')$ at $f_i' = 0$.

Notes

The central theme of tomography is to use the projection-slice theorem to reconstruct images. This was proposed by Radon in 1917 and was rediscovered independently in many fields. Bracewell (1956), and Bracewell and Riddle (1967), developed the idea in conjunction with the inversion of multiple radio-telescope scans. DeRosier and Klug (1968) developed the idea in conjunction with the inversion of multiple electron microscope images. Cormack (1963) proposed using the method for the reconstruction of medical images from X-ray data. Hounsfield (1972, 1973) built the first tomographic scanner for medical X-ray imaging. Systems for X-ray tomography, now in widespread use, can image three-dimensional objects by imaging many two-dimensional slices. More advanced illumination geometries are more difficult to process. Algorithms that form three-dimensional images directly by using cone-beam illumination have been studied by Tuy (1983), Smith (1985), and others. A tutorial on algebraic reconstruction techniques was written by Gordon (1974).

The phenomenon of nuclear magnetic resonance was discovered in 1946 by Bloch and Purcell. Thereafter, it was well understood that different tissues respond differently to magnetic excitation, but it was not immediately understood how to use this effect to form an image. Lauterbur (1974) contributed the idea of using gradient magnetic fields to encode the spatial information and then recovering the image from the received signal by tomographic processing. Lauterbur was also the first to demonstrate his technique. The important technology of magnetic-resonance imaging grew out of that work. Phase encoding methods were introduced by Kumar, Welti, and Ernst (1975). Phase contrast methods were extended to magnetic-resonance imaging by Moran (1982) and based on work in other fields, by Hahn (1960) and Singer (1971).

There have been many tutorial treatments of tomography such as those by Scudder (1978), and Mersereau and Oppenheim (1974). The topic of (polar coordinate) sampling in tomography was developed by Rattey and Lindgren (1981), following earlier work by Snyder and Cox (1977). A tomographic formulation of the synthetic-aperture principle may be found in Munson, O'Brien, and Jenkins (1983). An alternative tomographic imaging radar, using multiple chirp slopes, was proposed by Bernfeld (1984), and by Feig and Grünbaum (1986). Three-dimensional tomographic formulations of radar have been discussed by Jakowitz and Thompson (1992), and by Coble (1992). The use of the Wigner distribution to obtain an alternative method of forming images from radar data – akin to tomography – was proposed by Barbarosa and Farina (1990), and by Chen (1994). The Lewis–Bojarski (1969, 1982) equation describes the relationship between inverse scattering and object shape, and is closely related to the projection-slice theorem.

Diffraction tomography can be traced back to the work of Wolf (1969). The paper by Mueller, Kaveh, and Wade (1979) was also influential. Techniques that solve the inverse scattering problem for diffusing waves were developed by Schotland (1997) based on work by Ishii, Leigh, and Schotland (1995). These methods can be modified to solve the near-field diffraction inverse problem. Geophysical tomography was discussed in a review article by Dines and Lytle (1979). A striking application is the article by Dziewonski and Woodhouse (1987). Optical-coherence tomography was discussed in a review article by Schmitt (1999).

11 Likelihood and information methods

Now we shall introduce a broader view of the topic of image formation. Rather than think of finding the single "correct" image, we shall consider a set of possible images. The task of image formation, then, is to choose one image from the set of all possible images. The chosen image is the one that best accounts for a given set of data, usually a noisy and incomplete set of data. This set of images may be the set of all real-valued, two-dimensional functions on a given support, or the set of all nonnegative real-valued, two-dimensional functions on that given support. For this prescription to be followed, one must define criteria upon which to decide which image best accounts for a given set of data. A powerful class of optimality criteria is the class of information-theoretic criteria. These are optimality criteria based on a probabilistic formulation of the image-formation problem.

Elementary imaging techniques may be built on the idea of estimating the value of each image pixel separately. Within each cell of the image, the data are processed to estimate the signal within that cell without consideration of the signal in all other cells. This criterion does have some intuitive appeal and leads to relatively straightforward and satisfactory computational procedures. However, the criterion of estimating the signal in each pixel independently cannot be defended at a fundamental level because other structure is ignored. Fundamental approaches can be developed by maximizing more global performance measures such as the likelihood function or the entropy. These methods, in principle, give better images and often are preferred to the more conventional methods. However, maximum-entropy and maximum-likelihood algorithms are nonlinear and their behavior is sometimes unsatisfactory.

The maximum-likelihood principle is a general information-theoretic principle for problems of decision and estimation. This principle is best formulated first for the decision problem under the satisfying criterion of minimizing the probability of error. It can also be used for the estimation problem, but then without the same compelling justification. For the task of estimation the maximum-likelihood principle can be justified by its performance. The maximum-likelihood principle will be developed first for the case of a finite set of measurements. To apply the method to waveforms, we shall approximate the continuous-time waveform $v(t)$ by a finite set of discrete-time samples to which we will apply the maximum-likelihood principle. Then we will take the limit when the number of samples of $v(t)$ goes to infinity to obtain the maximum-likelihood principle for the waveform measurement $v(t)$. This leads to our desired maximum-likelihood approach to image formation as an estimation problem.

11.1 Likelihood functions and decision rules

The M-ary decision problem is the task of deciding between M hypotheses, one and only one of which is true. Corresponding to each of M hypotheses, indexed $m = 0, \ldots, M-1$, we are given the probability density function $p_m(x_1, \ldots, x_n)$ on the data space of n data points. Then, when given a set of n measured data values, one of the hypotheses must be selected. The natural rule is to choose the hypothesis so that the probability of error is as small as possible.

In the usual role of $p_m(x_1, \ldots, x_n)$, m is considered fixed, denoting a specific hypothesis, and $x_1, \ldots, x_n$ correspond to real-valued random variables $X_1, \ldots, X_n$. Then $p_m(x_1, \ldots, x_n)$ is called a *probability density function*. However, when the measurement $x_1, \ldots, x_n$ is known but m is unknown, $p_m(x_1, \ldots, x_n)$ is viewed as a function of m. Then $p_m(x_1, \ldots, x_n)$ is known as a *likelihood function*. Because of its important role in Theorem 11.1.1 that follows, the likelihood function $p_m(x_1, \ldots, x_n)$ will arise frequently. We shall often find it more convenient, however, to work with the likelihood function in the form of the *loglikelihood function*, given by $\Lambda(m) = \log p_m(x_1, \ldots, x_n)$.

Given a finite set of data, $(x_1, \ldots, x_n)$, we need to decide which probability density function was used to generate the data. A *decision rule* is a function of the data, denoted $\widehat{m}(x_1, \ldots, x_n)$, where $\widehat{m}$ is the estimated value of m when given the data vector $(x_1, \ldots, x_n)$. A decision error occurs if the mth probability density function was used to generate the data, but the decision $\widehat{m}(x_1, \ldots, x_n)$ is not equal to m. The conditional probability of error, given that m is true, is denoted $p_{e|m}$. The criterion we will use for defining a decision rule to be optimum is that the average probability of decision error is minimized over all decision rules, assuming that, prior to the measurement, each m is equally likely to be true. This means that each hypothesis occurs with the prior probability $1/M$.

Theorem 11.1.1 *For the M-ary decision problem with equiprobable hypotheses:*

1. *Given a vector measurement, $(x_1, \ldots, x_n)$, the optimum decision rule is*

$$\widehat{m}(x_1, \ldots, x_n) = \operatorname{argmax}_m p_m(x_1, \ldots, x_n)$$

where the right side denotes that m for which $p_m(x_1, \ldots, x_n)$ is largest.

2. *If, moreover, $x_1, \ldots, x_n$ is a gaussian vector random variable for each m with independent components of identical variances σ^2 and means $c_{m_1}, \ldots, c_{m_n}$, then the optimum decision rule is to choose that m for which the euclidean distance $\sum_{\ell=1}^{n}(x_\ell - c_{m_\ell})^2$ is smallest.*

3. *If, moreover,* $\sum_{\ell=1}^{n} c_{m_\ell}^2$ *is independent of* m*, then the optimum decision rule is to choose that* m *for which the correlation coefficient* $\sum_{\ell=1}^{n} x_\ell c_{m_\ell}$ *is largest.*

Proof: Let $\mathcal{U}_m$ be the set of data vectors $(x_1, \ldots, x_n)$ for which we decide that the mth probability density function was used. The probability of decision error, given that m is true, is

$$p_{e|m} = \int_{\mathcal{U}_m^c} p_m(x_1, \ldots, x_n)\,\mathrm{d}x_1 \ldots \mathrm{d}x_n.$$

The average probability of decision error is

$$\begin{aligned} p_e &= \sum_{m=0}^{M-1} \frac{1}{M} \int_{\mathcal{U}_m^c} p_m(x_1, \ldots, x_n)\,\mathrm{d}x_1 \ldots \mathrm{d}x_n \\ &= \sum_{m=0}^{M-1} \frac{1}{M}\left[1 - \int_{\mathcal{U}_m^c} p_m(x_1, \ldots, x_n)\,\mathrm{d}x_1 \ldots \mathrm{d}x_n\right]. \end{aligned}$$

Consequently, to minimize p_e, the measurement $(x_1, \ldots, x_n)$ should be assigned to the set $\mathcal{U}_m$ for which $p_m(x_1, \ldots, x_n)$ is larger than $p_{m'}(x_1, \ldots, x_n)$ for all $m' \neq m$. (A tie for the largest probability can be broken by any arbitrary rule, say, break a tie in favor of the smallest index.) This concludes the proof of statement **1** of the theorem.

Suppose that, for each m, $(x_1, \ldots, x_n)$ is a gaussian vector random variable with independent components of identical variances σ^2 and means $c_{m_1}, \ldots, c_{m_n}$, then

$$\begin{aligned} p_m(x_1, \ldots, x_n) &= \prod_{\ell=1}^{n} \frac{1}{\sqrt{2\pi\sigma^2}} \mathrm{e}^{-(x_\ell - c_{m_\ell})^2/2\sigma^2} \\ &= \frac{1}{(\sqrt{2\pi\sigma^2})^n} \mathrm{e}^{-\sum_\ell (x_\ell - c_{m_\ell})^2/2\sigma^2}. \end{aligned}$$

A maximum-likelihood decision rule decides on that m for which $p_m(x_1, \ldots, x_n)$ is largest. To maximize $p_m(x_1, \ldots, x_n)$ over m, one should minimize $\sum_\ell (x_\ell - c_{m_\ell})^2$ over m. This completes the proof of statement **2** of the theorem.

Statement **3** of the theorem follows from expanding the square and noting that the term $\sum_\ell x_\ell^2$ and, by assumption of that statement, the term $\sum_\ell c_{m_\ell}^2$ are independent of m. This completes the proof of the theorem. □

In general, the unknown index m can be replaced by a vector of parameters, γ, and the probability density function on the vector of measurements $p(x_1, \ldots, x_n|\gamma)$ is conditional on the parameter γ. We do not now assume that γ is a random variable so there need not be a prior on γ. Possibly, some components of the vector γ are discrete parameters and some components are continuous parameters.

As a function of the vector γ, the loglikelihood function is defined as

$$\Lambda(\gamma) = \log p(x_1, \ldots, x_n|\gamma).$$

Usually, we are interested in a loglikelihood function only for the purpose of finding where it achieves its maximum, but not in the actual value of the maximum. The value at which the maximum occurs is written

$$\widehat{\gamma} = \text{argmax}_\gamma \Lambda(\gamma).$$

Constants that are added to or multiplying the loglikelihood function do not affect the location of the maximum, so it is common practice to suppress such constants when they occur by redefining $\Lambda(\gamma)$. For example, the gaussian distribution with the unknown mean $\overline{x}$ has the loglikelihood function

$$\Lambda(\overline{x}) = -\log\sqrt{2\pi\sigma^2} - (x - \overline{x})^2/2\sigma^2.$$

For economy, we will commonly discard the constants and simply write

$$\Lambda(\overline{x}) = -(x - \overline{x})^2.$$

We often call this abbreviated function a *likelihood statistic*, which serves as a reminder that it is not precisely the loglikelihood function. A likelihood statistic is an example of a sufficient statistic. In general, a *statistic* is any function of the received data, and a *sufficient statistic* is a statistic from which the likelihood function can be recovered; no essential information is missing from a sufficient statistic.

We shall be interested in problems in which the number of measurements n tends toward infinity. In such a case, the limit as $n \to \infty$ of $\Lambda(\gamma)$ may be infinite for all (or many) values of γ. Then it would be meaningless to deal with the maximum over γ of the limit of $\Lambda(\gamma)$. We are not really interested in the value of the maximum, but rather in the value of γ where the maximum occurs, or in the limit as n goes to infinity of the sequence of values of γ that achieve the maximum for each n. Precisely, we want

$$\widehat{\gamma} = \lim_{n\to\infty} \text{argmax}_\gamma \log p_n(x_1, \ldots, x_n|\gamma),$$

provided the limit makes sense. The more appealing definition

$$\widehat{\gamma} = \text{argmax}_\gamma \lim_{n\to\infty} \log p_n(x_1, \ldots, x_n|\gamma)$$

may be nonsensical because the limit may diverge.

To avoid divergence of the limit, we also define a normalized version of the loglikelihood function, which can be done in several ways. One choice is to define

$$\Lambda(\gamma) = \log[B_n p(x_1, \ldots, x_n|\gamma)]$$

where B_n is any convenient constant, independent of γ, that will cause $\Lambda(\gamma)$ to have a finite limit. Alternatively, one can use a form called the *loglikelihood ratio*, also denoted by the symbol Λ, and given by the general form

$$\Lambda(\gamma, \gamma') = \log \frac{p(x_1, \ldots, x_n \mid \gamma)}{p(x_1, \ldots, x_n|\gamma')},$$

or alternatively,

$$\Lambda(\gamma) = \log \frac{p(x_1, \ldots, x_n | \gamma)}{p(x_1, \ldots, x_n)}$$

where $p(x_1, \ldots, x_n)$ is a convenient reference probability distribution, perhaps corresponding to noise only. The purpose in introducing either form of the loglikelihood ratio is to have a form that remains finite as n goes to infinity. The maximum-likelihood estimate is then that value of γ maximizing $\Lambda(\gamma)$, or for which $\Lambda(\gamma, \gamma')$ is nonnegative for all γ'.

Likelihood of waveforms

We are now ready to consider the likelihood function of the waveform $v(t)$, given by

$$v(t) = c(t, \gamma) + n(t),$$

where γ is an unknown vector parameter, and $n(t)$ is additive white gaussian noise with the power density spectrum $N_0/2$. The loglikelihood function for this waveform will often be written formally in the form

$$\Lambda(\gamma) = -\frac{1}{N_0} \int_{-\infty}^{\infty} |v(t) - c(t, \gamma)|^2 \, dt.$$

However, as it is written, the integral is infinite because the noise has infinite energy. Therefore the formula with infinite limits can be understood only symbolically. We shall derive the formula only for a finite observation time and only for white noise. Nonwhite noise can be converted to white noise by the use of a whitening filter, but not without violating the assumption of a finite observation interval. Nevertheless, we shall accept the use of a whitening filter under the assumption that the observation interval is very long in comparison with the response time of the whitening filters, so the transient effects at the edges of the interval have a negligible effect. A more elegant and more advanced treatment of the case of nonwhite noise would use the methods of functional analysis to reach the same conclusion rigorously.

We shall want to replace the waveform $v(t)$ by a finite-dimensional vector of samples having independent noise so that Theorem 11.1.1 can be used; this is why we suppose that $v(t)$ has been passed through a whitening filter to whiten the noise.

Now we will find an expression for the limit of $\Lambda(\gamma)$ as n goes to infinity for a waveform in additive white gaussian noise.

Proposition 11.1.2 *For the complex signal*

$$v(t) = c(t, \gamma) + n(t)$$

received on the interval $[-T_0/2, T_0/2]$ *in complex white gaussian noise* $n(t)$, *where* γ

is a vector parameter, the loglikelihood function for γ is

$$\Lambda(\gamma) = -\frac{1}{N_0}\int_{-T_0/2}^{T_0/2} |v(t) - c(t,\gamma)|^2\, dt.$$

Proof: Consider the Fourier series expansion of

$$v(t) = c(t,\gamma) + n(t)$$

restricted to the interval $[-T_0/2, T_0/2]$. The complex Fourier expansion coefficients $\ldots, V_{-1}, V_0, V_1, \ldots$ form an infinite set of independent, complex gaussian random variables. The means C_k are the Fourier coefficients of $c(t,\gamma)$ on the interval $[-T_0/2, T_0/2]$. Because the noise is white, the random variables are independent and identically distributed. The probability density function on the block $(V_{-K}, \ldots, V_K)$ is

$$p(V_{-K}, \ldots, V_K | c(t,\gamma)) = [2\pi\sigma^2]^{-(2K+1)} \prod_{k=-K}^{K} \mathrm{e}^{-|V_k - C_k|^2/2\sigma^2},$$

and $\sigma^2 = N_0/2$. This is conditional on the expected complex value $c(t,\gamma)$ of the received signal, which depends on the vector parameter γ. For a set of complex samples, we use as the loglikelihood function

$$\Lambda(\gamma) = \log Bp(V_{-K}, \ldots, V_K | c(t,\gamma)),$$

where we have used the normalizing constant

$$B = [2\pi\sigma^2]^{2K+1}.$$

With this choice of normalizing constant,

$$\Lambda(\gamma) = -\sum_{k=-K}^{K} |V_k - C_k|^2/N_0.$$

Now let K go to infinity and use the energy theorem for the Fourier series[1] to write

$$\lim_{K\to\infty} \Lambda(\gamma) = -\frac{1}{N_0}\int_{-T_0/2}^{T_0/2} |v(t) - c(t,\gamma)|^2\, dt,$$

which completes the proof of the proposition. □

[1] The energy theorem for the Fourier series

$$\int_{-T_0/2}^{T_0/2} |v(t)|^2\, dt = \sum_{k=-\infty}^{\infty} |V_k|^2$$

is a special case of Parseval's identity for the Fourier series

$$\int_{-T_0/2}^{T_0/2} v(t)u^*(t)\, dt = \sum_{k=-\infty}^{\infty} V_k U_k^*.$$

11.2 The maximum-likelihood principle

The estimation of an image from noisy data can be treated by the principle of maximum likelihood. Let $\boldsymbol{v}$ denote the set of noisy data, and let $\boldsymbol{\gamma}$ denote the image to be estimated. Let $p(\boldsymbol{v}|\boldsymbol{\gamma})$ be the probability density function on the space of possible data sets $\boldsymbol{v}$, given that the image is $\boldsymbol{\gamma}$. The loglikelihood function is

$$\Lambda(\boldsymbol{\gamma}) = \log p(\boldsymbol{v}|\boldsymbol{\gamma}).$$

The data vector $\boldsymbol{v}$ is regarded as fixed after the measurement takes place. From this fixed data vector, the unknown image $\boldsymbol{\gamma}$ is to be estimated. The maximum-likelihood estimate of the image is defined as

$$\widehat{\boldsymbol{\gamma}} = \text{argmax}_{\boldsymbol{\gamma}} \Lambda(\boldsymbol{\gamma}).$$

Thus the maximum-likelihood image $\widehat{\boldsymbol{\gamma}}$ is defined as the choice of $\boldsymbol{\gamma}$ that maximizes $\Lambda(\boldsymbol{\gamma})$.

For an example of the use of the maximum-likelihood principle, consider a measurement of a gaussian signal in gaussian noise, given by

$$v = s + n,$$

where v is the measurement, s is zero-mean gaussian noise whose variance σ_s^2 is unknown, and n is independent, zero-mean gaussian noise whose variance σ_n^2 is known. The elementary task is to estimate σ_s^2 when given the measurement v. The estimator consists of the function

$$\widehat{\sigma_s^2} = \widehat{\sigma_s^2}(v, \sigma_n^2),$$

which takes the measurement v into the estimate $\widehat{\sigma_s^2}$.

The measurement v, as the sum of two independent, gaussian random variables, is itself a gaussian random variable. It has the probability density function

$$p(v) = \frac{1}{\sqrt{2\pi(\sigma_s^2 + \sigma_n^2)}} \mathrm{e}^{-v^2/2(\sigma_s^2+\sigma_n^2)}.$$

The loglikelihood function is the logarithm of the density function, regarded as a function of the unknown σ_s^2, and ignoring additive constants. Thus

$$\Lambda(\sigma_s^2) = -\frac{1}{2}\log_e(\sigma_s^2 + \sigma_n^2) - \frac{v^2}{2(\sigma_s^2 + \sigma_n^2)}$$

where the term $\log\sqrt{2\pi}$ has been ignored as not affecting the subsequent maximization. The maximum-likelihood estimate $\widehat{\sigma_s^2}$ is found by setting the derivative with respect to

σ_s^2 equal to zero,

$$-\frac{1}{2(\sigma_s^2+\sigma_n^2)}+\frac{v^2}{2(\sigma_s^2+\sigma_n^2)^2}=0,$$

or

$$\sigma_s^2+\sigma_n^2=v^2.$$

Checking the second derivative shows that this is a maximum. Because σ_s^2 must be nonnegative, we conclude that the estimate

$$\widehat{\sigma_s^2}=\max\left[0,\, v^2-\sigma_n^2\right]$$

is the maximum-likelihood estimate.

This example can be elaborated. Suppose now that σ_s^2 is to be estimated by making N independent measurements of the random variable s in the presence of additive noise. The N measurements are

$$v_i = s_i + n_i \quad i = 1, \ldots, N$$

where the s_i are independent, zero-mean, gaussian random variables, each with variance σ_s^2, and the n_i are independent – and independent of the s_i – zero-mean, gaussian random variables, each with the known variance σ_n^2. The signal variance σ_s^2 is unknown and is to be estimated from the measurements.

The loglikelihood function is

$$\Lambda(\sigma_s^2)=\sum_{i=1}^{N}\log\left[\frac{1}{\sqrt{2\pi(\sigma_s^2+\sigma_n^2)}}\mathrm{e}^{-v_i^2/2(\sigma_s^2+\sigma_n^2)}\right].$$

To find the maximum, set the derivative with respect to σ_s^2 equal to zero. This gives

$$\sum_{i=1}^{N}\left[-\frac{1}{2(\sigma_s^2+\sigma_n^2)}+\frac{v_i^2}{2(\sigma_s^2+\sigma_n^2)^2}\right]=0,$$

which reduces to the estimate

$$\widehat{\sigma_s^2}=\max\left[0,\, \frac{1}{N}\sum_{i=1}^{N}v_i^2-\sigma_n^2\right].$$

11.3 Alternating maximization

Many maximization problems, including many problems of maximum-likelihood image formation, cannot be solved analytically; numerical methods are necessary. On the other hand, the problems are usually very large and often cannot be solved by general numerical methods either. Fortunately it has been possible to devise a mixture of analytical and numerical methods that take the form of a partial analytical solution that can be numerically iterated. This is the method of alternating maximization (or alternating minimization) which will be described in this section by means of several examples. A more formal, and more general, development appears in Section 11.11. There we will discuss the general method of Dempster, Laird, and Rubin for developing such alternating maximization algorithms.

Variance estimation

Alternating maximization will be introduced by a simple example. We will use it to form an algorithm to estimate the variance of a gaussian signal in gaussian noise. Alternating maximization is really not necessary to solve this problem because the problem can be solved analytically, as was done in Section 11.2. To get our desired algorithm for this simple example, we pretend that an analytical solution is not available. The received signal is

$$v = s + n$$

where s is an unknown random signal with a zero-mean gaussian probability density function of unknown variance σ_s^2 and n is an unknown noise of known variance σ_n^2. From v and σ_n^2, the maximum-likelihood estimate of σ_s^2 is to be computed.

To form the algorithm, we shall imagine that we have measured a different set of data from which the maximum-likelihood estimate can be found analytically. Specifically, we shall imagine that we have measured s and n individually, and analytically solve the problem under this assumption. The data set (s, n) is called the *complete data*. The actual data $s + n$, then, are called the *incomplete data*.

Figure 11.1 shows how the actual data space $v = s + n$ fits into the complete data space (s, n). Another possibility that could be chosen for the complete data space is the data set (v, s). Although either choice for the complete data space may be suitable, we

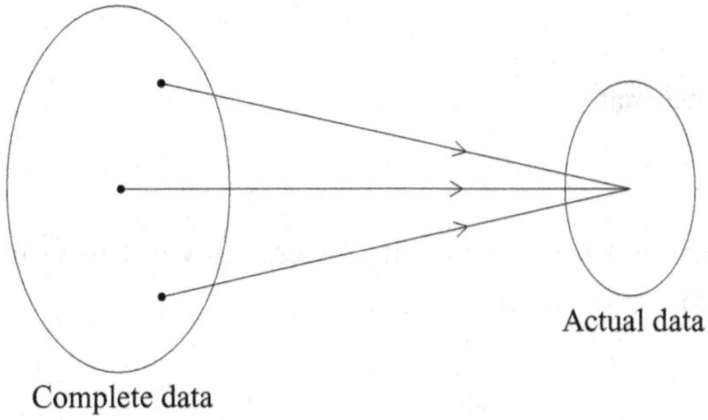

Figure 11.1 The complete data space

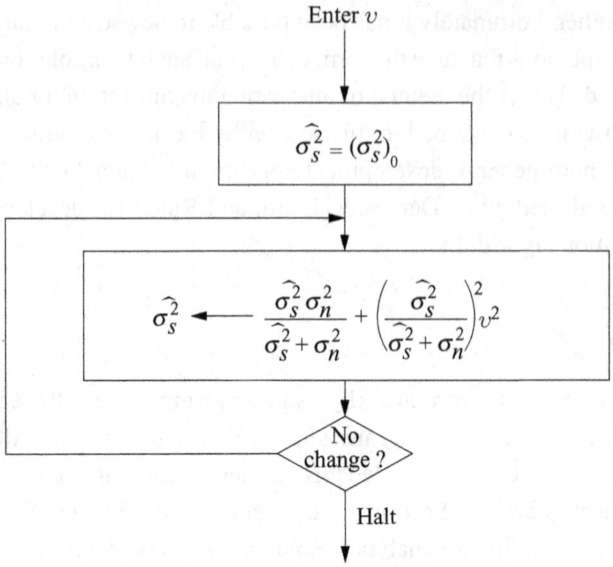

Figure 11.2 Iterative algorithm for estimating variance of signal in noise

will choose (s, n) as the complete data space, and we will derive the iterative algorithm shown in Figure 11.2.

The complete data space (s, n) is a two-dimensional random variable with the probability density function

$$p(s, n) = \left(\frac{1}{\sqrt{2\pi\sigma_s^2}} e^{-s^2/2\sigma_s^2}\right)\left(\frac{1}{\sqrt{2\pi\sigma_n^2}} e^{-n^2/2\sigma_n^2}\right).$$

The loglikelihood function for the complete data space is

$$\Lambda(\sigma_s^2) = -\tfrac{1}{2}\log\sigma_s^2 - \frac{s^2}{2\sigma_s^2} - \tfrac{1}{2}\log\sigma_n^2 - \frac{n^2}{2\sigma_n^2}.$$

Then σ_s^2 is chosen to maximize $\Lambda(\sigma_s^2)$. Differentiation with respect to σ_s^2 shows that the maximum occurs where

$$-\frac{1}{2\sigma_s^2} + \frac{s^2}{2(\sigma_s^2)^2} = 0.$$

Solving this for σ_s^2 gives the estimate

$$\widehat{\sigma_s^2} = s^2.$$

This would provide the desired estimate of σ_s^2 if s^2 were known. But s^2 is actually unknown, so it is replaced by its expectation

$$\widehat{s^2} = \mathrm{E}[s^2 | v, \widehat{\sigma_s^2}].$$

This step of computing the right side is called the expectation step. Perhaps surprisingly, even though σ_s^2 is actually the unknown being estimated, in this step the estimate $\widehat{\sigma_s^2}$ appears as a conditioning parameter. It is for this reason that numerical methods will be used eventually to piece together analytical expressions. An alternative way to set up the expectation step, which may be necessary in some other applications, is to compute $\mathrm{E}[\Lambda(\sigma_s^2)|v, \sigma_s^2]$ (now, for simplicity suppressing the 'hat' on σ_s^2). This expectation is defined by

$$\mathrm{E}[\Lambda(\sigma_s^2)|v, \sigma_s^2] = \mathrm{E}\left[\left(-\tfrac{1}{2}\log\sigma_s^2 - \frac{s^2}{2\sigma_s^2} - \tfrac{1}{2}\log\sigma_n^2 - \frac{n^2}{2\sigma^2}\right)|v, \sigma_s^2\right].$$

Everything in the likelihood function is known except for s and n. The only term that depends on s is the second term. Because the expectation operator can be distributed across addition, the task reduces to the task of evaluating $\mathrm{E}[s^2|v, \sigma_s^2]$. This calculation is provided by the following proposition.

Proposition 11.3.1 *Let $v = s + n$ where s and n are independent, gaussian random variables with the variances σ_s^2 and σ_n^2, respectively. Then*

$$\mathrm{E}\left[s^2|v, \sigma_s^2, \sigma_n^2\right] = \frac{\sigma_s^2\sigma_n^2}{\sigma_s^2 + \sigma_n^2} + \left(\frac{\sigma_s^2}{\sigma_s^2 + \sigma_n^2}\right)^2 v^2.$$

Proof: Let

$$\widehat{s^2} = \mathrm{E}[s^2|v, \sigma_s^2, \sigma_n^2] = \int_{-\infty}^{\infty} s^2 p(s|v)\,ds,$$

and

$$\widehat{s} = \mathrm{E}[s|v, \sigma_s^2, \sigma_n^2] = \int_{-\infty}^{\infty} sp(s|v)\,ds$$

where $p(s|v)$ is the appropriate gaussian probability density function conditioned on v, σ_s^2, and σ_n^2. By the standard properties of moments,

$$\begin{aligned}\widehat{s^2} &= \mathrm{E}[(s - \widehat{s})^2 + 2s\widehat{s} - \widehat{s}^2|v, \sigma_s^2] \\ &= \mathrm{E}[(s - \widehat{s})^2|v, \sigma_s^2] + \widehat{s}^2.\end{aligned}$$

To evaluate the two terms on the right, we first compute $p(s|v)$. By the Bayes rule,

$$\begin{aligned}p(s|v) &= \frac{p(s, v)}{p(v)} = \frac{p(s)p(v|s)}{p(v)} \\ &= \frac{\frac{1}{\sqrt{2\pi\sigma_s^2}}\mathrm{e}^{-s^2/2\sigma_s^2}\frac{1}{\sqrt{2\pi\sigma_n^2}}\mathrm{e}^{-(v-s)^2/2\sigma_n^2}}{\frac{1}{\sqrt{2\pi(\sigma_s^2+\sigma_n^2)}}\mathrm{e}^{-v^2/2(\sigma_s^2+\sigma_n^2)}} \\ &= \frac{1}{\sqrt{2\pi}}\sqrt{\frac{\sigma_s^2 + \sigma_n^2}{\sigma_s^2\sigma_n^2}}\,\mathrm{e}^{-\left(s - \frac{\sigma_s^2}{\sigma_s^2+\sigma_n^2}v\right)^2/2[\sigma_s^2\sigma_n^2/(\sigma_s^2+\sigma_n^2)]}.\end{aligned}$$

This expression now has the form of a gaussian distribution, so the mean and the variance can be recognized to be

$$E[s|v, \sigma_s^2] = \frac{\sigma_s^2}{\sigma_s^2 + \sigma_n^2} v$$

$$E[(s - \widehat{s})^2|v, \sigma_s^2] = \frac{\sigma_s^2 \sigma_n^2}{\sigma_s^2 + \sigma_n^2}.$$

Therefore

$$\widehat{s^2} = \frac{\sigma_s^2 \sigma_n^2}{\sigma_s^2 + \sigma_n^2} + \left(\frac{\sigma_s^2}{\sigma_s^2 + \sigma_n^2} \right)^2 v^2,$$

as was to be proved. □

Now we are ready to give a recursion to compute the maximum-likelihood estimate of σ_s^2. By combining the conclusion of the proposition with the likelihood maximization step, given by $\sigma_s^2 = s^2$, the recursion

$$\sigma_s^2 \leftarrow \frac{\sigma_s^2 \sigma_n^2}{\sigma_s^2 + \sigma_n^2} + \left(\frac{\sigma_s^2}{\sigma_s^2 + \sigma_n^2} \right)^2 v^2$$

is obtained. The final estimate $\widehat{\sigma_s^2}$ is the limit point of this recursion. Thus

$$\widehat{\sigma_s^2} = \frac{\widehat{\sigma_s^2} \sigma_n^2}{\widehat{\sigma_s^2} + \sigma_n^2} + \left(\frac{\widehat{\sigma_s^2}}{\widehat{\sigma_s^2} + \sigma_n^2} \right)^2 v^2,$$

which can be reduced to the direct solution given in Section 11.2. Because this problem can be solved analytically, it is not necessary to use this recursive algorithm. Our reason for describing the recursive algorithm is to give an example of a technique that is important in more complicated problems that cannot be solved analytically.

There is one more small point to be made. The recursion can be reworked into the form

$$\sigma_s^2 \leftarrow \sigma_s^2 - \left(\frac{\sigma_s^2}{\sigma_s^2 + \sigma_n^2} \right)^2 (\sigma_s^2 + \sigma_n^2 - v^2),$$

which some may find more attractive because it is in the form of a main term and a correction term that goes to zero. In this form of the recursion, the limit

$$\sigma_s^2 = \max[0, v^2 - \sigma_n^2]$$

is easy to see.

Covariance estimation

Next, we consider the vector version of this problem. We consider the vector measurement

$$\boldsymbol{v} = \boldsymbol{s} + \boldsymbol{n}$$

where $\boldsymbol{s}$ and $\boldsymbol{n}$ are gaussian, real vector random variables with zero means and with covariance matrices $\boldsymbol{\Sigma}_s$ and $\boldsymbol{\Sigma}_n$, respectively. The probability density functions of $\boldsymbol{s}$ and $\boldsymbol{n}$ are[2]

$$p(\boldsymbol{s}) = \frac{1}{\sqrt{\det 2\pi \boldsymbol{\Sigma}_s}} e^{-\frac{1}{2}\boldsymbol{s}^\dagger \boldsymbol{\Sigma}_s^{-1}\boldsymbol{s}} \quad p(\boldsymbol{n}) = \frac{1}{\sqrt{\det 2\pi \boldsymbol{\Sigma}_n}} e^{-\frac{1}{2}\boldsymbol{n}^\dagger \boldsymbol{\Sigma}_n^{-1}\boldsymbol{n}}.$$

Our task is to estimate $\boldsymbol{\Sigma}_s$ from $\boldsymbol{v}$ when $\boldsymbol{\Sigma}_n$ is known. Because covariances add under the addition of independent random vectors, we have

$$p(\boldsymbol{v}) = \frac{1}{\sqrt{\det 2\pi (\boldsymbol{\Sigma}_s + \boldsymbol{\Sigma}_n)}} e^{-\frac{1}{2}\boldsymbol{v}^\dagger (\boldsymbol{\Sigma}_s + \boldsymbol{\Sigma}_n)^{-1}\boldsymbol{v}},$$

and the loglikelihood statistic is

$$\Lambda(\boldsymbol{\Sigma}_s) = -\tfrac{1}{2}\log\det(\boldsymbol{\Sigma}_s + \boldsymbol{\Sigma}_n) - \tfrac{1}{2}\boldsymbol{v}^\dagger(\boldsymbol{\Sigma}_s + \boldsymbol{\Sigma}_n)^{-1}\boldsymbol{v}.$$

It is not tractable to maximize this loglikelihood over $\boldsymbol{\Sigma}_s$ by the elementary method of setting partial derivatives to zero. Instead, it is solved numerically by an iterative algorithm.

To set up the iterative algorithm we first replace the actual data $\boldsymbol{v} = \boldsymbol{s} + \boldsymbol{n}$ with the complete data $(\boldsymbol{s}, \boldsymbol{n})$. The complete-data loglikelihood $\Lambda(\boldsymbol{\Sigma}_s)$ is

$$\begin{aligned}\Lambda(\boldsymbol{\Sigma}_s) &= \log[p(\boldsymbol{s}, \boldsymbol{n})] \\ &= \log\left[\frac{1}{\sqrt{\det 2\pi \boldsymbol{\Sigma}_s}} e^{-\frac{1}{2}\boldsymbol{s}^\dagger \boldsymbol{\Sigma}_s^{-1}\boldsymbol{s}} \frac{1}{\sqrt{\det 2\pi \boldsymbol{\Sigma}_n}} e^{-\frac{1}{2}\boldsymbol{n}^\dagger \boldsymbol{\Sigma}_n^{-1}\boldsymbol{n}}\right].\end{aligned}$$

We maximize $\Lambda(\boldsymbol{\Sigma}_s)$ by the choice of the matrix $\boldsymbol{\Sigma}_s$ subject to the condition that $\boldsymbol{\Sigma}_s$ is positive definite. By discarding terms that do not depend on $\boldsymbol{\Sigma}_s$, this maximization step reduces to

$$\begin{aligned}\widehat{\boldsymbol{\Sigma}}_s &= \operatorname{argmax}_{\boldsymbol{\Sigma}_s}\left[-\log\det \boldsymbol{\Sigma}_s - \boldsymbol{s}^\dagger \boldsymbol{\Sigma}_s^{-1}\boldsymbol{s}\right] \\ &= \operatorname{argmin}_{\boldsymbol{\Sigma}_s}\left[\log\det \boldsymbol{\Sigma}_s + \boldsymbol{s}^\dagger \boldsymbol{\Sigma}_s^{-1}\boldsymbol{s}\right].\end{aligned}$$

Lemma 11.3.2 *Let $\boldsymbol{A}$ and $\boldsymbol{B}$ be positive-definite matrices. Then the minimum of the function*

$$f(\boldsymbol{A}) = \log\det \boldsymbol{A} + \operatorname{trace} \boldsymbol{A}^{-1}\boldsymbol{B}$$

occurs at $\boldsymbol{A} = \boldsymbol{B}$.

[2] The dagger † denotes transpose for real vectors and matrices and denotes conjugate transpose for complex vectors and matrices.

Proof: Let $C = A^{-1}B$. Then

$$f(B) - f(A) = \log\det B + \operatorname{trace} B^{-1}B - \log\det A - \operatorname{trace} A^{-1}B$$
$$= \log\det C - \operatorname{trace} C + \operatorname{trace} I.$$

But det C is the product of the eigenvalues of C and trace C is the sum of these eigenvalues. Because the matrix C is full rank, all its eigenvalues c_ℓ are positive, so

$$f(B) - f(A) = \Sigma_\ell[\log c_\ell - c_\ell + 1]$$
$$\leq 0$$

using the inequality $\log_e x \leq x - 1$. □

To use this lemma for the problem at hand, use the general relationship $x^\dagger Ax = \operatorname{trace} Axx^\dagger$ to write

$$\widehat{\Sigma}_s = \operatorname{argmin}_{\Sigma_s}\left[\log\det\Sigma_s + \operatorname{trace}\left(\Sigma_s^{-1}ss^\dagger\right)\right].$$

The lemma does not quite apply because $B = ss^\dagger$ is not full rank. However, there is always a full rank matrix arbitrarily close to $ss^\dagger$ to which the lemma does apply. In this sense, the lemma applies as well to $ss^\dagger$. Consequently, we can conclude that

$$\Sigma_s = ss^\dagger$$

is the maximum-likelihood estimate of Σ_s in terms of s, though s is unknown. This completes the maximization step.

To complete the setup of the iterative algorithm, we will need to compute the expectation

$$\mathrm{E}[ss^\dagger] = \mathrm{E}[ss^\dagger \mid v, \Sigma_s, \Sigma_n]$$

which will serve as a surrogate for the unknown $ss^\dagger$. This is the substance of the following theorem.

Theorem 11.3.3 *Let $v = s + n$ where s and n are independent, zero-mean, gaussian, real vector random variables with covariance matrices Σ_s and Σ_n, respectively. Then $p(s|v)$ is a multivariate gaussian probability density function with mean*

$$\bar{s} = (\Sigma_s^{-1} + \Sigma_n^{-1})^{-1}\Sigma_n^{-1}v$$

and covariance matrix

$$\Sigma_{s|v} = (\Sigma_s^{-1} + \Sigma_n^{-1})^{-1}.$$

Proof: By the Bayes formula,

$$p(s|v) = \frac{p(s)p(v|s)}{p(v)}.$$

Because the terms on the right are multivariate gaussian probability density functions, the term on the left is also a multivariate gaussian probability density function, and so has the form

$$p(s\,|\,v) = \frac{1}{\sqrt{\det(2\pi\,\boldsymbol{\Sigma}_{s|v})}}\,\mathrm{e}^{-\frac{1}{2}(s-\bar{s})^{\dagger}\boldsymbol{\Sigma}_{s|v}^{-1}(s-\bar{s})}$$

with covariance matrix $\boldsymbol{\Sigma}_{s|v}$ and mean $\bar{s}$. To find expressions for $\boldsymbol{\Sigma}_{s|v}$ and $\bar{s}$, it is enough to inspect the Bayes formula by writing only the terms in the exponent. This gives

$$p(s\,|\,v) \sim \frac{\mathrm{e}^{-\frac{1}{2}s^{\dagger}\boldsymbol{\Sigma}_s^{-1}s}\,\mathrm{e}^{-\frac{1}{2}(v-s)^{\dagger}\boldsymbol{\Sigma}_n^{-1}(v-s)}}{\mathrm{e}^{-\frac{1}{2}v^{\dagger}(\boldsymbol{\Sigma}_s+\boldsymbol{\Sigma}_n)^{-1}v}}.$$

We can identify terms to find the mean

$$\bar{s} = (\boldsymbol{\Sigma}_s^{-1} + \boldsymbol{\Sigma}_n^{-1})^{-1}\boldsymbol{\Sigma}_n^{-1}v,$$

and the variance

$$\boldsymbol{\Sigma}_{s|v} = \mathrm{E}[(s-\bar{s})^2|v] = (\boldsymbol{\Sigma}_s^{-1} + \boldsymbol{\Sigma}_n^{-1})^{-1},$$

as was to be proved. □

Corollary 11.3.4

$$\mathrm{E}[ss^{\dagger}|v] = \left(\boldsymbol{\Sigma}_s^{-1} + \boldsymbol{\Sigma}_n^{-1}\right)^{-1} + \bar{s}\bar{s}^{\dagger}.$$

Proof:

$$\begin{aligned}\mathrm{E}[ss^{\dagger}|v] &= \mathrm{E}[(s-\bar{s})(s-\bar{s})^{\dagger}|v] + \bar{s}\bar{s}^{\dagger}\\ &= \left(\boldsymbol{\Sigma}_s^{-1} + \boldsymbol{\Sigma}_n^{-1}\right)^{-1} + \bar{s}\bar{s}^{\dagger}\end{aligned}$$

as was to be proved. □

We can now describe the iterative algorithm. The maximization step is

$$\boldsymbol{\Sigma}_s \leftarrow \mathrm{E}[ss^{\dagger}\mid v, \boldsymbol{\Sigma}_s, \boldsymbol{\Sigma}_n]$$

and the expectation step, as given in Theorem 11.3.3 and Corollary 11.3.4 is

$$\mathrm{E}[ss^{\dagger}|v, \boldsymbol{\Sigma}_s, \boldsymbol{\Sigma}_n] = (\boldsymbol{\Sigma}_s^{-1} + \boldsymbol{\Sigma}_n^{-1})^{-1} + \bar{s}\bar{s}^{\dagger}$$

where

$$\bar{s} = \left(\boldsymbol{\Sigma}_s^{-1} + \boldsymbol{\Sigma}_n^{-1}\right)^{-1}\boldsymbol{\Sigma}_n^{-1}v.$$

Taken together, these give the iteration

$$\boldsymbol{\Sigma}_s \leftarrow (\boldsymbol{\Sigma}_s^{-1} + \boldsymbol{\Sigma}_n^{-1})^{-1} + (\boldsymbol{\Sigma}_s^{-1} + \boldsymbol{\Sigma}_n^{-1})^{-1}\boldsymbol{\Sigma}_n^{-1}vv^{\dagger}\boldsymbol{\Sigma}_n^{-1}(\boldsymbol{\Sigma}_s^{-1} + \boldsymbol{\Sigma}_n^{-1})^{-1}.$$

The iteration can be manipulated into the alternative form

$$\Sigma_s \leftarrow \Sigma_s - \Sigma_s(\Sigma_s + \Sigma_n)^{-1}(\Sigma_s + \Sigma_n - vv^\dagger)(\Sigma_s + \Sigma_n)^{-1}\Sigma_s$$

which may be preferred because it shows the iteration by means of a main term and a correction term. This correction term is zero if

$$\Sigma_s = vv^\dagger - \Sigma_n.$$

While this is a sufficient condition for a fixed point, it is a valid solution only if Σ_s is a positive-definite matrix. It is not a necessary condition for a fixed point. For Σ_s to be a fixed point, it is enough for the entire correction term be zero.

11.4 Other principles of inference

A set of noisy or incomplete data can be interpreted if one has a model that relates the data to the underlying problem. Time sampling, for example, represents a continuous-time function by its values at a discrete set of time instants. The samples are useful for recovering the continuous-time signal only if one accepts a model that imposes appropriate conditions on $s(t)$. One condition is the Nyquist condition, which is the condition that the Fourier transform $S(f)$ equals zero for $|f|$ greater than $1/2T$, where T is the sampling interval. Then, using the Nyquist–Shannon interpolation formula, $s(t)$ can be recomputed from the samples for any t. By requiring $S(f)$ to satisfy such a condition, we are imposing a prior model onto the data that the data by itself may not justify. Sometimes, an adequate model of this kind is not available; then the model must be construed by using the data and some general principle of inference.

We have already discussed the maximum-likelihood principle in some detail in the last section as a principle of inference. This principle assumes that a conditional probability distribution on the data is known. This conditional probability distribution $p(x|\gamma)$ depends on the unknown parameter γ. Then, when x is measured, an estimate of γ is formed as $\widehat{\gamma} = \text{argmax}_\gamma\, p(x|\gamma)$.

The maximum-likelihood principle is, perhaps, the most popular principle of inference. In this section, we shall introduce other principles of inference. While these principles are quite different, we will show later that they can often be reinterpreted to be seen as equivalent in some sense to the maximum-likelihood principle.

The maximum-likelihood principle assumes only that γ is unknown. It takes no position regarding whether γ is a random variable. If, however, γ is a random variable and has a known prior $p(\gamma)$, then one can write the Bayes formula

$$p(\gamma|x) = \frac{p(x|\gamma)p(\gamma)}{p(x)}$$

and use the estimate $\widehat{\gamma} = \text{argmax}_\gamma p(\gamma|x)$. This alternative estimator is known as the *maximum-posterior estimator* or the *bayesian estimator*.

The *Jaynes maximum-entropy principle* is a principle of data reduction that says when reducing a set of data into the form of an underlying model, one should be maximally noncommittal with respect to missing data. The measure of uncertainty to be used is a function known as the entropy. The *Shannon entropy* of the probability distribution $\boldsymbol{p}$ is defined as

$$H(\boldsymbol{p}) = -\sum_{j=0}^{J-1} p_j \log p_j.$$

The log base can be either two, in which case the entropy is measured in units of *bits*, or e in which case the entropy is measured in units of *nats*. The maximum-entropy principle says that if one must estimate a probability distribution, $\boldsymbol{p}$, on a data source, satisfying certain known constraints on $\boldsymbol{p}$ such as

$$\sum_j p_j f_j = t,$$

then, of those distributions that are consistent with the constraints, one should choose as the estimate of $\boldsymbol{p}$ the probability distribution $\boldsymbol{p}$ that has the maximum value of the entropy. The constraints may be imposed to ensure fidelity to various observations regarding the measurement data.

By using the inequality $\log_e x \leq x - 1$, it is easy to show that

$$H(\boldsymbol{p}) \leq \log J$$

and this inequality holds with equality if and only if $p_j = 1/J$ for all j. This means that one can define the difference $\log J - H(\boldsymbol{p})$ as a measure of the "distance" between $\boldsymbol{p}$ and the equiprobable distribution. Because $\log J = \sum_j p_j \log J$, this leads to the expression

$$\sum_{j=0}^{J-1} p_j \log \frac{p_j}{1/J} = \log J - H(\boldsymbol{p})$$

as a measure of the distance from probability distribution $\boldsymbol{p}$ to the equiprobable distribution. More generally, the *Kullback discrimination*, or *Kullback distance*, defined as

$$L(\boldsymbol{p}, \boldsymbol{q}) = \sum_{j=0}^{J-1} p_j \log \frac{p_j}{q_j}$$

is a measure of the "distance" from probability distribution $\boldsymbol{q}$ to probability distribution $\boldsymbol{p}$, both on an alphabet of size J, which is not the same as the "distance" from probability distribution $\boldsymbol{q}$ to probability distribution $\boldsymbol{p}$. The discrimination is nonnegative and equals zero if and only if $\boldsymbol{p} = \boldsymbol{q}$. The discrimination plays a role in the space of probability distributions on J symbols similar to the role that euclidian distance plays in the euclidean space $\boldsymbol{R}^n$. The discrimination can also be defined for probability density functions. Thus

$$L(p(x), q(x)) = \int_{-\infty}^{\infty} p(x) \log \frac{p(x)}{q(x)} \, dx$$

is the appropriate form for probability density functions.

The *Kullback minimum-discrimination principle* is a principle of data reduction that says when reducing a set of data to a probability distribution $\boldsymbol{p}$ in the presence of a prior $\boldsymbol{q}$ of that distribution, one should choose the admissible distribution $\boldsymbol{p}$ that minimizes $L(\boldsymbol{p}, \boldsymbol{q})$. The admissible probability distributions are those that satisfy some set of appropriate side conditions. These side conditions may take the form of equality constraints, which typically have the form of a constraint on expectation of the form

$$\sum_j p_j t_j = T.$$

A side condition may also have the form of an inequality constraint of the form

$$\sum_j p_j t_j \geq T.$$

One way to accommodate an inequality constraint is by solving two problems, one with the inequality constraint replaced by an equality constraint and one with the constraint ignored, then choosing between the two solutions appropriately. Thus, we need only study equality constraints.

A single equality constraint is treated by introducing a Lagrange multiplier s and minimizing the augmented object function as follows

$$\frac{\partial}{\partial p_j} \left[\sum_j p_j \log \frac{p_j}{q_j} + \lambda \left(\sum_j p_j - 1 \right) + s \left(\sum_j p_j t_j - T \right) \right] = 0$$

which leads to

$$p_j = \frac{q_j \, e^{-st_j}}{\sum_j q_j \, e^{-st_j}}$$

where the Lagrange multiplier s is to be chosen so that $\sum_j p_j t_j = T$.

This procedure is essentially the same if instead of only one constraint, there are many constraints of the form

$$\sum_j p_j t_{j\ell} = T_\ell \qquad \ell = 1, \ldots, L.$$

Then the solution is

$$p_j = \frac{q_j \, e^{-\sum_\ell s_\ell t_{j\ell}}}{\sum_j q_j \, e^{-\sum_\ell s_\ell t_{j\ell}}}$$

where s_ℓ is a Lagrange multiplier corresponding to the ℓth constraint. If the constraints are not consistent, then it will not be possible to choose the Lagrange multipliers to satisfy the constraints. In this case, the problem can be solved only if some of the constraints are relaxed to inequality constraints.

11.5 Nonnegativity constraints and discrimination

We introduced the Richardson–Lucy algorithm in Section 9.2 as an iterative algorithm, giving a heuristic development, but not a formal development. Now, in this section, we are ready to develop the Richardson–Lucy algorithm using formal information-theoretic methods. We start with the appropriate discrepancy measure on the space of nonnegative real functions. This discrepancy measure is the *Csiszár discrimination*, or *Csiszár distance*, which is defined as

$$L(a(x), b(x)) = \int_{-\infty}^{\infty} a(x) \log \frac{a(x)}{b(x)} \, dx - \int_{-\infty}^{\infty} [a(x) - b(x)] \, dx$$

where $a(x)$ and $b(x)$ are elements of the space of nonnegative real functions. The Csiszár discrimination is an appropriate way to measure the separation between nonnegative functions. It plays a role in the space of nonnegative, real functions analogous to the role played by euclidean distance in the space of real functions. It is not invariant under the interchange of $a(x)$ and $b(x)$.

The Csiszár discrimination is nonnegative and equal to zero if and only if $a(x) = b(x)$. This is a consequence of the elementary inequality $\log x \leq x - 1$ applied as follows:

$$\begin{aligned} -L(a(x), b(x)) &= \int_{-\infty}^{\infty} a(x) \log \frac{b(x)}{a(x)} \, dx + \int_{-\infty}^{\infty} [a(x) - b(x)] \, dx \\ &\leq \int_{-\infty}^{\infty} a(x) \left[\frac{b(x)}{a(x)} - 1 \right] dx + \int_{-\infty}^{\infty} [a(x) - b(x)] \, dx \\ &= \int_{-\infty}^{\infty} b(x) \, dx - \int_{-\infty}^{\infty} a(x) \, dx + \int_{-\infty}^{\infty} a(x) \, dx - \int_{-\infty}^{\infty} b(x) \, dx \\ &= 0. \end{aligned}$$

This means that $L(a(x), b(x)) \geq 0$, which is referred to as the *discrimination inequality*.

The Csiszár discrimination is closely related to the Kullback discrimination, which is defined on the space of probability functions as

$$L(p(x), q(x)) = \int_{-\infty}^{\infty} p(x) \log \frac{p(x)}{q(x)} \, dx$$

where $p(x)$ and $q(x)$ are nonnegative functions that integrate to one. The Kullback discrimination does not satisfy the triangle inequality, so it is not a metric. Nevertheless, it is the appropriate notion of distance in the space of probability density functions.

The Kullback discrimination is used in this section to give a formal development of the Richardson–Lucy algorithm, and to show that the algorithm converges. We shall treat only the case in which the integral of $h(x|y)$ over x is independent of y. This condition can be removed, but the development is then more cluttered. With this condition, the proof uses the Kullback discrimination. Otherwise, it would use the Csiszár discrimination.

The task is as follows: given the nonnegative functions $h(x|y)$ and $v(x)$ such that $\int_{-\infty}^{\infty} v(x)\,dx = 1$ and $\int_{-\infty}^{\infty} h(x|y)\,dx = 1$, find a nonnegative function, $s(y)$, that integrates to one such that

$$\widehat{v}(x) = \int_{-\infty}^{\infty} h(x|y)s(y)\,dy$$

minimizes the discrimination $L(v(x), \widehat{v}(x))$ over the choice of such $s(y)$. That is, the function

$$L\left(v(x), \int_{-\infty}^{\infty} h(x|y)s(y)\,dy\right) = \int_{-\infty}^{\infty} v(x)\log\frac{v(x)}{\int_{-\infty}^{\infty} h(x|y)s(y)\,dy}\,dx$$

is to be minimized by choice of $s(y)$. The function that achieves the minimum is written as

$$\widehat{s}(y) = \text{argmin}_{s(y)} \int_{-\infty}^{\infty} v(x)\log\frac{v(x)}{\int_{-\infty}^{\infty} h(x|y)s(y)\,dy}\,dx.$$

This minimization problem can be studied by temporarily ignoring the condition that $s(y)$ is nonnegative, introducing the Lagrange multiplier λ to constrain the integral of $s(y)$ to be one, and then using the calculus of variations to treat the augmented object function

$$\int_{-\infty}^{\infty} v(x)\log\frac{v(x)}{\int_{-\infty}^{\infty} h(x|y)s(y)\,dy}\,dx + \left[\lambda\int_{-\infty}^{\infty} s(y)\,dy - 1\right].$$

The calculus of variations states that because $s(y)$ is arbitrary, it can be replaced by the alternative arbitrary function $s(y) + \epsilon\eta(y)$ where ϵ is an arbitrary scalar parameter and $\eta(y)$ is arbitrary. Then compute the derivative with respect to ϵ and set the derivative equal to zero at the point where ϵ is equal to zero. This gives

$$\frac{\partial}{\partial\epsilon}\left[\int_{-\infty}^{\infty} v(x)\log\frac{v(x)}{\int_{-\infty}^{\infty} h(x|y)[s(y)+\epsilon\eta(y)]\,dy}\,dx \right.$$
$$\left. + \lambda\left[\int_{-\infty}^{\infty} (s(y)+\epsilon\eta(y))\,dy - 1\right]\right]_{\epsilon=0} = 0,$$

which leads to the equation

$$\int_{-\infty}^{\infty}\left[v(x)\frac{\int_{-\infty}^{\infty} h(x|y)\eta(y)\,dy}{\int_{-\infty}^{\infty} h(x|y)s(y)\,dy}\right]dx - \lambda\int_{-\infty}^{\infty}\eta(y)\,dy = 0$$

where $\eta(y)$ is arbitrary. This can be rewritten as

$$\int_{-\infty}^{\infty}\left[\int_{-\infty}^{\infty}\frac{h(x|y)}{\int_{-\infty}^{\infty}h(x|y)s(y)\,dy}v(x)\,dx-\lambda\right]\eta(y)\,dy=0.$$

This equation must hold for every function $\eta(y)$. Unless the bracketed term is zero, we could make the left side nonzero by choosing $\eta(y)$ to be ±1 according to the sign of the bracketed term. Thus we conclude that the bracketed term must be zero and

$$\int_{-\infty}^{\infty}\frac{h(x|y)}{\int_{-\infty}^{\infty}h(x|y)s(y)\,dy}v(x)\,dx=\lambda.$$

To evaluate the constant λ, multiply by $s(y)$ and integrate with respect to y, which leads to the conclusion that $\lambda = 1$.

Our development is incomplete because it ignored the requirement that $s(x)$ is nonnegative. A complete development leads to the statement that

$$\int_{-\infty}^{\infty}\frac{h(x|y)}{\int_{-\infty}^{\infty}h(x|y)s(y)\,dy}v(x)\,dx\leq 1$$

with equality at all y for which $s(y)$ is nonzero. This statement, known as a Kuhn–Tucker condition, is a necessary condition on any $s(y)$ that achieves the desired minimum. Moreover, because the object function is a convex function, any local minimum is a global minimum, so the Kuhn–Tucker condition is both a necessary and a sufficient condition.

In particular, by multiplying through by $s(y)$, the Kuhn–Tucker condition becomes the statement that for all y,

$$s(y)=s(y)\int_{-\infty}^{\infty}\frac{h(x|y)}{\int_{-\infty}^{\infty}h(x|y)s(y)\,dy}v(x)\,dx.$$

This expression suggests the Richardson–Lucy algorithm. This is the recursion

$$s^{(r+1)}(y)=s^{(r)}(y)\int_{-\infty}^{\infty}\frac{h(x|y)}{\int_{-\infty}^{\infty}h(x|y)s^{(r)}(y)\,dy}v(x)\,dx.$$

Before deriving the Richardson–Lucy recursion more formally we will develop a lemma as a technical device that is often useful to lift a certain kind of problem to a larger setting where it may be easier to solve. This lemma deals with nonnegative functions that integrate to one, so it is suggestive to use the notation of probability theory. Let $q(y;\gamma)$ be a probability density function, dependent on the parameter γ, that can be written in the form

$$q(y;\gamma)=\int_{-\infty}^{\infty}p(x;\gamma)Q(y|x)\,dx.$$

Thus, $q(y;\gamma)$ can be understood as the output of the channel $Q(y|x)$ when $p(x;\gamma)$ is the input. This situation also suggests that a backwards channel, as shown in Figure 11.3,

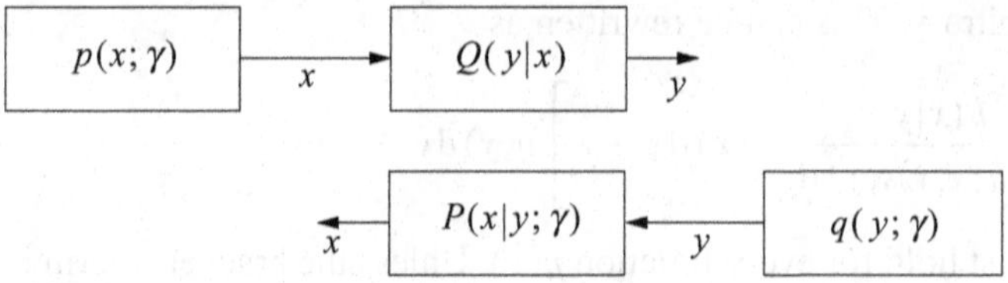

Figure 11.3 The forward channel and the backward channel

can be defined by the Bayes formula as

$$P(x|y;\gamma) = \frac{p(x;\gamma)Q(y|x)}{\int_{-\infty}^{\infty} p(x';\gamma)Q(y|x')\,dx'}.$$

Suppose that it is desired to work with the logarithm of such an expression. The denominator will then lead to the logarithm of an integral, which is difficult to work with. The following lemma expresses the logarithm of an integral in terms of the integral of a logarithm, but at the cost of a gratuitous minimization.

Lemma 11.5.1 (Convex Decomposition Lemma)

$$-\log \int_{-\infty}^{\infty} p(x;\gamma)Q(y|x)\,dx = \min_{P(x|y)} \int_{-\infty}^{\infty} P(x|y) \log \frac{P(x|y)}{p(x;\gamma)Q(y|x)}\,dx.$$

Proof: To minimize the right side over conditional probability densities, introduce the Lagrange multiplier λ. The augmented object function to be minimized is

$$\int_{-\infty}^{\infty} P(x|y) \log \frac{P(x|y)}{p(x;\gamma)Q(y|x)}\,dx + \lambda \left(\int_{-\infty}^{\infty} P(x|y)\,dx - 1 \right).$$

Now use the calculus of variations to differentiate, in effect, with respect to $P(x|y)$. Setting the derivative with respect to $P(x|y)$ to zero gives

$$\log \frac{P(x|y)}{p(x;\gamma)Q(y|x)} + 1 + \lambda = 0$$

where λ is to be chosen so that $P(x|y)$ is a probability density function. Therefore

$$P(x|y) = \frac{p(x;\gamma)Q(y|x)}{\int_{-\infty}^{\infty} p(x';\gamma)Q(y|x')\,dx'}.$$

The conclusion of the lemma now follows directly. □

To obtain the Richardson–Lucy algorithm, alternate between computing the estimates $\widehat{p}^{(r)}(x;\gamma)$ and $\widehat{P}^{(r)}(x|y)$. Use Lemma 11.5.1 to write

$$\widehat{p}^{(r+1)}(x;\gamma) = \text{argmax}_{p(x;\gamma)} \left[-\int_{-\infty}^{\infty} \widehat{P}^{(r)}(x|y) \log \frac{\widehat{P}^{(r)}(x|y)}{p(x;\gamma)Q(y|x)}\, dx \right].$$

Then, with $\widehat{p}^{(r)}(x;\gamma)$ fixed, maximizing with respect to $\widehat{P}^{(r)}(x|y)$ gives the Bayes formula

$$\widehat{P}^{(r)}(x|y) = \frac{\widehat{p}^{(r)}(x;\gamma)Q(y|x)}{\int_{-\infty}^{\infty} \widehat{p}^{(r)}(x';\gamma)Q(y|x')\, dx'}.$$

Theorem 11.5.2 (Richardson–Lucy Algorithm) *The iterated sequence*

$$s^{(r+1)}(y) = s^{(r)}(y) \int_{-\infty}^{\infty} \frac{h(x|y)}{\int_{-\infty}^{\infty} h(x|y)s^{(r)}(y)\, dy} v(x)\, dx$$

satisfies

$$\lim_{r\to\infty} s^{(r)}(y) = \text{argmin}_{s(x)} L(v(x), \widehat{v}(x))$$

where

$$\widehat{v}(x) = \int_{-\infty}^{\infty} h(x|y)\widehat{s}(y)\, dy.$$

Proof: To prove convergence of the Richardson–Lucy algorithm, we temporarily write it as

$$s^{(r+1)}(y) = s^{(r)}(y) f^{(r)}(y)$$

where

$$f^{(r)}(y) = \int_{-\infty}^{\infty} \frac{h(y|x)}{\int_{-\infty}^{\infty} h(x|y)s^{(r)}(y)\, dy} v(x)\, dx.$$

Next, let

$$L_{\min} = \min_{s(y)} L\left(v(x), \int_{-\infty}^{\infty} h(x|y)s(y)\, dy \right),$$

and let

$$L^{(r)} = L\left(v(x), \int_{-\infty}^{\infty} h(x|y)s^{(r)}(y)\, dy \right).$$

We first argue that the $L^{(r)}$ are monotonically decreasing and nonnegative so the sequence of $L^{(r)}$ must converge. Then we argue that the limit must be $L_{\min}$ and, finally, that the $s^{(r)}(x)$ must converge to the desired function.

Let $v^{(r)}(x) = \int_{-\infty}^{\infty} h(x\,|\,y)s^{(r)}(y)\,dy$. With the aid of Jensen's inequality,[3] we have the following:

$$\begin{aligned}
L^{(r)} - L^{(r+1)} &= \int_{-\infty}^{\infty} v(x) \log \frac{\int_{-\infty}^{\infty} h(x\,|\,y)s^{(r+1)}(y)\,dy}{v^{(r)}(x)}\,dx \\
&= \int_{-\infty}^{\infty} v(x) \log \int_{-\infty}^{\infty} \left[\frac{h(x\,|\,y)s^{(r)}(y)}{\int_{-\infty}^{\infty} h(x\,|\,y)s^{(r)}(y)\,dy} \right] f^{(r)}(y)\,dy\,dx \\
&\geq \int_{-\infty}^{\infty} v(x) \int_{-\infty}^{\infty} \frac{h(x\,|\,y)s^{(r)}(y)}{\int_{-\infty}^{\infty} h(x\,|\,y)s^{(r)}(y)\,dy} \log f^{(r)}(y)\,dy\,dx \\
&= \int_{-\infty}^{\infty} s^{(r)}(y) f^{(r)}(y) \log \frac{s^{(r+1)}(y)}{s^{(r)}(y)}\,dy \\
&= L(s^{(r)}(y), s^{(r+1)}(y)) \\
&\geq 0.
\end{aligned}$$

Therefore $L^{(r+1)} \leq L^{(r)}$. This means that the sequence of terms $L^{(1)}, L^{(2)}, L^{(3)}, \ldots$ is nonincreasing and the terms are nonnegative, and so the sequence of the terms has a limit. A reasonable conjecture is that the limit is $L_{\min}$ because no alternative conjecture is in evidence. The proof, however, is not so trivial. The above inequality can be written

$$\int_{-\infty}^{\infty} v(x) \log \frac{v^{(r+1)}(x)}{v^{(r)}(x)}\,dx \geq L(s^{(r+1)}(y), s^{(r)}(y)).$$

By summing both sides over r, we can write

$$\begin{aligned}
\sum_{r=0}^{R} L(s^{(r+1)}(y), s^{(r)}(y)) &\leq \int_{-\infty}^{\infty} v(x) \log \frac{v^{(R+1)}(x)}{v^{(0)}(x)}\,dx \\
&= \int_{-\infty}^{\infty} v(x) \log \frac{v(x)}{v^{(0)}(x)}\,dx - \int_{-\infty}^{\infty} v(x) \log \frac{v(x)}{v^{(R+1)}(x)}\,dx \\
&\leq \int_{-\infty}^{\infty} v(x) \log \frac{v(x)}{v^{(0)}(x)}\,dx
\end{aligned}$$

where the second inequality uses the discrimination inequality. Because the final term on the right side is a constant, independent of R, and R is arbitrary, we conclude that the summands $L(s^{(r+1)}(y), s^{(r)}(y))$ go to zero as r goes to infinity. Finally, because the discrimination is continuous, it is apparent that, in some sense, the sequence of $s^{(r)}(y)$ must converge and that the Kuhn–Tucker conditions are satisfied in the limit, so the convergence of the $s^{(r)}(y)$ must be to $\widehat{s}(y)$. □

[3] Jensen's inequality states that

$$f(\mathrm{E}[x]) \leq \mathrm{E}[f(x)]$$

for any convex function $f(x)$.

In the special case of a one-dimensional, time-invariant blur function, the integrals become convolutions and the Richardson–Lucy algorithm has the form

$$s^{(r+1)}(t) = s^{(r)}(t)\left[\frac{h(t)}{h(t) * s^{(r)}(t)} \quad * \quad v(-t)\right]$$

by using the notation of the convolution.

In the case of a two-dimensional, space-invariant, pointspread function, the blurred function is given by the convolution

$$v(x, y) = h(x, y) \quad ** \quad s(x, y).$$

The Richardson–Lucy algorithm for a two-dimensional, space-invariant, pointspread function takes the special form given in the following theorem.

Theorem 11.5.3 *The iterated sequence*

$$s^{(r+1)}(x, y) = s^{(r)}(x, y)\left[\frac{v(x, y)}{s^{(r)}(x, y) \quad ** \quad h(x, y)} \quad ** \quad h(-x, -y)\right]$$

satisfies

$$\lim_{r\to\infty} s^{(r)}(x, y) = \text{argmin}_{s(x,y)} L[v(x, y), \widehat{v}(x, y)]$$

where

$$\widehat{v}(x, y) = \int_{-\infty}^{\infty}\int_{-\infty}^{\infty} h(x - \xi, y - \eta)\widehat{s}(\xi, \eta)\, d\xi\, d\eta.$$

Proof: The proof of the two-dimensional case is the same as the proof of the one-dimensional case. □

11.6 Likelihood methods in blind image deconvolution

A more difficult topic of image restoration is the topic of two-dimensional blind deconvolution. The standard task of two-dimensional blind deconvolution is to estimate $h(x, y)$ and $s(x, y)$, both nonnegative, satisfying the stochastic equation

$$v(x, y) = h(x, y) \quad ** \quad s(x, y) + n(x, y)$$

when given the realization $v(x, y)$ where $n(x, y)$ is a zero-mean, white gaussian-noise process whose power density spectrum is known. One procedure for making estimates of $h(x, y)$ and $s(x, y)$ was given in Section 9.4 without any theory or formal justification. In this section, we will develop a more thoughtful procedure.

The likelihood function is

$$\Lambda(h(x, y), s(x, y)) = \int_{-\infty}^{\infty}\int_{-\infty}^{\infty} (v(x, y) - h(x, y) \quad ** \quad s(x, y))^2\, dx\, dy.$$

If the condition that the functions are nonnegative were not imposed, then we would use the maximum-likelihood principle and write

$$(\widehat{h}(x, y), \widehat{s}(x, y)) = \operatorname{argmin}_{(h(x,y),s(x,y))} \int_{-\infty}^{\infty}\int_{-\infty}^{\infty} (v(x, y) - h(x, y) ** s(x, y))^2 \, dx \, dy.$$

However, the solutions $h(x, y)$ and $s(x, y)$, in general, would take negative values and hence would not satisfy the constraint. Simply setting those negative values to zero would give a solution, though not one satisfying an evident optimality criterion. Instead, as an alternative to maximizing likelihood, we will choose the Kullback discrimination

$$L(v(x, y), h(x, y) ** s(x, y)) = \int_{-\infty}^{\infty}\int_{-\infty}^{\infty} v(x, y) \log \frac{v(x, y)}{h(x, y) ** s(x, y)} \, dx \, dy$$

as a measure of discrepancy and choose the pair of functions that minimizes the discrimination. This criterion can be regarded as closely related to the maximum-likelihood criterion.

The minimum-discrimination solution then consists of the pair $h(x, y)$ and $s(x, y)$ that minimizes the right side of this equation. This is a more general version of the problem that led to the Richardson–Lucy algorithm in Section 11.5. In this case, however, there are two functions, $s(x, y)$ and $h(x, y)$, that must be found to minimize the object function. This leads to two iterates in the generalization of the Richardson–Lucy algorithm.

The blind deconvolution algorithm for nonnegative functions is as follows:

$$s^{new}(x, y) = s^{old}(x, y)\left[\frac{v(x, y)}{s^{old}(x, y) ** h(x, y)} ** h(-x, -y)\right]$$

$$h^{new}(x, y) = h^{old}(x, y)\left[\frac{v(x, y)}{h^{old}(x, y) ** s(x, y)} ** s(-x, -y)\right].$$

This iteration can be motivated in the same way as the Richardson–Lucy algorithm. Alternate one step of the iteration for $s(x, y)$ with one step of the iteration for $h(x, y)$.

Theorem 11.6.1 *The pair of alternating iterations*

$$s^{(r+1)}(x, y) = s^{(r)}(x, y)\left[\frac{v(x, y)}{s^{(r)}(x, y) ** h(x, y)} ** h(-x, -y)\right]$$

$$h^{(r+1)}(x, y) = h^{(r)}(x, y)\left[\frac{v(x, y)}{h^{(r)}(x, y) ** s(x, y)} ** s(-x, -y)\right]$$

satisfies

$$\lim_{r\to\infty} (s^{(r)}(y), h^{(r)}(y)) = \operatorname{argmin}_{(h(x,y),s(x,y))} L(v(x, y), h(x, y) ** s(x, y)).$$

Proof: The proof of this convergence theorem is not given. It is an extension of the proof of the Richardson–Lucy algorithm. □

11.7 Likelihood methods in photon imaging

Although it is easy to define the maximum-likelihood image, it may not be easy to find this image. It can be extremely difficult to compute or even to test whether a candidate image is indeed the maximum-likelihood image. To develop the maximum-likelihood image for a given application, we must first derive the probability function $p(v|\gamma)$. Then we must present an algorithm for finding argmax $\Lambda(\gamma)$. Our example in this section will use the Dempster–Laird–Rubin method to derive an iterative algorithm to compute the maximum-likelihood image.

In Section 9.5, we described how to estimate images in weak light by forming histograms of photon position differences, and then we used a general phase-retrieval procedure. We now reconsider that problem, using the maximum-likelihood principle to obtain a better estimator of the image.

Suppose, as in Section 9.5, that $2N$ photoconversions have been detected and parameterized. The raw data consist of the sequence of space-time points (x_i, y_i, t_i) for $i = 1, \ldots, 2N$. To remove the effect of uncertainty in the spatial origin, we will choose to preprocess the data by replacing the $2N$ spatial coordinates by N pairs of position differences $(x_{2\ell} - x_{2\ell-1}, y_{2\ell} - y_{2\ell-1})$. This step simply removes the unknown position displacement from the data set. We do not justify this step in any deeper way.

The times of the photon arrivals t_i intuitively seem to have very little to do with the unknown $\lambda(x, y)$, and it almost seems that t_i can be discarded. However, this is a little rash because the integral of $\lambda(x, y)$ does depend on the average arrival rate, so at least $T = t_{2N} - t_1$ should be retained. We will take a middle position that allows the loglikelihood function to work out cleanly. We will preprocess the times of arrival to retain in the data set only the time interval between every second arrival. The preprocessed data set now has the form

$$(x_1 - x_2, y_1 - y_2, t_3 - t_1)$$
$$(x_3 - x_4, y_3 - y_4, t_5 - t_3)$$
$$(x_5 - x_6, y_5 - y_6, t_7 - t_5).$$

Thus the data set consists of photon-pair position differences and photon-pair intervals. Retaining the times of arrival in the data set as photon-pair intervals yields an analytically tractable form for the likelihood function. Specifically, because the arrival times are a poisson random process with the parameter λ, the probability density function on the interval is $P_{w_2}(t_{2\ell+1} - t_{2\ell-1})$ where

$$P_{w_n}(\tau) = \frac{(\lambda\tau)^{n-1}}{(n-1)!}\lambda e^{-\lambda\tau}$$

is the probability density function on the nth occurrence time for the homogeneous poisson process. In particular, the second occurrence time is

$$P_{w_2}(\tau) = \lambda^2 \tau \, \mathrm{e}^{-\lambda\tau}.$$

When we form the likelihood function, the term λ^2 will cancel a like term arising from the spatial data.

The probability density function for the entire set of preprocessed data can now be stated. Because the N points $(x_{2\ell} - x_{2\ell-1}, y_{2\ell} - y_{2\ell-1}, t_{2\ell+1} - t_{2\ell-1})$ for $\ell = 1, \ldots, N$ are independent, the probability density function is a product. Thus the likelihood function is the same product.

Let u, v, and $\boldsymbol{\tau}$ be the vectors with the components $u_\ell = x_{2\ell} - x_{2\ell-1}$, $v_\ell = y_{2\ell} - y_{2\ell-1}$, and $\tau_\ell = t_{2\ell+1} - t_{2t-1}$, respectively. Then the loglikelihood function is

$$\begin{aligned} \Lambda(\lambda(x, y)) &= \log p((\mathrm{u}, v, \boldsymbol{\tau})|\lambda(x, y)) \\ &= \log \prod_{\ell=1}^{N} P_D(u_\ell, v_\ell) P_{w_2}(\tau_\ell) \end{aligned}$$

where

$$P_D(u, v) = \frac{1}{\lambda^2} \int_{-\infty}^{\infty} \int_{-\infty}^{\infty} \lambda(x, y)\lambda(x + u, y + v)\, \mathrm{d}x\, \mathrm{d}y$$

and $P_{w_2}(\tau)$ is as given previously. Consequently,

$$\begin{aligned} \Lambda(\lambda(x, y)) &= \sum_{\ell=1}^{N} \log \left[\lambda^2 \tau_\ell \, \mathrm{e}^{-\lambda\tau_\ell} \frac{1}{\lambda^2} \int_{-\infty}^{\infty} \int_{-\infty}^{\infty} \lambda(x, y)\lambda(x + u_\ell, y + v_\ell)\, \mathrm{d}x\, \mathrm{d}y \right] \\ &= \sum_{\ell=1}^{N} \log \tau_\ell - \lambda \sum_{\ell=1}^{N} \tau_\ell + \sum_{\ell=1}^{N} \log \int_{-\infty}^{\infty} \int_{-\infty}^{\infty} \lambda(x, y)\lambda(x + u_\ell, y + v_\ell)\, \mathrm{d}x\, \mathrm{d}y. \end{aligned}$$

We suppress the first term because it does not depend on $\lambda(x, y)$, and so does not affect the location of the maximum. Replace the sum of intervals by T in the second term and replace λ by its definition. Thus our task is to find

$$\begin{aligned} \widehat{\lambda}(x, y) = \operatorname{argmin}_{\lambda(x,y)} \Bigg[-T \int_{-\infty}^{\infty} \int_{-\infty}^{\infty} \lambda(x, y)\, \mathrm{d}x\, \mathrm{d}y \\ + \sum_{\ell=1}^{N} \log \int_{-\infty}^{\infty} \int_{-\infty}^{\infty} \lambda(x, y)\lambda(x + u_\ell, y + v_\ell)\, \mathrm{d}x\, \mathrm{d}y \Bigg]. \end{aligned}$$

Because N can be very large, finding the function $\widehat{\lambda}(x, y)$ is a formidable computational problem. Fortunately, the Dempster–Laird–Rubin method can be used to derive an iterative algorithm for its solution.

Before developing the algorithm, the next theorem states a necessary condition on $\lambda(x, y)$ to achieve the maximum. To find a nonnegative $\lambda(x, y)$ that satisfies this condition is far from trivial.

Theorem 11.7.1 *A necessary condition for* $\lambda(x, y)$ *to maximize the photon difference loglikelihood function is*

$$1 = \frac{1}{T}\sum_{\ell=1}^{N} \frac{\lambda(x-u_\ell, y-v_\ell) + \lambda(x+u_\ell, y+v_\ell)}{\int_{-\infty}^{\infty}\int_{-\infty}^{\infty} \lambda(x, y)\lambda(x+u_\ell, y+v_\ell)\,\mathrm{d}x\,\mathrm{d}y}.$$

Proof: To find this condition on $\lambda(x, y)$, replace $\lambda(x, y)$ by $\lambda(x, y) + \epsilon\eta(x, y)$ and use the methods of variational calculus. For any well-behaved function, $\eta(x, y)$, if $\lambda(x, y)$ maximizes $\Lambda(\lambda(x, y))$, then for any $\eta(x, y)$

$$\frac{\partial}{\partial\epsilon}\Lambda(\lambda(x, y) + \epsilon\eta(x, y))\Big|_{\epsilon=0} = 0$$

where

$$\Lambda(\lambda(x, y)) = -T\int_{-\infty}^{\infty}\int_{-\infty}^{\infty} \lambda(x, y)\,\mathrm{d}x\,\mathrm{d}y + \sum_{\ell=1}^{N} \log \int_{-\infty}^{\infty}\int_{-\infty}^{\infty} \lambda(x, y)\lambda(x+u_\ell, y+v_\ell)\,\mathrm{d}x\,\mathrm{d}y.$$

Substituting $\lambda(x, y) + \epsilon\eta(x, y)$ in place of $\lambda(x, y)$, and differentiating with respect to ϵ, gives

$$0 = -T\int_{-\infty}^{\infty}\int_{-\infty}^{\infty} \eta(x, y)\,\mathrm{d}x\,\mathrm{d}y + \sum_{\ell=1}^{N} \frac{\int_{-\infty}^{\infty}\int_{-\infty}^{\infty}[\lambda(x, y)\eta(x+u_\ell, y+v_\ell) + \eta(x, y)\lambda(x+u_\ell, y+v_\ell)]\,\mathrm{d}x\,\mathrm{d}y}{\int_{-\infty}^{\infty}\int_{-\infty}^{\infty} \lambda(x, y)\lambda(x+u_\ell, y+v_\ell)\,\mathrm{d}x\,\mathrm{d}y}$$

at $\epsilon = 0$. The integrand can be expressed as a multiple of the arbitrary function $\eta(x, y)$ as follows:

$$0 = \int_{-\infty}^{\infty}\int_{-\infty}^{\infty}\left[-T + \frac{\lambda(x-u_\ell, y-v_\ell) + \lambda(x+u_\ell, y+v_\ell)}{\int_{-\infty}^{\infty}\int_{-\infty}^{\infty} \lambda(x, y)\lambda(x+u_\ell, y+v_\ell)\,\mathrm{d}x\,\mathrm{d}y}\right]\eta(x, y)\,\mathrm{d}x\,\mathrm{d}y.$$

The standard methods of variational calculus assert at this point that the term multiplying $\eta(x, y)$ must equal zero when $\lambda(x, y)$ maximizes $\Lambda(\lambda(x, y))$. Otherwise, a function, $\eta(x, y)$, could be constructed making $\Lambda(\lambda(x, y))$ larger. Thus the necessary condition follows, and the proof is complete. □

It is instructive to relate the condition of Theorem 11.7.1 first to the phase-retrieval problem and then to the iterative computational algorithm to be derived below. To connect the condition with phase retrieval, notice that, by the law of large numbers, the histogram of photoconversions will converge with increasing data and shrinking bin size to $NR(u, v)$ where

$$R(u, v) = \int_{-\infty}^{\infty}\int_{-\infty}^{\infty} \lambda(x, y)\lambda(x+u, y+v)\,\mathrm{d}x\,\mathrm{d}y,$$

and T will be approximately $2N\lambda$. Therefore, with $\widehat{\lambda}(x, y)$ denoting the estimated distribution, the condition on $\widehat{\lambda}(x, y)$ is approximated by

$$1 = \frac{1}{2\lambda} \int_{-\infty}^{\infty} \int_{-\infty}^{\infty} \frac{\widehat{\lambda}(x-u, y-v) + \widehat{\lambda}(x+u, y+v)}{\widehat{R}(u, v)} R(u, v)\, du\, dv.$$

If $\widehat{\lambda}(x, y)$ is chosen to equal $\lambda(x, y)$, then $\widehat{R}(u, v) = R(u, v)$, and this equation is satisfied. Thus solving the equation for $\lambda(x, y)$ is equivalent to finding $\lambda(x, y)$, given the autocorrelation $R(x, y)$. This is the problem of phase retrieval. Evidently, phase-retrieval algorithms approximate a maximum-likelihood solution when the data set is very large, but fail to approximate a maximum-likelihood solution when the data are inadequate to closely estimate the autocorrelation function.

To restate the necessary condition on $\lambda(x, y)$, given in Theorem 11.7.1 in a form that suggests an iterative algorithm, multiply that equation by $\lambda(x, y)$. Then

$$\lambda(x, y) = \frac{\lambda(x, y)}{T} \sum_{\ell=1}^{N} \frac{\lambda(x-u_\ell, y-v_\ell) + \lambda(x+u_\ell, y+v_\ell)}{\int_{-\infty}^{\infty} \int_{-\infty}^{\infty} \lambda(x, y)\lambda(x+u_\ell, y+v_\ell)\, dx\, dy}.$$

This equation suggests an iterative procedure to compute $\lambda(x, y)$. Using the left side as an updated iterative based on the right side will give the Schulz–Snyder iteration

$$\lambda^{(r+1)}(x, y) = \frac{\lambda^{(r)}(x, y)}{T} \sum_{\ell=1}^{N} \frac{\lambda^{(r)}(x-u_\ell, y-v_\ell) + \lambda^{(r)}(x+u_\ell, y+v_\ell)}{\int_{-\infty}^{\infty} \int_{-\infty}^{\infty} \lambda^{(r)}(x, y)\lambda^{(r)}(x+u_\ell, y+v_\ell)\, dx\, dy}.$$

If the sequence $\lambda^{(r)}(x, y)$ defined here converges to a limit point, then that limit point is an extremum (in fact, a maximum). Although we have introduced this algorithm only in a heuristic way, it can be derived more formally by using the Dempster–Laird–Rubin method.

11.8 Radar imaging of diffuse reflectors

Radar reflectors, as discussed in Section 7.1, may be modeled as either specular or diffuse. A diffuse model for a radar target is one in which there are a large number of reflectors in a typical resolution cell such that no single reflector dominates that cell and the multiple reflectors have some kind of statistical fluctuation in amplitude and phase. Constructive and destructive interference from the multitude of reflectors in each cell will create a speckle pattern that will appear as multiplicative noise. A very slight change in the imaging geometry can cause a large change in the reflected signal. In such a case, the reflectivity density may be modeled as a two-dimensional random process with zero mean and variance[4] $\sigma^2(x, y) = \mathrm{E}[\rho(x, y)^2]$. The expected value of

[4] Unfortunately in this situation, there is a clash in traditional notation. The variance, which is usually denoted σ^2, is equal to the expected radar cross section $\mathrm{E}[\rho(x, y)^2]$, which is usually denoted $\sigma(x, y)$.

the reflectivity density $E[\rho(x, y)]$ is equal to zero, so there is no need to estimate it. Instead, one estimates the variance $\Sigma(x, y) = E|\rho(x, y)|^2$ as a function of x and y.

The distinction between specular and diffuse reflectors may be understood by considering a reflectivity density, $\rho(x, y)$, viewed through a pointspread function, $h(x, y)$, to produce the image

$$r(x, y) = h(x, y) ** \rho(x, y),$$

which has the intensity

$$|r(x, y)|^2 = |h(x, y) ** \rho(x, y)|^2.$$

To the extent that $h(x, y)$ approximates an impulse, we can approximate this by

$$|r(x, y)|^2 = |h(x, y)|^2 ** |\rho(x, y)|^2.$$

Although we are not able to observe $|\rho(x, y)|^2$, the approximation enables us to use $|r(x, y)|^2$ as a blurred approximation of $|\rho(x, y)|^2$.

Now suppose that $\rho(x, y)$ is multiplied by a *texture function*, $u(x, y)$, to give

$$\rho'(x, y) = u(x, y)\rho(x, y).$$

The texture function $u(x, y)$ is a complex, stationary random field with zero mean and unit variance. The particular realization of the texture is usually not regarded as an informative part of the reflectivity density, though the statistical parameters of the texture may be of interest. Thus only $\rho(x, y)$ is of interest – $\rho'(x, y)$ is not – and the covariance function of $u(x, y)$ may be of interest.

The elementary problem of estimating the variance of a gaussian signal in gaussian noise was discussed in Section 11.3. One is given a measurement of the form

$$v = s + n$$

where s is a gaussian random variable of unknown variance σ_s^2, and n is a gaussian random variable of known variance σ_n^2. The task is to estimate the variance σ_s^2. The loglikelihood is

$$\Lambda(\sigma_s^2) = -\tfrac{1}{2}\log_e(\sigma_s^2 + \sigma_n^2) - \frac{v^2}{2(\sigma_s^2 + \sigma_n^2)}.$$

The estimate is

$$\widehat{\sigma_s^2} = \max(0, v^2 - \sigma_n^2).$$

More generally, if there are M independent measurements, then the maximum-likelihood estimate of σ_s^2 is

$$\widehat{\sigma_s^2} = \max\left[0, \widehat{\sigma_v^2} - \sigma_n^2\right]$$

where

$$\widehat{\sigma_v^2} = \frac{1}{M} \sum_{\ell=1}^{M} v_\ell^2.$$

We are interested in the problem of estimating the variance function $\sigma^2(x, y) = E|\rho(x, y)|^2$ of the diffuse reflectivity density from a reflected radar signal. This is a more general version of the problem of estimating variance. No simple analytic solution is evident for the maximum-likelihood estimate. Therefore we will resort to an iterative computational algorithm, which is the main purpose of this section.

So that we may regularize the problem and describe it in terms of matrix and vector operations, we will pixelate the diffuse reflectivity density $\rho(x, y)$ to the array $\rho_{i'i''}$. The (i', i'')th pixel has a reflectivity, $\rho_{i'i''}$, which is modeled as a zero-mean, gaussian random variable of variance $\sigma_{i'i''}^2$ that is independent of all other pixels. We shall regard the array of reflectivity samples as arranged into a single vector, denoted $\boldsymbol{\rho}$, and we shall regard the array of variances as arranged into a single vector, denoted $\boldsymbol{\sigma}^2$. The covariance matrix $\boldsymbol{\Sigma}$ is defined as the diagonal matrix with the elements of the vector $\boldsymbol{\sigma}^2$ on the diagonal. Thus our model of the reflectivity is that every sample of the reflectivity has its own variance and these samples are uncorrelated.

The vector of the measured data is

$$\boldsymbol{v} = \boldsymbol{\Gamma}^\dagger \boldsymbol{\rho} + \boldsymbol{n},$$

where $\boldsymbol{\Gamma}$ is a matrix and $\boldsymbol{n}$ is a white gaussian-noise vector with covariance matrix $N_0\boldsymbol{I}$ and $\boldsymbol{\rho}$ is a gaussian random vector with diagonal covariance matrix $\boldsymbol{\Sigma}_\rho$. The vector $\boldsymbol{v}$ has the form of a sum of a gaussian signal and gaussian noise but, because of the matrix $\boldsymbol{\Gamma}$, it is not the case studied in Section 11.3. Because $\boldsymbol{\rho}$ and $\boldsymbol{n}$ are both gaussian, the vector $\boldsymbol{v}$ is gaussian with the covariance matrix $\boldsymbol{K} = \boldsymbol{\Gamma}^\dagger \boldsymbol{\Sigma}_\rho \boldsymbol{\Gamma} + N_0\boldsymbol{I}$. Our assumption that the pixel values are independent random variables implies that $\boldsymbol{\Sigma}_\rho$ is diagonal.

By setting $\boldsymbol{s} = \boldsymbol{\Gamma}^\dagger \boldsymbol{\rho}$, we have

$$\boldsymbol{v} = \boldsymbol{s} + \boldsymbol{n}$$

which is the problem studied in Section 11.2. Thus, if $\boldsymbol{\Gamma}$ were invertible, we could use the methods of that section to compute $\boldsymbol{\Sigma}_s$ and then write $\boldsymbol{\Sigma}_s = \boldsymbol{\Gamma}^\dagger \boldsymbol{\Sigma}_\rho \boldsymbol{\Gamma}$. However, we do not require that $\boldsymbol{\Gamma}$ is invertible. Both because of the inclusion of $\boldsymbol{\Gamma}$ and because $\boldsymbol{\Sigma}_\rho$ is constrained to be diagonal, the problem differs from the problem studied in Section 11.3.

The probability density function is

$$p(\boldsymbol{v}) = \frac{1}{\sqrt{\det(2\pi \boldsymbol{K})}} e^{-\frac{1}{2}\boldsymbol{v}^\dagger \boldsymbol{K}^{-1} \boldsymbol{v}}.$$

This leads to the loglikelihood function

$$\Lambda(\boldsymbol{\Sigma}) = -\log \det \boldsymbol{K} - \boldsymbol{v}^\dagger \boldsymbol{K}^{-1} \boldsymbol{v}.$$

The maximum-likelihood estimate then is

$$\widehat{\boldsymbol{\Sigma}}_\rho = \operatorname{argmax}_{\boldsymbol{\Sigma}} \left[-\log\det \boldsymbol{K} - \boldsymbol{v}^\dagger \boldsymbol{K}^{-1} \boldsymbol{v} \right],$$

and the range of the argmax is over all $\boldsymbol{\Sigma}$ that are diagonal.

Because this argmax problem apparently has no closed-form solution, we will describe an iterative algorithm developed by using the method of alternating maximization. To form this algorithm, the actual data $\boldsymbol{v}$ is replaced by the complete data $(\boldsymbol{\rho}, \boldsymbol{n})$. The mapping from the complete data to the actual data is given by

$$\boldsymbol{v} = \boldsymbol{\Gamma}^\dagger \boldsymbol{\rho} + \boldsymbol{n}.$$

The probability density function for the complete data is

$$p(\boldsymbol{\rho}, \boldsymbol{n}) = \frac{1}{\sqrt{\det(2\pi \boldsymbol{\Sigma}_\rho)}} \mathrm{e}^{-\frac{1}{2}\boldsymbol{\rho}^\dagger (\boldsymbol{\Sigma}_\rho)^{-1} \boldsymbol{\rho}} \frac{1}{\sqrt{\det(2\pi N_0 \boldsymbol{I})}} \mathrm{e}^{-\frac{1}{2}\boldsymbol{n}^\dagger \boldsymbol{\Sigma}_n^{-1} \boldsymbol{n}}.$$

Where $\boldsymbol{\Sigma}_m = N_0 I_0$. The complete-data loglikelihood statistic is

$$\Lambda_{cd}(\boldsymbol{\Sigma}_\rho) = -\log\det \boldsymbol{\Sigma}_\rho - \boldsymbol{\rho}^\dagger \boldsymbol{\Sigma}_\rho^{-1} \boldsymbol{\rho}$$

where constants have been dropped. Because $\boldsymbol{\Sigma}_\rho$ is a diagonal matrix, this reduces to

$$\Lambda_{cd}(\boldsymbol{\Sigma}_\rho) = -\sum_{i=1}^{I} \log \sigma_i^2 - \sum_{i=1}^{I} \frac{|\rho_i|^2}{\sigma_i^2}.$$

Suppose that an estimate, $\boldsymbol{\Sigma}^{old}$, is given for the unknown diagonal matrix $\boldsymbol{\Sigma}_\rho$. The expectation of the complete-data loglikelihood, given the actual data and $\boldsymbol{\Sigma}^{old}$, is

$$\mathrm{E}\left[\Lambda_{cd}(\boldsymbol{\Sigma}_\rho) | \boldsymbol{v}, \boldsymbol{\Sigma}^{old}\right] = -\sum_{i=1}^{I} \log \sigma_i^2 - \sum_{i=1}^{I} \frac{1}{\sigma_i^2} \mathrm{E}[|\rho_i|^2 | \boldsymbol{v}, \boldsymbol{\Sigma}^{old}].$$

The expectation on the right side will take some work to evaluate. This is a standard estimation-theoretic calculation similar to the one studied in Section 11.3 in Theorem 11.3.3 and Corollary 11.3.4. We state it as a general proposition.

Proposition 11.8.1 *Let $\boldsymbol{v} = \boldsymbol{\Gamma}^\dagger \boldsymbol{\rho} + \boldsymbol{n}$ where $\boldsymbol{\rho}$ and $\boldsymbol{n}$ are independent, gaussian vector random variables with covariance matrices $\boldsymbol{\Sigma}_\rho$ and $N_0 \boldsymbol{I}$, respectively. Then*

$$\mathrm{E}[|\rho_i|^2 | \boldsymbol{v}, \boldsymbol{\Sigma}_\rho] = \sigma_i^2 - \sigma_i^4 [\boldsymbol{\Gamma} \boldsymbol{K}^{-1} \boldsymbol{\Gamma}^\dagger - \boldsymbol{\Gamma} \boldsymbol{K}^{-1} \boldsymbol{S} \boldsymbol{K}^{-1} \boldsymbol{\Gamma}^\dagger]_{ii}$$

where

$$\boldsymbol{K} = \boldsymbol{\Gamma}^\dagger \boldsymbol{\Sigma}_\rho \boldsymbol{\Gamma} + N_0 \boldsymbol{I}$$

and

$$\boldsymbol{S} = \boldsymbol{v}\boldsymbol{v}^\dagger.$$

Proof: As in the proof of Theorem 11.3.3, the method of proof is to use the Bayes formula

$$p(\boldsymbol{\rho}|\boldsymbol{v}) = \frac{p(\boldsymbol{\rho}, \boldsymbol{v})}{p(\boldsymbol{v})} = \frac{p(\boldsymbol{\rho})p(\boldsymbol{v}|\boldsymbol{\rho})}{p(\boldsymbol{v})}$$

to compute the probability density function $p(\boldsymbol{\rho}|\boldsymbol{v})$ and to then recognize terms. Because $p(\boldsymbol{\rho}|\boldsymbol{v})$ is gaussian, it must have the form

$$p(\boldsymbol{\rho}|\boldsymbol{v}) = \frac{1}{\det(2\pi\boldsymbol{\Sigma}_{\rho|v})} \mathrm{e}^{-\frac{1}{2}(\boldsymbol{\rho}-\overline{\boldsymbol{\rho}})^{\dagger}\boldsymbol{\Sigma}_{\rho|v}^{-1}(\boldsymbol{\rho}-\overline{\boldsymbol{\rho}})}.$$

Recall that

$$p(\boldsymbol{v}) = \frac{1}{\sqrt{\det(2\pi\boldsymbol{K})}} \mathrm{e}^{-\frac{1}{2}\boldsymbol{v}^{\dagger}\boldsymbol{K}^{-1}\boldsymbol{v}}$$

$$p(\boldsymbol{\rho}) = \frac{1}{\sqrt{\det(2\pi\boldsymbol{\Sigma}_{\rho})}} \mathrm{e}^{-\frac{1}{2}\boldsymbol{\rho}^{\dagger}\boldsymbol{\Sigma}_{\rho}^{-1}\boldsymbol{\rho}}$$

and

$$p(\boldsymbol{v}|\boldsymbol{\rho}) = \frac{1}{\sqrt{\det(2\pi\boldsymbol{\Sigma}_{n})}} \mathrm{e}^{-\frac{1}{2}(\boldsymbol{v}-\boldsymbol{\Gamma}\boldsymbol{\rho})^{\dagger}\boldsymbol{\Sigma}_{n}^{-1}(\boldsymbol{v}-\boldsymbol{\Gamma}\boldsymbol{\rho})}$$

where $\boldsymbol{\Sigma}_n = N_0\boldsymbol{I}$. Therefore,

$$p(\boldsymbol{\rho}|\boldsymbol{v}) \sim \frac{\mathrm{e}^{-\frac{1}{2}\boldsymbol{\rho}^{\dagger}\boldsymbol{\Sigma}_{\rho}^{-1}\boldsymbol{\rho}}\,\mathrm{e}^{-\frac{1}{2}(\boldsymbol{v}-\boldsymbol{\Gamma}\boldsymbol{\rho})^{\dagger}\boldsymbol{\Sigma}_{n}^{-1}(\boldsymbol{v}-\boldsymbol{\Gamma}\boldsymbol{\rho})}}{\mathrm{e}^{-\frac{1}{2}\boldsymbol{v}^{\dagger}(\boldsymbol{\Gamma}^{\dagger}\boldsymbol{\Sigma}_{\rho}\boldsymbol{\Gamma}+\boldsymbol{\Sigma}_{n})^{-1}\boldsymbol{v}}}.$$

There is no need to compute the proportionality constant separately because $p(\boldsymbol{\rho}|\boldsymbol{v})$ must match the standard form of a gaussian probability density function. Next, we collect terms in the exponent to write

$$\begin{aligned}(\boldsymbol{\rho}-\bar{\boldsymbol{\rho}})^{\dagger}\boldsymbol{\Sigma}_{\rho|v}^{-1}(\boldsymbol{\rho}-\bar{\boldsymbol{\rho}}) = \boldsymbol{\rho}^{\dagger}(\boldsymbol{\Sigma}_{\rho}^{-1} + \boldsymbol{\Gamma}^{\dagger}\boldsymbol{\Sigma}_{n}^{-1}\boldsymbol{\Gamma})\boldsymbol{\rho}& \\ + \boldsymbol{v}^{\dagger}\left(\boldsymbol{\Sigma}_{n}^{-1} - (\boldsymbol{\Gamma}^{\dagger}\boldsymbol{\Sigma}_{\rho}\boldsymbol{\Gamma} + \boldsymbol{\Sigma}_{n})^{-1}\right)\boldsymbol{v}& \\ - \boldsymbol{v}^{\dagger}\boldsymbol{\Sigma}_{n}^{-1}\boldsymbol{\Gamma}\boldsymbol{\rho} - \boldsymbol{\rho}^{\dagger}\boldsymbol{\Gamma}^{\dagger}\boldsymbol{\Sigma}_{n}^{-1}\boldsymbol{v}&\end{aligned}$$

from which the terms $\boldsymbol{\Sigma}_{\rho|v}$ and $\bar{\boldsymbol{\rho}}$ can be readily recognized to be

$$\boldsymbol{\Sigma}_{\rho|v} = (\boldsymbol{\Sigma}_{\rho}^{-1} + \boldsymbol{\Gamma}^{\dagger}\boldsymbol{\Sigma}_{n}^{-1}\boldsymbol{\Gamma})^{-1}$$

and

$$\bar{\boldsymbol{\rho}} = \boldsymbol{\Sigma}_{s|v}\boldsymbol{\Gamma}^{\dagger}\boldsymbol{\Sigma}_{n}^{-1}\boldsymbol{v} = -(\boldsymbol{\Sigma}_{\rho}^{-1} + \boldsymbol{\Gamma}^{\dagger}\boldsymbol{\Sigma}_{n}^{-1}\boldsymbol{\Gamma})^{-1}\boldsymbol{\Gamma}^{\dagger}\boldsymbol{\Sigma}_{n}^{-1}\boldsymbol{v}.$$

Alternatively these can be rewritten as

$$\begin{aligned}\boldsymbol{\Sigma}_{\rho|v} &= \boldsymbol{\Sigma}_{n}\left(\boldsymbol{\Sigma}_{n}^{-1} - (\boldsymbol{\Sigma}_{s} + \boldsymbol{\Sigma}_{n})^{-1}\right)\boldsymbol{\Sigma}_{n} \\ \bar{\boldsymbol{\rho}} &= -\boldsymbol{\Sigma}_{s}\boldsymbol{\Gamma}^{\dagger}\left(\boldsymbol{\Gamma}\boldsymbol{\Sigma}_{s}\boldsymbol{\Gamma}^{\dagger} + \boldsymbol{\Sigma}_{n}\right)^{-1}\boldsymbol{v}.\end{aligned}$$

The alternative forms can be related by using $\boldsymbol{\Sigma}_s = \boldsymbol{\Gamma}^\dagger \boldsymbol{\Sigma}_\rho \boldsymbol{\Gamma}$. By the standard properties of moments,

$$\begin{aligned}
\mathrm{E}[\rho_i^2 | \boldsymbol{v}, \boldsymbol{\Sigma}_n] &= \mathrm{E}[|\rho_i - \overline{\rho}_i|^2 + 2\rho_i\overline{\rho}_i - \overline{\rho}_i^2 | \boldsymbol{v}, \boldsymbol{\Sigma}] \\
&= \mathrm{E}[|\rho_i - \overline{\rho}_i|^2 | \boldsymbol{v}, \boldsymbol{\Sigma}] + \overline{\rho}_i^2 = \boldsymbol{\Sigma}_{\rho|v} + \overline{\rho}_i^2 \\
&= (\boldsymbol{\Sigma}_\rho^{-1} + \boldsymbol{\Gamma}^\dagger \boldsymbol{\Sigma}_n^{-1} \boldsymbol{\Gamma})^{-1} + (\boldsymbol{\Sigma}_\rho^{-1} + \boldsymbol{\Gamma}^\dagger \boldsymbol{\Sigma}_n^{-1} \boldsymbol{\Gamma})^{-1} \boldsymbol{v}\boldsymbol{v}^\dagger \left(\boldsymbol{\Sigma}_\rho^{-1} + \boldsymbol{\Gamma}^\dagger \boldsymbol{\Sigma}_n^{-1} \boldsymbol{\Gamma}\right)^{-1}.
\end{aligned}$$

We now are ready to describe the algorithm by the following proposition. □

Proposition 11.8.2 *The maximum-likelihood variance image is computed from* $\boldsymbol{S} = \boldsymbol{v}\boldsymbol{v}^\dagger$ *by the iteration*

$$\sigma_i^{2(new)} = \sigma_i^{2(old)} - (\sigma_i^{2(old)})^2 \left[\boldsymbol{\Gamma} \boldsymbol{K}^{-1} \boldsymbol{\Gamma}^\dagger - \boldsymbol{\Gamma} \boldsymbol{K}^{-1} \boldsymbol{S} \boldsymbol{K}^{-1} \boldsymbol{\Gamma}^\dagger\right]_{ii}$$

where

$$\boldsymbol{K} = \boldsymbol{\Gamma}^\dagger \boldsymbol{\Sigma}_\rho^{old} \boldsymbol{\Gamma} + N_0 \boldsymbol{I},$$

and $\boldsymbol{\Sigma}_\rho^{old}$ *is the diagonal matrix with* $\sigma_i^{2(old)}$ *as the* i*th diagonal element.*

Proof. The maximum-likelihood estimate from the complete data $(\boldsymbol{\rho}, \boldsymbol{n})$ is

$$\widehat{\sigma_i^2} = [\boldsymbol{\rho}\boldsymbol{\rho}^\dagger]_{ii}.$$

Because $\boldsymbol{\rho}$ is unknown, the right side of this equation is replaced by the update for expectation $\mathrm{E}[\rho_i^2 | \boldsymbol{v}, \boldsymbol{\Sigma}_\rho]$. Now use the identity

$$(\boldsymbol{\Sigma}_\rho^{-1} + \boldsymbol{\Gamma}^\dagger \boldsymbol{\Sigma}_n^{-1} \boldsymbol{\Gamma})^{-1} = \boldsymbol{\Sigma}_\rho - \boldsymbol{\Sigma}_\rho \boldsymbol{\Gamma} (\boldsymbol{\Gamma}^\dagger \boldsymbol{\Sigma}_\rho \boldsymbol{\Gamma} + \boldsymbol{\Sigma}_n)^{-1} \boldsymbol{\Gamma}^\dagger \boldsymbol{\Sigma}_\rho$$

to write

$$\boldsymbol{\Sigma}_{\rho|v} + \bar{\boldsymbol{\rho}} \bar{\boldsymbol{\rho}}^\dagger = \boldsymbol{\Sigma}_\rho - \boldsymbol{\Sigma}_\rho (\boldsymbol{\Gamma} \boldsymbol{K}^{-1} \boldsymbol{\Gamma}^\dagger - \boldsymbol{\Gamma} \boldsymbol{K}^{-1} \boldsymbol{S} \boldsymbol{K}^{-1} \boldsymbol{\Gamma}^\dagger) \boldsymbol{\Sigma}_\rho$$

where $\boldsymbol{K} = \boldsymbol{\Gamma}^\dagger \boldsymbol{\Sigma}_\rho \boldsymbol{\Gamma} + \boldsymbol{\Sigma}_n$ and $\boldsymbol{S} = \boldsymbol{v}\boldsymbol{v}^\dagger$. Thus, because $\boldsymbol{\Sigma}_\rho$ is constrained to be diagonal, the update for the new diagonal element is

$$\sigma_s^2 \leftarrow \sigma_s^2 - \sigma_s^2 [\boldsymbol{\Gamma} \boldsymbol{K}^{-1} \boldsymbol{\Gamma}^\dagger - \boldsymbol{\Gamma} \boldsymbol{K}^{-1} \boldsymbol{S} \boldsymbol{K}^{-1} \boldsymbol{\Gamma}^\dagger]_{ii} \sigma_i^2$$

as was to be proved. □

11.9 Regularization

The maximum-likelihood principle is a powerful principle of inference, and the expectation-maximization method of Dempster, Laird, and Rubin provides a powerful tool to implement this principle. As is often the case with powerful tools, the tool may be too strong for a problem at hand. Consequently, maximum-likelihood image formation has a potential flaw that is subtle. We subjectively expect an image to have a satisfactory combination of crispness and smoothness, and we usually regard less detail in an image to be more likely than more detail, but this presumption is difficult to quantify. Whereas most computational problems become more precise as the variables are more finely quantized, maximum-likelihood imaging can become less precise because the algorithm will create unnatural details in the image to account for small inaccuracies in the data. The task of augmenting the computation with prior expectations in order to impose some degree of order or smoothness in the image is called *regularization*. When the prior expectation is in the form of a prior probability distribution, this becomes maximum posterior estimation as discussed in Section 11.4.

An image on continuous space $\boldsymbol{R}^2$ has an infinite number of degrees of freedom and cannot be determined by a finite set of data. An elementary and commonly-used way to regularize an image-formation computation is by *pixelization*. The size of the pixels should be small enough to obtain a good resolution, yet not so small that artificial details are created.

A standard method of regularization is the method of *penalties*. The method of penalties adds an empirical penalty term to the negative loglikelihood and then finds the minimum of this sum. Thus, if $\Lambda(\gamma)$ is the negative loglikelihood, the constrained maximum-likelihood estimate is

$$\widehat{\gamma} = \text{argmin}_{\gamma}[\Lambda(\gamma) + c(\gamma)]$$

where $c(\gamma)$ is the empirical penalty term. A quadratic constraint, sometimes called an energy constraint, is often considered to be a good choice for a penalty constraint. Thus, the maximum-likelihood estimate under the quadratic penalty constraint is

$$\widehat{\gamma} = \text{argmin}_{\gamma}[\Lambda(\gamma) + \alpha^2\|\gamma\|^2]$$

where $\alpha^2\|\gamma\|^2$ is the quadratic penalty term, written in terms of a constant α^2 and the energy in the image, denoted $\|\gamma\|^2$.

An elementary example of the use of a penalty constraint is the estimation of the value of a vector $\boldsymbol{x}$ when given a vector measurement $\boldsymbol{y}$ satisfying

$$\boldsymbol{y} = \boldsymbol{c}\boldsymbol{x} + \boldsymbol{n}$$

where $\boldsymbol{n}$ is a gaussian random vector of zero mean and covariance matrix $\boldsymbol{\Sigma}$. The conditional gaussian probability density function is

$$p(\boldsymbol{y}|\boldsymbol{x}) = \frac{1}{\sqrt{\det(2\pi\boldsymbol{\Sigma})}}\,\mathrm{e}^{-\frac{1}{2}(\boldsymbol{y}-\boldsymbol{c}\boldsymbol{x})^{\dagger}\boldsymbol{\Sigma}^{-1}(\boldsymbol{y}-\boldsymbol{c}\boldsymbol{x})}$$

and the (negative) loglikelihood function is

$$\Lambda(\boldsymbol{x}) = (\boldsymbol{y} - \boldsymbol{c}\boldsymbol{x})^{\dagger}\boldsymbol{\Sigma}^{-1}(\boldsymbol{y} - \boldsymbol{c}\boldsymbol{x}).$$

Then

$$\hat{x} = \text{argmin}_x \left[(y - cx)^\dagger \Sigma^{-1}(y - cx) + \alpha^2 \|x\|^2\right]$$

under a quadratic penalty constraint.

This likelihood estimate for a deterministic vector parameter x under a quadratic penalty constraint is identical to a bayesian estimator with a gaussian prior, which arises in a much different way. The bayesian estimator is discussed in Section 11.4. For this estimator, consider x to be a random vector with an independent gaussian distribution of variance σ^2 as a prior. For this formulation,

$$\begin{aligned}\hat{x} &= \text{argmin}_x[-\log p(x|y)] \\ &= \text{argmin}_x[-\log p(y|x)p(x)] \\ &= \text{argmin}_x \left[(y - cx)^\dagger \Sigma^{-1}(y - cx) + \sigma^2 \|x\|^2\right]\end{aligned}$$

which is the same as before with the variance σ^2 replacing the free parameter α^2. Hence, choosing a quadratic penalty function gives the same estimate as was obtained by placing a prior on x so, in this sense, the two estimators are equivalent.

Another method, called the method of sieves, can be used to suppress the noise artifacts. The method constrains the set of allowed images to be from a chosen sequence of smooth images called a *Grenander sieve*. The sieve is defined as a sequence of sets of functions that depends on the size of the data set. For a larger data set, the sieve contains more functions. Because the size of the sieve is allowed to grow with the number of data measurements, it is hoped that the image estimate converges to the true image as the size of the data set goes to infinity.

11.10 Notions of equivalence

It is natural to ask how the maximum-likelihood problem is related to the minimum-discrimination problem (or the maximum-entropy problem). Often the same probability distribution solves both criteria. In this sense, the maximum-likelihood principle and the minimum-discrimination principle (or the maximum-entropy principle) are equivalent. These problems seem to be very different, and are justified by very different arguments, yet we will find at a deeper level that there are similarities. We will first show that the minimum-discrimination principle and the maximum-likelihood principle are connected by the law of large numbers and, to this extent, are equivalent.

Let x be a discrete random variable described by the prior probability vector q so that x takes value x_j with probability q_j. The random variable x is the complete data. Let y be the actual data obtained by a measurement described by $y = h(x)$. The actual data y is related to the complete data x by the function $y = h(x)$. The task is to estimate the value of x when given the measurement y. Because $h(x)$ is a many-to-one function, the measurement $\boldsymbol{y}$ does not determine a unique $\boldsymbol{x}$, but only a set of possible $\boldsymbol{x}$, so $\boldsymbol{y}$ is also called the incomplete data. Let $p(\boldsymbol{x})$ be a posterior probability distribution on $\boldsymbol{x}$. Then we have the constraint on $p(\boldsymbol{x})$ that

$$\sum_{\boldsymbol{x}:h(\boldsymbol{x})=\boldsymbol{y}} p(\boldsymbol{x}) = 1.$$

Theorem 11.10.1 *Let $\boldsymbol{q}$ be a prior on a set $\mathcal{X}$ and let $\boldsymbol{p}$ be any probability distribution on $\mathcal{X}$ that is nonzero only on $\boldsymbol{x}$ satisfying $h(\boldsymbol{x}) = \boldsymbol{y}$. The discrimination $L(\boldsymbol{p}, \boldsymbol{q})$ is maximized over such $\boldsymbol{p}$ by*

$$p(\boldsymbol{x}) = \frac{q(\boldsymbol{x})}{\sum_{\boldsymbol{x}:h(\boldsymbol{x})=\boldsymbol{y}} q(\boldsymbol{x})}.$$

Proof: Let $\mathcal{X}'$ be the set of $\boldsymbol{x}$ satisfying the constraint $h(\boldsymbol{x}) = \boldsymbol{y}$. Then

$$\widehat{p}(\boldsymbol{x}) = \operatorname{argmax}_{p(x)} \sum_{\boldsymbol{x}\in\mathcal{X}'} p(\boldsymbol{x}) \log \frac{p(\boldsymbol{x})}{q(\boldsymbol{x})} + \lambda \left[\sum_{\boldsymbol{x}\in\mathcal{X}'} p(\boldsymbol{x}) - 1 \right].$$

Setting the partial derivatives equal to zero shows that $p(\boldsymbol{x})/q(\boldsymbol{x})$ must be a constant for $\boldsymbol{x} \in \mathcal{X}'$, and the constraint requires that $p(\boldsymbol{x}) = 0$ for $\boldsymbol{x} \notin \mathcal{X}'$. The conclusion of the theorem follows. □

Figure 11.4 illustrates an example of the theorem for a scalar measurement on the real line. On the set $\mathcal{X}'$, $p(x)$ is a scaled version of $q(x)$. Elsewhere it is zero.

The significance of the theorem is that the minimizing $p(\boldsymbol{x})$ is the same as the conditional on $\boldsymbol{x}$ when given $\boldsymbol{y}$. To verify this formally, recall that

$$q(\boldsymbol{x})p(\boldsymbol{y}|\boldsymbol{x}) = p(\boldsymbol{y})q(\boldsymbol{x}|\boldsymbol{y}).$$

If $\boldsymbol{y} = h(\boldsymbol{x})$, then $p(\boldsymbol{y}|\boldsymbol{x}) = 1$ for that $\boldsymbol{y}$ and $p(\boldsymbol{y}|\boldsymbol{x}) = 0$ for all other $\boldsymbol{y}$. Because $p(\boldsymbol{y}) = \sum_{\boldsymbol{x}:h(\boldsymbol{x})=\boldsymbol{y}} q(\boldsymbol{x})$ we conclude that

$$q(\boldsymbol{x}|\boldsymbol{y}) = \frac{q(\boldsymbol{x})}{\sum_{\boldsymbol{x}:h(\boldsymbol{x})=\boldsymbol{y}} q(\boldsymbol{x})}.$$

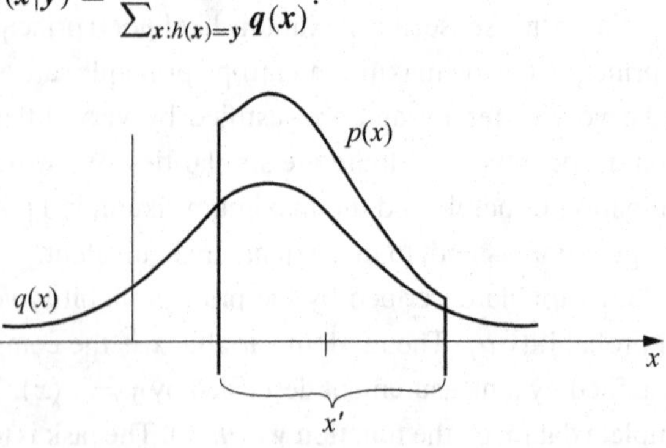

Figure 11.4 Constraining a probability distribution

This is the same as $p(\boldsymbol{x})$ as is given in Theorem 11.10.1. Once again, as in the example of Figure 11.4, the probability distribution q(x), after imposing the condition that x $\in \mathcal{X}$, is replaced by a new probability $p(x)$ which is simply a multiple of $q(x)$ on that interval and zero elsewhere.

Theorem 11.10.2 *Let* $\boldsymbol{q}$ *be a prior on a set* $\mathcal{X}$ *and let* $\boldsymbol{p}$ *be any probability distribution on* $\mathcal{X}$ *satisfying the constraint* $E[h(\boldsymbol{x})] = \boldsymbol{y}$. *Then* $L(\boldsymbol{p}, \boldsymbol{q})$ *is minimized over such p by*

$$p(\boldsymbol{x}) = \frac{q(\boldsymbol{x})\mathrm{e}^{-\sum_j s_j h_j(\boldsymbol{x})}}{\sum_{\boldsymbol{x}} q(\boldsymbol{x})\mathrm{e}^{-\sum_j s_j h_j(\boldsymbol{x})}}$$

where the lagrange multipliers s_j are chosen to satisfy the constraint.

Proof: Introduce appropriate lagrange multipliers and write

$$\frac{\partial}{\partial p(\boldsymbol{x})}\left[\sum_{\boldsymbol{x}} p(\boldsymbol{x}) \log \frac{p(\boldsymbol{x})}{q(\boldsymbol{x})} + \lambda\left(\sum_{\boldsymbol{x}} p(\boldsymbol{x}) - 1\right) + \sum_j s_j \left(\sum_{\boldsymbol{x}} p(\boldsymbol{x}) h_j(\boldsymbol{x}) - y_j\right)\right] = 0.$$

which leads to the expression stated in the theorem. □

A second notion of equivalence is an implicit equivalence between probability distributions and distance measures. Thus, it is common practice to use euclidean distance to measure the discrepancy between two vectors or functions. This criterion is regarded as quite reasonable. It is also common practice to use a gaussian probability density function as a surrogate when the true probability density function is not known. This is perhaps regarded as a bit less reasonable. It is often partially justified by the loose statement that many probability densities are actually gaussian, often as a consequence of the central limit theorem. It might also be justified by the heuristic statement that this choice is maximally noncommittal or by the formal statement that the gaussian probability density function maximizes the entropy for a specified mean and variance. We shall provide a third justification, one that may be more compelling. We shall show that accepting that the minimum euclidean distance as a criterion of estimation optimality, which is a widely accepted criterion, is equivalent to assuming a gaussian probability density function. We already saw an instance of this in Section 11.9. There we saw that using a quadratic penalty for regularization is equivalent to assuming a gaussian prior.

Suppose that we are given a polynomial $y = \sum_{i=0}^{I-1} a_i x^i$ where the coefficients are unknown and are to be estimated from a set of noisy measurements (x_j, y_j) where $y_j = \sum_{i=0}^{I-1} a_i x_j^i + n_i$ where the n_i are random variables with unknown joint distributions. Suppose that for want of a better model, one assumes random variables are gaussian, with independent noise components of equal variance. Then the maximum of the loglikelihood function is the same as minimum euclidean distance. Thus, choosing to minimize euclidean distance is the same as assuming a memoryless gaussian distribution.

A parallel sequence of statements can be made for the poisson distribution. It is common practice to use a poisson probability distribution for a point process when the true distribution is not known. In particular, we shall argue that accepting the Csiszár discrimination as an appropriate measure of distance in the space of nonnegative functions is tantamount to accepting a poisson distribution as appropriate for the data.

Thus, the Csiszár discrimination is written

$$L(y(x), \lambda(x)) = \sum_x \left[y(x) \log \frac{y(x)}{\lambda(x)} + \lambda(x) - y(x) \right]$$

$$= \sum_x [y(x) \log y(x) - \lambda(x)] - \sum_x [y(x) \log \lambda(x) - y(x)].$$

Then

$$\mathrm{e}^{-L(y(x),\lambda(x))} = \frac{1}{Z} \prod_x \mathrm{e}^{-\lambda(x)} \lambda(x)^{y(x)}$$

where Z does not depend on $\lambda(x)$. This formula is to be compared with the poisson distribution

$$p_\lambda(y) = \frac{\mathrm{e}^{-\lambda} \lambda^y}{y!}$$

which has the same dependence on λ. Essentially the same formula arises in two different ways. The Csiszár discrimination is appropriate when treating nonnegative functions, in general, in a nonprobabilistic space. The poisson distribution is appropriate for a point process in an probabilistic setting. Because the same formulas arise, the same computational procedures may apply. For example, the Richardson–Lucy algorithm, in addition to minimizing the Csiszár discrimination, can be seen as an algorithm of the Dempster–Laird–Rubin type.

11.11 The Dempster–Laird–Rubin method

The maximum-likelihood principle is well accepted in estimation theory for estimating a vector parameter, γ. However, except in very simple problems, finding the maximum of the likelihood function $\Lambda(\gamma)$ is not analytically tractable. Maximum-likelihood problems in image formation can be very complex. For example, γ may represent an array of 500 by 500 pixels constituting an image to be estimated. Then $\Lambda(\gamma)$ is a function of 500^2 variables. The *Dempster–Laird–Rubin method*[5] (also known as the *expectation-maximization algorithm*) constructs an iterative algorithm that determines a maximum-likelihood estimate of the vector parameter γ from measured data. In the

[5] We prefer to call this a method for constructing algorithms rather than itself an algorithm. Each algorithm constructed by this method is a computational procedure for a unique application and is specific to that application.

usual formulation, the method evaluates an expectation, then computes a maximization, and then repeats this sequence of steps a number of times. In most cases, after the equations of the expectation step and the maximization step are derived, they can easily be merged into a single equation for the combined iteration. We have already discussed examples of such iterative algorithms, in Sections 11.3 and 11.8. Here we recapitulate these methods in a more formal discussion.

The development of the Dempster–Laird–Rubin method takes place in the context of two spaces. The space in which the available data is defined is called the *actual data space* or the *incomplete data space*, the latter name referring to the fact that the available data are not suitable to determine an answer. A larger and somewhat arbitrary data space in which a more tractable set of data can be considered is called the *complete data space*. The idea is that, if the data in the complete data space were available, it would be analytically tractable to maximize the likelihood function. Thus one pretends that data in the complete data space are available. In practice, an estimate of the complete data is available from a previous iteration. One first estimates the likelihood function as a function of the complete data, conditional on the actual data and the solution of the previous iteration. In other words, one extrapolates from the likelihood function of the actual data space to a likelihood function on the complete data space.

In this way, a recursive algorithm is derived. First, estimate the loglikelihood function from the currently computed value of the unknown parameter γ and the data in the actual data space, then maximize the likelihood to find a new γ, which is a simple computation in the complete data space. This pair of steps is repeated until a stable point is reached.

To use the Dempster–Laird–Rubin method, one must define the complete data space. To find a suitable choice we may call for insight gained by experience. There may be several reasonable choices that work, but one choice may be considerably easier computationally or analytically. On the other hand, it may be that no suitable choice is evident. The formalism of the Dempster–Laird–Rubin method also allows some general statements about convergence to be made.

Take the logarithm of Bayes' formula

$$p(r;\gamma) = \frac{p(r|c)p(c;\gamma)}{p(c|r;\gamma)}$$

which we have written with an unknown parameter γ. Then

$$\log p(r;\gamma) = \log p(r|c) + \log p(c;\gamma) - \log p(c|r;\gamma).$$

Then take the expectation of both sides of this equation with probability distribution $p(c|r;\widetilde{\gamma})$ to obtain

$$\mathrm{E}[\log p(r;\gamma)] = \mathrm{E}[\log p(r|c)] + \mathrm{E}[\log p(c;\gamma)] - \int_{-\infty}^{\infty} p(c|r;\widetilde{\gamma}) \log p(c|r;\gamma)\,\mathrm{d}c$$

where the parameter $\widetilde{\gamma}$ is arbitrary. The left side does not depend on c, so it is equal to

$\log p(r;\gamma)$. The second term on the right will be denoted $Q(\gamma|\widetilde{\gamma})$. Therefore,

$$\log p(r,\gamma) = \mathrm{E}[\log p(r|c)] + Q(\gamma|\widehat{\gamma}) - \int_{-\infty}^{\infty} p(c|r;\widetilde{\gamma}) \log p(c|r;\gamma)\, \mathrm{d}c.$$

During the iterations, the right side changes as follows

$$\log p(r,\gamma^{(k+1)}) - \log p(r,\gamma^{(k)}) = Q(\gamma^{(k+1)}|\widetilde{\gamma}) - Q(\gamma^{(k)}|\widetilde{\gamma}) + \int_{-\infty}^{\infty} p(c|r,\gamma^{(k)}) \log \frac{p(c|r,\gamma)}{p(c|r,\gamma^{(k+1)})}\, \mathrm{d}c.$$

The task of maximum-likelihood estimation is to maximize the left side by choice of γ. The method is to alternate between maximizing the right side with respect to γ and estimating c from the current value of γ.

Problems

11.1 Prove that the Kullback discrimination

$$L(p(x), q(x)) = \int_{-\infty}^{\infty} p(x) \log \frac{p(x)}{q(x)}\, \mathrm{d}x$$

is convex in the first variable and convex in the second variable.

11.2 **a.** Prove that $\log_e x \leq x - 1$.

b. Prove that the Csiszár discrimination satisfies

$$L(a(x), b(x)) \geq 0.$$

11.3 A term of the form $\log\left(\sum_i q_i\right)$ is often difficult to handle analytically because the sum is inside the logarithm. Prove that, if the q_i are all positive,

$$-\log\left(\sum_i q_i\right) = \min_{p} \sum_i p_i \log \frac{p_i}{q_i}$$

where $\boldsymbol{p}$ must be a probability vector. This formula allows one to pull the sum outside the logarithm at the cost of a superfluous minimization. This is a discrete form of the convex decomposition lemma.

11.4 A poisson process is a point process with the probability distribution

$$P_N = \frac{(\lambda T)^{N(T)}}{N(T)!} \mathrm{e}^{-\lambda T}$$

where $N(T)$ is the number of events observed in the observation interval T. If λ is unknown, then P_N becomes the likelihood function

$$\Lambda(\lambda) = \frac{(\lambda T)^{N(T)}}{N(T)!} \mathrm{e}^{-\lambda T}$$

in the parameter λ. The unknown λ can be estimated by maximizing this function.

Two independent poisson processes have the arrival rates a and b. The arrival rate b is known, and the arrival rate a is unknown. The points of the two processes are pooled to form the single poisson process with the arrival rate $a+b$. The occurrence times $(t_1, t_2, t_3, \ldots, t_N)$ of the pooled process are the observed data.

a. Find analytically the maximum-likelihood estimator of parameter a when given the observed data $(t_1, t_2, t_3, \ldots, t_N)$.

b. Use the Dempster–Laird–Rubin method to construct an iterative algorithm to estimate a. What is the limit point of this iteration?

11.5 Show that the Csiszár discrimination does not satisfy the triangle inequality.

11.6 Let

$$p(v) = \frac{1}{\sqrt{\det 2\pi \boldsymbol{\Sigma}}} e^{-(v-\bar{v})^{\dagger}\boldsymbol{\Sigma}^{-1}(v-\bar{v})}.$$

Calculate the mean $\mathrm{E}[v]$ and the covariance matrix $\mathrm{E}[vv^T] - \mathrm{E}[v]\mathrm{E}[v^T]$.

11.7 A maximum-likelihood estimate is known to be *asymptotically unbiased* in the size or quality of the data set. This means that the expected error goes to zero as the amount of data goes to infinity. Show that the problem of estimating the variance of a single gaussian random variable, measured in gaussian noise, is asymptotically unbiased with respect to the signal-to-noise ratio. It will be helpful to recall that the square of a zero-mean, gaussian random variable with variance σ^2 is an exponential random variable with mean σ^2.

11.8 The discrete noisy deblurring problem is the task of estimating a deterministic image $\boldsymbol{s}$ when given a noisy blurred image, possibly complex,

$$v = \boldsymbol{Hs} + \boldsymbol{n},$$

where $\boldsymbol{n}$ is a white gaussian noise vector with the covariance matrix $\boldsymbol{N} = N_0\boldsymbol{I}$, and $\boldsymbol{H}$ is a known blurring matrix.

a. Show that the loglikelihood function for this problem is

$$\Lambda(\boldsymbol{s}) = 2\mathrm{Re}[\boldsymbol{n}^{\dagger}\boldsymbol{N}^{-1}\boldsymbol{Hs}] - \boldsymbol{s}^{\dagger}\boldsymbol{H}^{\dagger}\boldsymbol{N}^{-1}\boldsymbol{Hc}.$$

b. Give an algorithm for computing the maximum-likelihood deblurred image $\hat{\boldsymbol{s}}$.

11.9 The discrete noisy deblurring problem is the task of estimating a random image, $\boldsymbol{s}$, when given a noisy blurred image, possibly complex, given by

$$v = \boldsymbol{Hs} + \boldsymbol{n}$$

where $\boldsymbol{s}$ is a zero-mean, gaussian random process with covariance matrix $\boldsymbol{\Sigma}_s$,

$\boldsymbol{n}$ is white gaussian noise with the covariance matrix $\boldsymbol{\Sigma}_n = N_0\boldsymbol{I}$, and $\boldsymbol{H}$ is a known blurring matrix.

a. Show that the loglikelihood function for this problem is

$$\Lambda(\boldsymbol{\Sigma}_s) = -\log\det(\boldsymbol{H}^\dagger\boldsymbol{\Sigma}_s\boldsymbol{H} + N_0\boldsymbol{I}) - \boldsymbol{v}^\dagger(\boldsymbol{H}^\dagger\boldsymbol{\Sigma}_s\boldsymbol{H} + N_0\boldsymbol{I})^{-1}\boldsymbol{v}.$$

b. Given the constraint that $\boldsymbol{\Sigma}_s$ is diagonal, give an algorithm for computing the maximum-likelihood deblurred image.

11.10 Reorganize the computations of the algorithm for radar imaging of diffuse targets to improve computational efficiency. How many multiplications are required in each iteration?

11.11 **a.** Prove the *Pinsker inequality* for discrimination

$$L(p_1, p_2) \geq |p_1 - p_2|^2$$

for both discrete and continuous probability distributions.

b. Prove the inequality

$$L(p_1(x), p_2(x)) \geq \tfrac{1}{2}\int_{-\infty}^{\infty}\left(\sqrt{p_1(x)} - \sqrt{p_2(x)}\right)^2 dx.$$

11.12 **a.** Let

$$\boldsymbol{C} = \begin{bmatrix} \boldsymbol{A} & \boldsymbol{0} \\ \boldsymbol{0} & \boldsymbol{B} \end{bmatrix}$$

where $\boldsymbol{A}$ and $\boldsymbol{B}$ are square matrices. Show that $\det \boldsymbol{C} = \det \boldsymbol{A} \det \boldsymbol{B}$.

b. Let $\boldsymbol{v} = \boldsymbol{s} + \boldsymbol{n}$ where $\boldsymbol{s}$ and $\boldsymbol{n}$ are gaussian vector random processes with covariance matrices $\boldsymbol{\Sigma}_s$ and $\boldsymbol{\Sigma}_n$, respectively. Find the joint probability density function of $(\boldsymbol{s}, \boldsymbol{v})$ by evaluating $p(\boldsymbol{s})p(\boldsymbol{v}|\boldsymbol{s})$.

c. Verify this conclusion by directly computing

$$\boldsymbol{\Sigma} = \begin{bmatrix} \mathrm{E}[\boldsymbol{s}\boldsymbol{s}^\dagger] & \mathrm{E}[\boldsymbol{s}\boldsymbol{v}^\dagger] \\ \mathrm{E}[\boldsymbol{v}\boldsymbol{s}^\dagger] & \mathrm{E}[\boldsymbol{v}\boldsymbol{v}^\dagger] \end{bmatrix}.$$

11.13 Let $\boldsymbol{x}$ and $\boldsymbol{y}$ each be a gaussian vector random variable of blocklength n with the joint probability density function

$$p(\boldsymbol{x}, \boldsymbol{y}) = \frac{1}{\sqrt{\det(2\pi\boldsymbol{\Sigma})}} e^{-\frac{1}{2}|\boldsymbol{x},\boldsymbol{y}|\boldsymbol{\Sigma}^{-1}|\boldsymbol{x},\boldsymbol{y}|^\dagger}$$

where $\boldsymbol{\Sigma} = \begin{bmatrix} \boldsymbol{\Sigma}_x & \boldsymbol{R}_{xy} \\ \boldsymbol{R}_{xy} & \boldsymbol{\Sigma}_y \end{bmatrix}$, which is defined by $\boldsymbol{\Sigma} = \mathrm{E}\left[\begin{vmatrix} \boldsymbol{x} \\ \boldsymbol{y} \end{vmatrix} \begin{vmatrix} \boldsymbol{x} & \boldsymbol{y} \end{vmatrix}\right]$. From $p(\boldsymbol{x}, \boldsymbol{y})$, find expressions for the marginals $p(\boldsymbol{x})$ and $p(\boldsymbol{y})$ and the conditionals $p(\boldsymbol{x}|\boldsymbol{y})$ and $p(\boldsymbol{y}|\boldsymbol{x})$.

11.14 Develop the alternative forms

$$\Sigma_s \leftarrow \Sigma_s - (\Sigma_s^{-1} + \Sigma_n^{-1})^{-1}\Sigma_n^{-1}(\Sigma_s + \Sigma_n - vv^\dagger)\Sigma_n^{-1}(\Sigma_s^{-1} + \Sigma_n^{-1})^{-1}$$
$$\Sigma_s \leftarrow \Sigma_s - \Sigma_s(\Sigma_s + \Sigma_n)(\Sigma_s + \Sigma_n - vv^\dagger)(\Sigma_s + \Sigma_n)^{-1}\Sigma_s$$

for the iterative algorithm for estimating covariance matrices given in Section 11.3.

11.15 Let $v = s + n$ where s and n are gaussian vector random variables with zero mean and covariance matrices Σ_s and Σ_n, respectively. Find the conditional density function

$$p_{s|v}(s\,|\,v) = \frac{1}{\sqrt{\det(2\pi\Sigma_{s|v})}}\,e^{-(s-\bar{s})^\dagger\Sigma_{s|v}^{-1}(s-\bar{s})}$$

from Bayes' formula. Verify every term.

11.16 Prove the matrix identities

a.

$$\Sigma_s\Gamma^\dagger(\Gamma\Sigma_s\Gamma^\dagger + \Sigma_n)^{-1} = (\Gamma^\dagger\Sigma_n^{-1}\Gamma + \Sigma_s^{-1})^{-1}\Gamma^\dagger\Sigma_n^{-1}$$

b.

$$(\Sigma_s^{-1} + \Gamma^\dagger\Sigma_n^{-1}\Gamma)^{-1} = \Sigma_s - \Sigma_s\Gamma(\Gamma^\dagger\Sigma_s\Gamma + \Sigma_n)^{-1}\Gamma^\dagger\Sigma_s.$$

11.17 Develop the Richardson–Lucy algorithm for the case of Csiszár discrimination.

11.18 The fixed-point conditions for blind deconvolution can be developed by the calculus of variations. Replace $h(x, y)$ and $s(x, y)$ by $h'(x, y) = h(x, y) + \epsilon\eta(x, y)$ and $s'(x, y) = s(x, y) + \epsilon'\eta'(x, y)$, respectively. Then

$$L(v(x, y), h'(x, y) ** s'(x, y))$$
$$= \int_{-\infty}^{\infty}\int_{-\infty}^{\infty} v(x, y)\log\frac{v(x, y)}{h'(x, y) ** s'(x, y)}\,dx\,dy.$$

Now, set the partial derivatives with respect to ϵ and ϵ' equal to zero at the point at which both ϵ and ϵ' are equal to zero to establish the iteration.

11.19 Verify convergence of the iterative algorithm for blind deconvolution given in Section 11.6. Is there a unique fixed point? Must the algorithm converge to the correct solution?

11.20 An *Ali–Silvey distance* between two probability distributions is defined as

$$d(p_0, p_1) = f(\mathrm{E}[c(\Lambda(x))])$$

where $f(\cdot)$ is any nondecreasing function and $c(\cdot)$ is any convex function. The function $\Lambda(x)$ denotes the likelihood ratio $p_1(x)/p_0(x)$.

Show that the Kullback discrimination is an Ali–Silvey distance.

Notes

The maximum-likelihood principle is a long-standing and time-honored principle of statistics. Similarly, the entropy is an important function in the literature of informatics, having been introduced by Shannon (1948). It already had a prominent place in statistical physics. The discrimination was introduced into the statistics literature by Kullback (1959) and studied, under other names, by Shore and Johnson (1980) and Csiszár (1991). In his work, Csiszár argued that the only reasonable choices for distance measures are: quadratic distance in euclidian space, Csiszár discrimination in the space of nonnegative n-tuples, and Kullback discrimination in probability space. Miller and Snyder (1987) observed that maximum likelihood and minimum discrimination in many situations give the same answer and, in this sense, are equivalent. An important method of developing computational algorithms for obtaining maximum-likelihood images was presented by Dempster, Laird, and Rubin (1977). Convergence of the algorithms was established under general conditions by Wu (1983). A stronger statement regarding convergence under certain special conditions was established by Vardi, Shepp, and Kaufman (1985).

The Richardson (1972)–Lucy (1974) algorithm for deblurring nonnegative images was first developed in a heuristic manner, without any formal motivation or proof, and was justified experimentally. The Richardson–Lucy algorithm was rediscovered in a probabilistic formulation by Shepp and Vardi (1982) in the context of medical imaging. It was placed into a formal framework, based on minimizing the discrimination, by Snyder, Schulz, and O'Sullivan (1992). Similar ideas were reached independently by Vardi and Lee (1993). O'Sullivan (1998, 2002) presented the formal development of the Richardson–Lucy algorithm as an alternating minimization. The blind version of the deconvolution algorithm along lines similar to the Richardson–Lucy algorithm was discussed by Ayers and Dainty (1982). By introducing the convex decomposition lemma, O'Sullivan (2002) completed the fusion of alternating minimization algorithms for likelihood problems and information theory. The maximum-entropy principle was formulated by Jaynes (1957) based on the observation that any probability density function consistent with a set of constraints but with less entropy than the maximum entropy so achievable is atypical of the data.

Imaging from weak light has been studied by Snyder and Schulz (1990). Schulz and Snyder (1992) developed a maximum-likelihood iteration for the photon-differencing problem by using the Dempster, Laird, and Rubin (1977) method. Motivated by the particular characteristic of reflection at optical frequencies, Shapiro, Capron, and Horney (1981) proposed that lidar systems use a diffuse model of reflectivity in which the two-dimensional reflectivity density is a zero-mean random variable, and the variance density of this random variable is the desired image. A similar model of reflectivity had been proposed by Van Trees (1971) for microwave frequencies, with earlier work in this direction by Gaarder (1968). The use of maximum-likelihood methods for radar imaging of diffuse reflector scenes was proposed by Snyder, O'Sullivan, and Miller (1989). Iterative algorithms for this imaging problem, as formed by the expectation-maximization method, were studied by Moulin, O'Sullivan, and Snyder (1992), and by Lanterman (2000). Their method of imaging reflectivity variance was a radical departure from traditional methods for forming radar images.

The topic of imaging from point event data was first formulated for emission tomography by Rockmore and Macovski (1976) using the maximum-likelihood principle, but their formulation was intractable because they had no way to compute the maximum-likelihood estimates. A computationally tractable form was developed by Shepp and Vardi (1982), and Lange and Carson (1984), when they applied the Dempster–Laird–Rubin method to this problem. Politte and Snyder (1991) described two algorithms based on two choices of the complete data space that can be developed by using the expectation-maximization method.

The method of sieves was suggested by Grenander (1981) as a general tool of statistical inference. The method of sieves was applied to the problem of noncoherent imaging by Snyder and Miller (1985).

12 Radar search systems

A radar processor consists of a preprocessor, a detection and estimation function, and a postprocessor. On entering the receiver, a received radar signal first encounters the preprocessor, which often can be viewed as the computation of the sample cross-ambiguity function, though perhaps in a crudely approximated form. A search radar is one whose sample cross-ambiguity function typically consists of isolated peaks that are examined to detect objects in the environment and to estimate parameters associated with those objects.

A reflecting object may be made up of many individual reflecting elements, such as corners, edges, and so on. When the resolution of the radar is coarse compared with the size of the individual reflecting elements, then the single reflecting object is regarded as a point and appears as a single peak in the sample cross-ambiguity function. The search radar detects that peak, and the delay and doppler coordinates of the peak form an estimate of the delay and doppler coordinates of the reflector considered as a single object.

When the resolution of the radar is fine compared with the size of an individual reflector, there will be structure in the processed image. Then the search radar begins to take on the character of an imaging radar. Thus a somewhat loose distinction between a search radar and an imaging radar can be made based on the relationship between the resolution of the radar and the spacing between reflecting elements within a local scene. In this chapter, we shall study the tasks of detection and estimation as applied to the specialized problems of radar and sonar search systems. These tasks are special cases of the general theories of detection and estimation, which are topics of the subject of mathematical statistics.

12.1 The radar range equation

For a search radar, the resolution of the radar is usually coarse compared to the size of the target elements. The reflectivity density is then modeled as a sparse set of discrete point reflectors, each represented by an impulse. The ith point reflector has a complex reflectivity, ρ_i, at the delay-doppler coordinates (τ_i, ν_i). Then the reflectivity density

$\rho(\tau, \nu)$ is the sum of impulses given by

$$\rho(\tau, \nu) = \sum_i \rho_i \delta(\tau - \tau_i, \nu - \nu_i),$$

with one impulse for each reflecting object. The separation between any two impulses in the τ, ν plane is assumed to be large in comparison with the resolution of the search radar.

It may be more convenient to deal with the radar cross section of a reflector because then we can deal with reflected signal energy rather than reflected signal amplitude. Thus

$$\begin{aligned}\sigma(\tau, \nu) &= |\rho(\tau, \nu)|^2 \\ &= \sum_i \sigma_i \delta(\tau - \tau_i, \nu - \nu_i)\end{aligned}$$

where $\sigma_i = |\rho_i|^2$ is the radar cross section of the ith target. Because each target is modeled as a point target, it can be characterized simply by its radar cross section, which is a real number.

The performance of a search radar depends in large measure on the amount of energy E_{pr} in the pulse that reaches the receiver in comparison with the internal thermal noise inevitably generated within the receiver. The equation that expresses E_{pr} in terms of the transmitted pulse energy E_{pt} is known as the *radar range equation.* One form of the radar range equation[1] is given by

$$E_{pr} = \left(\frac{\lambda^2}{4\pi} G_r\right)\left(\frac{1}{4\pi R_r^2}\right)\sigma\left(\frac{1}{4\pi R_t^2}\right) G_t E_{pt}$$

where G_t and G_r are the gains of the transmitting and receiving antennas in the direction of propagation; σ is the radar cross section of the reflector; R_t and R_r are the ranges from transmitter to reflector and from reflector to receiver, and λ is the wavelength of the radiation.

In the case of a one-way, point-to-point antenna link, the corresponding expression is

$$E_{pr} = \left(\frac{\lambda^2}{4\pi} G_r\right)\left(\frac{1}{4\pi R^2}\right) G_t E_{pt}.$$

This expression is called the *Friis formula.*

The radar range equation has been introduced with the terms arranged from right to left to tell the story of the energy bookkeeping. The transmitted energy E_{pt} appears at the output of the antenna as the effective radiated energy $G_t E_{pt}$. The energy spreads in spherical waves. The term $4\pi R_t^2$ is the area of a sphere of radius R_t. By dividing $G_t E_{pt}$ by this term, we have the energy per unit area that passes through a surface

[1] Replacing E_{pr} and E_{pt} with P_r and P_t restates the radar range equation in terms of power rather than energy. The energy in a pulse is the power integrated over time, and it is usually the more convenient quantity for our purposes.

at the distance R_t from the transmitter. A reflector reradiates a portion of this energy, as described by the cross section σ, in the direction of the receiver. Of this reflected energy, a fraction is captured by the receiving antenna. By dividing the reflected energy by $4\pi R_r^2$, we have the energy per unit area that passes through a spherical surface at the distance R_r from the reflector. The energy captured by the receiving antenna is determined by the effective area A_e of the receiving antenna. It is related to the receiving antenna gain by antenna theory as

$$A_e = G_r \frac{\lambda^2}{4\pi}.$$

The product of all factors leads to the radar range equation. When $R_t = R_r = R$, the equation can be collapsed into the form

$$E_{pr} = G_r G_t \frac{\lambda^2 \sigma}{(4\pi)^3 R^4} E_{pt}.$$

The radar range equation is written in many alternative forms by grouping the terms differently, or by renaming groups of terms. Another form of the equation is

$$E_{pr} = G_r \left(\frac{\lambda^2}{4\pi R_r^2} \right) \left(\frac{\sigma}{4\pi \lambda^2} \right) \left(\frac{\lambda^2}{4\pi R_t^2} \right) G_t E_{pt}.$$

This is a somewhat artificial rearrangement of terms that normalizes range by wavelength, and normalizes σ by the area of a sphere of radius λ.

12.2 Coherent detection of pulses in noise

The radar-detection problem in its most elementary form consists of a received signal, $v(t)$, which is either a pulse, $s(t)$, contaminated by stationary noise, $n(t)$, or noise alone. We shall allow $s(t)$ to be a complex pulse and $n(t)$ to be a complex noise process. The pulse $s(t)$ is known. The noise process $n(t)$ is known stochastically by its mean and autocorrelation function. The pulse may be received in the form $s(t - \tau_0)\, e^{j(2\pi \nu_0 t + \theta_0)}$ in which form it includes a known arrival time, τ_0, a known phase shift, θ_0, and a known doppler shift, ν_0. For the present purposes, we can easily redefine $s(t)$ to include all of these known parameters. We do not need to display them explicitly at this time.

The task is to detect the presence of the pulse $s(t)$. In this section, because the phase shift θ_0 is known, the task is called *coherent detection*. When the phase shift θ_0 is not known, as is the case studied in Section 12.5, the task is called *noncoherent detection*.

The detection of a pulse in noise is a special case of the general theory of binary hypothesis testing. In general, this problem of testing hypotheses consists of two hypotheses, called H_0 and H_1. One and only one of these two hypotheses is true, and the problem is to decide which is true. Hypothesis H_0 is called the *null hypothesis*, and

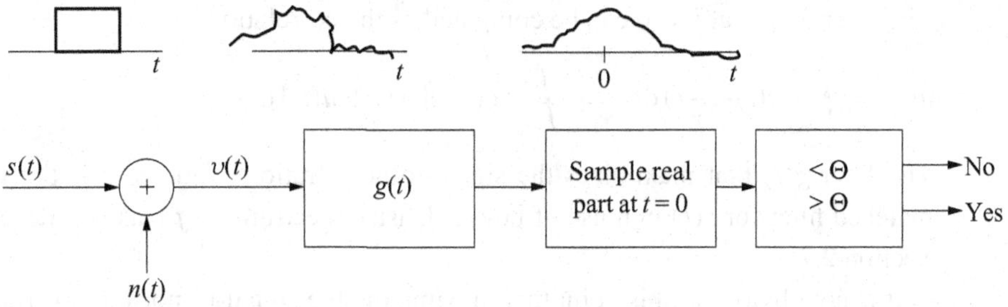

Figure 12.1 Detection of a pulse in noise

hypothesis H_1 is called the *alternative hypothesis.* They are defined as

$$H_0 : v(t) = n(t) \quad \text{(noise only)}$$
$$H_1 : v(t) = s(t) + n(t) \quad \text{(signal plus noise)}$$

where the signal $s(t)$ is a known pulse of finite energy E_p. In general, the noise $n(t)$ is a complex, stationary, zero-mean, gaussian-noise process, given by $n(t) = n_R(t) + jn_I(t)$. The real and imaginary noise components $n_R(t)$ and $n_I(t)$ are independent, identically distributed, random processes; each component has the known autocorrelation function $\phi(\tau) = \mathrm{E}[n(t)n^*(t+\tau)]/2$ and the power density spectrum $N(f)$, which is the Fourier transform of $\phi(\tau)$. Commonly, $n(t)$ is gaussian noise, but we need not make this assumption at this time. Only the second-order properties of $n(t)$ will be used initially.

The detection problem weak decide between H_0 and H_1 based on an observation of $v(t)$. This amounts to formulating a rule that assigns every possible $v(t)$ to either of two sets: $\mathcal{U}_0$ or $\mathcal{U}_1 = \mathcal{U}_0^c$. If $v(t)$ is in $\mathcal{U}_0$, then the decision is in favor of H_0; that there is no pulse in the received signal. If $v(t)$ is in $\mathcal{U}_1$, the decision is in favor of H_1; that there is a pulse, $s(t)$, in the received signal. Of course, we cannot describe $\mathcal{U}_0$ and $\mathcal{U}_1$ by enumerating all of the $v(t)$ that belong to each; that would be an infinite list. The sets must be described by a rule called a *decision rule*. The decision rule will decide whether the observed $v(t)$ is in $\mathcal{U}_0$ or in $\mathcal{U}_1$. Thus the decision rule will detect whether the received signal $v(t)$ consists of a pulse in noise or of noise only.

The decision rule we will choose is to pass the received signal $v(t)$ through a filter with the impulse response $g(t)$, as shown in Figure 12.1, and then to test the amplitude of the real part at the output of the filter at $t = 0$ by comparing it to a threshold, Θ. We do not require the filter to be causal. If the amplitude at time zero is larger than the threshold, the decision is that the pulse is present in the received signal. The filter output is the convolution

$$u(t) = \int_{-\infty}^{\infty} v(\xi)g(t-\xi)\,\mathrm{d}\xi.$$

The filter output at $t = 0$ can be computed as the correlation

$$u(0) = \int_{-\infty}^{\infty} v(t)g(-t)\,dt \;=\; \int_{-\infty}^{\infty} g(-t)[s(t) + u(t)]\,dt.$$

The filter $g(t)$ that maximizes the signal-to-noise ratio at time zero is the whitened matched filter for $s(t)$ in noise of power density spectrum $N(f)$, as was developed in Section 2.7.

It is not obvious at this point that maximizing the signal-to-noise ratio at the single time instant $t = 0$ is the optimum thing to do for the detection problem; in general, it is not. We shall argue at the end of the next section that, if the noise is gaussian, the matched filter followed by a threshold test is the optimum detection procedure. Even if the noise is not gaussian, the matched-filter detector is still a very practical and popular detector. Then it is optimal only in the restricted class of detectors that have the form of a linear filter followed by a threshold test.

To determine the performance of the matched-filter detector, we must evaluate the two probabilities of error. For the detection problem, the probability that the real part of the filter output exceeds the threshold Θ when the input is noise only is called the *probability of false alarm* and is denoted α or p_{FA}. Similarly, the probability that the filter output does not exceed the threshold when the input is signal plus noise is called the *probability of missed detection* and is denoted β or p_{MD}.

The matched filter collapses the received waveform $v(t)$ into a single number, $\mathrm{Re}[u(0)]$, also denoted $\Lambda(v(t))$ when we wish to emphasize that it is a function of the received signal $v(t)$. We need to determine the probability distribution of $\Lambda(v(t))$ under each hypothesis.

Under the hypothesis H_0, the mean of $\Lambda(v(t))$ is zero. Under the hypothesis H_1, the mean is

$$\mathrm{E}[\Lambda(v(t))] = \mathrm{E}[\mathrm{Re}[u(0)]].$$

As we have seen when deriving the whitened matched filter, the mean is always real so

$$\begin{aligned}\mathrm{E}[\mathrm{Re}[u(0)]] &= \mathrm{E}[u(0)]\\ &= \int_{-\infty}^{\infty} g(-t)s(t)\,dt\\ &= \int_{-\infty}^{\infty} G(f)S(f)\,df.\end{aligned}$$

Under either hypothesis, the variance (per component when the noise is complex) is the real number given by

$$\begin{aligned}N &= \tfrac{1}{2}\,\mathrm{var}[u(0)] = \mathrm{var}[\mathrm{Re}[u(0)]] = \sigma^2\\ &= \int_{-\infty}^{\infty} N(f)|G(f)|^2\,df.\end{aligned}$$

In particular, as we saw in Section 2.7 when using the whitened matched filter, $G(f) = S^*(f)/N(f)$, the mean value of the filter output is the real number

$$E[u(0)] = \int_{-\infty}^{\infty} \frac{|S(f)|^2}{N(f)} \, df,$$

and the variance per component is

$$N = \int_{-\infty}^{\infty} \frac{|S(f)|^2}{N(f)} \, df.$$

If the noise is white, it has power density spectrum

$$N(f) = \frac{N_0}{2}$$

and the signal-to-noise ratio is

$$\frac{S}{N} = \frac{2E_p}{N_0}.$$

This holds for any noise probability density function.

To numerically evaluate the probability of false alarm p_{FA} and the probability of missed detection p_{MD}, we must specify the noise $n(t)$ more fully. If the matched filter has stationary gaussian noise at the input, then the output at $t = 0$ must be a gaussian random variable, and we have found its mean and variance under each hypothesis. Therefore the probability density functions on the output of the matched filter under hypotheses H_0 and H_1 are

$$p_0(x) = \frac{1}{\sqrt{2\pi}\,\sigma} e^{-x^2/2\sigma^2}$$

$$p_1(x) = \frac{1}{\sqrt{2\pi}\,\sigma} e^{-(x-A)^2/2\sigma^2}$$

where $A = E[u(0)]$. The probability of a false alarm and the probability of a missed detection are given by

$$p_{FA} = \int_{\Theta}^{\infty} p_0(x) \, dx$$

$$p_{MD} = \int_{-\infty}^{\Theta} p_1(x) \, dx.$$

Make the change in the variables $z = x/\sigma$ in the first integral and $z = (x - A)/\sigma$ in the second integral. Then, using the symmetry of the gaussian function, the integrals can be written in terms of the error function

$$Q(x) = \int_{x}^{\infty} \frac{1}{\sqrt{2\pi}} e^{-z^2/2} \, dz$$

as

$$p_{FA} = Q\left(\frac{\Theta}{\sigma}\right)$$

$$p_{MD} = Q\left(\frac{A-\Theta}{\sigma}\right).$$

The error probabilities also can be written in terms of the ratio E_p/N_0 as

$$p_{FA} = Q\left(\lambda\sqrt{\frac{2E_p}{N_0}}\right)$$

$$p_{MD} = Q\left((1-\lambda)\sqrt{\frac{2E_p}{N_0}}\right),$$

where $\lambda = \Theta/A$ is a parameter between $-\infty$ and ∞, and normally between 0 and 1. These expressions only depend on the ratio E_p/N_0 and the parameter λ. By crossplotting $p_{FA}(\lambda)$ and $p_{MD}(\lambda)$ for each value of λ, the parameter λ can be eliminated.

The performance of the detector of a pulse in gaussian noise is shown in Figure 12.2 with E_p/N_0 expressed in decibels. The performance of the coherent detector should be compared with the performance of the noncoherent detector, shown in Figure 12.3, which will be analyzed in Section 12.3.

It is important to observe that, for both the coherent detector and the noncoherent detector, the performance of the matched-filter detector in gaussian noise only depends on the ratio E_p/N_0 but not on the detailed structure of $s(t)$. Only the energy of the pulse matters; the shape may be chosen for reasons of convenience.

12.3 The Neyman–Pearson theorem

Now that we have studied the matched filter followed by a threshold as one kind of detector of a signal in noise, we shall turn to the larger task of finding the optimal such detector of a signal in noise. Our conclusion will be that, if the noise is white gaussian noise, the optimal detector is indeed the matched filter followed by a threshold.

If the noise is nonwhite gaussian noise, the optimal detector is the whitened matched filter followed by a threshold. If the noise is not gaussian, or if it is correlated with the signal (as is clutter), then the optimal detector does not have this structure of a filter followed by a threshold. Generally, it will have other nonlinearities besides the threshold.

The Neyman–Pearson theorem is a general theorem in the subject of binary hypothesis testing. It deals with the case where there is no prior probability on the validity of the two hypotheses, H_0 and H_1. In many applications, it may be unreasonable to insist that such a probability distribution could be known, even in principle. For example,

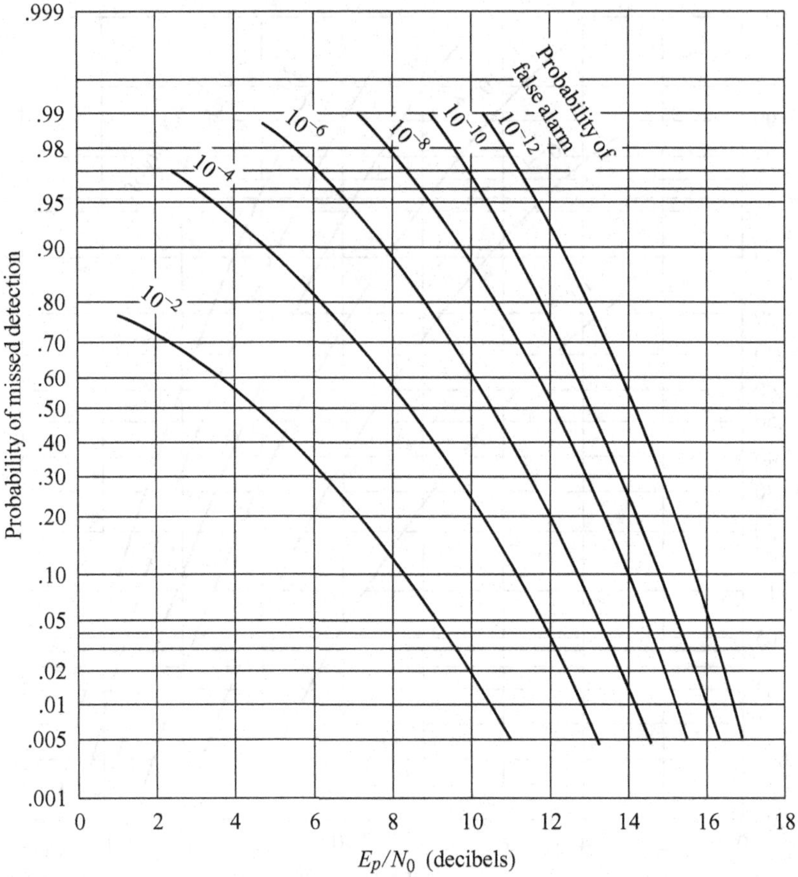

Figure 12.2 Performance of a coherent detector in gaussian noise

one would not like to insist that a meaningful probability model could be constructed for the event that there will be an aircraft at a particular point of space at a given time. The Neyman–Pearson point of view sidesteps this question by refusing to introduce the notion of a prior probability on the two hypotheses.

To start out, we shall suppose that we are summarizing the real signal $v(t)$ by the single number x, called a *statistic* or *measurement*. To emphasize that it is a function of $v(t)$, the statistic x is also denoted $\Lambda(v(t))$. Perhaps the statistic is a sample of $v(t)$ at some time t_0, so that $x = v(t_0)$, or perhaps a linear combination of samples such as $x = \sum_i v(t_i)$. First, we shall determine how to detect the pulse, given only the statistic x. Then we will determine how to form the optimum statistic from the received signal $v(t)$.

The statistic $x = \Lambda(v(t))$ is a real random variable. Associated with each of the two simple hypotheses H_0 and H_1 is a probability distribution on the statistic x. If H_0 is true, then $p_0(x)$ is the probability density function on the statistic; if H_1 is true, then $p_1(x)$ is this probability density function.

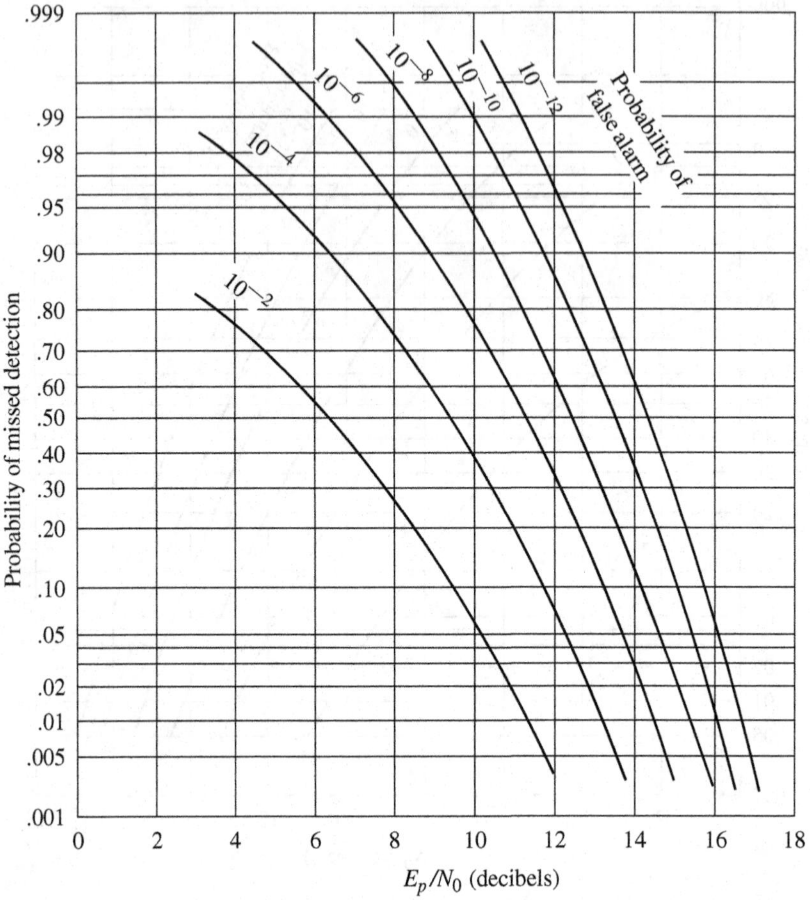

Figure 12.3 Performance of a noncoherent detector in gaussian noise

As for the case of pulse detection, a simple measurement x consists of an observation of one realization of the statistic x, and for some values of x, the decision will be that hypothesis H_0 is true; for other values of x, the decision will be that hypothesis H_1 is true. A hypothesis-testing rule is a partition of the measurement space into two disjoint sets, $\mathcal{U}_0$ and $\mathcal{U}_1$. We decide that H_0 or H_1 is true according to whether x is an element of $\mathcal{U}_0$ or of $\mathcal{U}_1$. Each partition is a different hypothesis-testing rule and, except for randomized rules, which we do not consider, there are no other rules. We should use the best hypothesis-testing rule, but first we must agree on what is meant by the "best" rule.

Accepting hypothesis H_1 when H_0 is actually true is called a *type I error*[2] (or *false alarm*), and the probability of this event is denoted by α. Accepting hypothesis H_0 when H_1 is actually true is called a *type II error* (or *missed detection*), and the probability of

[2] The convention is unfortunate in that the errors are denoted type I and type II, while the hypotheses are subscripted 0 and 1.

this event is denoted by β. Obviously,

$$\alpha = \int_{\mathcal{U}_1} p_0(x) \qquad \beta = \int_{\mathcal{U}_0} p_1(x).$$

The problem is to specify $(\mathcal{U}_0, \mathcal{U}_1)$ so that α and β are as small as possible. This is not yet a well-defined problem because α generally can be made smaller by reducing $\mathcal{U}_1$, although β thereby is increased. The Neyman–Pearson point of view assumes that a maximum value of β is specified, and $(\mathcal{U}_0, \mathcal{U}_1)$ must be determined to minimize α subject to the constraint that β is not larger than this maximum.

Theorem 12.3.1 (Neyman–Pearson Theorem) *For any real number, Θ, let*

$$\mathcal{U}_0(T) = \{x : p_1(x) \le p_0(x)\,\mathrm{e}^{-\Theta}\}$$
$$\mathcal{U}_1(T) = \{x : p_1(x) > p_0(x)\,\mathrm{e}^{-\Theta}\},$$

and let α^* and β^* be the probabilities of type I and type II error corresponding to this choice of decision regions. Suppose α and β are the probabilities of type I and type II error corresponding to some other choice of decision regions, and suppose $\alpha < \alpha^*$. Then $\beta > \beta^*$.

Proof: Let $(\mathcal{U}_0', \mathcal{U}_1')$ be any other decision procedure such that $\alpha < \alpha^*$. Define the indicator functions on x as: $\phi(x) = 1$ if $x \in \mathcal{U}_1(\Theta)$, and otherwise $\phi(x) = 0$; and $\phi'(x) = 1$ if $x \in \mathcal{U}_1'$, and otherwise $\phi'(x) = 0$. Then $(\phi(x) - \phi'(x))(p_1(x) - p_0(x)\,\mathrm{e}^{-\Theta}) \ge 0$ for all x, which can be verified by separately examining the cases $x \in \mathcal{U}_0(\Theta)$ and $x \in \mathcal{U}_1(\Theta)$. Therefore

$$\int_{-\infty}^{\infty} (\phi(x) - \phi'(x))(p_1(x) - p_0(x)\,\mathrm{e}^{-\Theta})\,\mathrm{d}x \ge 0$$

$$\int_{-\infty}^{\infty} p_1(x)\phi(x)\,\mathrm{d}x - \int_{-\infty}^{\infty} p_1(x)\phi'(x)\,\mathrm{d}x \ge$$
$$\mathrm{e}^{-\Theta}\int_{-\infty}^{\infty} p_0(x)\phi(x)\,\mathrm{d}x - \mathrm{e}^{-\Theta}\int_{-\infty}^{\infty} p_0(x)\phi'(x)\,\mathrm{d}x.$$

Hence

$$(1 - \beta^*) - (1 - \beta) \ge \mathrm{e}^{-\Theta}(\alpha^* - \alpha).$$

Because by assumption $\alpha^* - \alpha > 0$, we conclude that $\beta - \beta^* > 0$, as was to be proved. □

Notice that, because the theorem does not explicitly define α^* as a function of β^*, but rather expresses the optimum pair parametrically as $\alpha^*(\Theta)$ and $\beta^*(\Theta)$, many values of the threshold must be examined to find the smallest α^* satisfying the constraint on the maximum permissable β^*. An efficient procedure is to compute $\alpha^*(\Theta)$ and $\beta^*(\Theta)$ for a range of Θ, then to crossplot them to construct a graph of α^* versus β^*.

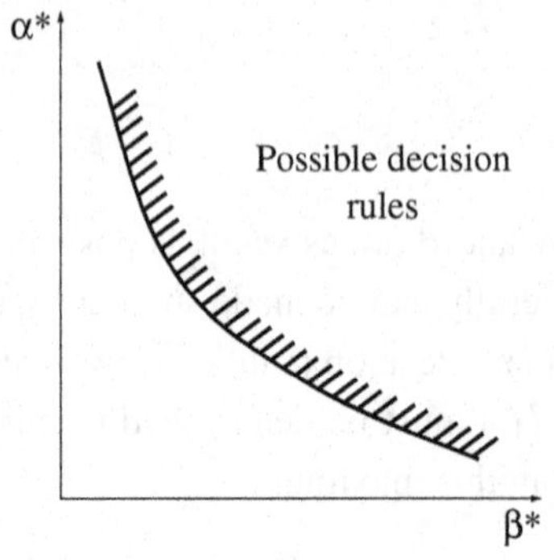

Figure 12.4 The region of possible decision rules

Figure 12.4 presents a typical graph of α^* versus β^*. This graph provides a visual portrayal of the Neyman–Pearson theorem. The theorem says that every possible decision rule has a performance lying above or on the curve, but no decision rule has a performance lying below the curve. The Neyman–Pearson threshold rules have their performance lying on the curve.

The Neyman–Pearson decision regions can be rewritten in terms of the loglikelihood ratio, given by

$$\Lambda(x) = \log \frac{p_0(x)}{p_1(x)}$$

as the sets

$$\mathcal{U}_0(T) = \mathcal{U}_1(T)^c = \{x : \Lambda(x) \geq \Theta\}.$$

We will decide that H_0 is true if the loglikelihood ratio is at least as large as the threshold Θ. It is clear that $\Lambda(x)$ (or any monotonic function of $\Lambda(x)$) is the significant function in the Neyman–Pearson theorem rather than the probability distributions individually.

Suppose that the single measurement x is replaced by a block of independent, identically distributed measurements. We will consider the block as a vector of measurements of length n. Then we will write the vector measurement as an n-tuple of simple measurements,

$$\boldsymbol{v} = (x_1, \ldots, x_n),$$

where x_ℓ is the value of the ℓth measurement. The theory holds just as before. However, the vector structure enables us to divide the problem into sections. The loglikelihood ratio for the vector $\boldsymbol{v}$ is

$$\Lambda(\boldsymbol{v}) = \log \frac{p_0(\boldsymbol{v})}{p_1(\boldsymbol{v})}.$$

Because the measurements are independent, the probability of a block is the product of the probabilities of the individual measurements. This means that the logarithm becomes a sum,

$$\Lambda(\boldsymbol{v}) = \sum_{\ell=1}^{n} \log \frac{p_{0\ell}(x_\ell)}{p_{1\ell}(x_\ell)}$$

where $p_{0\ell}(x)$ and $p_{1\ell}(x)$ are probability density functions on the ℓth component

$$\Lambda(\boldsymbol{v}) = \sum_{\ell=1}^{n} \Lambda(x_\ell).$$

The loglikelihood ratio of a sum of independent measurements is the sum of the loglikelihood ratios. If $p_{0\ell}(x)$ is gaussian with zero mean and variance σ^2 and $p_{1\ell}(x)$ is gaussian with mean A_ℓ and variance σ^2, then

$$\begin{aligned}\Lambda(x_\ell) &= \frac{(x_\ell - A_\ell)^2}{2\sigma^2} - \frac{x_\ell^2}{2\sigma^2} \\ &= -x_\ell A_\ell/\sigma^2 + A_\ell^2/2\sigma^2.\end{aligned}$$

The second term does not depend on the measurement and so is an uninformative constant, which can be ignored, as can the denominator of the first term. Then, if the components are independent,

$$\Lambda(\boldsymbol{v}) = \sum_{\ell=1}^{n} x_\ell A_\ell.$$

If now $\boldsymbol{v}$ is obtained by sampling the pulse $v(t)$ in white noise, where $v(t) = s(t) + n(t)$, then this becomes

$$\Lambda(v(t_\ell), \ell = 1, \ldots, L) = \sum_{\ell=1}^{L} v(t_\ell)s(t_\ell).$$

Even without a formal proof it is apparent that, in the limit,

$$\Lambda(v(t)) = \int_{-\infty}^{\infty} v(t)s(t)\,dt.$$

Thus for white gaussian noise, the matched filter is not only optimal in the class of linear decision rules, it is optimal in general. We can conclude from this that the threshold detector in Section 12.2 is optimal for gaussian noise because it is the loglikelihood detector. If the noise is not white, it can be made white by a whitening filter. This means that the whitened matched filter followed by a threshold is optimal for stationary gaussian noise with power density spectrum $N(f)$.

12.4 Rayleigh and ricean probability distributions

This section studies the noncoherent detection of pulses in noise. The detection statistic is the magnitude of the complex matched-filter output because the carrier phase is unknown. This corresponds to taking the square root of the sum of the squares of the real and imaginary components. When the input is a signal in additive gaussian noise, the output of each component of the matched filter is a gaussian random variable. To analyze the probability of error of noncoherent detection, we must study what happens to gaussian noise under the computation of the square root of the sum of the squares. This can be viewed as a transformation of the noise from rectangular coordinates (in-phase and quadrature components) to polar coordinates (amplitude and phase).

The two-dimensional gaussian probability density function is defined in terms of two random variables, say X and Y. If the random variables X and Y are independent, zero mean, and have equal variance, then the two-dimensional gaussian probability density function is

$$p(x, y) = \frac{1}{2\pi\sigma^2} \mathrm{e}^{-(x^2+y^2)/2\sigma^2},$$

as shown in Figure 12.5. If the mean is nonzero, the two-dimensional gaussian probability density function is

$$p(x, y) = \frac{1}{2\pi\sigma^2} \mathrm{e}^{-[(x-\overline{x})^2+(y-\overline{y})^2]/2\sigma^2}.$$

The transformation from rectangular coordinates to polar coordinates is

$$r = \sqrt{x^2 + y^2}$$
$$\phi = \tan^{-1}\frac{x}{y}.$$

We wish to find an expression for the probability density function of the amplitude r.

The probability of the measurement lying within any region $\mathcal{A}$ must be the same whether the measurement is expressed in rectangular coordinates or in polar

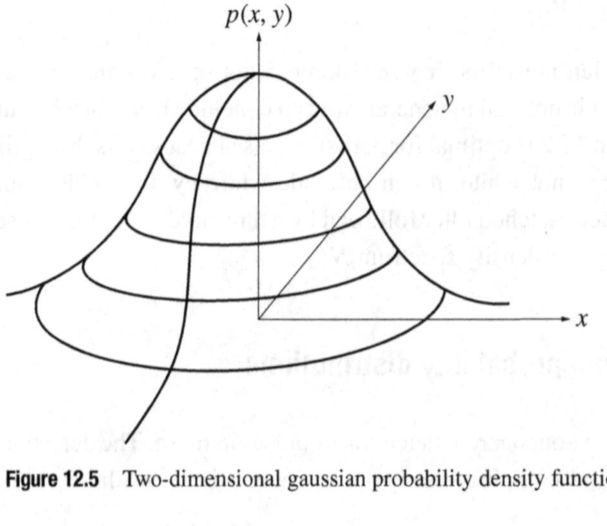

Figure 12.5 Two-dimensional gaussian probability density function

coordinates. That is, for any region $\mathcal{A}$,

$$\int_{\mathcal{A}} p(r,\phi)\,\mathrm{d}r\,\mathrm{d}\phi = \int_{\mathcal{A}} \frac{1}{2\pi\sigma^2}\,\mathrm{e}^{-(x^2+y^2)/2\sigma^2}\,\mathrm{d}x\,\mathrm{d}y$$

where $p(r,\phi)$ is the probability density function in polar coordinates. On the right side, substitute

$$\begin{aligned} x^2 + y^2 &= r^2 \\ \mathrm{d}x\,\mathrm{d}y &= r\,\mathrm{d}\phi\,\mathrm{d}r. \end{aligned}$$

This gives

$$\int_{\mathcal{A}} p(r,\phi)\,\mathrm{d}r\,\mathrm{d}\phi = \int_{\mathcal{A}} \frac{1}{2\pi\sigma^2}\,\mathrm{e}^{-r^2/2\sigma^2} r\,\mathrm{d}r\,\mathrm{d}\phi,$$

from which it follows that

$$p(r,\phi) = \frac{r}{2\pi\sigma^2}\,\mathrm{e}^{-r^2/2\sigma^2} \qquad 0 \le \phi < 2\pi,$$

which is uniform in ϕ. Integrating over the range of ϕ gives the marginal

$$p(r) = \frac{r}{\sigma^2}\,\mathrm{e}^{-r^2/2\sigma^2} \qquad r \ge 0.$$

This probability density function is known as a *rayleigh probability density function*, and the corresponding random variable is called a *rayleigh random variable*. It is the probability density function for the envelope of unbiased, complex gaussian noise. A rayleigh random variable has the mean $\sigma\sqrt{\pi/2}$ and variance $\sigma\sqrt{2}$.

To write the rayleigh probability density function in a standardized form, make the change in variables

$$z = \frac{r}{\sigma},$$

and define

$$p_{Ra}(z) = z\,\mathrm{e}^{-z^2/2} \qquad z \ge 0$$

so that

$$p(r) = \frac{1}{\sigma}\,p_{Ra}\left(\frac{r}{\sigma}\right).$$

For a signal plus noise, we must use a two-dimensional, gaussian probability density function with a nonzero mean, $(\overline{x}, \overline{y})$. Write the mean in the form

$$\begin{aligned} \overline{x} &= A\cos\theta \\ \overline{y} &= A\sin\theta \end{aligned}$$

where θ is an unknown phase angle. Then

$$p_R(x) = \frac{1}{\sqrt{2\pi}\,\sigma}\,\mathrm{e}^{-(x-A\cos\theta)^2/2\sigma^2}$$

$$p_I(y) = \frac{1}{\sqrt{2\pi}\,\sigma}\,\mathrm{e}^{-(y-A\sin\theta)^2/2\sigma^2}.$$

Carrying through the transformation of variables, as was done previously, gives

$$p(r,\phi) = \frac{r}{2\pi\sigma^2}\,\mathrm{e}^{-(r^2-2Ar\cos(\theta-\phi)+A^2)/2\sigma^2}.$$

Integrating over ϕ gives

$$p(r) = \frac{r}{2\pi\sigma^2}\,\mathrm{e}^{-r^2/2\sigma^2}\,\mathrm{e}^{-A^2/2\sigma^2}\int_{-\pi}^{\pi}\mathrm{e}^{Ar\cos(\theta-\phi)/\sigma^2}\,\mathrm{d}\phi.$$

The integral is clearly independent of θ because the integral is periodic and extends over one period whatever the value of θ. The integral can be expressed in terms of the standard function known as the *modified, zero-order Bessel function* of the first kind and order zero. It is defined by the integral

$$I_0(\xi) = \frac{1}{2\pi}\int_{-\pi}^{\pi}\mathrm{e}^{\xi\cos\theta}\,\mathrm{d}\theta.$$

The modified Bessel function $I_0(x)$ is shown in Figure 12.6. Now we can write

$$p(r) = \frac{r}{\sigma^2}\,\mathrm{e}^{-(r^2+A^2)/2\sigma^2}\,I_0\left(\frac{Ar}{\sigma^2}\right) \qquad r > 0.$$

To re-express this function in terms of a single parameter, let

$$p_{Ri}(z,a) = z\,\mathrm{e}^{-(z^2+a^2)/2}\,I_0(az).$$

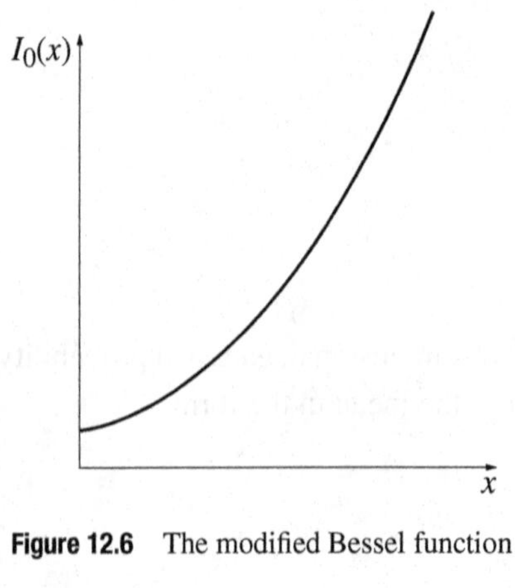

Figure 12.6 The modified Bessel function

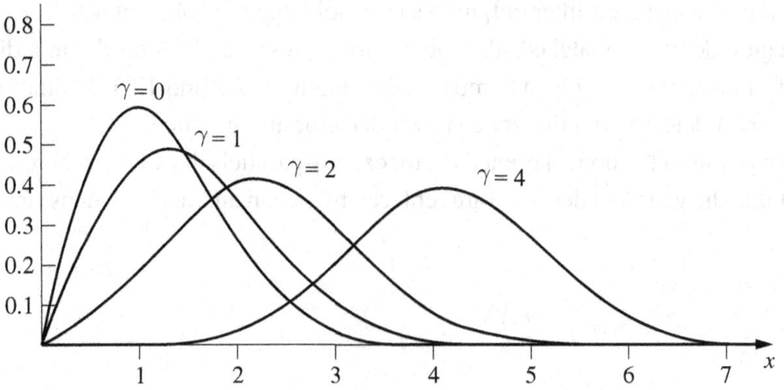

Figure 12.7 Ricean probability density functions

Therefore, with

$$z = \frac{r}{\sigma}$$
$$a = \frac{A}{\sigma},$$

we have

$$p(r) = \frac{1}{\sigma} \, p_{Ri}\left(\frac{r}{\sigma}, \frac{A}{\sigma}\right).$$

The probability density $p_{Ri}(z, a)$ is known as a *ricean probability density function* with the parameter a, and the random variable is called a *ricean random variable*. The parameter a defines a family of probability densities, one for each value of a, as shown in Figure 12.7. The ricean probability density function reduces to the rayleigh probability density function when a is equal to zero. When a is large, the ricean probability density function resembles a gaussian probability density function. Of course, the ricean density is never identical to a gaussian density because it is zero when its argument is negative, while the gaussian density is not zero for any value of its argument.

12.5 Noncoherent detection of pulses in noise

We will now treat another problem of hypothesis testing in which the pulse has an unknown phase, but is otherwise known:

$$H_0 : v(t) = n(t)$$
$$H_1 : v(t) = s(t)\,e^{-j\theta} + n(t).$$

This model corresponds to the usual radar detection problem in which a known passband pulse of unknown phase is received in noise. This is called a *noncoherently received pulse*, or more loosely, a *noncoherent pulse*. Whereas the coherent detector compares the real part of the matched-filter output to a threshold, the noncoherent detector compares the magnitude of the matched-filter output to a threshold. This results in a difference in performance indicated by a comparison of Figures 12.2 and 12.3. In other respects, the coherent detector and the noncoherent detector are the same.

The analysis of the noncoherent detector exactly parallels that of the coherent detector except that the gaussian densities are replaced by ricean and rayleigh densities. That is,

$$p_0(r) = \frac{r}{\sigma^2} e^{-r^2/2\sigma^2} \qquad r \geq 0$$

$$p_1(r) = \frac{r}{\sigma^2} e^{-(r^2+A^2)/2\sigma^2} I_0\left(\frac{rA}{\sigma^2}\right) \qquad r \geq 0$$

where $A = \mathrm{E}[u(0)]$. The loglikelihood ratio is

$$\log \frac{p_0(r)}{p_1(r)} = \frac{A^2}{2\sigma^2} - \log I_0\left(\frac{rA}{\sigma^2}\right).$$

The loglikelihood ratio should be compared to a threshold, but, because it is monotonic in r, it is enough to compare r itself directly to a threshold, again denoted Θ. As before, the probabilities of error are

$$\alpha = \int_{\Theta}^{\infty} p_0(r)\,dr$$

$$\beta = \int_{-\infty}^{\Theta} p_1(r)\,dr.$$

However, from the theory of the matched filter, $(A/\sigma)^2 = 2E_p/N_0$. Redefine the variable of integration as $z = r/\sigma$ and the threshold as $\Theta' = \Theta/\sigma$. Then

$$\alpha(\Theta') = \int_{\Theta'}^{\infty} z\,e^{-z^2/2}\,dz = e^{-\Theta'^2/2}$$

$$\beta(\Theta') = \int_{0}^{\Theta'} z\,e^{-(z^2+2E_p/N_0)/2} I_0\left(z\sqrt{\frac{2E_p}{N_0}}\right) dz$$

where the lower limit in the second integral is set equal to zero because the integrand is zero for negative z.

These expressions only depend on the ratio E_p/N_0 and the normalized threshold Θ'. By crossplotting $\alpha(\Theta')$ and $\beta(\Theta')$ for each value of Θ', the threshold Θ' can be eliminated. The performance of the noncoherent detector in gaussian noise is shown in Figure 12.3. This figure should be compared to Figure 12.2 to conclude that the noncoherent detector needs at most one decibel more E_p/N_0 in the region of high E_p/N_0.

One may also wish instead to detect not just a single pulse, but an entire pulse train. If the pulse train is coherent, it can itself be regarded as a pulse, and the discussion in

Section 12.2 applies without change. The detector performance in gaussian noise is the same as in Figure 12.2, only now E_p refers to the energy in the pulse train.

If the pulse train is coherent from pulse to pulse, but noncoherently detected – that is, the same unknown phase applies to every pulse – then, again, the pulse train is treated simply as one pulse. It is matched-filtered, and the magnitude of the matched filter output is applied to a threshold. The performance of this noncoherent detector is the same as in Figure 12.7, only now E_p refers to the energy in the pulse train.

In the more general case, to be studied in Section 14.1, the pulses each have an unknown and independent phase. Then the pulses are individually matched filtered, and the magnitudes of the matched-filter outputs are added to provide a detection statistic. The detection statistic is applied to a threshold to make a detection decision. More elaborate situations, such as when each pulse has both an unknown and an independent amplitude fluctuation, may also arise.

12.6 Detection under unknown parameters

Thus far we have studied only the detection of a pulse that is fully known or is known except for an unknown phase angle. In this section, we shall study the detection of a pulse in the presence of other unknown parameters. The problem of detection of a pulse of unknown amplitude is the hypothesis-testing problem given by

$$H_0 : v(t) = n(t)$$
$$H_1 : v(t) = As(t) + n(t)$$

where A is an unknown parameter. If A were known, then a matched filter followed by a threshold would allow the best possible compromise between the probability of false alarm p_{FA} and the probability of missed detection p_{MD}. If p_{FA} is specified, then a threshold Θ that achieves this probability of false alarm is implied. The threshold that achieves a desired probability of false alarm is independent of the actual value of A, but the probability of missed detection depends on the actual value of A. In this way, one obtains p_{MD} as a function of A given by

$$p_{MD} = Q\left(\frac{A - \Theta}{\sigma}\right).$$

Likewise, if H_1 is replaced by $v(t) = As(t)\mathrm{e}^{\mathrm{j}\theta_0} + n(t)$, with θ_0 unknown, then a noncoherent detector is needed. The detector is a complex matched filter whose output magnitude is applied to a threshold. Again, for a fixed p_{FD}, a threshold is implied. Then, again, p_{MD} is a function of A.

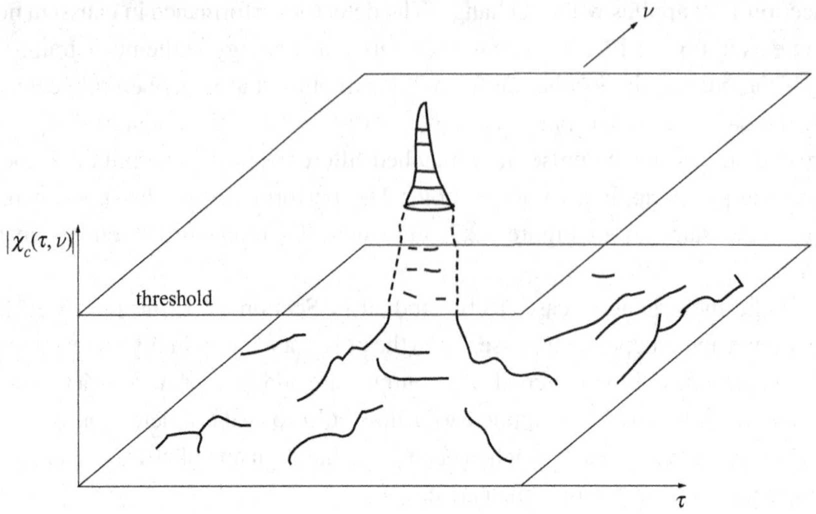

Figure 12.8 Thresholding the sample ambiguity function

Now consider

$$H_0 : v(t) = n(t)$$
$$H_1 : v(t) = As(t - \tau_0)\,e^{j(2\pi\nu_0 t + \theta_0)} + n(t)$$

wherein the pulse has an unknown amplitude, A, an unknown delay, τ_0, an unknown phase, θ_0, and an unknown doppler, ν_0. The pulse $s(t)$ itself is known. The problem is still a binary hypothesis-testing problem, but now hypothesis H_1 is a composite hypothesis. If τ_0 and ν_0 were known, the Neyman–Pearson theorem says filter $v(t)$ with the filter matched to $s(t - \tau_0)\,e^{j2\pi\nu_0 t}$. But this amounts to the computation of the sample cross-ambiguity function $\chi_c(\tau, \nu)$ at the point (τ_0, ν_0). Because τ_0 and ν_0 are actually unknown, they can be estimated from the location of the peak of $|\chi_c(\tau, \nu)|$, assuming H_1 is true. Then the decision between H_0 and H_1 is based on the magnitude of the peak.

Figure 12.8 illustrates the detection of a single pulse in noise by thresholding the sample cross-ambiguity function. A false alarm will occur if noise alone causes the sample cross-ambiguity function to break the threshold, as shown in Figure 12.9.

In practice, the sample cross-ambiguity function may be computed on a discrete delay-doppler grid from sampled values of $v(t)$, and the magnitude at each grid point compared to a threshold. This will usually be quite satisfactory, but it does not necessarily detect all crossings of the threshold by $|\chi_c(\tau, \nu)|$, even if the data samples satisfy the Nyquist criterion. Figure 12.10 shows a case in which the threshold is crossed between two samples of the matched filter output. The samples themselves can all be below

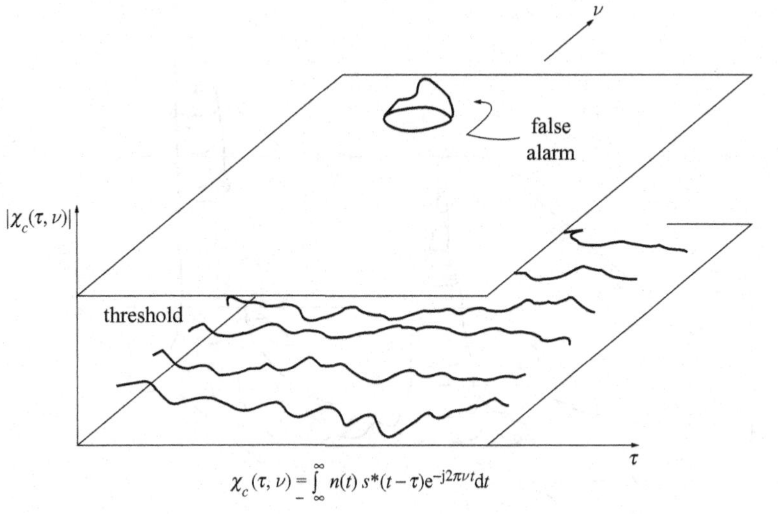

Figure 12.9 False detection of a pulse with unknown delay and doppler

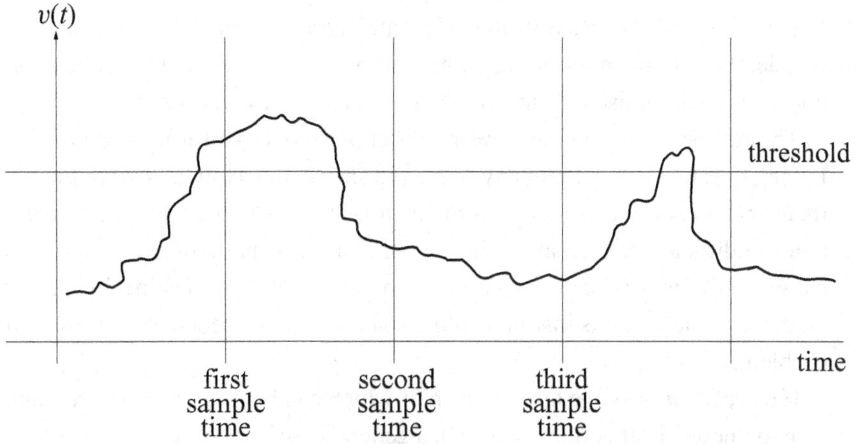

Figure 12.10 An effect of sampling on threshold detection

the threshold even though the sampling rate satisfies the Nyquist sampling theorem. To detect a threshold crossing at an intermediate point requires interpolation of the sampled data.

Compound hypotheses

The detection problem in which there may be more than one pulse is much harder and is usually handled in an ad hoc way by thresholding the ambiguity surface as shown in

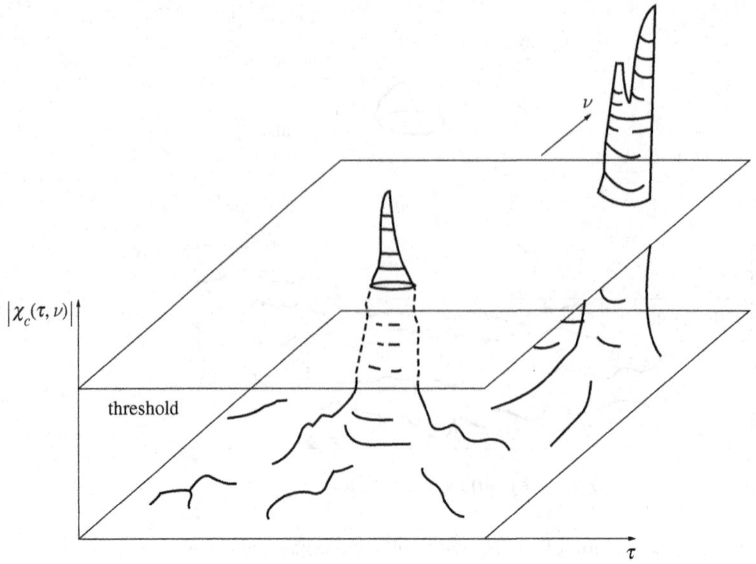

Figure 12.11 Illustration of multiple detections

Figure 12.11. This is usually quite adequate in practice, but it does create the possibility of sidelobes of one return hiding other returns by destructive interference, or sidelobes combining with noise or other sidelobes to create false alarms.

The full detection problem we might choose to solve formally consists of the following questions: Are there any targets in the scene? How many are there? What are their (τ, ν) coordinates? The problem in this form is usually considered too vague and too difficult to seek an optimal procedure, and perhaps not worth the effort because suboptimal procedures based on binary hypothesis testing in each pixel work well. Nevertheless, it is instructive to consider what is needed to treat the more general problem.

If there are n possible targets, each of which can be present or absent, then there are 2^n hypotheses. Rather than treat such a general case, we will treat the case with at most two targets. We shall consider the four-ary hypothesis-testing problem consisting of a possible pair of pulses in white gaussian noise $n(t)$,

$$\begin{aligned}
H_0 &: v(t) = n(t)\\
H_1 &: v(t) = s(t-\tau_1) + n(t)\\
H_2 &: v(t) = s(t-\tau_2) + n(t)\\
H_3 &: v(t) = s(t-\tau_1) + s(t-\tau_2) + n(t)
\end{aligned}$$

where τ_1 and τ_2 are known parameters. The problem is to decide which hypothesis is correct when given a realization of $v(t)$. In this formulation, the notion of a prior probability distribution on the hypotheses is not admitted. The maximum-likelihood principle, studied in Chapter 11, can be used to set up a general method for addressing such problems. For gaussian noise, maximum likelihood reduces to minimum euclidean distance. For each hypothesis, the euclidean distance between the received signal and the hypothesized noise-free signal is

$$H_0 : \int_{-\infty}^{\infty} v(t)^2 \, dt$$

$$H_1 : \int_{-\infty}^{\infty} [v(t) - s(t - \tau_1)]^2 \, dt$$

$$H_2 : \int_{-\infty}^{\infty} [v(t) - s(t - \tau_2)]^2 \, dt$$

$$H_3 : \int_{-\infty}^{\infty} [v(t) - s(t - \tau_1) - s(t - \tau_2)]^2 \, dt$$

and the decision is to be based on the minimum of these four euclidean distances. Because the integral of the squared noise will be infinite over infinite time, the integral is actually restricted to a finite interval large enough to capture almost all of the signal. To compare the four terms, it is enough to compute $\lambda_1 = \int v(t)s^*(t - \tau_1)\, dt$ and $\lambda_2 = \int v(t)s^*(t - \tau_2)\, dt$. The terms $E = \int_{-\infty}^{\infty} |s(t)|^2 \, dt$ and $\rho = \int_{-\infty}^{\infty} s(t - \tau_1)s(t - \tau_2)\, dt$ also arise, but these can be considered constants of the problem and can be precomputed. The term $\int |v(t)|^2 \, dt$ is common to all four hypotheses and need not be computed. The preprocessed data consist of the pair of real numbers λ_1 and λ_2, and the decision is based on the values of these two numbers. The likelihoods of the four hypotheses can now be collapsed to the following

$$H_0 : 0$$

$$H_1 : E - 2\lambda_1$$

$$H_2 : E - 2\lambda_2$$

$$H_3 : 2E - 2\lambda_1 - 2\lambda_2 + 2\rho.$$

The decision rule can be described as the partition of the λ_1, λ_2 plane into *decision regions* such as shown in Figure 12.12. The boundaries of these decision regions can be defined by applying the difference between two statistics to a threshold. For example, the boundary between H_2 and H_3 will have the form

$$(E - 2\lambda_2) - (2E - 2\lambda_1 - 2\lambda_2 + 2\rho) = \Theta$$

which reduces to

$$2\lambda_1 - 2\rho - E = \Theta$$

where Θ is arbitrary. The other boundaries can be found in a similar way. Because a prior on the hypotheses is not available, the maximum-likelihood principle cannot

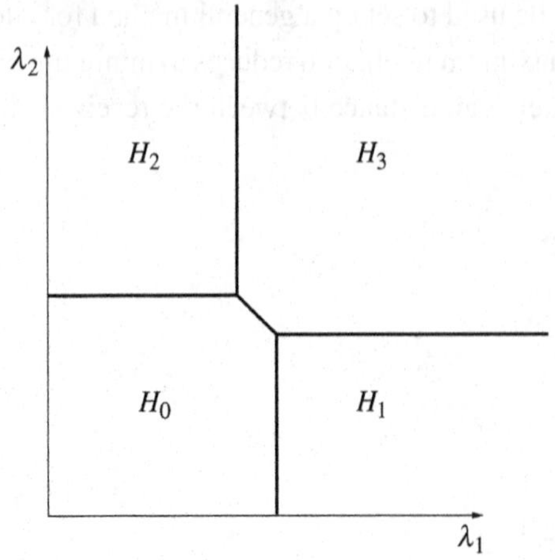

Figure 12.12 Decision regions for hypotheses testing

determine the values of the thresholds. These can be chosen by trading the probabilities of decision error. Rather than only two probabilities of error p_{FA} and p_{MD}, as in the case of two hypotheses, there will be an array of error probabilities with entry $p_{i|j}$ denoting the probability of choosing hypothesis i when hypothesis j is true. These are given by

$$p_{i|j} = \int\int_{\mathcal{U}_i} p_j(\lambda_1, \lambda_2)\, d\lambda_1\, d\lambda_2$$

where $p_j(\lambda_1, \lambda_2)$ is the probability density function on (λ_1, λ_2). The total error probability, given that H_j is true is

$$p_{e|j} = \sum_{i \neq j} p_{i|j}.$$

If λ_1 and λ_2 are gaussian random variables, the error probability $p_{e|j}$ for any given set of thresholds can be computed by integrating the appropriate two-dimensional gaussian density function $p_j(\lambda_1, \lambda_2)$ over the complimentary decision region $\mathcal{U}_j^c$. The various thresholds can be varied to simultaneously adjust the several probabilities of error.

If the cross-correlation coefficient ρ is zero, then this problem can be decoupled into two independent binary hypothesis-testing problems. Thus, it is enough to deal only with binary hypothesis testing unless ρ is not negligible. Even then, it is common practice to treat the problem as two independent binary hypothesis-testing problems, accepting a loss in performance in return for a simple procedure.

If τ_1 and τ_2 are not known, but the pulse $s(t)$ is known then one can first compute the two largest values of the matched-filter output as candidates for τ_1 and τ_2, and then apply the above decision rule to determine whether these are actually signals. Furthermore, if unknown dopplers are also introduced into the problem as well, then one can compute the sample cross-ambiguity surface $|\chi_c(\tau, \nu)|$ and predetect the largest two peaks as candidates for the above decision rule.

12.7 Clutter

The cumulative signal caused by radar reflections from background objects other than those of interest is called *clutter*. Usually, the term clutter carries the connotation that the objects causing the clutter are a dense set of small, unresolvable elements, perhaps moving elements such as raindrops, birds, swaying tree branches, or ocean waves; or perhaps fixed elements such as buildings or trees. The effect of clutter on the radar ambiguity function is similar to thermal noise in that it creates a background intensity above which the main lobes of sample ambiguity functions of targets must rise if they are to be detected. Clutter is different from thermal noise, however, in that it is dependent on the transmitted signal, and it has statistical properties that may be exploited to improve performance.

The most elementary and naive way of dealing with clutter is to treat it as nonwhite gaussian noise and to design "whitening" filters into the system that improve performance by de-emphasizing spectral regions where the clutter is strong. A better approach is to recognize that the clutter is not gaussian noise. In fact, the clutter is caused by the transmitted signal; therefore it is related to the signal reflected from targets of interest. We shall consider three techniques for clutter reduction: clutter-matched filtering, delay-doppler filtering, and clutter notching.

The quality of a radar in clutter is judged by a notion called the *subclutter visibility*. This term refers to the amount by which the maximum allowable clutter-to-signal ratio at the receiver input exceeds the maximum allowable noise-to-signal ratio at the receiver input.

The simplest clutter is stationary clutter, consisting of a large number of echoes from small, randomly placed, stationary reflectors. Let $s(t)$ be the transmitted pulse, let τ_i be a set of time delays distributed randomly on a segment of the time axis, and let a_i be a set of complex gaussian random variables with $\mathrm{E}[a_i a_j] = 0$ for $i \neq j$. Let a_i denote the reflected signal from a reflector with the delay τ_i. The clutter from a set of stationary reflectors is

$$n(t) = \sum_i a_i s(t - \tau_i).$$

Figure 12.13 shows a clutter signal from a sparse set of stationary reflectors that are sparse enough to appear as individual interfering reflectors. Figure 12.14 shows a clutter signal from a dense set of reflectors. These reflectors cannot be resolved; therefore one may prefer to treat the clutter signal as noise. If the clutter reflectors are moving, then

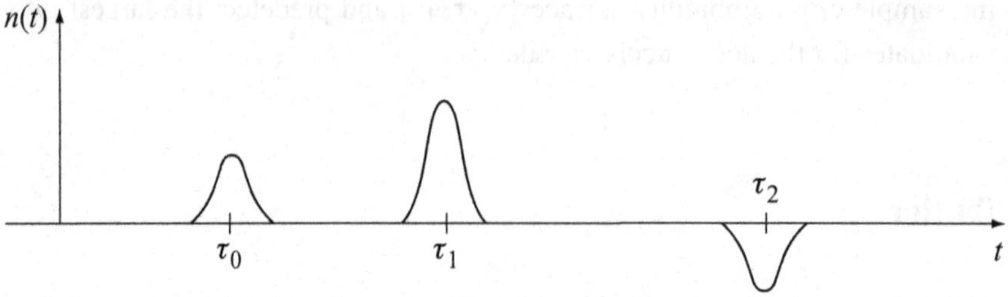

Figure 12.13 Illustrating sparse clutter

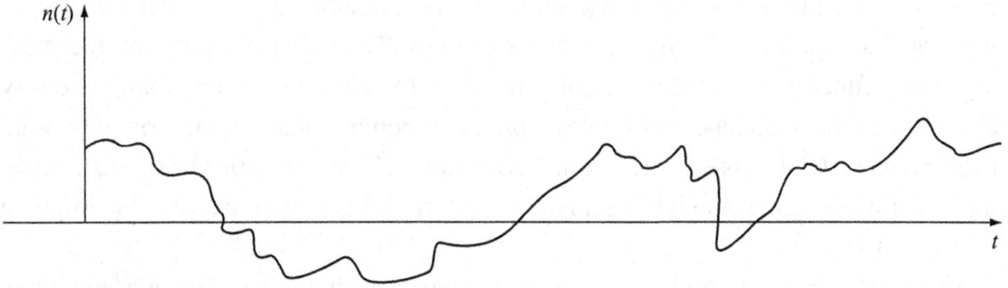

Figure 12.14 Illustrating dense clutter

the clutter signal has the form

$$n(t) = \sum_i a_i s(t - \tau_i)\, e^{j2\pi \nu_i t}$$

which, again, looks like noise if the individual terms are not resolved.

The first, and most simplistic, approach in dealing with clutter is to treat it as equivalent noise. The sample correlation function of the clutter is

$$\phi_c(\tau) = \int_{-\infty}^{\infty} n(t) n^*(t - \tau)\, dt.$$

If the reflectors are not moving, the correlation function is

$$\phi_c(\tau) = \sum_i \sum_j a_i a_j^* \int_{-\infty}^{\infty} s(t - \tau_i) s^*(t - \tau - \tau_j)\, dt.$$

The Fourier transform is obtained by using the convolution theorem as

$$\begin{aligned}\Phi_c(f) &= \sum_i \sum_j a_i a_j^* [S(f)\, e^{j2\pi \tau_i}][S(f)\, e^{j2\pi \tau_j}]^* \\ &= |S(f)|^2 \sum_i \sum_j a_i a_j^*\, e^{j2\pi(\tau_i - \tau_j)}.\end{aligned}$$

The double summation is a real number because the ij term is the complex conjugate of the ji term. Define the clutter power density spectrum as the expectation

$$\begin{aligned} N(f) &= \mathrm{E}[\Phi_c(f)] \\ &= |S(f)|^2 \mathrm{E}\left[\sum_i \sum_j a_i a_j^* \, \mathrm{e}^{\mathrm{j}2\pi(\tau_i - \tau_j)}\right]. \end{aligned}$$

If the reflector amplitudes are modeled as independent random variables, then

$$N(f) = \gamma |S(f)|^2$$

for some real constant γ. This is the power density spectrum of stationary clutter, which depends on the signal spectrum. For this simple model, one ignores the fact that the clutter originates with the signal, and models the clutter as random noise that is independent of the signal, but which has a power density spectrum that is the same as the signal power spectrum.

Suppose that we have a signal in independent additive noise with the power density spectrum $N(f) = \gamma |S(f)|^2$. The matched filter is

$$\begin{aligned} G(f) &= \frac{S^*(f)}{N(f)} \\ &= \frac{1}{\gamma S(f)}. \end{aligned}$$

This filter, clearly, is not realizable because $S(f)$ goes to zero as $|f|$ goes to infinity, which implies that $|S(f)|^{-1}$ goes to infinity. Indeed, the output of the filter would be an impulse because

$$G(f)S(f) = 1/\gamma,$$

which is independent of frequency, and the output signal-to-noise ratio would be

$$\begin{aligned} \frac{S}{N} &= \int_{-\infty}^{\infty} \frac{|S(f)|^2}{N(f)} \, \mathrm{d}f \\ &= \int_{-\infty}^{\infty} \frac{1}{\gamma} \, \mathrm{d}f, \end{aligned}$$

which is infinite. Thus, this model, while informative, is not realistic.

An approximation to the clutter-matched filter is the *Urkowitz filter* given by

$$G(f) = \begin{cases} 1/S(f) & |f| \leq W/2 \\ 0 & \text{otherwise.} \end{cases}$$

Therefore

$$G(f)S(f) = \begin{cases} 1 & |f| \leq W/2 \\ 0 & \text{otherwise,} \end{cases}$$

so the filter output is

$$\begin{aligned} u(t) &= \int_{-W/2}^{W/2} e^{j2\pi ft}\, df \\ &= W \operatorname{sinc} Wt. \end{aligned}$$

The value at the origin can be made arbitrarily large by making W large.

The Urkowitz filter has been stated for a somewhat artificial problem because it ignores additive thermal noise, which is independent of the signal and which will always be present. If white thermal noise is included, the total power density spectrum for the clutter plus the noise is given by $\gamma|S(f)|^2 + N(f)$. Then the clutter-matched filter is replaced by the whitened matched filter

$$G(f) = \frac{S^*(f)}{\gamma|S(f)|^2 + N(f)}.$$

This filter behaves like the whitened matched filter at frequencies where thermal noise dominates clutter, and it behaves like the Urkowitz filter at frequencies where clutter dominates thermal noise. Between these is a gradual transition.

It is interesting to notice that the discussion suggests that, if the clutter is regarded as a set of sparse reflectors in thermal noise, we should use the filter

$$G(f) = \frac{S^*(f)}{N(f)},$$

while if the clutter is regarded as dense, we might use the modified Urkowitz filter above. We cannot give a general answer as to which of these is correct. Both methods have been developed in a somewhat ad hoc way – not to minimize the probability of error but to maximize some measure of signal-to-noise ratio.

12.8 Detection of moving objects

A radar waveform that is designed to have an ambiguity function with good resolution in the doppler direction can be used to detect moving objects even though there may be stationary (or slowly moving) reflectors that are many orders of magnitude larger. If only the moving objects are of interest, then the cumulative reflected signal from the stationary objects is regarded as clutter. A *moving-target detection* (MTD) radar is one that is designed to detect the presence of moving objects in the presence of such clutter by using a waveform with an appropriate ambiguity function. The term MTD refers to radars that detect moving reflectors and measure one or more components of the velocity vector, as well as radars that measure only the range and angle from the radar to the moving reflector.

An MTD radar may be stationary or it may be moving. If the radar is moving, the detection of moving reflectors is much more difficult, as is the estimation of their

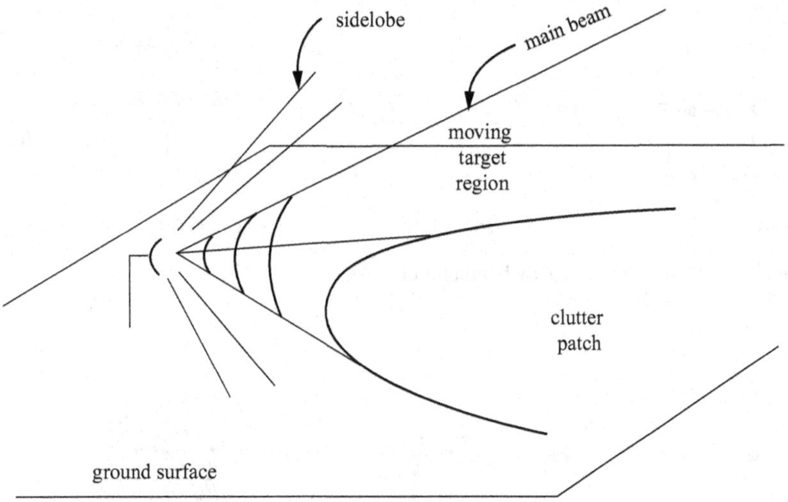

Figure 12.15 Moving-target detection from stationary radar

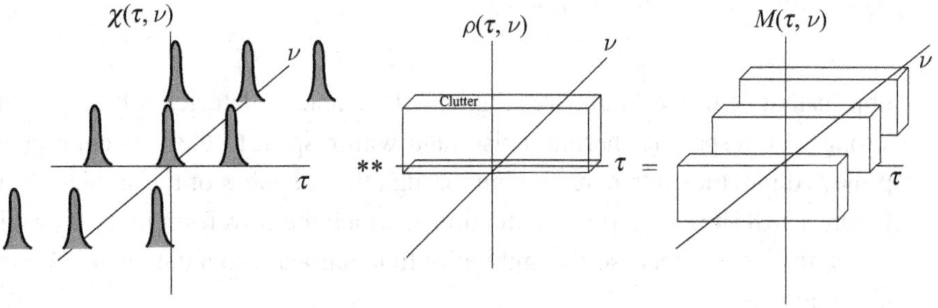

Figure 12.16 Illustrating the effect of clutter

parameters. For a stationary MTD radar fixed on the ground, shown in Figure 12.15, the targets are in motion with respect to the radar; any object with a doppler shift in an appropriate interval can be detected and declared to be a target.

The ambiguity function can be used to study the effect of clutter on a linear receiver. The radar return from moving targets is cluttered by echoes from stationary or slowly moving reflectors in the antenna main beam or sidelobes. The sample cross-ambiguity function computed from the received signal will be cluttered by the effect of the echoes. We want to determine when the clutter masks the signal in the sample cross-ambiguity function. Figure 12.16 shows an example in which the clutter reflectivity density $\rho(\tau, \nu)$ is confined to small values of ν. This clutter reflectivity density $\rho(\tau, \nu)$ is (two-dimensionally) convolved with the ambiguity function of the radar waveform to produce clutter in the image. Figure 12.17 shows a plan view which illustrates that the clutter, when confined to small values of ν, may contaminate certain regions of the sample ambiguity function while leaving other regions free of clutter. Moving

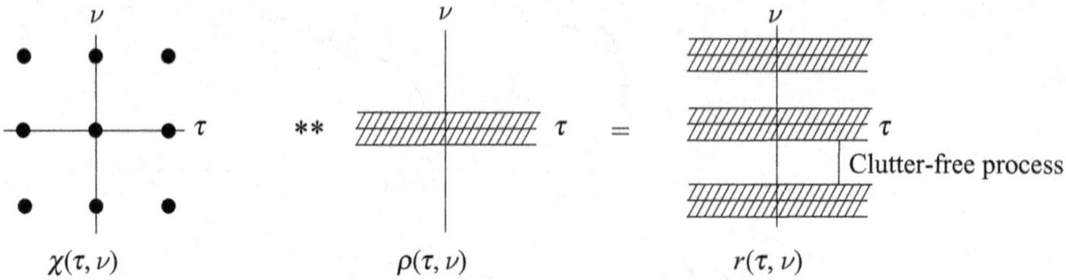

Figure 12.17 Plan view for narrowband clutter

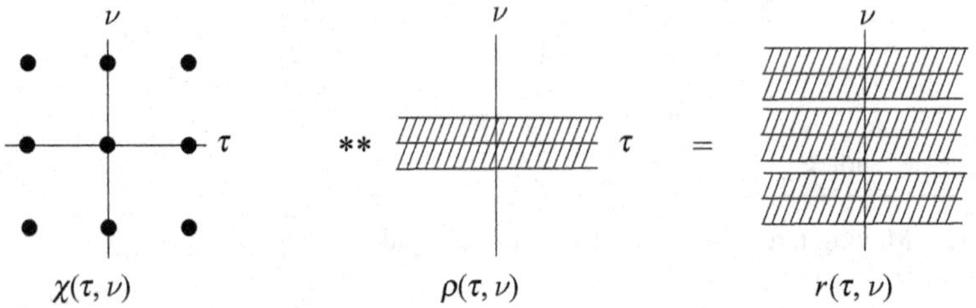

Figure 12.18 Plan view for wideband clutter

objects that lie in the clutter-free region will be readily detected if they are sufficiently strong with respect to thermal noise (and with respect to the incidental clutter noise projected into the clutter-free region through the sidelobes of the ambiguity function). Figure 12.18 shows a different situation in which the waveform may be unsuitable for the clutter model because the ambiguity function leads to a clutter-free region that is too small.

For a moving MTD radar, as shown in Figure 12.19, clutter suppression can be much more difficult because the stationary objects in the background will be in relative motion with respect to the radar. Therefore signals reflected from stationary objects will have a doppler shift and the clutter reflectivity density is not confined to small values of ν. If the doppler of the clutter is comparable to that of the moving target, then doppler alone is not enough to separate the targets from the clutter.

A moving object in the air and a stationary object on the ground with the same doppler can both be within the illumination beam and a signal from one can mask the other. Therefore, a simple MTD radar might not be able to detect slowly moving objects in the presence of clutter from stationary objects. There may also be slowly moving objects on the ground which means that the doppler of objects on the ground cannot be predicted only from knowledge of position.

An alternative method of suppressing clutter is by using multiple antennas to form a null in the antenna beam. This technique is based on the fact that the many reflectors in the same range-doppler cell have different azimuth angles. Hence the signal contained in a single range-doppler cell can be sorted out by interferometric phase information.

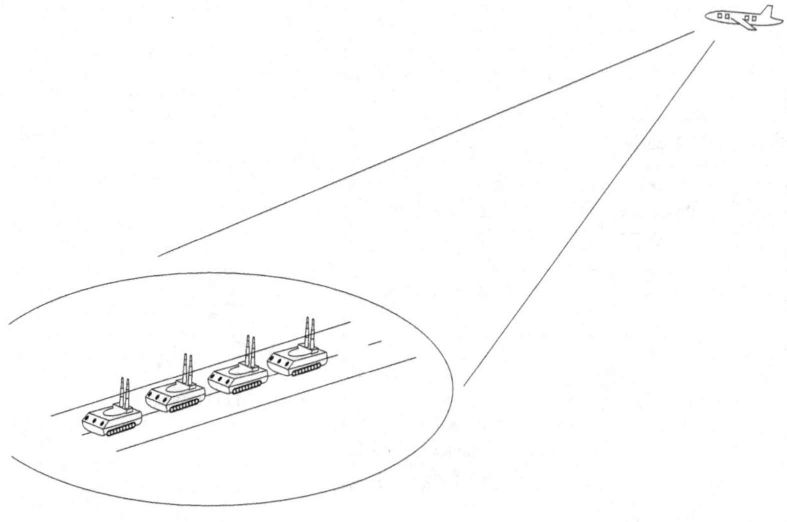

Figure 12.19 Moving target detection

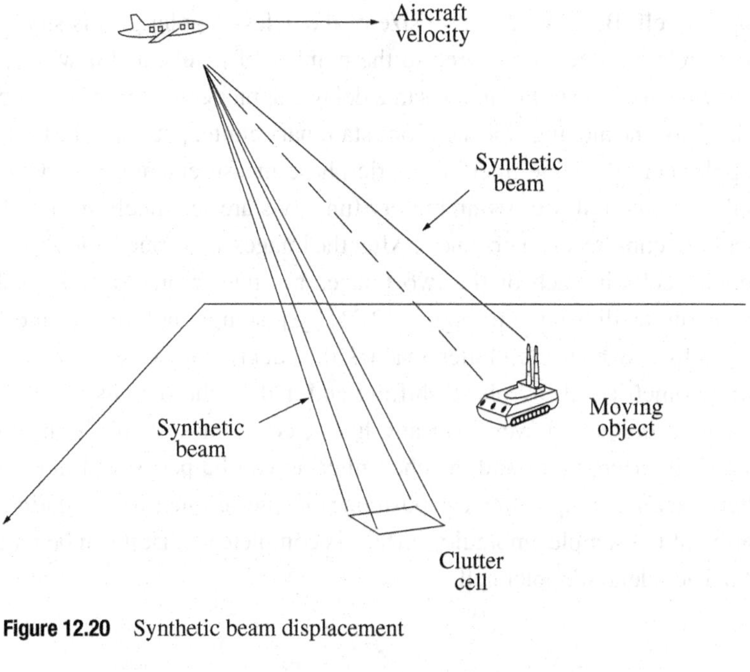

Figure 12.20 Synthetic beam displacement

For a radar, shown in Figure 12.20, with velocity V_t and a reflecting object with velocity V_o, the relative velocity is

$$V_r = V_t \sin\theta + V_o \sin\phi$$

where θ and ϕ are the angles between the transmitter and reflector velocity vectors and the range vector.

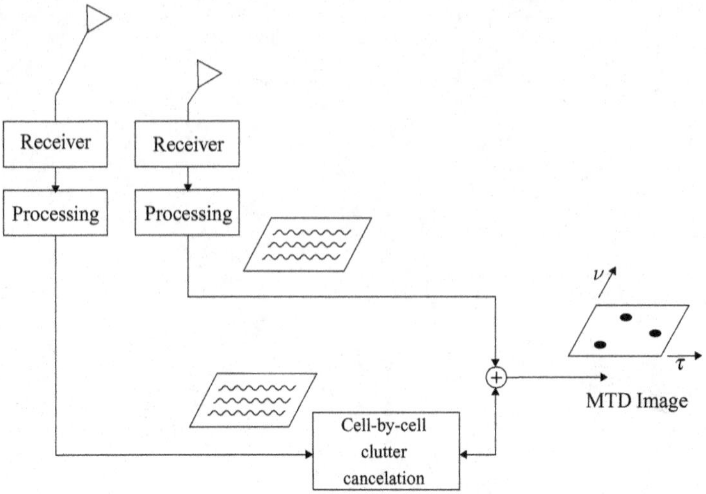

Figure 12.21 Clutter notching

All reflectors at the same range R and relative velocity V_r will fall in the same delay-doppler cell. Because the mesh size of the delay-doppler grid is small, moving vehicles are relatively few compared to the number of pixels, and it will be unusual for two moving vehicles to fall in the same delay-doppler cell. Hence, with occasional exceptions, only one moving object and one stationary clutter patch will be found in each delay-doppler cell. A single interferometric phase measurement can sort those. This requires that two complex cross-ambiguity functions are separately formed from two side-by-side antenna-receiver channels. After the images are formed, the corresponding delay-doppler cells in each of the two images are interferometrically combined to suppress clutter, as illustrated in Figure 12.21, by placing a null of the antenna beam at the angle from which the clutter reaches that delay-doppler cell. The formation of an interferometric null by phase shifting and adding the outputs of two antennas is a familiar technique. However, in any linear, coherent system, the mathematical operations of interferometry and matched filtering can be performed in either order. The clutter-notching computation can postpone the interferometric calculation until the computation of the sample ambiguity surface is complete so that it can be individually adjusted in each delay-doppler cell.

12.9 Coherent estimation of pulse parameters

The task of coherent estimation of pulse parameters is stated as follows: we are given a received complex baseband signal of the form

$$v(t) = s(t - \tau_0)\, e^{j2\pi \nu_0 t} + n(t)$$

where $s(t)$ is a known waveform with finite energy, and

$$n(t) = n_R(t) + jn_I(t)$$

is a stationary, complex noise process whose components $n_R(t)$ and $n_I(t)$ are independent and have known and identical correlation functions $\phi(\tau)$ and power density spectrum $N(f)$. The parameters τ_0 and ν_0 are unknown.

The task of noncoherent estimation of pulse parameters is an alternative problem to be studied in Section 12.10, in which the phase of the received signal is not known. In that case, the received signal is

$$v(t) = s(t - \tau_0)\, e^{j2\pi\nu_0 t}\, e^{j2\pi\theta_0} + n(t).$$

The new feature is the inclusion of an unknown phase θ_0.

In either case, we wish to estimate the unknown arrival time τ_0, or the unknown arrival frequency ν_0, of the received pulse, or both. Intuitively, we may simply compute the cross-ambiguity function – either the real part or the magnitude – and set the coordinates of the peak in the (τ, ν) plane as the estimates $(\widehat{\tau}_0, \widehat{\nu}_0)$ of the parameters (τ_0, ν_0). In this section and the next section, we shall study the coherent and noncoherent estimators.

We shall begin our treatment with a simpler estimation problem, which is the estimation of the arrival time of a real-valued pulse in noise. The received baseband waveform to be considered is

$$v(t) = s(t - \tau_0) + n(t).$$

The noise is covariance stationary noise with the known correlation function $\phi(\tau)$ and power density spectrum $N(f)$. Higher-order moments of the noise are not given; the noise $n(t)$ need not be gaussian noise. Accordingly, we will restrict ourselves to a class of estimators, called *linear estimators*, that can be judged by using only the second-order properties of the noise. The linear estimator passes $v(t)$ through a linear filter, $g(t)$, and determines the estimate $\widehat{\tau}_0$ from the time at which the filter output is maximum. Without loss of generality, we can require that $g(t) * s(t)$ takes its maximum value at the origin. (Otherwise, replace $g(t)$ by an appropriate translation.) In the absence of noise, the filter output is $g(t) * s(t - \tau_0)$. This means that $g(t) * s(t - \tau_0)$ has its maximum value at τ_0. The linear estimator chooses the time of the maximum value of $g(t) * v(t)$ as the estimate $\widehat{\tau}_0$.

The linear estimator whose filter $g(t)$ satisfies

$$G(f) = \frac{S^*(f)}{N(f)}$$

is called a *matched-filter estimator.*

Proposition 12.9.1 *The matched-filter estimator of pulse arrival time in additive stationary gaussian noise of power density spectrum $N(f)$ has the smallest variance of*

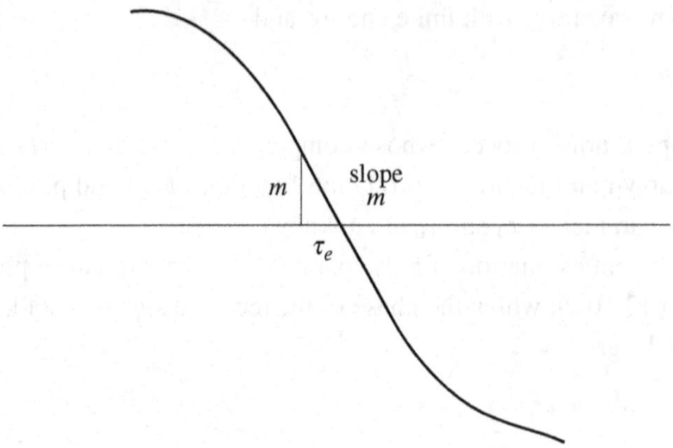

Figure 12.22 Linearized triangle

any linear estimator. The variance satisfies

$$\sigma_{\tau_0}^2 \simeq \left[(2\pi)^2 \int_{-\infty}^{\infty} f^2 \frac{|S(f)|^2}{N(f)}\,\mathrm{d}f\right]^{-1},$$

and the approximation is tight asymptotically as the bracketed term becomes large.

Proof: Let the random variable $u(t)$ denote the filter output, and let $\dot{u}(t)$ be the time derivative of $u(t)$, which can be written

$$\dot{u}(t) = \int_{-\infty}^{\infty} \dot{g}(t-\xi)v(\xi)\,\mathrm{d}\xi \quad = \quad \int_{-\infty}^{\infty} \dot{g}(\xi)v(t-\xi)\,\mathrm{d}\xi$$

where $\dot{g}(t)$ is the time derivative of $g(t)$. The estimate τ_0 is taken as the peak of $u(t)$. This is the value $\widehat{\tau_0}$ such that $\dot{u}(\widehat{\tau_0}) = 0$. We will find the variance of the error τ_e by using a linearized analysis based on the triangle shown in Figure 12.22. The linearized analysis is only valid when the signal from the matched filter is large compared with the noise because, otherwise, second-order terms become important and the linearized approximation breaks down.

Based on the linearized triangle, we have for the error in τ_0,

$$m\tau_e = \dot{u}(\tau_0),$$

and the variance is

$$\sigma_{\tau_0}^2 = \frac{\mathrm{var}[\dot{u}(\widehat{\tau_0})]}{m^2}$$

where the slope m is given by

$$m = \frac{\mathrm{d}}{\mathrm{d}t}\mathrm{E}[\dot{u}(t)]_{t=\widehat{\tau_0}} = (2\pi)^2 \int_{-\infty}^{\infty} f^2 S(f)G(f)\,\mathrm{d}f.$$

The mean of $\dot{u}(t)$ at any point t is given by

$$\begin{aligned} \mathrm{E}[\dot{u}(t)] &= \mathrm{E}\int_{-\infty}^{\infty} \dot{g}(\xi)v(t-\xi)\,\mathrm{d}\xi \\ &= \int_{-\infty}^{\infty} \dot{g}(\xi)s(t-\xi)\,\mathrm{d}\xi. \end{aligned}$$

At $t = 0$, Parseval's formula gives

$$\mathrm{E}[\dot{u}(0)] = \int_{-\infty}^{\infty} \mathrm{j}2\pi f G(f)S(f)\,\mathrm{d}f$$

where $G(f)$ is the Fourier transform of $g(t)$.

The variance in $\dot{u}(t)$ is due to the noise only:

$$\begin{aligned} \mathrm{var}[\dot{u}(t)] &= \int_{-\infty}^{\infty}\int_{-\infty}^{\infty} \dot{g}(\xi)\dot{g}(\xi')\mathrm{E}[n(t-\xi)n(t-\xi')|\,\mathrm{d}\xi\,\mathrm{d}\xi' \\ &= \int_{-\infty}^{\infty}\int_{-\infty}^{\infty} \dot{g}(\xi)\dot{g}(\xi')\phi(\xi-\xi')\,\mathrm{d}\xi\,\mathrm{d}\xi'. \end{aligned}$$

This is independent of t. Now make the change in variables $\eta = \xi - \xi'$ so that

$$\mathrm{var}[\dot{u}(t)] = \int_{-\infty}^{\infty} \phi(\xi)\int_{-\infty}^{\infty} \dot{g}(\xi)\dot{g}(\xi-\eta)\,\mathrm{d}\xi\,\mathrm{d}\eta.$$

To transform this formula into the frequency domain, use Parseval's theorem and note that the convolution $\int \dot{g}(\xi)\dot{g}(\xi-\eta)\,\mathrm{d}\xi$ has the transform $(2\pi f)^2|G(f)|^2$. Then

$$\mathrm{var}[\dot{u}(t)] = (2\pi)^2\int_{-\infty}^{\infty} f^2 N(f)|G(f)|^2\,\mathrm{d}f.$$

Therefore

$$\sigma_{\tau_0}^2 = \left(\frac{1}{2\pi}\right)^2 \frac{\int_{-\infty}^{\infty} f^2 N(f)|G(f)|^2\,\mathrm{d}f}{\left[\int_{-\infty}^{\infty} f^2 S(f)G(f)\,\mathrm{d}f\right]^2}.$$

To minimize the error variance by the choice of $G(f)$, we manipulate the denominator into an appropriate form and use the Schwarz inequality (noting that $N(f)$ is real and nonnegative)

$$\begin{aligned} \left|\int_{-\infty}^{\infty} f^2 S(f)G(f)\,\mathrm{d}f\right|^2 &= \left|\int_{-\infty}^{\infty} [fG(f)N^{1/2}(f)]\left[\int_{-\infty}^{\infty} \frac{S(f)}{N^{1/2}(f)}\,\mathrm{d}f\right]\right|^2 \\ &\le \int_{-\infty}^{\infty} f^2|G(f)|^2 N(f)\,\mathrm{d}f \int_{-\infty}^{\infty} f^2\frac{|S(f)|^2}{N(f)}\,\mathrm{d}f \end{aligned}$$

with equality if and only if

$$G(f) = \frac{S(f)^*}{N(f)}.$$

Therefore, for this choice of filter, we have the conclusion of the proposition. □

How should the estimator be changed if the pulse is a complex baseband pulse? We know that the signal output of the matched filter is real (and equal to E_p) at the peak, so the estimator should find the peak of the real part of the output signal. Otherwise, the estimator is the same.

For any covariance stationary process, the matched-filter estimator is the best linear estimator, at least to the extent that the linearized model is valid, but there may be better nonlinear estimators. Consequently, if one knows only the power density spectrum of covariance stationary noise, but not higher-order moments, one may choose to use this estimator. If the noise is gaussian, this estimator is actually the optimum estimator asymptotically in the signal-to-noise ratio. If the noise is nongaussian, the linear estimator still has this performance, though it is then not necessarily the best estimator. This partially justifies the common assumption of gaussian noise when the true noise is not known or is intractable.

Corollary 12.9.2 *The asymptotic variance of the matched-filter estimate of the arrival time of a pulse in white noise is*

$$\sigma_{\tau_0}^2 = \frac{1}{(2\pi)^2(2E_p/N_0)\overline{f^2}}$$

where

$$\overline{f^2} = \frac{1}{E_p}\int_{-\infty}^{\infty} f^2|S(f)|^2 \, df.$$

Proof: Set $N(f) = N_0/2$ in Proposition 12.9.1. □

The conclusion of Corollary 12.9.2 can be expressed in terms of the Gabor bandwidth as

$$\sigma_{\tau_0}^2 = \frac{1}{(2\pi)^2(2E_p/N_0)(B_G^2 + \overline{f}^2)}.$$

This expression should be contrasted with the like expression for the noncoherent estimator given in the next section.

A task that is dual to the coherent estimation of pulse arrival time is the coherent estimation of pulse arrival frequency. This is a somewhat artificial problem because it assumes that phase is known even though frequency is unknown. By analogy with the asymptotic variance of pulse arrival time, we can suppose that the error of a coherent matched-filter estimator in white noise of pulse arrival frequency has an asymptotic variance given by

$$\sigma_{\nu_0}^2 = \frac{1}{(2\pi)^2(2E_p/N_0)(T_G^2 + \overline{t}^2)}$$

where $N(f) = N_0/2$ is the power density spectrum of the noise. The matched-filter estimator passes $v(t)\,e^{j2\pi ft}$ through a filter matched to $s(t)$ and then searches over f for the maximum filter output.

12.10 Noncoherent estimation of pulse parameters

We shall show that the asymptotic error variance of the noncoherent estimator of the arrival time of a pulse of known frequency is

$$\sigma_{\tau_0}^2 = \frac{1}{(2\pi)^2(2E_p/N_0)B_G^2},$$

where B_G is the Gabor bandwidth. The asymptotic error variance of the noncoherent estimator of the arrival frequency of a pulse of known arrival time is

$$\sigma_{\nu_0}^2 = \frac{1}{(2\pi)^2(2E_p/N_0)T_G^2}$$

where T_G is the Gabor timewidth. The difference between the performance of the coherent and noncoherent estimators in white noise is due ultimately to the fact that, for small τ,

$$\mathrm{Re}[\chi(\tau, 0)] = E_p[1 - 2\pi^2(\overline{f^2}\tau^2 + 2\mathrm{Re}[\overline{tf}]\tau\nu + \overline{t^2}\nu^2)],$$

while

$$|\chi(\tau, 0)| = E_p[1 - 2\pi^2(B_G^2\tau^2 + 2T_G B_G \rho\tau\nu + T_G^2\nu^2)],$$

as illustrated in Figure 12.23. Because the peak is sharper in the first case, it can be located more accurately.

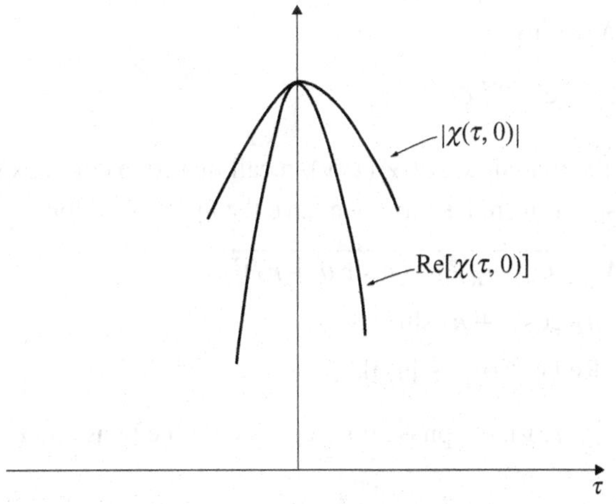

Figure 12.23 Illustrating the curvature of $\chi(\tau, 0)$

For a pulse of known frequency in white noise, the matched filter is equivalent to computing the sample cross-ambiguity function at the known frequency, which we may take to be zero. Thus, in the language of the ambiguity function, to estimate τ_0 from the received signal $v(t)$, we should first compute the cross-ambiguity function $\chi_c(\tau, \nu)$ with $\nu = 0$. Then, if $v(t)$ is real, find the peak of $\chi_c(\tau, 0)$, and if $v(t)$ is complex, find the peak of $\text{Re}[\chi_c(\tau, 0)]$. Thus, the sample cross-ambiguity function $\chi_c(\tau, \nu)$ is a statistic for estimating τ_0.

The noncoherent waveform with an unknown arrival time is

$$v(t) = s(t - \tau_0)\,\mathrm{e}^{\mathrm{j}\theta} + n(t)$$

where now $s(t)$ and $n(t)$ are taken to be complex. We have decided to estimate τ_0 by first computing $\chi(\tau, \nu)$, setting $\nu = 0$, and finding the peak. However, because of the unknown phase, we find the peak of $|\chi(\tau, 0)|$, or equivalently, of $|\chi(\tau, 0)|^2$.

More generally, let

$$v(t) = s(t - \tau_0)\,\mathrm{e}^{\mathrm{j}2\pi\nu_0 t}\,\mathrm{e}^{\mathrm{j}\theta} + n(t).$$

To estimate τ_0 when ν_0 is known, we find the peak of $|\chi(\tau, \nu_0)|$. This is a noncoherent estimator of τ_0. To estimate ν_0 when τ_0 is known, we find the peak of $|\chi(\tau_0, \nu)|$ as a function of ν. This is a noncoherent estimator of ν_0. To simultaneously estimate τ_0 and ν_0, we find the peak of $|\chi(\tau, \nu)|$ in both variables. This is a noncoherent estimator of the pair (τ_0, ν_0).

The variance of the matched-filter estimator does not depend on the particular values assumed by τ_0 and ν_0, so we can choose the special case in which $\tau_0 = 0$ and $\nu_0 = 0$ to simplify the analysis. The sample cross-ambiguity function is

$$\chi_c(\tau, \nu) = \chi(\tau, \nu)\,\mathrm{e}^{\mathrm{j}\theta} + \chi^{(n)}(\tau, \nu)$$

where the noise term on the ambiguity function is

$$\begin{aligned}\chi^{(n)}(\tau, \nu) &= N_R(\tau, \nu) + \mathrm{j}N_I(\tau, \nu)\\ &= \int_{-\infty}^{\infty} s(t - \tau)n(t)\,\mathrm{e}^{-\mathrm{j}2\pi\nu t}\,\mathrm{d}t.\end{aligned}$$

To analyze the noise in the magnitude of $\chi_c(\tau, \nu)$, recall that for a complex signal in noise, for large values of signal-to-noise ratio, we have the approximation

$$\begin{aligned}|A\,\mathrm{e}^{\mathrm{j}\theta} + n_R + \mathrm{j}n_I| &= \sqrt{(A\cos\theta + n_R)^2 + (A\sin\theta + n_I)^2}\\ &\approx A + n_R\cos\theta + n_I\sin\theta\\ &= A + \text{Re}\,[\,\mathrm{e}^{-\mathrm{j}\theta}(n_R + \mathrm{j}n_I)].\end{aligned}$$

To apply this reasoning to $\chi_c(\tau, \nu)$, the phase θ is replaced by the phase of $\chi(\tau, \nu)\,\mathrm{e}^{\mathrm{j}\theta}$. Thus

$$|\chi_c(\tau, \nu)| \approx |\chi(\tau, \nu)| + N_e(\tau, \nu)$$

where

$$N_e(\tau,\nu) = \mathrm{Re}\left[\frac{\chi(\tau,\nu)}{|\chi(\tau,\nu)|}\,\mathrm{e}^{\mathrm{j}\theta}\chi^{(n)}(\tau,\nu)\right]$$
$$= \mathrm{Re}\left[\frac{\chi(\tau,\nu)}{|\chi(\tau,\nu)|}\,\mathrm{e}^{\mathrm{j}\theta}\int_{-\infty}^{\infty} s(t-\tau)n^*(t)\,\mathrm{e}^{\mathrm{j}2\pi\nu t}\,\mathrm{d}t\right].$$

Because white noise is statistically invariant under a phase shift, we can absorb the phase term θ by redefining $n(t)$.

Now that we have an expression for the noise on the ambiguity surface, we must find an expression for its magnitude.

Lemma 12.10.1 *Let* $\mathrm{e}^{\mathrm{j}\phi(\tau,\nu)} = \chi(\tau,\nu)/|\chi(\tau,\nu)|$. *Then*

$$\mathrm{E}[N_e(\tau,\nu)N_e(\tau',\nu')| = \frac{N_0}{2}\,\mathrm{Re}[\,\mathrm{e}^{\mathrm{j}[\phi(\tau',\nu')-\phi(\tau,\nu)]}\,\mathrm{e}^{\mathrm{j}2\pi(\nu-\nu')}\chi(\tau-\tau',\nu-\nu')].$$

Proof: For any complex numbers, z and z', we have the useful identity on the real part

$$\mathrm{Re}[z]\mathrm{Re}[z'] = \tfrac{1}{2}\mathrm{Re}[zz' + z^*z'],$$

and

$$\mathrm{E}[\mathrm{Re}[z]\mathrm{Re}[z']] = \tfrac{1}{2}\mathrm{Re}[\mathrm{E}[zz'] + \mathrm{E}[z^*z']].$$

For our problem,

$$z = \mathrm{e}^{\mathrm{j}\phi(\tau,\nu)}\int_{-\infty}^{\infty} s(\xi-\tau)n^*(\xi)\,\mathrm{e}^{\mathrm{j}2\pi\nu\xi}\,\mathrm{d}\xi.$$

In the computation of $\mathrm{E}[z\,z']$, the expectation can be brought inside the double integral to produce a term, $\mathrm{E}[n^*(\xi)n^*(\xi')]$, that is equal to zero for white complex noise that is invariant under phase change because

$$\mathrm{E}[n^*(t)n^*(t')] = \mathrm{E}[n_R(t)n_R(t')] - \mathrm{E}[n_I(t)n_I(t')] - \mathrm{j}\mathrm{E}[n_R(t)n_I(t')] - \mathrm{j}\mathrm{E}[n_I(t)n_R(t')]$$
$$= 0.$$

In the computation of $\mathrm{E}[z^*z']$, there will be the term $\mathrm{E}[n(\xi)n^*(\xi')]$ that equals $N_0\delta(\xi-\xi')$ for white noise. Therefore

$$\mathrm{E}[N_e(\tau,\nu)N_e(\tau',\nu')]$$
$$= \tfrac{1}{2}\mathrm{Re}[\,\mathrm{e}^{\mathrm{j}[\phi(\tau',\nu')-\phi(\tau,\nu)]}\int_{-\infty}^{\infty}\int_{-\infty}^{\infty} s^*(\xi-\tau)s(\xi'-\tau')N_0\delta(\xi-\xi')\,\mathrm{e}^{\mathrm{j}2\pi(\nu-\nu')t}\,\mathrm{d}t,$$

and the rest of the proof follows directly. □

Proposition 12.10.2 *The variance of the noncoherent estimator of* τ_0*, based on maximizing* $|\chi(\tau, \nu_0)|$*, is given by*

$$\sigma_{\tau_0}^2 = \frac{1}{(2\pi)^2(2E_p/N_0)B_G^2}$$

asymptotically for large E_p/N_0.

Proof: To find the peak of a function, set the derivative equal to zero. To determine the error in the estimate of the peak location due to small noise terms, use a first-order perturbation analysis. This analysis requires the partial derivatives of $|\chi(\tau, \nu)|$. They can be computed from the series approximation

$$|\chi(\tau, \nu)| = E_p[1 - 2\pi^2\nu^2 T_G^2 - 4\pi^2\rho\tau\nu T_G B_G - 2\pi^2\tau^2 B_G^2],$$

which was computed in Section 6.3. Consequently,

$$\left.\frac{\partial^2|\chi(\tau, \nu)|}{\partial\tau^2}\right|_{\tau=\nu=0} = -(2\pi)^2 B_G^2 E_p.$$

As in the proof of Proposition 12.9.1, we have

$$\sigma_{\tau_0}^2 = \frac{\mathrm{var}[\frac{\partial}{\partial\tau}|\chi_c(\tau_0, \nu_0)|]}{m^2}.$$

To evaluate the numerator, we first want to compute derivatives of the noise at the origin:

$$N_e(\tau, \nu) = \mathrm{Re}\left[\mathrm{e}^{\mathrm{j}\phi(\tau,\nu)} \int_{-\infty}^{\infty} s(t-\tau)n^*(t)\,\mathrm{e}^{\mathrm{j}2\pi\nu t}\,\mathrm{d}t\right].$$

Replace $\phi(\tau, \nu)$ in Lemma 12.10.1 by a linearized approximation:

$$\begin{aligned} &\mathrm{E}[N_e(\tau, \nu)N_e(\tau', \nu')] \\ &\approx \frac{N_0}{2}\mathrm{Re}[(1 + \mathrm{j}2\pi\overline{f}\tau + \mathrm{j}2\pi\overline{t}\nu)(1 - \mathrm{j}2\pi\overline{f}\tau' - \mathrm{j}2\pi\overline{t}\nu')\,\mathrm{e}^{\mathrm{j}2\pi(\nu-\nu')\tau}\chi(\tau - \tau', \nu - \nu')]. \end{aligned}$$

From this equation, it is straightforward to compute the numerator needed for the variance in the estimate. Set $\nu = \nu' = 0$, and take derivatives with respect to τ and τ' evaluated at $\tau = \tau' = 0$. This gives

$$\mathrm{E}\left[\left(\frac{\partial N_e}{\partial\tau}\right)^2\right] \approx \frac{N_0}{2}\mathrm{Re}\left[\frac{\partial}{\partial\tau}\frac{\partial}{\partial\tau'}(1 - \mathrm{j}2\pi\overline{f}\tau)(1 + \mathrm{j}2\pi\overline{f}\tau')\chi(\tau - \tau', 0)\right].$$

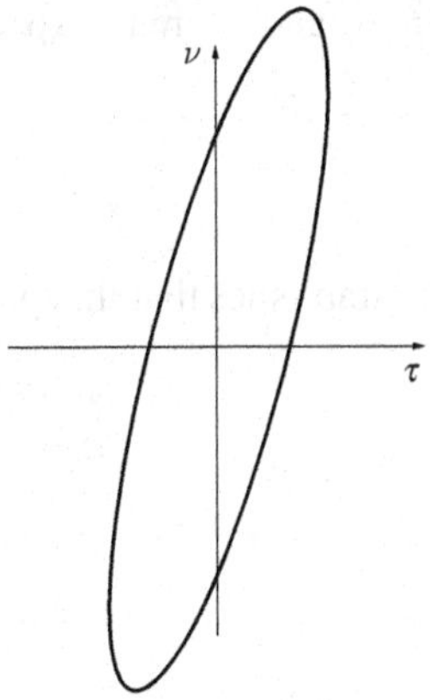

Figure 12.24 An uncertainty ellipse

Set $\tau = \tau' = 0$,

$$= \frac{N_0}{2} \operatorname{Re}\left[-(2\pi \overline{f})^2 \chi(0,0) - \frac{\partial^2 \chi(0,0)}{\partial \tau^2}\right]$$

$$= \frac{N_0}{2}(2\pi)^2 E_p[-(\overline{f})^2 + \overline{f^2}]$$

$$= \frac{N_0}{2} E_p(2\pi B_G)^2.$$

Consequently,

$$\sigma_{\tau_0}^2 = \frac{\frac{N_0}{2} E_p(2\pi B_G)^2}{[(2\pi B_G)^2 E_p]^2},$$

which reduces to the statement of the proposition, so the proof is complete. □

When both τ and ν are unknowns to be estimated, we may expect that the estimate in τ will be less accurate than it would be if ν were known. A rough argument may be formulated by looking at the uncertainty ellipse in Figure 12.24. When ν is known, the Gabor bandwidth B_G appears in the equation for σ_τ^2. When ν is unknown, we may expect that the projection of the uncertainty ellipse onto the axis replaces the role of B_G. To find the maximum excursion of the ellipse

$$\tau^2 B_G^2 + 2\tau\nu T_G B_G \rho + \nu^2 T_G^2 = 1$$

in the τ direction, rewrite it in the form

$$(B_G^2 - B_G^2 \rho^2)\tau^2 + (T_G \nu + \rho B_G \tau)^2 = 1.$$

The maximum value of τ can occur when ν is such that the second term on the left equals zero. Consequently, B_G^2 is replaced by $B_G^2 - B_G^2\rho^2$, and we may expect that

$$\sigma_\tau^2 = \frac{1}{(2\pi)^2(2E_p/N_0)B_G^2(1-\rho^2)}$$

is the modified expression for σ_τ^2. The next proposition establishes that this plausibility argument does indeed give the right formula.

Let $\boldsymbol{\Sigma}$ denote the covariance matrix of τ and ν errors

$$\boldsymbol{\Sigma} = \mathrm{E}\left[\begin{bmatrix}\tau_e \\ \nu_e\end{bmatrix}[\tau_e \;\; \nu_e]\right],$$

and let $\boldsymbol{A}$ denote the matrix of the second partial derivatives. Thus

$$\boldsymbol{A} = \begin{bmatrix}\frac{\partial^2|\chi|}{\partial\tau^2} & \frac{\partial^2|\chi|}{\partial\tau\partial\nu} \\ \frac{\partial^2|\chi|}{\partial\nu\partial\tau} & \frac{\partial^2|\chi|}{\partial\nu^2}\end{bmatrix} = (2\pi)^2 E_p \begin{bmatrix}B_G^2 & B_G T_G \rho \\ B_G T_G \rho & T_G^2\end{bmatrix}.$$

Proposition 12.10.3 *In the limit of a large signal-to-noise ratio, the covariance matrix satisfies*

$$\boldsymbol{\Sigma} = \frac{N_0}{2}\boldsymbol{A}^{-1}.$$

Proof: The peak of $|\chi_c(\tau, \nu)|$ occurs where the partial derivatives are equal to zero,

$$\frac{\partial|\chi|}{\partial\tau} = 0 \quad \frac{\partial|\chi|}{\partial\nu} = 0.$$

It is enough to consider $\tau_0 = 0$, $\nu_0 = 0$, and

$$|\chi_c(\tau, \nu)| = |\chi(\tau, \nu) + \chi_n(\tau, \nu)|,$$

where $\chi_n(\tau, \nu)$ is the contribution to $\chi_c(\tau, \nu)$ due to the noise. Because this is a rayleigh random variable for high signal-to-noise ratio and near the peak, we have the approximation

$$|\chi_c(\tau, \nu)| = |\chi(\tau, \nu)| + N(\tau, \nu)$$

where $N(\tau, \nu)$ is a gaussian random variable for each τ, ν. The conditions for a peak are now

$$\frac{\partial|\chi(\tau, \nu)|}{\partial\tau} + \frac{\partial N(\tau, \nu)}{\partial\tau} = 0$$

$$\frac{\partial|\chi(\tau, \nu)|}{\partial\nu} + \frac{\partial N(\tau, \nu)}{\partial\nu} = 0.$$

Because of the small noise terms, the leading terms must be perturbed from their peaks by error terms τ_e, ν_e. Thus

$$\tau_e \frac{\partial^2 |\chi|}{\partial \tau^2} + \nu_e \frac{\partial^2 |\chi|}{\partial \nu \partial \tau} + \frac{\partial N}{\partial \tau} = 0$$

$$\tau_e \frac{\partial^2 |\chi|}{\partial \tau \partial \nu} + \nu_e \frac{\partial^2 |\chi|}{\partial^2 \nu} + \frac{\partial N}{\partial \nu} = 0,$$

which can be abbreviated in matrix notation as

$$\begin{bmatrix} A_{11} & A_{12} \\ A_{21} & A_{22} \end{bmatrix} \begin{bmatrix} \tau_e \\ \nu_e \end{bmatrix} = - \begin{bmatrix} \frac{\partial N_e}{\partial \tau} \\ \frac{\partial N_e}{\partial \nu} \end{bmatrix}.$$

Then

$$\begin{bmatrix} \tau_e \\ \nu_e \end{bmatrix} = - \begin{bmatrix} A_{11} & A_{12} \\ A_{21} & A_{22} \end{bmatrix}^{-1} \begin{bmatrix} \frac{\partial N_e}{\partial \tau} \\ \frac{\partial N_e}{\partial \nu} \end{bmatrix}.$$

Let

$$\boldsymbol{\Sigma} = \mathrm{E}\left[\begin{bmatrix} \tau_e \\ \nu_e \end{bmatrix} [\tau_e \ \ \nu_e] \right]$$

$$= \begin{bmatrix} A_{11} & A_{12} \\ A_{21} & A_{22} \end{bmatrix}^{-1} \mathrm{E}\left[\begin{bmatrix} \frac{\partial N_e}{\partial \tau} \\ \frac{\partial N_e}{\partial \nu} \end{bmatrix} \begin{bmatrix} \frac{\partial N_e}{\partial \tau} & \frac{\partial N_e}{\partial \nu} \end{bmatrix} \right] \begin{bmatrix} A_{11} & A_{21} \\ A_{12} & A_{22} \end{bmatrix}^{-1}.$$

Recall from the proof of Proposition 12.10.2 that

$$\mathrm{E}\left[\left(\frac{\partial N_e}{\partial \tau} \right)^2 \right] = (2\pi)^2 \frac{N_0}{2} E_p B_G^2.$$

A similar analysis for the terms $\mathrm{E}\left[(\partial N_e / \partial \nu)^2 \right]$ and $\mathrm{E}\left[(\partial N_e / \partial \nu \; \partial N_e / \partial \tau) \right]$ begins with the τ equation of Lemma 12.10.1 with the phase terms replaced by linear approximations. First, set $\tau = \tau' = 0$ so that

$$\mathrm{E}\left[\left(\frac{\partial N_e}{\partial \nu} \right)^2 \right] = \frac{N_0}{2} \mathrm{Re}\left[\frac{\partial}{\partial \nu} \frac{\partial}{\partial \nu'} (1 + \mathrm{j}2\pi \bar{t} \nu)(1 + \mathrm{j}2\pi \bar{t} \nu') \chi(0, \nu - \nu') \right]$$

$$= (2\pi)^2 \frac{N_0}{2} E_p T_G^2.$$

Next, set $\tau = \nu' = 0$ so that

$$\mathrm{E}\left[\frac{\partial N_e}{\partial \tau} \frac{\partial N_e}{\partial \nu} \right] = \frac{N_0}{2} \mathrm{Re}\left[\frac{\partial}{\partial \nu} \frac{\partial}{\partial \tau'} (1 + \mathrm{j}2\pi \bar{t} \nu)(1 + \mathrm{j}2\pi \bar{f} \tau') \chi(-\tau', \nu) \right]$$

$$= (2\pi)^2 \frac{N_0}{2} E_p T_G B_G.$$

Therefore

$$\mathrm{E}\left[\begin{bmatrix}\dfrac{\partial N_e}{\partial \tau}\\ \dfrac{\partial N_e}{\partial \nu}\end{bmatrix}\begin{bmatrix}\dfrac{\partial N_e}{\partial \tau} & \dfrac{\partial N_e}{\partial \nu}\end{bmatrix}\right] = (2\pi)^2 \frac{N_0}{2} E_p \begin{bmatrix} B_G^2 & B_G T_G \rho \\ B_G T_G \rho & T_G^2 \end{bmatrix},$$

$$= \frac{N_0}{2} \boldsymbol{A}$$

and because $\boldsymbol{\Sigma} = \frac{1}{2} N_0 A^{-1} A A^{-1}$, the proof of the proposition is complete. □

Problems

12.1 **a.** Use a series expansion on a term of the form $\sqrt{(A+x)^2+y^2}$ to explain why the ricean probability density function looks like a gaussian probability density function when A is sufficiently large.

b. The modified Bessel function of the first kind, $I_0(x)$, can be approximated as

$$I_0(x) = \frac{\mathrm{e}^x}{\sqrt{2\pi x}}$$

when x is large. Using this approximation as a starting point, show again that the ricean probability density function looks like a gaussian probability density function when A is large.

12.2 **a.** Prove that, as long as enough energy exists in each segment so that the linearized analysis holds, with no loss in arrival-time estimation accuracy, a pulse train waveform can be chopped into segments, and the arrival time of the pulse train can be estimated by averaging the estimates of the arrival time of each segment. This justifies the common practice of estimating the arrival time of a pulse train by estimating the arrival time of each pulse individually, and then averaging (after a suitable correction for pulse offset time). For time-of-arrival estimation, it is not necessary to maintain coherence in long-duration waveforms, provided the signal-to-signal ratio of the individual segments is large.

b. Prove that the arrival frequency of a pulse train can be estimated much more accurately than by estimating and averaging the arrival frequencies of the individual pulses. Estimating the arrival frequency of a pulse train by averaging estimates made on individual pulses can result in excessive degradation in accuracy.

12.3 The *rayleigh probability density function* (rayleigh pulse) is defined as the radial marginal $p(r)$ of the two-dimensional gaussian probability density function

$$p(x, y) = \frac{1}{2\pi} \mathrm{e}^{-(x^2+y^2)/2}.$$

The *maxwellian probability density function* (maxwellian pulse) is defined as the radial marginal $p(r)$ of the three-dimensional gaussian probability density function

$$p(x, y, z) = \frac{1}{(2\pi)^{3/2}} e^{-(x^2+y^2+z^2)/2}.$$

Find the rayleigh pulse and the maxwellian pulse and their Fourier transforms.

12.4 Consider the trapezoid pulse

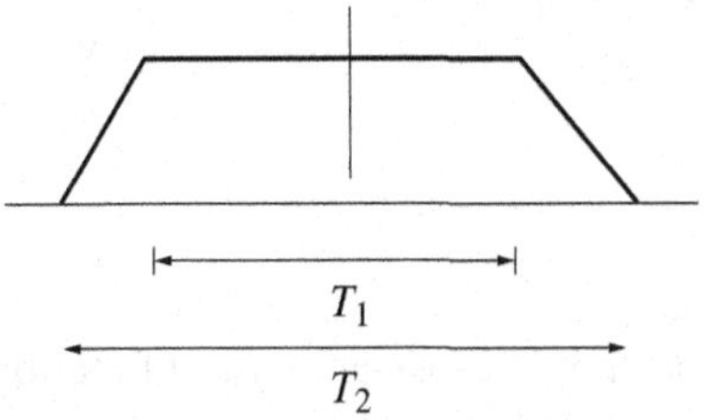

a. Supposing that $2E_p/N_0 = 10$ decibels, $T_1 = 1$ microsecond, and $T_2 = 1.1$ microseconds, what is the variance of the error of the best noncoherent estimator of pulse arrival time?

b. Now let the pulse width be increased so that $T_1 = 2$ microseconds, and $T_2 = 2.1$ microseconds. The amplitude of the pulse is unchanged. How will the error variance of the estimator change?

12.5 Give an example of a signal for which the maximum Nyquist sample is smaller than the maximum value of the signal. For this signal, how does the probability of missed detection depend on the offset between the location of the maximum and the nearest sample? Give approximate numerical values for the probability of missed detection.

12.6 The Fourier transform of the ranging waveform $s(t)$ is

$$S(f) = \text{rect}\left(\frac{f - f_0}{B}\right) + \text{rect}\left(\frac{f + f_0}{B}\right)$$

where $f_0 > B$.

a. What is $\chi(\tau, 0)$?

b. Evaluate the variance of a noncoherent estimator of pulse arrival time in white gaussian noise, based on the assumption that a linearized analysis in the vicinity of the main lobe is adequate.

c. With reference to the following figure, and using the fact that the noise output of a matched filter has the autocorrelation function $(N_0/2)\chi(\tau, 0)$ when the input is white noise, give an expression for the probability that, in white gaussian noise, the first sidelobe in noise on the right or the left is larger than the main lobe.

d. Give a refined expression for the variance developed in part b, based on the approximation that noise causes either a small perturbation to the position of the main lobe, or a false detection of either of the principal sidelobes, but with no further error.

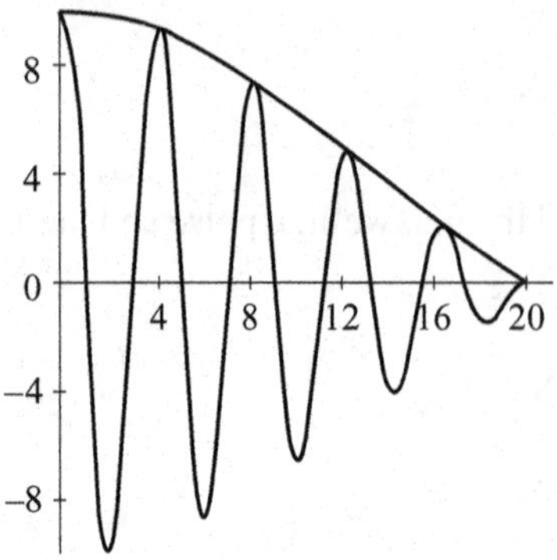

12.7 The digital computation of a sample cross-ambiguity function must use a discrete (τ, ν) grid consistent with the spatial frequency content of $\chi(\tau, \nu)$. To simplify the processing, a two-step estimator, consisting of a coarse estimate followed by a fine estimate, is proposed for estimating the arrival time of the pulse given in the previous problem. A coarse estimate of τ_0 is computed by using only a single "window" of $S(f)$. Then $|\chi(\tau, 0)|$ is computed by using both windows, but only in the vicinity of the coarse estimate. Specifically, the received signal is first filtered by $S_0(f) = \text{rect}((f - f_0)/B)$ to reject the lower sideband. What is the variance of the coarse estimate? Given that the noise is gaussian, at $E_p/N_0 = 10$ dB and $f_0/B = 10$, what is the probability that the coarse estimate will be closer to the main lobe than to a sidelobe? How does this compare with the previous problem?

12.8 A radar with the carrier frequency f_0 and a narrow beam fixed at an up-looking angle of 45° is used to measure the altitude of satellites in low circular orbits, moving right to left, as shown in the following figure.

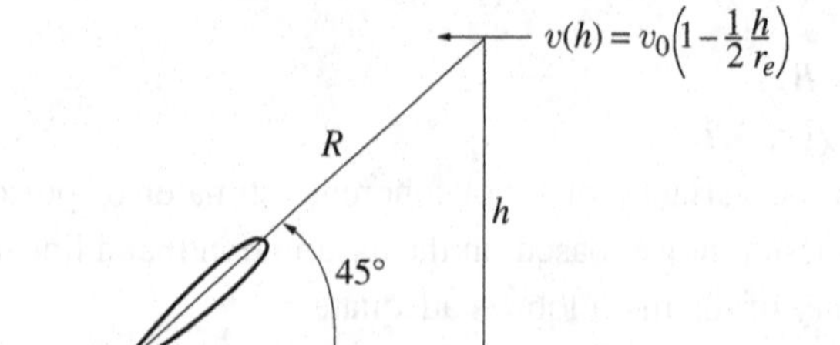

Because the satellite is known to be in a circular orbit, its velocity and altitude are interrelated by the laws of mechanics. For h small compared to the earth's radius r_e, the relationship is approximated by

$$v(h) = v_0\left(1 - \frac{1}{2}\frac{h}{r_e}\right).$$

a. Express this equation as a constraint in the τ, ν plane of the form

$$a\nu + b\tau = 1$$

for some appropriate constants, a and b. Show the location of a typical cross-ambiguity function by sketching the uncertainty ellipse of the cross-ambiguity function in the (τ, ν) plane.

b. Given that the uncertainty ellipse of the waveform $s(t)$ has the form

$$T_G^2\nu^2 + B_G^2\tau^2 = C,$$

and taking advantage of the constraint, give an expression for the variance of a noncoherent estimator of τ_0.

c. If the pulse $s(t)$ is replaced by $s(t)\,e^{j\pi\alpha t^2}$, how will the variance change? Is there a best choice for α?

12.9 A moving-target detection radar uses the waveform

$$s(t) = \sum_{n=0}^{N-1} p(t - nT_r).$$

A primitive processing algorithm passes the received echo $v(t)$ through the filter

$$g(t) = p(t) - p(t - T_r).$$

a. What is the filter output due to stationary clutter?

b. What is the filter output due to a moving reflector?

c. Let E_p be the energy of pulse $p(t)$. In the presence of white gaussian noise of power density spectrum $N_0/2$, what is the signal-to-noise ratio of a moving reflector at the output of the filter?

12.10 **a.** A pulse train consisting of N identical pulses is noncoherently received in white gaussian noise with a single random phase that is the same on all pulses of the pulse train. What is the Neyman–Pearson detector of the pulse train?

b. A pulse train consisting of N identical pulses is noncoherently received in white gaussian noise with the phase of each pulse random, independent, and uniformly distributed. What is the Neyman–Pearson detector of this pulse train?

c. A pulse train consisting of N real identical pulses is received in white gaussian noise, and the amplitude of each pulse is random, independent, and rayleigh-distributed. What is the Neyman–Pearson detector of this pulse train?

12.11 Given the hypothesis-testing problem with $v(t) = a_1 s(t - \tau_1) + a_2 s(t - \tau_2) + n(t)$, where $a_1, a_2 \in \{0, 1\}$, sketch the minimum-distance decision regions for the case in which $\mathrm{Re}\,[\int s(t - \tau_1)s^*(t - \tau_2)\,dt]$ is negative. Give expressions for the conditional probabilities of error if $n(t)$ is white gaussian noise.

Notes

The detection of objects by means of reflected radio waves became an obvious possibility as soon as Heinrich Hertz demonstrated the transmission and reflection of radio waves in 1887. A formal theory came much later. Detection is an instance of the problem of the testing of binary hypotheses, as was studied by Neyman and Pearson (1933), and also by Zadah and Ragazzini (1952), Marcum (1960), Swerling (1957, 1960), and many others.

Detection of a signal with an unknown delay or unknown frequency is a more difficult problem. By quantizing delay to one of M discrete values, the detection of a signal with an unknown arrival time becomes an M-ary detection problem and was treated as such by Peterson and Birdsall (1953), and also by Middleton and Van Meter (1955). Selin (1965) employed an averaging technique to treat the detection of a signal with an unknown doppler.

Detection of multiple, overlapping echo signals, each with an unknown delay and unknown doppler, is a yet more difficult detection problem. The resolution of multiple targets was treated by Helstrom (1960) by using the method of maximum likelihood. A direct approach to the combined problem of multitarget detection and estimation of their parameters was due to Nilsson (1961), who introduced a hybrid loss function that assigned penalties to both detection and estimation errors; the loss was then minimized with respect to a posterior probability measure. The solution requires a computer to perform a multidimensional maximization; as the number of targets grows, the computational requirements grow in difficulty.

Detection of moving targets against a stationary background has a somewhat different history because of the problem of clutter. Moving-target detection had its origins during World War II. This early work was treated in the reports of Emslie (1946) and Emerson (1954). Detection in the presence of clutter has been studied from many points of view, as in the papers of Capon (1964); Manasse (1961); and Sekine, Ohtani, and Muska (1981). Urkowitz (1953) proposed the use of a model of clutter as statistically independent of the signal, but showing the same spectrum.

13 Passive and baseband surveillance systems

A propagating medium may be teeming with a multitude of weak signals even when it appears superficially to be empty. For example, an acoustic medium such as a lake may appear quite still, and yet it may contain numerous faint pressure waves originating in various submerged objects and reflecting off other submerged objects. A passive sonar system can intercept these waves and extract useful information from the raw received data. Indeed, these invisible pressure waves can be used in principle to form images of submerged objects. Likewise, a seismographic sensor or an array of such sensors on the surface of the earth can measure tiny vibrations of the earth's surface and deduce the location of distant earthquakes and other geophysical disruptions or can form images of geological structures. Even the electromagnetic environment in which we are immersed contains immense quantities of information. For example, some of these electromagnetic signals can be intercepted by suitable large apertures and formed into detailed images of far distant galaxies. We need provide no illumination, nor can we provide illumination in such an application. We need only gather the data with appropriate passive sensors and process the sensed data into the images that allow us to observe these galaxies.

Passive surveillance systems include systems for radio astronomy, seismic data analysis, electromagnetic surveillance, and sonar surveillance. A passive surveillance system collects a propagating signal that the system itself does not generate. The signal originates in the environment, usually in the scene being observed. At one or more places within the propagation medium, the information-bearing signals that are incident on a set of receivers are collected. From this set of received signals, contaminated by noise and perhaps other interference, we can estimate parameters such as the locations of radiation sources or the properties of the propagation medium, or we can obtain images of the radiation source or the environment of that source.

We shall discuss the function of imaging in a passive surveillance system primarily in the context of radio astronomy. We shall discuss the functions of detection and estimation in a passive surveillance system primarily in the context of systems for the location of radar or sonar sources.

13.1 Radio astronomy

Our everyday environment is rich with invisible signals of many kinds, including electromagnetic waves in the radio bands. Some of these radio waves are emitted naturally by radio sources in distant galaxies. The tasks of radio astronomy are to detect and image the astronomical – usually extragalactic – sources of these waves. The spatial (angular) distribution of the received energy provides an image of the distribution of radiation sources in the galaxy. In addition to this image of radiation intensity, one can also image both the radial velocity distribution and the temperature distribution of the radiation source by measurements on spectral lines. Perhaps the most important spectral line for this purpose is the hydrogen line at 1.420405 GHz. The doppler effect will move the observed frequency (or wavelength) of this line by an amount proportional to the radial velocity. The doppler effect will also cause this line to be broadened by an amount proportional to the temperature at the source because of the thermal motion of the hydrogen atoms. By such methods, an intensity image can be annotated with additional information. We will not consider such refinements further. We will study only the formation of the intensity image.

Figure 13.1 shows several elements of an array that forms a radio telescope. Each element of the array is a large, high-gain antenna. The antenna beams are very narrow and are steered together so that all beams point in parallel to the same point at infinity. The wavelengths in the microwave band are large, which would lead to poor resolution for a small array, but one can create an array as large as the diameter of the earth, thereby obtaining good resolution. Some present-day radio telescopes use such an array of antennas that spans an entire continent. Consequently, in recent years, radio astronomy has outperformed classical optical astronomy in many applications.

A radio telescope treats each pair of antennas as a simple interferometer. Consider any pair of antennas of the array, as shown in Figure 13.2, and consider only the signal $s(t)$ arriving at angle ϕ from a point source at infinity. With the phase reference chosen so that the phase delay at the first element is zero, the complex baseband signals

$$v_1(t) = s(t)$$
$$v_2(t) = s(t)e^{j2\pi(d/\lambda)\sin\phi}$$

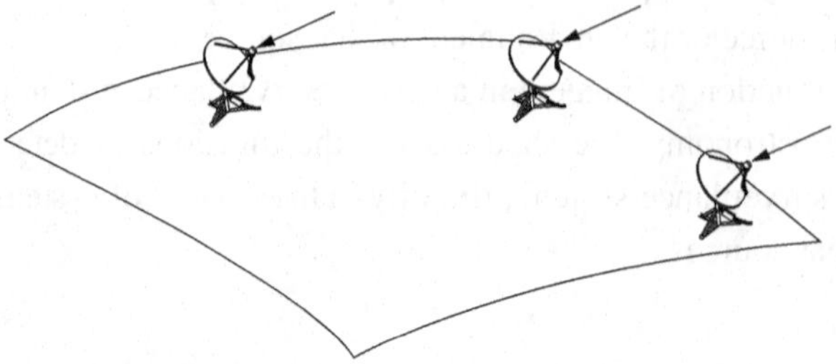

Figure 13.1 Radio astronomy

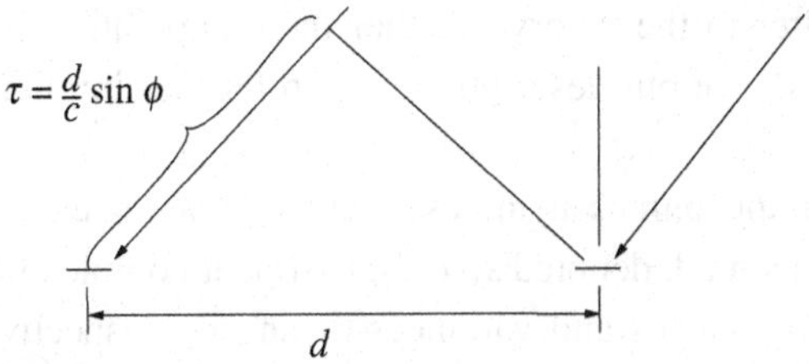

Figure 13.2 Illustrating the phase delay

are received during an observation interval of duration T. The transmitted signal $s(t)$ from the point source at infinity is spatially coherent across the array of antennas. Because the signal $s(t)$ is generated by a great many independent and random radiation sources in the resolution cell at angle ϕ, it is accurately modeled temporally as a complex, white gaussian stationary random process of mean zero and variance c. The imaging task is to estimate both the signal power c and the direction ϕ. The accepted procedure is to compute the sample cross correlation, given by

$$\begin{aligned}\frac{1}{T}\int_0^T v_1(t)v_2^*(t)\,\mathrm{d}t &= \mathrm{e}^{-\mathrm{j}2\pi(d/\lambda)\sin\phi}\frac{1}{T}\int_0^T |s(t)|^2\,\mathrm{d}t,\\ &= \widehat{c}\mathrm{e}^{-\mathrm{j}2\pi(d/\lambda)\sin\phi}\end{aligned}$$

where the estimate $\widehat{c}$ is given by

$$\widehat{c} = \frac{1}{T}\int_0^T |s(t)|^2\,\mathrm{d}t.$$

The estimate $\widehat{c}$ has the expected value c.

An equivalent computation of this estimate can be formulated at passband. At passband, the received signals are

$$\begin{aligned}\widetilde{v}_1(t) &= A(t)\cos 2\pi(f_0 t + \theta)\\ \widetilde{v}_2(t) &= A(t)\cos 2\pi(f_0 t + \theta + (d/\lambda)\sin\phi).\end{aligned}$$

Consequently, using a standard trigonometric identity,

$$2\int_0^T \widetilde{v}_1(t)\widetilde{v}_2(t)\,\mathrm{d}t = \cos(2\pi(d/\lambda)\sin\phi)\int_0^T |s(t)|^2\,\mathrm{d}t$$

where the negligible contribution of the double-frequency term at $2f_0$ to the integral has been dropped. When regarded as a function of ϕ, the term $\cos(2\pi(\mathrm{d}/\lambda)\sin\phi)$ is referred to as the *fringe pattern*. An observation of the phase of the fringe pattern provides a measurement of ϕ. To measure the phase of the fringe pattern, one may phase-shift the first of the two passband inputs $\widetilde{v}_1(t)$ by 90° to form $\widetilde{v}_1'(t)$ and then compute

$$2\int_0^T \widetilde{v}_1'(t)\widetilde{v}_2(t)\,\mathrm{d}t = \sin(2\pi(d/\lambda)\sin\phi)\int_0^T |s(t)|^2\,\mathrm{d}t.$$

The phase of the fringe is then computed as an arc tangent of the ratio of the two integrals. Thus, it does not matter to the theory whether the computation is done at complex baseband or at passband. For our description, we prefer the simplicity of the complex baseband formulation.

Next, suppose there are two point sources at angles ϕ_1 and ϕ_2 that are each emitting an independent gaussian random signal, denoted $s_1(t)$ and $s_2(t)$, at complex baseband. The transmitted signals have zero means and variances c_1 and c_2, respectively. The received signals are

$$v_1(t) = s_1(t) + s_2(t)$$
$$v_2(t) = s_1(t)\mathrm{e}^{\mathrm{j}2\pi(d/\lambda)\sin\phi_1} + s_2(t)\mathrm{e}^{\mathrm{j}2\pi(d/\lambda)\sin\phi_2}.$$

Now the computational task is to estimate the parameters c_1, c_2, ϕ_1, and ϕ_2. Again, the accepted procedure is to compute the sample cross-correlation:

$$\frac{1}{T}\int_0^T v_1(t)v_2^*(t)\,\mathrm{d}t = \widehat{c}(\phi_1)\mathrm{e}^{-\mathrm{j}2\pi(d/\lambda)\sin\phi_1} + \widehat{c}(\phi_2)\mathrm{e}^{-\mathrm{j}2\pi(d/\lambda)\sin\phi_2} + n_e(t)$$

where

$$\widehat{c}(\phi_1) = \frac{1}{T}\int_0^T |s_1(t)|^2\,\mathrm{d}t$$
$$\widehat{c}(\phi_2) = \frac{1}{T}\int_0^T |s_2(t)|^2\,\mathrm{d}t,$$

and the so-called *self-noise* term is

$$n_e(t) = \mathrm{e}^{\mathrm{j}2\pi(d/\lambda)\sin\phi_2}\frac{1}{T}\int_0^T s_1(t)s_2^*(t)\,\mathrm{d}t + \mathrm{e}^{\mathrm{j}2\pi(d/\lambda)\sin\phi_1}\frac{1}{T}\int_0^T s_2(t)s_1^*(t)\,\mathrm{d}t.$$

For large enough T, the self-noise term can be considered negligible in comparison with the other terms. This statement can be justified by computing the variance of the self-noise terms, which is inversely proportional to T.

Now consider a continuum of infinitesimal signals distributed in the angle ϕ. The infinitesimal signal at angle ϕ is denoted $s(t,\phi)\,\mathrm{d}\phi$, and $s(t,\phi)$ as a function of ϕ is the signal density. The signal densities arriving from two different angles could be modeled as independent, but this condition would be more than we need. For our needs, it is enough that the signal densities at different angles are uncorrelated. Then we can write

$$\mathrm{E}[s(t,\phi)s^*(t,\phi')] = c(\phi)\delta(\phi - \phi').$$

The received signals are

$$v_1(t) = \int_0^{2\pi} s(t,\phi)\,\mathrm{d}\phi$$
$$v_2(t) = \int_0^{2\pi} s(t,\phi)\mathrm{e}^{\mathrm{j}2\pi(d/\lambda)\sin\phi}\,\mathrm{d}\phi.$$

The sample cross-correlation coefficient is computed from the received signals as the integral

$$\int_0^T v_1(t)v_2^*(t)\,\mathrm{d}t = \int_0^T \int_0^{2\pi} \int_0^{2\pi} s(t,\phi)s^*(t,\phi')\mathrm{e}^{-\mathrm{j}2\pi(d/\lambda)\sin\phi'}\,\mathrm{d}\phi\,\mathrm{d}\phi'\,\mathrm{d}t.$$

The expected value of the integral is

$$\begin{aligned}\mathrm{E}\int_0^T v_1(t)v_2^*(t)\,\mathrm{d}t &= \int_0^T \int_0^{2\pi} \int_0^{2\pi} c(\phi)\delta(\phi-\phi')\mathrm{e}^{-\mathrm{j}2\pi(d/\lambda)\sin\phi'}\,\mathrm{d}\phi\,\mathrm{d}\phi'\,\mathrm{d}t \\ &= T\int_0^{2\pi} \int_0^{2\pi} c(\phi)\delta(\phi-\phi')\mathrm{e}^{-\mathrm{j}2\pi(d/\lambda)\sin\phi'}\,\mathrm{d}\phi\,\mathrm{d}\phi' \\ &= T\int_0^{2\pi} c(\phi)\mathrm{e}^{-\mathrm{j}2\pi(d/\lambda)\sin\phi}\,\mathrm{d}\phi.\end{aligned}$$

If, moreover, $c(\phi)$ is negligible except near $\phi = 0$, the approximation $\sin\phi \approx \phi$ allows this to be written

$$\begin{aligned}\frac{1}{T}\mathrm{E}\int_0^T v_1(t)v_2^*(t)\,\mathrm{d}t &\approx \int_{-\infty}^{\infty} c(\phi)\mathrm{e}^{-\mathrm{j}2\pi(d/\lambda)\phi}\,\mathrm{d}\phi \\ &= C\left(\frac{d}{\lambda}\right).\end{aligned}$$

Consequently, the expected value of the cross-correlation coefficient is approximately equal to the Fourier transform $C(f)$ at the single point $f = d/\lambda$. For large T, the sample cross-correlation coefficient, computed from the observations, approaches its expectation which is the true cross-correlation coefficient. Therefore, the sample cross-correlation coefficient provides an estimate of the Fourier transform of $c(\phi)$ at the single point $f = d/\lambda$. To obtain acceptable accuracy in this estimate, very large values of T will be needed.

A real radio source is two-dimensional, as shown in Figure 13.3. Provided the scene is small, with the z axis pointed at the scene, the angular coordinates can be described by the small angles ϕ_x and ϕ_y. Then the source can be described as the function $c(\phi_x, \phi_y)$. Just as before, each sample cross-correlation coefficient provides a sample of the two-dimensional Fourier transform at the single point (f_x, f_y) provided T is sufficiently large. To form an image, we shall need to determine $C(f_x, f_y)$ for many values of (f_x, f_y). These samples must satisfy the Nyquist criteria or the data set must be supplemented in some way by prior information.

Consider three colinear antennas separated by d_1, d_2, and $d_1 + d_2$. By computing the correlation using two antennas at a time, estimates of three samples of the two-dimensional Fourier transform can be obtained. If the three antennas lie along the x axis, then the three values of the Fourier transform are $C(d_1/\lambda, 0)$, $C(d_2/\lambda, 0)$, and $C((d_1 + d_2)/\lambda, 0)$. If the alignment of the three antennas is at the angle ψ with respect to the x axis, then the three values of the Fourier transform are at $C(d_1 \cos\psi/\lambda, d_1 \sin\psi/\lambda)$,

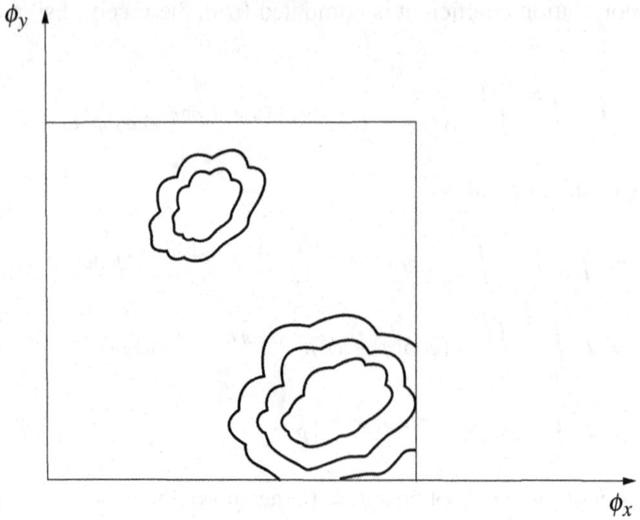

Figure 13.3 A density profile of radio sources

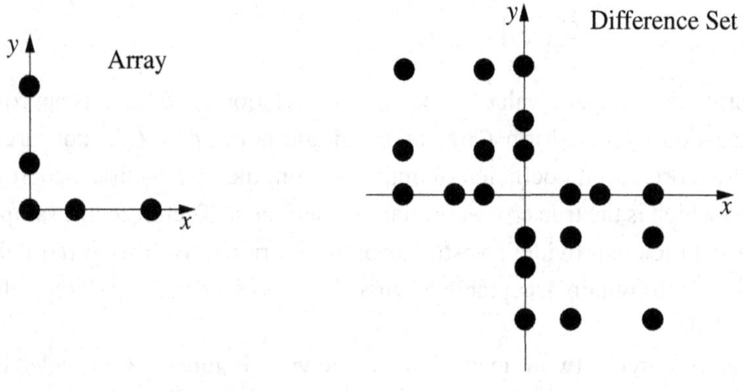

Figure 13.4 An array and its difference set

$C(d_2 \cos \psi/\lambda, d_2 \sin \psi/\lambda)$, and $C((d_1 + d_2)\cos \psi/\lambda, (d_1 + d_2)\sin \psi/\lambda)$. If the three antennas are not colinear, then the three points of the Fourier transform will not be colinear.

With M antennas, the M received signals can be correlated two at a time to produce estimates of many samples of the Fourier transform. There are $M(M-1)/2$ pairs. One should design the antenna placements in the array so that the pairwise (vector) distances are all different. Then $M(M-1)/2$ samples of $C(f_x, f_y)$ can be computed. Figure 13.4 shows an array of five antennas, as well as the twenty vector differences between pairs of antennas. These vector differences determine the twenty points of the f_x, f_y plane where the values of $C(f_x, f_y)$ are obtained. Because $c(x, y)$ is real, $C(f_x, f_y) = C(-f_x, -f_y)$. Thus only ten of the samples of $C(f_x, f_y)$ are unique. This is in accord with the statement that $M(M-1)/2$ samples of the Fourier transform can be obtained.

These twenty samples provide only partial knowledge of $c(x, y)$. To form a high-resolution image of $c(x, y)$, one needs many more samples of $C(f_x, f_y)$. A good image uses between 10^5 and 10^6 samples of $C(f_x, f_y)$. To obtain these by pairwise correlation would require something on the order of 1000 radio antennas, each consisting of a very large and expensive high-gain dish.

In practice, one obtains the effect of many antennas by reusing a smaller number of antennas. One method is to mount one or more antennas on a track so that one antenna can be physically repositioned with respect to the other antennas after each set of data is collected. A second method is simply to allow the normal rotation of the earth to rotate the array with respect to the source. A new set of data is taken after the earth rotates a small amount, and these data consists of samples of $C(f_x, f_y)$ in a new reference frame that is rotated with respect to the first reference frame. Of course, such observation methods produce Fourier transform samples that are not on a rectangular grid. This means that computing the inverse Fourier transform is not a straightforward task.

Even with all of these methods for enlarging the data set, the number of measured components of the Fourier transform may still be much smaller than the number of pixels in a high-resolution image. The Fourier samples may fail to satisfy the Nyquist condition. Moreover, there may be some regions of the Fourier plane that are only sparsely sampled, or not observed at all. For these reasons, methods have been developed that can supplement the measured data with prior knowledge about the images. Two elementary facts are of great value: one is that the image is real and nonnegative; the other is that the object being imaged is compact and a large part of that image is black, a so-called mostly-black image. Various principles of inference may be invoked to infer the missing Fourier data from the condition that the image is nonnegative. The clean algorithm, which was discussed in Section 9.1, uses the prior information that the image is mostly black.

13.2 Estimation of direction

Interferometry was discussed in Section 5.6. The interferometric technique using a pair of antennas can be used to measure the direction of arrival in a great variety of situations. In a situation in which several receiving apertures are closely spaced in comparison with the distance to a source of radiation, the incoming wave may be considered to be a plane wave, and the difference in the time of arrival is a statistic to measure the direction to that radiation source. In particular, the direction of arrival may be measured by the difference in the carrier phase at two apertures. Any such system of measurement is called an interferometer.

In this section, in order to deepen our understanding of an interferometer, we shall take a more fundamental view. We begin with the observation that the distribution of the phase of an incoming waveform across an aperture depends on the direction of arrival of the waveform. We have already seen how the aperture can be divided into two or more subapertures to estimate the direction of arrival by a phase-comparison interferometer. This is a widely used and practical approach, but we did not develop this phase comparison method from any condition of optimality. We did not distinguish the noise distribution across the aperture from the noise introduced by the receivers because in that section the way to partition the full aperture was not a part of the consideration. In this section, we shall continue the study of elementary methods for the estimation of the direction of arrival of a waveform, both against an external noise background and with noise introduced by the receivers. The maximum-likelihood methods of Chapter 11 can also be applied to this problem.

The simplest approach to the problem, as studied in Section 5.6, consists of partitioning the aperture into two disjoint subapertures with the two received signals given by

$$v_1(t) = s(t) + n_1(t)$$
$$v_2(t) = s(t)e^{j\Delta\theta} + n_2(t),$$

where $n_1(t)$ and $n_2(t)$ are independent, complex, white gaussian-noise processes, each with variance σ^2, and where $\Delta\theta$ is the difference in carrier phase due to path length difference to the two phase centers. The usual way to estimate $\Delta\theta$ is to pass each received signal through a filter matched to $s(t)$ and to sample the filter outputs at the expected peak. The two complex matched-filter samples are:

$$\begin{aligned} u_1 &= A + n_1 \\ &= A + n_{1R} + jn_{1I} \\ u_2 &= Ae^{j\Delta\theta} + n_2 \\ &= Ae^{j\Delta\theta} + n_{2R} + jn_{2I} \end{aligned}$$

where n_1 and n_2 are independent, complex gaussian random variables of variance σ^2 per component. The difference in the phase angles of u_1 and u_2 provides an estimate of $\Delta\theta$. This estimate will include an error, denoted θ_e, because of the noise terms.

We shall derive a simple expression for the direction-of-arrival error based on the received noise in the two subapertures. Consider a noisy complex number u given by

$$\begin{aligned} u &= Ae^{j\theta} + n \\ &= (A_R + n_R) + j(A_I + n_I) \end{aligned}$$

where $Ae^{j\theta}$ denotes the complex signal and n denotes the complex noise. The phase angle of u provides an estimate of the phase angle θ. This estimate will be in error because of the noise. To find the phase error θ_e due to the noise, choose a new coordinate system in which the imaginary part of the signal is zero. In this coordinate system, the real part of the signal is A, and

$$u' = (A + n'_R) + jn'_I.$$

The phase error is

$$\theta_e = \tan^{-1}\frac{n'_I}{A + n'_R},$$

which is also the phase error in the original coordinate system. If the noise is small in comparison to the signal,

$$\theta_e \approx \frac{n'_I}{A}.$$

Hence

$$\sigma_\theta^2 = \frac{\sigma_n^2}{A^2}.$$

In the phase-difference interferometer, each signal u_1 and u_2 has a phase error with this variance, and these phase errors are modeled as independent. Hence the differential phase error in the phase difference $\theta_1 - \theta_2$ has the variance

$$\sigma_\theta^2 = 2\frac{\sigma_n^2}{A^2}.$$

The phase difference between the two antennas is related to the angle of arrival by

$$\Delta\theta = 2\pi\frac{d}{\lambda}\sin\phi$$

where d is the separation in the phase centers of the antennas. The error in angle of arrival, for small values of the error, satisfies

$$\theta_e \approx 2\pi\frac{d}{\lambda}\cos\phi\phi_e.$$

Within this approximation, the variance in θ_e satisfies

$$\left(2\pi\frac{d}{\lambda}\cos\phi\right)^2\sigma_\phi^2 = 2\frac{\sigma_n^2}{A^2}.$$

In particular, near broadside,

$$\sigma_\phi^2 = 2\left(\frac{\lambda}{2\pi d}\right)^2\frac{\sigma_n^2}{A^2}.$$

Thus, the variance in measured angle depends on signal and noise only through their power ratio. If the signal power and the noise power both depend directly on the size of the two apertures, as when the noise is external, then the aperture size does not matter. If only the signal power depends on the size of the apertures, but the noise power does not, as when the noise arises in the receivers, then the aperture size does matter.

Suppose now that the total aperture, which is fixed in size, is sampled by dividing it into multiple subapertures that together make up the total aperture. Also, suppose that the signal received in each subaperture is contaminated by an additive noise term $n_i(t)$ that is independent of other noise terms. The received signal at the ith subaperture is

$$v_i(t) = \alpha_i s(t) e^{j\theta_i} + n_i(t)$$

where the $n_i(t)$ are independent, identically distributed, white gaussian-noise processes, each with the power density spectrum $N_0/2$. Because the total energy received in the full aperture is unchanged,

$$\sum_{i=0}^{I-1} \int_{-\infty}^{\infty} \alpha_i^2 |s(t)|^2 \, dt = \int_{-\infty}^{\infty} |s(t)|^2 \, dt.$$

Therefore $\sum_i \alpha_i^2 = 1$. The θ_i depend on the positions of the phase centers of the subapertures. In each subaperture, the phase is measured with an independent phase error. At the phase center of the ith aperture, the phase θ_i is measured with the variance

$$\sigma_{\theta_i}^2 = \frac{\sigma_{ni}^2}{A^2 \alpha_i^2}.$$

If all subapertures are the same size with the same noise variance,

$$\sigma_{\theta_i}^2 = \frac{\sigma_n^2}{A^2/I}.$$

We can fit a straight line to these I phase measurements. The phase depends linearly on position, so it has the form

$$\theta = ax + b.$$

Because

$$a = \frac{d\theta}{dx} = \frac{2\pi}{\lambda} \sin\phi,$$

an estimate of the coefficient a gives the desired estimate of ϕ.

The coefficients a and b of a straight line that fits a set of data points can be estimated by the standard method of least-squares. The standard least-squares straight-line fit, $y = ax + b$, to any set of points (x_i, y_i), $i = 1, \ldots, I$, is derived from the fact that for any parameters a and b, the squared error in the y direction at the ith point is $(y_i - ax_i - b)^2$. The total squared error is

$$\Sigma_i (y_i - ax_i - b)^2.$$

To find the least-squares fit, set the partial derivatives with respect to a and b equal to zero to give

$$\Sigma_i x_i (y_i - ax_i - b) = 0$$
$$\Sigma_i (y_i - ax_i - b) = 0.$$

This can be written as

$$\begin{bmatrix} \overline{x^2} & \overline{x} \\ \overline{x} & 1 \end{bmatrix} \begin{bmatrix} a \\ b \end{bmatrix} = \begin{bmatrix} \overline{xy} \\ \overline{y} \end{bmatrix}$$

where $\overline{x} = \Sigma_i x_i/I$, $\overline{y} = \Sigma_i y_i/I$, $\overline{x^2} = \Sigma_i x_i^2/I$, and $\overline{xy} = \Sigma_i x_i y_i/I$. In the simplest case, the x_i are chosen so that $\overline{x} = 0$. Then the estimates are $\widehat{a} = \overline{xy}/\overline{x^2}$ and $\widehat{b} = \overline{y}$. Futhermore

$$\frac{1}{I}\Sigma_i \left(y_i - \widehat{a}x_i - \widehat{b}\right)^2 = (\overline{y^2} - \overline{y}^2) - \frac{\overline{xy}^2}{\overline{x^2}}$$

is the residual sample error variance of the straight-line fit.

The variance in the estimate of the slope a is

$$\begin{aligned} \sigma_a^2 &= \left(\mathrm{E}[\overline{xy}]^2 - [\mathrm{E}[\overline{xy}]]^2\right)/(\overline{x^2})^2 \\ &= \Sigma_i \Sigma_j x_i x_j \left(\mathrm{E}[y_i y_j] - \mathrm{E}[y_i]\mathrm{E}[y_j]\right)/(\overline{x^2})^2. \end{aligned}$$

Suppose that the errors in y are independent, zero-mean, and identically distributed. Then this reduces to $\sigma_a^2 = \sigma_y^2/I\overline{x^2}$.

Therefore, adjusting notation for the application to phase interferometry,

$$\sigma_a^2 = \frac{\sigma_\theta^2}{I\overline{x^2}} = \frac{\sigma_n^2}{A^2\overline{x^2}}.$$

The phase slope is related to spatial angle by

$$a = \frac{2\pi}{\lambda}\sin\phi.$$

Therefore

$$\sigma_a^2 = \left(\frac{2\pi}{\lambda}\right)^2 \cos^2\phi\,\sigma_\phi^2.$$

At boresight, $\phi = 0$, and

$$\sigma_\phi^2 = \left(\frac{\lambda}{2\pi}\right)^2 \frac{\sigma_n^2}{\overline{x^2}A^2}.$$

This is similar to the formula with only two subapertures spaced by d except that $d^2/2$ is replaced by $\overline{x^2}$. If the full aperture has length $2d$, and the number of subapertures is large, then

$$\overline{x^2} \approx \frac{2}{3}d^2.$$

Thus, using multiple samples of the aperture gives an improvement over using only two subapertures if the noise is external, but only a modest improvement. If the noise is internal, and enters in each sample, the performance can be worse.

13.3 Passive location of emitters

The passive location of the source of received radiation is a task encountered in electromagnetic surveillance, passive sonar, and seismic data analysis. The task of passive location is usually understood to employ receivers at several observation stations. At each of several widely separated receivers, possibly moving as shown in Figure 13.5, we can observe an attenuated and delayed version of a signal, $s(t)$, embedded in receiver noise. The signal $s(t)$ itself is unknown. Only the received noisy version of the signal is observed. The source of the signal $s(t)$ is at an unknown location and that location is to be estimated from the received signals.

A formal statement of the problem for the case of nonmoving receivers is as follows: given the received signals

$$v_i(t) = s(t - R_i/c) + n_i(t) \quad i = 1, \dots, I$$

form an estimate, $(\widehat{x}, \widehat{y})$, of the source location (x, y). In a two-dimensional problem, the range R_i has the form

$$R_i = \sqrt{(x - x_i)^2 + (y - y_i)^2}.$$

We shall consider only the simple case where the propagation velocity c is a known constant and the receiver locations (x_i, y_i) are known. We shall study a two-step approach to the task of location that first estimates the differential delays

$$\tau_{ij} = (R_i - R_j)/c$$

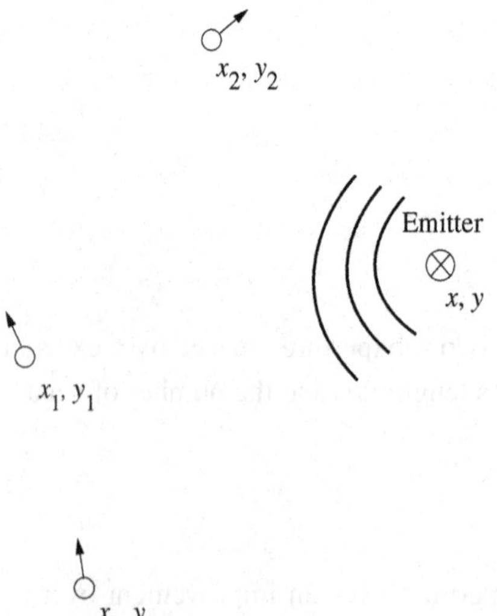

Figure 13.5 Geometry for passive source location

as intermediate parameters. These estimates $\widehat{\tau}_{ij}$ then constitute the input to a geometric computation to determine the location from which the signal has emanated. The set of differential delays forms a statistic for the task of location, though in general this set is not a sufficient statistic. In general, the original waveforms $v_i(t)$ contain information useful for estimating (x, y) that is not preserved in the intermediate variables τ_{ij}.

More generally, a signal transmitted at the center frequency f_0 by a radiation source, possibly moving, has undergone a time delay and a frequency shift when received at a moving point. The ith received signal can be written in complex form as

$$v_i(t) = s(t - R_i(t)/c)e^{j2\pi(f_0/c)\dot{R}_i(t)t}e^{j\theta_i} + n_i(t)$$

where $s(t)$ is the complex representation of the transmitted signal; where $R_i(t)$, given by

$$R_i(t) = \sqrt{(x(t) - x_i(t))^2 + (y(t) - y_i(t))^2},$$

is the distance from the emitter to the receiver; where $\dot{R}_i(t)$, given by

$$\dot{R}_i(t) = \frac{(x(t) - x_i(t))(\dot{x}(t) - \dot{x}_i(t)) + (y(t) - y_i(t))(\dot{y}(t) - \dot{y}_i(t))}{\sqrt{(x(t) - x_i(t))^2 + (y(t) - y_i(t))^2}},$$

is the rate of change of these distances (range rates); and θ_i is the constant phase shift. We shall only treat the case in which the radiation source is not moving, so $\dot{x}(t) = \dot{y}(t) = 0$. We can also write the received signal as

$$v_i(t) = s(t - \tau_i(t))e^{j2\pi\nu_i(t)}e^{j\theta_i} + n_i(t)$$

where $\tau_i(t) = R_i(t)/c$ and $\dot{\tau}_i(t) = (f_0/c)\dot{R}_i(t)$. In the first approximation, $\tau_i(t)$ is taken to be the constant τ_i and $\nu_i(t)$ to be the constant ν_i.

The estimator that we shall study is based on the sample cross-ambiguity surface

$$\left|\chi_{ij}(\tau, \nu)\right| = \left|\int_{-\infty}^{\infty} v_i(t)v_j^*(t + \tau)e^{j2\pi\nu t}\,dt\right|.$$

This differs from the problem studied in Chapter 12 in that now both terms, $v_i(t)$ and $v_j(t)$, in the integrand are measured and noise enters through both $v_i(t)$ and $v_j(t)$. In Chapter 12, one of the signals is a local replica generated within the receiver and so does not introduce noise.

We will next introduce differential delay and differential doppler as the intermediate variables

$$\tau_{ij} = (R_i - R_j)/c$$
$$\nu_{ij} = (f_0/c)(\dot{R}_i - \dot{R}_j).$$

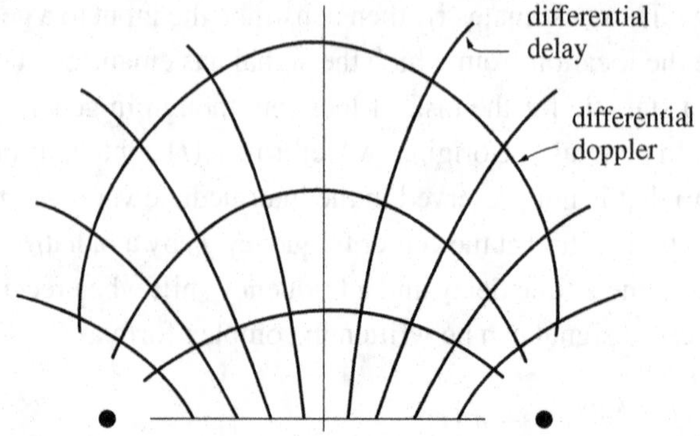

Figure 13.6 Lines of constant differential delay and differential doppler

In the absence of noise, the sample cross-ambiguity surface is

$$
\begin{aligned}
|\chi_{ij}(\tau, \nu)| &= \left| \int_{-\infty}^{\infty} s(t-\tau_i)\mathrm{e}^{\mathrm{j}2\pi\nu_i t} s^*(t-\tau_j-\tau)\mathrm{e}^{-\mathrm{j}2\pi\nu_j t}\mathrm{e}^{\mathrm{j}2\pi\nu t}\,\mathrm{d}t \right| \\
&= \left| \int_{-\infty}^{\infty} s(t)s^*(t-\tau+\tau_{ij})\mathrm{e}^{\mathrm{j}2\pi(\nu-\nu_{ij})t}\,\mathrm{d}t \right|,
\end{aligned}
$$

which has a peak at $\tau = \tau_{ij}$, $\nu = \nu_{ij}$. Thus, by computing the sample cross-ambiguity surface from $v_i(t)$ and $v_j(t)$, we have a way of estimating τ_{ij} and ν_{ij}.

The estimates $\widehat{\tau}_{ij}$ and $\widehat{\nu}_{ij}$ satisfy the two equations

$$
\begin{aligned}
\widehat{\tau}_{ij} &= (R_i(x, y) - R_j(x, y))/c \\
\widehat{\nu}_{ij} &= (\dot{R}_i(x, y) - \dot{R}_j(x, y)) f_0/c,
\end{aligned}
$$

which we wish to invert to obtain position estimates $\widehat{x}$ and $\widehat{y}$. Each equation defines a curve in the x, y plane on which the radiation source must lie. Figure 13.6 shows a family of differential delay curves in the x, y plane, each consisting of all points for which $\tau_{12} = (R_1(x, y) - R_2(x, y))/c$ is a constant. The figure also shows a family of differential doppler loci in the x, y plane, each consisting of all points for which $\nu_{12} = (\dot{R}_1(x, y) - \dot{R}_2(x, y)) f_0/c$ is a constant. The loci of constant differential delay are easy to describe: they are hyperbola. The loci of constant differential doppler have no such simple description: they must be computed numerically. The highlighted curves correspond to a specific pair of measurements. The intersection of these two curves determines the location of the radiation source. For some configurations, the curves may have two intersections. The solution then is ambiguous.

When there are three or more receivers, one can compute the sample cross-ambiguity surface for each pair. For example, when there are three receivers, one can compute $|\chi_{12}(\tau, \nu)|$, $|\chi_{23}(\tau, \nu)|$, and $|\chi_{31}(\tau, \nu)|$. The peaks of these surfaces provide the differential delay estimates $\widehat{\tau}_{12}$, $\widehat{\tau}_{23}$, $\widehat{\tau}_{31}$ and the differential doppler estimates $\widehat{\nu}_{12}$, $\widehat{\nu}_{23}$, and

$\widehat{\nu}_{31}$. The differential delay coordinates are not independent because they satisfy

$$\tau_{12} + \tau_{23} + \tau_{31} = 0.$$

The estimates $\widehat{\tau}_{12}$, $\widehat{\tau}_{13}$, and $\widehat{\tau}_{23}$, however, need not sum to zero because they contain errors.

When there is little or no motion, differential doppler is not useful for estimating position. Then the position estimate $(\widehat{x}, \widehat{y})$ may be found by using only the differential delay estimates. The simplest case estimates $(\widehat{x}, \widehat{y})$ from the pair of estimates $\widehat{\tau}_{12}$ and $\widehat{\tau}_{13}$. This is the task of computing the intersection of two hyperboli with one common focus: find (x, y) by solving the two equations

$$c\widehat{\tau}_{12} = \sqrt{(x - x_1)^2 + (y - y_1)^2} - \sqrt{(x - x_2)^2 + (y - y_2)^2}$$
$$c\widehat{\tau}_{13} = \sqrt{(x - x_1)^2 + (y - y_1)^2} - \sqrt{(x - x_3)^2 + (y - y_3)^2}$$

where (x_1, y_1), (x_2, y_2), and (x_3, y_3) are the three distinct foci of the two hyperboli. Perhaps surprisingly, these equations can be algebraically solved for (x, y) using nothing more complicated than the rooting of a quadratic equation. First, translate the coordinate system so that $x_1 = y_1 = 0$, and let $R = \sqrt{x^2 + y^2}$. The method of solution is first to set up a quadratic equation to solve for R. Once R is known, x and y can be computed easily.

We begin with a restatement of the problem, replacing two equations in the two unknowns x and y with three equations in three unknowns x, y, and R.

$$(R + c\tau_{12})^2 = (x_2 - x)^2 + (y_2 - y)^2$$
$$(R + c\tau_{13})^2 = (x_3 - x)^2 + (y_3 - y)^2$$
$$R^2 = x^2 + y^2.$$

Expand the first two equations and subtract the third,

$$2Rc\tau_{12} + c^2\tau_{12}^2 = -2x_2x - 2y_2y + R_2^2$$
$$2Rc\tau_{13} + c^2\tau_{13}^2 = -2x_3x - 2y_3y + R_3^2,$$

where

$$R_2^2 = x_2^2 + y_2^2$$
$$R_3^2 = x_3^2 + y_3^2.$$

Now solve for x and y in terms of R,

$$\begin{bmatrix} x \\ y \end{bmatrix} = \frac{1}{2} \begin{bmatrix} x_2 & y_2 \\ x_3 & y_3 \end{bmatrix}^{-1} \begin{bmatrix} -2Rc\tau_{12} - c^2\tau_{12}^2 - R_2^2 \\ -2Rc\tau_{13} - c^2\tau_{13}^2 - R_3^2 \end{bmatrix}.$$

This gives equations for x and y that are linear in R. Therefore, we may substitute for x and y in the equation $R^2 = x^2 + y^2$, to obtain a quadratic equation in R, which has two roots. The positive roots give possible values of R from which x and y are easily computed. A negative root is an impossible value of R and can be discarded. If both roots of the quadratic equation are positive, then the pair of measurements (τ_{12}, τ_{13}) is explained by either of two (x, y) locations. In this case, the solution has an ambiguity.

The solution for x, y involves a matrix inverse that fails to exist whenever the three foci lie on a straight line. This special case can be subsumed in the same set of equations by carefully removing the determinant D from the denominator of the equations. Let D denote the determinant $x_2y_3 - y_2x_3$. Then the above equation can be written in the form

$$\begin{bmatrix} Dx \\ Dy \end{bmatrix} = \begin{bmatrix} a_1R + b_1 \\ a_2R + b_2 \end{bmatrix}$$

for appropriate definitions of a_1, a_2, b_1, and b_2. Then the equation $D^2R^2 = D^2x^2 + D^2y^2$ can be written

$$D^2R^2 = (a_1R + b_1)^2 + (a_2R + b_2)^2$$

which again has two roots.

13.4 Magnetic anomaly detection

A metallic object distorts an ambient magnetic field. Hence, by measuring anomalies in the earth's magnetic field, one may detect a nearby metallic object. This is referred to as *magnetic anomaly detection*. Magnetic anomaly detection can be used for detecting submarines and land mines, as well as buried pipes.

Magnetic fields can also be created by electrical currents. By precisely measuring the magnetic field structure in one region of space, say the space surrounding the brain, one can hope to deduce something about the distribution of currents in another nearby region of space, say within the brain. This is referred to as *magnetic imaging*. We will not discuss magnetic imaging.

The problems of magnetic anomaly detection and magnetic imaging can be quite delicate, requiring extremely precise measurements of the structure of the local magnetic field. This section addresses the processing techniques that can be used for magnetic anomaly detection to extract the desired information from the background magnetic field, which is of no interest, and the sensor noise.

The magnetic field is a vector field, which is written as the vector function

$$\boldsymbol{H}(x, y, z) = (H_1(x, y, z), H_2(x, y, z), H_3(x, y, z)).$$

Each of the three field components has a derivative with respect to each of the three spatial coordinates. This is a set of nine derivatives that can be expressed compactly if the spatial coordinates are denoted by (x_1, x_2, x_3) instead of (x, y, z). Then the nine derivatives are denoted $\partial H_k/\partial x_j$ for $k = 1, 2, 3$ and $j = 1, 2, 3$. This three by three matrix of partial derivatives is not arbitrary. It must correspond to a field that satisfies Maxwell's equations. Whenever there are no currents or time-varying electrical fields, Maxwell's equations reduce to

$$\nabla \cdot \boldsymbol{H} = 0$$
$$\nabla \times \boldsymbol{H} = 0.$$

Because magnetic monopoles do not exist, the simplest static magnetic field is a dipole field. The magnetic field $\boldsymbol{H}(x, y, z)$ for a magnetic dipole, $\boldsymbol{m} = (m_1, m_2, m_3)$, at the origin is

$$\boldsymbol{H}(x, y, z) = \frac{3(\boldsymbol{m} \cdot \boldsymbol{R})\boldsymbol{R}}{R^5} - \frac{\boldsymbol{m}}{R^3}$$

where $\boldsymbol{R} = (x, y, z)$ and $R = |\boldsymbol{R}|$. The inverse dependence of the field on R^3 means that the magnetic field will be weak at large R. This implies that magnetic anomaly detection is useful only for short distances.

Differentiating the dipole field at the point (x_1, x_2, x_3) gives

$$\frac{\partial H_k}{\partial x_j} = -\sum_{i=1}^{3} \frac{3m_i}{R^4} N_{ijk}$$

where the nine quantities N_{ijk}, given by

$$N_{ijk} = 5\alpha_i\alpha_j\alpha_k - (\delta_{ki}\alpha_i + \delta_{kj}\alpha_j + \delta_{ji}\alpha_k),$$

are the nine elements of a third-rank tensor, expressed concisely in terms of the direction cosines $\alpha_i = x_i/|\boldsymbol{R}|$ of the range vector $\boldsymbol{R} = (x_1, x_2, x_3)$. It will be convenient to introduce the three quantities M_i, given by

$$M_i = 3m_i/R^4.$$

These quantities are called *scaled moments.*

The simplest problem of magnetic anomaly detection is to find (x_1, x_2, x_3) and (m_1, m_2, m_3) from measurements of the three by three matrix of the partial derivatives of $\boldsymbol{H}$ at the point (x_1, x_2, x_3). Any two of the direction cosines are independent and imply the third because the direction cosines satisfy $\alpha_1^2 + \alpha_2^2 + \alpha_3^2 = 1$. The three scaled moments, and any two of the three direction cosines, constitute a set of five independent unknowns that can be computed from the matrix of the partial derivatives of $\boldsymbol{H}$. This solution is best obtained in two steps: first compute the direction cosines, then compute the scaled moments.

The first step in the inversion is to solve the equations for the scaled moments so they can be eliminated. This can be easily done with the aid of another third-rank tensor,

$$\widetilde{N}_{jk\ell} = \frac{3}{2}\alpha_j\alpha_k\alpha_\ell - \frac{1}{2}(\delta_{j\ell}\alpha_k + \delta_{k\ell}\alpha_j),$$

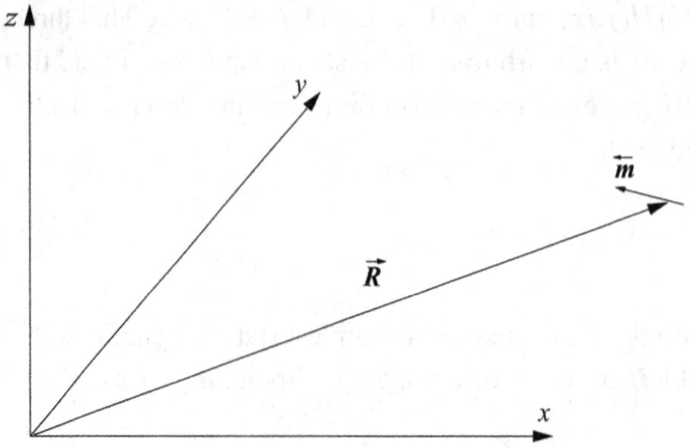

Figure 13.7 Location of an unknown magnetic dipole

which is an inverse of N_{ijk} in the sense that

$$\sum_j \sum_k N_{ijk} \widetilde{N}_{jk\ell} = \delta_{i\ell}.$$

Contracting both sides of the equations for the field derivative with $\widetilde{N}_{jk\ell}$ yields

$$M_\ell = -\sum_j \sum_k \frac{\partial H_k}{\partial x_j} \widetilde{N}_{jk\ell},$$

which gives the scaled moments in terms of the direction cosines and the measurements. It can be used to eliminate the scaled moments. Substituting this equation back into the equations for the field derivatives then yields

$$\frac{\partial H_k}{\partial x_j} = \sum_\ell \sum_m \sum_i \frac{\partial H_m}{\partial x} \widetilde{N}_{\ell mi} N_{ijk}.$$

The scaled moments have now been eliminated. The term $\widetilde{N}_{\ell mi} N_{ijk}$ depends only on the direction cosines. Therefore this is a system of equations relating the unknown direction cosines to the observed field derivatives. The solution of this system lies at the crux of the inversion problem. Once the direction cosines are known, the scaled moments can be obtained from the preceding expression.

We will begin with the problem, shown in Figure 13.7, of estimating the direction to a magnetic dipole, $\boldsymbol{m}$, at an unknown location, (x, y, z), when given the partial derivatives of the magnetic field $\boldsymbol{H}$ at the origin. The point dipole $\boldsymbol{m}$ and the location $\boldsymbol{R}$ are unknown; each unknown vector has three unknown components, resulting in six scalar unknowns. At the origin, we have the three by three array of partial derivatives $\partial H_k / \partial x_j$ for $k = 1, 2, 3$ and $j = 1, 2, 3$. Maxwell's equations require that the trace of this matrix equals zero (because $\nabla \cdot \boldsymbol{B} = 0$), and also that the matrix is symmetric. Therefore there are only five independent measurements, and the six independent unknowns cannot be estimated. To obtain a full solution, additional information, such as the scalar range $|\boldsymbol{R}|$ or the magnetic field magnitude $|\boldsymbol{H}|$, must be available.

To solve this system of equations, we assume that the coordinate system has been chosen so that the matrix of derivatives is a diagonal matrix. This entails no loss of generality because the general system of equations can always be rotated into a principal axis frame.

Therefore

$$\sum_{m=1}^{3} \frac{\partial H_k}{\partial x_m} \sum_{i=1}^{3} \widetilde{N}_{mmi} N_{ijk} = \begin{cases} \frac{\partial H_k}{\partial x_k} & j = k \\ 0 & j \neq k \end{cases}.$$

Because $\sum_i \alpha_i^2 = 1$, the sum on i can be collapsed as follows

$$\begin{aligned} \sum_{i=1}^{3} \widetilde{N}_{mmi} N_{ijk} &= \sum_{i=1}^{3} \left[\frac{3}{2}\alpha_m^2 \alpha_i - \delta_{mi}\alpha_m \right] \left[5\alpha_i\alpha_j\alpha_k - \delta_{ki}\alpha_j - \delta_{kj}\alpha_i - \delta_{ji}\alpha_k \right] \\ &= -\frac{1}{2}\alpha_m^2(\alpha_j\alpha_k + \delta_{kj}) + \alpha_m(\alpha_j\delta_{mk} + \alpha_k\delta_{mj}) \\ &= -\frac{1}{2}\alpha_m^2(\alpha_j\alpha_k + \delta_{kj}) + \alpha_k\alpha_j\delta_{mk} + \alpha_j\alpha_k\delta_{mj}. \end{aligned}$$

When $j = k$, this reduces to

$$\sum_{i=1}^{3} \widetilde{N}_{mmi} N_{ikk} = -\frac{1}{2}\alpha_m^2(\alpha_k^2 + 1) + 2\alpha_m\alpha_k\delta_{mk}.$$

When $j \neq k$, this reduces to

$$\sum_{i=1}^{3} \widetilde{N}_{mmi} N_{ijk} = \left(-\frac{1}{2}\alpha_m^2 + \delta_{mk} + \delta_{mj} \right) \alpha_k\alpha_j.$$

13.5 Estimation of differential parameters

The estimation of the difference in pulse arrival time is based on a pair of baseband signals of the form

$$v_1(t) = s(t - \tau_1) + n_1(t)$$
$$v_2(t) = s(t - \tau_2) + n_2(t)$$

received at two separated points, where $s(t)$ is an unknown waveform with finite energy, and $n_1(t)$ and $n_2(t)$ are independent, stationary noise processes with the identical and known correlation function $\phi(\tau)$ and power density spectrum $N(f)$. In general, $s(t)$ is complex. Then these conditions apply to both the real and the imaginary parts. A coherent estimator of τ_{12} computes the sample cross-ambiguity function of $v_1(t)$ and $v_2(t)$

$$\chi_{12}(\tau, \nu) = \int_{-\infty}^{\infty} v_1(t)v_2^*(t - \tau)\mathrm{e}^{-\mathrm{j}2\pi\nu t}\, \mathrm{d}t$$

and then finds the peak of $\text{Re}[\chi_{12}(\tau, 0)]$. Since ν is set to zero in this case, it is enough to compute the correlation

$$\chi_{12}(\tau, 0) = \int_{-\infty}^{\infty} v_1(t) v_2^*(t - \tau)\, dt.$$

The parameters τ_1 and τ_2 are unknown but fixed. The parameter estimation problem is to determine the delay difference τ_{12}, defined as

$$\tau_{12} = \tau_1 - \tau_2,$$

from the received waveforms $v_1(t)$ and $v_2(t)$.

If $s(t)$ were known, one approach would be to estimate τ_1 and τ_2 individually, using the estimation methods of Chapter 12, and to then take their difference. We are interested instead in the more general problem of estimating τ_{12} when $s(t)$ is not known.

Furthermore, we may allow the received signals to have the unknown phases θ_1 and θ_2 so that

$$v_1(t) = s(t - \tau_1)e^{j\theta_1} + n_1(t)$$
$$v_2(t) = s(t - \tau_2)e^{j\theta_2} + n_2(t).$$

A noncoherent estimator of τ_{12} first computes the sample cross-ambiguity function of $v_1(t)$ and $v_2(t)$,

$$\chi_{12}(\tau, \nu) = \int_{-\infty}^{\infty} v_1(t) v_2^*(t - \tau) e^{-j2\pi\nu t}\, dt,$$

and then finds the peak of $|\chi_{12}(\tau, \nu)|$ regarded as a function of τ. More generally we can consider the received signals to also have unknown frequency offsets

$$v_1(t) = s(t - \tau_1)e^{j2\pi\nu_1 t}e^{j\theta_1} + n_1(t)$$
$$v_2(t) = s(t - \tau_2)e^{j2\pi\nu_2 t}e^{j\theta_2} + n_2(t).$$

A noncoherent estimator of τ_{12} and ν_{12} computes the sample cross-ambiguity function of $v_1(t)$ and $v_2(t)$, then finds the peak of $|\chi_{12}(\tau, \nu)|$ over both τ and ν. To illustrate the analysis of error, we will discuss the error in the noncoherent estimation of differential pulse arrival time τ. Recall from the proof of Proposition 12.10.2, that the variance of the τ error can be written

$$\sigma_\tau^2 = \frac{\mathrm{E}\left[\left(\dfrac{\partial N_e(\tau, 0)}{\partial \tau}\right)^2\right]}{\left[\dfrac{\partial^2 |\chi(\tau, 0)|}{\partial \tau^2}\right]^2_{\tau=0}}$$

where $|\chi(\tau, 0)|$ and $N_e(\tau, 0)$ are determined by approximating the sample cross-ambiguity surface by the first two terms of a series expansion,

$$|\chi_{12}(\tau, \nu)| \approx |\chi(\tau, \nu)| + N_e(\tau, \nu),$$

where

$$N_e(\tau, \nu) = \mathrm{Re}\left[\frac{\chi(\tau, \nu)}{|\chi(\tau, \nu)|}(N_R(\tau, \nu) + \mathrm{j}N_I(\tau, \nu))\right].$$

The denominator in the expression for σ_τ^2 was shown in the proof of Proposition 12.10.2 to be equal to $[(2\pi)^2 T_G^2 E_p]^2$. Because the denominator does not depend on noise, it remains the same in the estimator of differential delay. The noise term, however, is not the same because noise enters the sample cross-ambiguity function in two ways.

We find the numerator by inspecting $N(\tau, \nu)$ at $\nu = 0$,

$$N(\tau, 0) = \int_{-\infty}^{\infty} s(t)n_2^*(t)\,\mathrm{d}t + \int_{-\infty}^{\infty} n_1(t)s^*(t)\,\mathrm{d}t + \int_{-\infty}^{\infty} n_1(t)n_2^*(t)\,\mathrm{d}t.$$

Whereas the analysis of the time-of-arrival error variance had one noise term at this point, the analysis of the differential time-of-arrival has three noise terms. The third is a noise times noise term and will be incidental if noise is smaller than the signal. Thus, in this sense, we can expect that the variance of the estimate of differential time of arrival is about twice as large as the estimate of time of arrival, when noise is small. When noise is not small, there is an additional noise-only term. A formal analysis involves expectations of products of random processes, and is quite tedious.

13.6 Detection of unknown waveforms

Let $s(t)$ be a fixed but unknown waveform from some appropriate class of waveforms. Given a received signal, $v(t)$, in a finite time interval, we are to decide between two hypotheses,

$$H_0 : v(t) = n(t)$$
$$H_1 : v(t) = s(t) + n(t),$$

where $n(t)$ is stationary noise whose power density spectrum $N(f)$ is known. This differs from the hypothesis-testing problems studied in Chapter 12 in that here $s(t)$ is unknown, though possibly known in some statistical sense. A common detector for this problem, called a *radiometer*, computes the energy in $v(t)$ and decides on H_1 if the energy is larger than some threshold. Setting the threshold requires knowledge of $N(f)$. If $N(f)$ is imprecisely known, the correct threshold will be uncertain.

A more interesting problem is the problem of detecting a known pulse in the pair of received signals $v_1(t)$ and $v_2(t)$. The null hypothesis is

$$H_0 : \begin{cases} v_1(t) = n_1(t) \\ v_2(t) = n_2(t), \end{cases}$$

and the alternative hypothesis is

$$H_1 : \begin{cases} v_1(t) = s(t) + n_1(t) \\ v_2(t) = s(t) + n_2(t) \end{cases}$$

where $n_1(t)$ and $n_2(t)$ are stationary noise processes whose autocorrelation function $\mathrm{E}[n_1(t)n_2^*(t-\tau)]$ is known, and the same $s(t)$ is common to both $v_1(t)$ and $v_2(t)$ under H_1. More generally, we can study the case consisting of multiple received signals

$$\begin{aligned} H_0 &: v_i(t) = n_i(t) & i &= 1, \ldots, I \\ H_1 &: v_i(t) = s(t) + n_i(t) & i &= 1, \ldots, I. \end{aligned}$$

When there are multiple received signals, the various copies of $s(t)$ may appear with different delays, dopplers, phase shifts, and amplitudes. Then the problem becomes

$$\begin{aligned} H_0 &: v_i(t) = n_i(t) & i &= 1, \ldots, I \\ H_1 &: v_i(t) = a_i s(t-\tau_i)\mathrm{e}^{\mathrm{j}2\pi\nu_i t}\mathrm{e}^{\mathrm{j}\theta_i} + n_i(t) & i &= 1, \ldots, I \end{aligned}$$

where the parameters τ_i and ν_i are unknown but not independent, while a_i and θ_i are random and independent. In such a case, we are to decide between H_0 and H_1 and, if we decide on H_1, we must estimate τ_i, ν_i, and perhaps a_i and θ_i as well.

A standard estimator for the case in which $I = 2$ is to compute the cross-ambiguity function of $v_1(t)$ and $v_2(t)$ and test the peak against a threshold. This is similar to the methods of Section 12.6 except that noise enters through both terms.

13.7 Lidar surveillance

A *lidar* (light radar) is a surveillance system that operates at optical or infrared frequencies. It may be used for image formation by measuring the optical signal scattered from an object of interest. The common form of lidar may be regarded as a baseband system because it uses only the intensity of the received signal. Because the wavelength is so small – of the order of a micron – a lidar has a narrow beam compared to a radar. A lidar imaging system may be a scanning-beam, real-aperture system with cross-beam resolution determined primarily by the structure of the beam. The beam is scanned in azimuth and elevation to produce an image of the reflectance density of a reflecting object. A lidar can also be used to image the three-dimensional density of a nearly transparent gas such as the atmosphere, or a particular species of atom within the gas. Images of pollutants in the upper atmosphere can be obtained in this way. The wavelength λ of the lidar is chosen to target a particular spectral line of an atomic species, such as sodium. An individual volume cell is isolated by the beamwidth and by range gating. Then the return signal in each single volume cell has the simple form $v = r + n$.

The scattering of light by an atomic species is a strong function of wavelength because the scattering involves the resonances within the atomic structure. This means that the image, in general, is a function of four variables: the three spatial variables and the wavelength. At a fixed spatial point, (x, y, z), the image $c(x, y, z)$ also becomes a function of λ. This function may now be denoted $c(x, y, z, \lambda)$. Near a known resonance, λ_0, $c(x, y, z, \lambda)$ gives a great deal of information about the composition of the gas in a volume cell at (x, y, z). The actual location of the peak of $c(\lambda)$ near λ_0 measures the radial velocity of that cell because of the doppler effect. By measuring this peak shift as a function of (x, y, z), one obtains a three-dimensional image of the radial velocity of the scattering gas.

The function $c(x, y, z, \lambda)$ near a known resonance λ_0 also gives information about the temperature within the reflecting cell. At any fixed spatial point (x, y, z), the temperature is caused by the thermal agitation of the gas molecules. This thermal agitation is manifested by a spread in velocity which leads to a spread in doppler shift, causing the spectral line to be broadened. By measuring the width of the resonance at position (x, y, z), the temperature at (x, y, z) is measured.

Problems

13.1 Suppose that a radio telescope has a resolution cell that is 0.1 arc second on a side. How many pixels (resolution cells) are there on the full celestial sphere? How long would it take to image the entire celestial sphere (at one radio frequency) if a scene of 1000 by 1000 pixels can be imaged in 1 hour? Repeat the calculation for a resolution cell that is 10^{-4} arc second on a side. Repeat that calculation for the situation where the process is to be repeated for each of N different frequency intervals.

13.2 Formulate the equations for a *weighted least-squares* straight-line fit to a set of data. Given that $y = ax + b$ and that the value y_i is measured with the error variance σ_i^2 at location x_i, determine the estimates of a and b to minimize

$$\Sigma_i \frac{(y_i - ax_i - b)}{\sigma_i^2}.$$

13.3 A plane wave at angle ϕ is incident on an aperture $\text{rect}(t/T)$. Show that the phase of the integrated received signal captured by the aperture has the phase angle of the signal received at the center of the aperture.

13.4 Which pair of apertures is better for estimating the direction of arrival of a plane wave in the presence of internal receiver gaussian noise:

a. $\text{rect}\left(x - \frac{1}{2}, y\right)$ and $\text{rect}\left(x + \frac{1}{2}, y\right)$, or

b. $\text{rect}\left(2x - \frac{3}{2}, y\right)$ and $\text{rect}\left(2x + \frac{3}{2}, y\right)$?

Does the conclusion change if the noise is external noise and so the noise power is dependent on aperture size?

13.5 An n by n array of antennas occupies a total aperture of size L by L.

a. What is the angular resolution of the antenna beam at wavelength λ?

b. Suppose that the antenna elements of a radio telescope are placed in a pattern of size L by L such that all correlation values in an n by n grid are obtained. What is the angular resolution of the image?

c. Is it true that correlation processing is equivalent to synthesizing an aperture from individual elements?

This comparison is a variation of the van Cittert–Zernike theorem, which states a parallel between the correlation function on an aperture and a Fourier transform on that aperture.

13.6 **(Phase Closure)** Phase errors (or gain errors) can be introduced into a received radio signal by flaws in the antenna or the receiver circuitry. A cross-correlation sample $c_{12} = \int_0^T v_1(t)v_2^*(t)\,dt$, will have a phase error caused by phase errors in both $v_1(t)$ and $v_2(t)$. Because the image is known to be real and nonnegative, and perhaps mostly black, it may be possible to remove the phase errors, in part, by insisting that the image be appropriately constrained. Suppose that the true cross-correlation coefficient c_{ij} has phase ϕ_{ij}, and the computed cross-correlation coefficient $\widehat{c_{ij}}$ has phase $\widehat{\phi_{ij}}$, which differs from ϕ_{ij} because of phase errors in the receivers. Show that

$$\widehat{\phi_{ij}} + \widehat{\phi_{jk}} + \widehat{\phi_{ki}} = \phi_{ij} + \phi_{jk} + \phi_{ki}.$$

That is, the sum of the correlation phases is correct even if the individual correlation phases are in error. Can this fact be used to improve a radio astronomy image? Is a similar statement true for amplitudes?

13.7 A radio telescope consists of an array of antennas of diameter D. It is decided to position the antennas so that all position differences are on a uniformly spaced grid. Given that two antennas cannot share the same space (or otherwise block each other), what can be said about grating lobes? How is the situation changed if the antennas are placed so that the grid is slightly perturbed from a uniform grid?

13.8 To account for missing data samples, let $H(f_x, f_y)$ be an indicator function for an "aperture" in the f_x, f_y plane. (The function $H(f_x, f_y)$ only takes values 0 and 1.) Suppose that a radio telescope uniformly samples $H(f_x, f_y)C(f_x, f_y)$ on a square grid in the f_x, f_y plane. How do the missing data, as modeled by $H(f_x, f_y)$, affect the image if simple inverse Fourier transform processing is used? Describe the degradation if $H(f_x, f_y)$ has the form of a plus sign (a square with the corners missing). Are there any signal-processing techniques that will recover the missing data? Why?

13.9 A source emits white gaussian noise of the power density spectrum $N_S(f) = N_{0S}/2$. At each of two surveillance receivers, with identical transfer functions $H(f)$, the radiation is contaminated by white gaussian noise of the power density spectrum $N_{01}/2$ and $N_{02}/2$. Suppose that $H(f)$ consists of a split frequency window, as shown below. Find the variance of a noncoherent estimator of the relative delay based on finding the peak of a cross-ambiguity surface.

$$\sigma_{\hat{\tau}}^2 = \frac{1}{8\pi} \frac{\gamma_1 + \gamma_2 + \gamma_1\gamma_2}{(2\Delta BT)\frac{1}{12}(3B^2 + \Delta B^2)}.$$

Notice that the denominator is written as a product of two factors. The first is a product of time and observation bandwidth: it is called the *processing gain.* The second factor is the square of the Gabor bandwidth of the filter. This factor can be made large by using a split window without increasing the observation bandwidth. In this way, the estimation accuracy is improved.

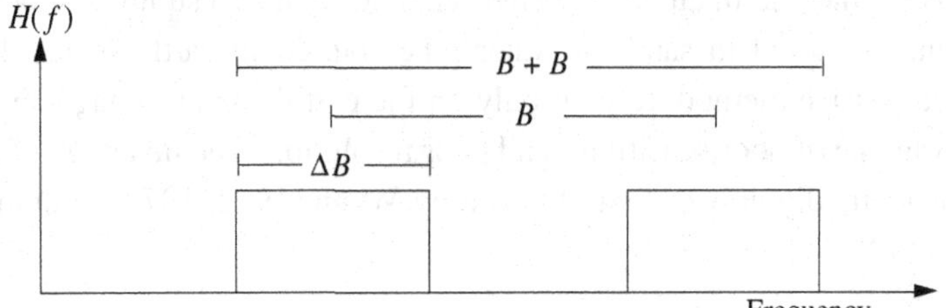

13.10 Under thermal agitation, the velocity distribution of the molecules of a gas that consists of a single species of molecule is a three-dimensional gaussian distribution. What is the distribution of radial velocity in the direction toward an observation point? Give an expression relating the temperature to the width of the three-dimensional gaussian distribution describing the velocity spread.

13.11 Prove the one-dimensional form of the *van Cittert–Zernike theorem*: the antenna pattern of a uniformly-illuminated one-dimensional aperture has the same functional form as the correlation function of a spatially white random process in that aperture. Extend this statement to two dimensions.

13.12 Three known points $(x_1, y_1), (x_2, y_2), (x_3, y_3)$ in the plane are given. Differential ditstances $\Delta_{12} = R_1 - R_2$ and $\Delta_{13} = R_1 - R_3$ are given where R_1, R_2, and R_3 are the distances from the three known points to an unknown point (x, y). Write out an explicit flow diagram for computing (x, y) from $(\Delta_{12}, \Delta_{13})$. Ensure that the computation does not break down when the three known positions are on a straight line.

Notes

Radio astronomy has been a spectacular success and is now the principal source of information about the distant universe. Following Jansky's discovery in 1931 of cosmic radiation in the radio bands, Ryle (1952) first proposed the idea of a radio telescope based on the interferometric techniques originally introduced into optical astronomy by Michelson (1890, 1921). The first radio telescope to use the earth's rotation was the Cambridge One-Mile Radio Telescope in 1964. Earth rotation had been used earlier to form images of the sun by Christiansen and Warburton (1955).

The clean algorithm was introduced into radio astronomy by Högbom (1974) as a way to accommodate the sparsity of Fourier data by exploiting the compactness and nonnegativity of the images of galaxies. The clean algorithm has had a very successful history in radio astronomy. The method of phase closure was introduced into radio astronomy by Jennison (1958), and was further discussed by Readhead and Wilkinson (1978).

Magnetic anomaly detection is a baseband technique for detecting large metallic objects submerged in an opaque environment, such as a submarine beneath the ocean, land mines buried in sand, or water pipes buried in earth. It has the advantage of being a passive method, relying only on the earth's natural magnetic field. It has the disadvantage of poor sensitivity and poor resolution. The inversion of the equations of the magnetic dipole was first obtained by Wynn (1972, 1975), Frahm (1972), and by Clem (1995).

14 Data combination and tracking

Some types of large data sets have a hierarchal substructure that leads to new kinds of surveillance algorithms for tasks such as data combination and tracking. The term *data combination* typically refers to a task in which several estimates based on partial data are combined. Perhaps several snapshots of the same scene or object are available, and multiple measurements or estimates are to be combined or averaged in some way. The topics of this chapter refer to partially processed data, and the methods may be used subsequent to correlation or tomographic processing.

Various sets of data may be combined either before detection or after detection. The combination of closely associated data prior to detection is called integration, and can be either *coherent integration* or *noncoherent integration*. The combination of data from multiple sensors, usually in the form of parameters that have been estimated from the data, is called *data fusion*. This term usually conveys an emphasis that there is a diversity of types of data. Sometimes, only a tentative detection is made before tracking, while a hard detection is deferred until after tracking. In some applications, this is called "track before detect." In this chapter, we shall study the interplay between the functions of detection, data fusion, and tracking, which leads us to think of sensor processing on a much longer time scale than in previous chapters.

A post-detection data record of a radar search system will usually contain a large number of detected data points. These samples will consist of multiple observations of each of a number of stationary or moving targets made by a single sensor, or multiple observations of several targets from multiple sensors. At each observation time, the detected targets will be described by a large number of points in space, possibly with each point labeled by one or more components of its velocity vector. The data may be contaminated by many kinds of impairments, such as measurement errors, false alarms, missing samples, or ambiguities. One task is to sort the data into individual targets, removing both ambiguities and false alarms. After the task of association is complete, the data are partitioned and smoothed into estimates of the individual trajectories of each target.

The trajectory of a moving point target is called a *track*. A radar or sonar system may estimate the tracks of targets from the postdetection data. This can be an easy problem if the target is continually under surveillance, or if the density of targets is low. There are many radars, however, such as those with scanning antennas, in which a target is only observed intermittently and the density of targets is high. Then a method is needed to sort the successive detections and string them together to form estimated tracks. This is the task of *multitarget tracking*. This is the same form of mathematical problem as the task of data fusion. In one case, one type of sensor is used many times; in the other case, many kinds of sensors are used.

The usual multitarget environment consists of a set of trajectories in three-dimensional space, denoted $(x_m(t), y_m(t), z_m(t))$ for $m = 1, \ldots, M$. A measurement of a trajectory may provide partial knowledge of the trajectory positions and velocities at some instant of time but need not fully measure the position or velocity. For example, a radar with a scanning antenna will produce a bearing angle to the Mth trajectory, $(x_m(t_{km}), y_m(t_{km}), z_m(t_{km}))$, at each time, t_{km}, that the antenna beam sweeps over the mth target. The range might be unmeasured. If the scan rate is slow, the target may move a considerable distance between samples, in which case the samples of the trajectory will be sparse.

14.1 Noncoherent integration

A binary hypothesis-testing problem may involve more than one measurement. We shall consider an instance of this problem that leads us to the topic of noncoherent integration. Let I be the number of received waveforms and consider the formal hypothesis-testing problem given by

$$\begin{aligned} H_0 : v_i(t) &= n_i(t) & i &= 0, \ldots, I-1 \\ H_1 : v_i(t) &= s_i(t)e^{j\theta_i} + n_i(t) & i &= 0, \ldots, I-1. \end{aligned}$$

The θ_i are independent random variables on the interval $[0, 2\pi]$, and the $n_i(t)$ are independent, complex gaussian random processes. When I equals one, the optimal detector is simply the noncoherent detector studied in Section 12.5. When I is larger than one and the $s_i(t)$ are all copies of the same pulse $s(t)$, this binary hypothesis-testing problem is referred to as the detection of a noncoherent, nonfluctuating pulse train.

Because the noise and the phase errors are independent, the probability density function for the composite measurement is the product of the probability density functions for the I component measurements. This means that the loglikelihood function or loglikelihood ratio for the composite measurement is the sum of the individual loglikelihood functions or loglikelihood ratios. The decision rule, analyzed in the next theorem, is to compare the loglikelihood ratio to a threshold Θ.

Theorem 14.1.1 *The Neyman–Pearson decision rule for the nonfluctuating, noncoherent pulse train with matched-filter samples r_i is:*

Decide H_0 if:

$$-\sum_{i=0}^{I-1} \log I_0\left(\frac{r_i A}{\sigma^2}\right) < \Theta.$$

Decide H_1 if:

$$-\sum_{i=0}^{I-1} \log I_0\left(\frac{r_i A}{\sigma^2}\right) \geq \Theta$$

where $(A/\sigma)^2 = 2E_p/N_0$.

Proof: The likelihood function for the ith measurement, maximized over θ_i, is given in terms of the ricean and rayleigh probability density functions

$$p_0(r) = \mathrm{e}^{-r^2/2\sigma^2} \qquad r \geq 0$$

$$p_1(r) = \frac{r}{\sigma^2}\mathrm{e}^{-(r^2+A^2)/2\sigma^2} I_0\left(\frac{rA}{\sigma^2}\right) \qquad r \geq 0.$$

The loglikelihood function for the ith measurement r_i is

$$\Lambda_0(r_i) = \log \frac{r_i}{\sigma^2}\mathrm{e}^{-r_i^2/2\sigma^2}$$

$$\Lambda_1(r_i) = \log \frac{r_i}{\sigma^2}\mathrm{e}^{-(r_i^2+A^2)/2\sigma^2} I_0\left(\frac{r_i A}{\sigma^2}\right).$$

The loglikelihood function for the block of I measurements is

$$\Lambda_0(\boldsymbol{r}) = \log \prod_{i=0}^{I-1} \frac{r_i}{\sigma^2}\mathrm{e}^{-r_i^2/2\sigma^2}$$

$$\Lambda_1(\boldsymbol{r}) = \log \prod_{i=0}^{I-1} \frac{r_i}{\sigma^2}\mathrm{e}^{-(r_i^2+A^2)/2\sigma^2} I_0\left(\frac{r_i A}{\sigma^2}\right).$$

The loglikelihood ratio is

$$\Lambda(\boldsymbol{r}) = \log \frac{\prod_{i=0}^{I-1} \frac{r_i}{\sigma^2}\mathrm{e}^{-r_i^2/2\sigma^2}}{\prod_{i=0}^{I-1} \frac{r_i}{\sigma^2}\mathrm{e}^{-(r_i^2+A^2)/2\sigma^2} I_0\left(\frac{r_i A}{\sigma^2}\right)}$$
$$= \sum_{i=0}^{I-1}\left[\frac{A^2}{2\sigma^2} - \log I_0\left(\frac{r_i A}{\sigma^2}\right)\right].$$

We compare $\Lambda(\boldsymbol{r})$ with the threshold Θ to decide between H_0 and H_1. Because the term $A^2/2\sigma^2$ is a constant, it can be absorbed into the choice of threshold. Therefore we will redefine the loglikelihood ratio as

$$\Lambda(\boldsymbol{r}) = -\sum_{i=0}^{I-1} \log I_0\left(\frac{r_i A}{\sigma^2}\right).$$

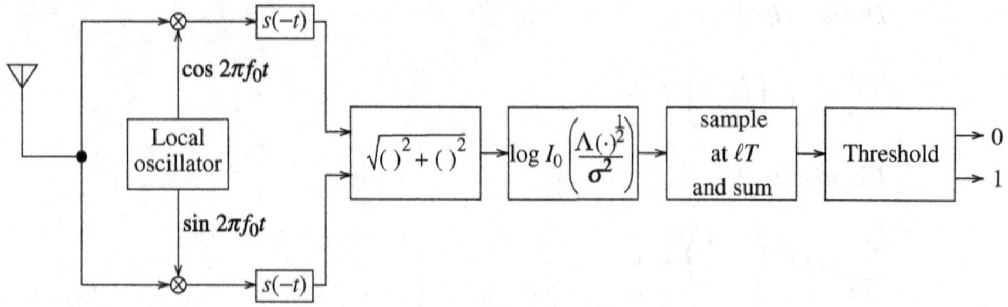

Figure 14.1 Detection of a nonfluctuating, noncoherent pulse train

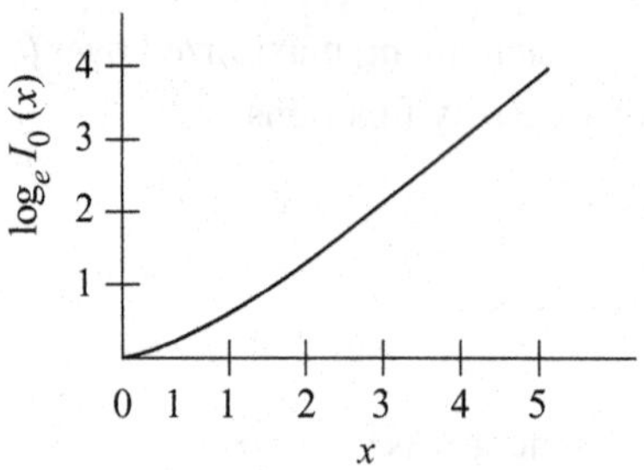

Figure 14.2 The function $\log I_0(x)$

The Neyman–Pearson rule simply applies $\Lambda(\boldsymbol{r})$ to a threshold Θ. By the choice of Θ, the probability of a false alarm, p_{FA}, can be traded against the probability of a missed detection, p_{MD}. □

An optimal receiver for a nonfluctuating, noncoherent pulse train is shown in Figure 14.1. The structure of this receiver is an immediate consequence of Theorem 14.1.1. The function $\log I_0(x)$ that appears in the receiver is shown in Figure 14.2.

This decision rule requires the computation of the modified Bessel function $I_0(x)$. This is much different from the optimal noncoherent detection of a single pulse, which did not require the computation of a Bessel function. In that situation, the Bessel function was only needed to compute the performance curves for the optimal detector. In contrast, for noncoherent integration, every pulse of a received pulse train is passed through the log-Bessel function $\log I_0(x)$ at the output of the matched filter.

For small values of x, we may consider the series expansion

$$\log I_0(x) = \tfrac{1}{4}x^2 - \tfrac{1}{64}x^4 + \dots$$

Therefore the function $\log I_0(x)$ can be well approximated by the quadratic function

$$\log I_0(x) \approx \tfrac{1}{4}x^2$$

when x is small. Because of this approximation, the detector will be suboptimal at large values of x, but this may be only a minor concern. The approximation has another important consequence in the structure of the receiver. The detection rule now has the form

$$\left(\frac{A}{\sigma^2}\right)\sum_{i=0}^{I-1} r_i^2 \geq \Theta.$$

By redefining the threshold, this becomes

$$\sum_{i=0}^{I-1} r_i^2 \geq \Theta.$$

The detector structure now does not depend on A. Only the performance of the detector depends on A.

14.2 Sequential detection

The Neyman–Pearson detection rule processes a set of data to form the likelihood statistic, then decides between H_0 and H_1 (target present or target absent) based on whether the likelihood statistic exceeds a fixed threshold. The probability of false alarm and the probability of missed detection both depend on the threshold. To make the probability of false alarm smaller, increase the threshold. To make the probability of missed detection smaller, decrease the threshold. To simultaneously do both, use two thresholds. This then leaves a region between the thresholds which corresponds to no decision. It is straightforward to calculate the probability of no decision under each of the two hypotheses, as well as the probability of missed detection and the probability of false alarm.

A sequential detection rule introduces an alternative way to fill this gap between the two thresholds. This is to collect more data. Thus the sequential rule will decide H_0 if the statistic is below the lower threshold, and will decide H_1 if the statistic is above the upper threshold. If the statistic is between the two thresholds, the decision will be left open, pending the collection of more data. As more data are gathered, the statistic is reapplied to two thresholds in order to detect target present or target absent. This process will continue until one of the two thresholds is crossed. Normally, the two thresholds will be changed as the amount of data collected increases.

A sequential decision rule will almost surely make a decision eventually. However, the amount of time that it takes to make a decision is not predetermined; it varies with the actual data realization. This may be unacceptable in some applications.

14.3 Multitarget sorting

Suppose that a detection sensor, such as a radar or sonar, observes the same scene several times in succession; each such observation of a common scene is called a *scan.* During each scan, the radar detects and locates those targets that lie within that scene, but with some position error. It may be that a single target has a unique characteristic, such as size or shape, that allows it to be recognized within each scan. However, in many cases, the detected targets are not distinguishable. Then one has the task of target association which is the task of matching the targets from one scan with targets in the next scan. In this section we are interested in associating targets using only consistency of position.

In the general case, the targets may be densely distributed and moving, and there may be false alarms and missed detections, whereby the problem of target association by consistency of position can appear formidable. We shall begin our discussion with a simple case with no target motion, no false alarms, and no missed detections. Even then, the problem can be quite difficult.

Figure 14.3 shows a two-dimensional static situation in which there are four stationary targets and three scans. The four targets in the same scan are all coded by the same symbol. Because of measurement errors, each target appears displaced from its true position. The goal is to sort the targets into bins, one target from each scan to a bin. Because the targets are widely separated compared to the error dispersion, this case is easy to solve, and the solution is visually apparent in Figure 14.3. After the targets are partitioned into bins, the cluster of three measurements in each bin can be averaged or smoothed in some way to give a better estimate of the target location, but this step is

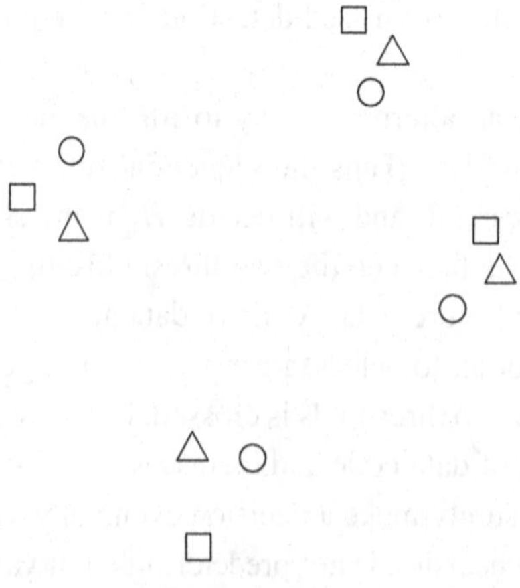

Figure 14.3 A simple association problem

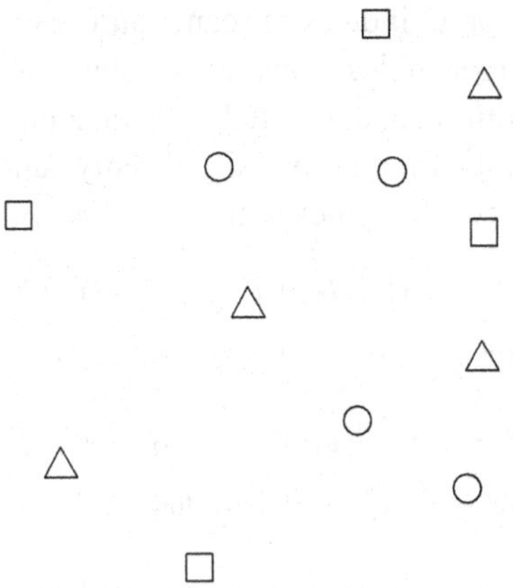

Figure 14.4 A more difficult association problem

not our main interest in this section. Our purpose here is to study the assignment of targets to bins.

Figure 14.4 shows another instance of the problem, now with a larger dispersion in the measurements. Now the solution is not apparent, nor are the criteria apparent for judging a particular solution to be the best solution. We shall see that if the errors are gaussian distributed, then the optimal association for two scans minimizes the sum of pairwise euclidean distances between associated points.

This kind of problem is called a *three-dimensional assignment problem*. The term "three-dimensional" here refers to the fact that there are three scans; it does not refer to the two physical dimensions of the spatial problem. In general, an n-dimensional assignment problem refers to n scans of a scene, which itself may be either a two-dimensional or three-dimensional scene.

The two-dimensional assignment problem with n targets in a two-dimensional space is treated as follows. The n targets are at true positions (x_i, y_i) for $i = 1, \ldots, n$. During each scan, the positions are measured. The measurements of the positions in each scan have no preferred order. The true positions in the x, y plane during the first scan are denoted $(x_i^{(1)}, y_i^{(1)})$ for $i = 1, \ldots, n$. Thus $(x_i^{(1)}, y_i^{(1)}) = (x_{\pi_1(i)}, y_{\pi_1(i)})$, where $\pi_1(i)$ denotes a permutation of the indices. The actual locations during the second scan are $(x_i^{(2)}, y_i^{(2)})$ for $i = 1, \ldots, n$. Thus $(x_i^{(2)}, y_i^{(2)}) = (x_{\pi_2(i)}, y_{\pi_2(i)})$. If there are only two scans, then there are only two hypotheses:

$$H_0 : \begin{cases} (x_1^{(2)}, y_1^{(2)}) = (x_1^{(1)}, y_1^{(1)}), \\ (x_2^{(2)}, y_2^{(2)}) = (x_2^{(1)}, y_2^{(1)}), \end{cases}$$

$$H_1 : \begin{cases} (x_1^{(2)}, y_1^{(2)}) = (x_2^{(1)}, y_2^{(1)}), \\ (x_2^{(2)}, y_2^{(2)}) = (x_1^{(1)}, y_1^{(1)}). \end{cases}$$

In other words, hypothesis H_0 says that the target indices in scan 1 and scan 2 are the same, while hypothesis H_1 says that the target indices have been interchanged. The measured datum $(u_i^{(j)}, v_i^{(j)})$ is the actual position contaminated by measurement error. If the errors are gaussian random variables, then the probability density functions on the data on scan one and scan two under the two hypotheses are

$$H_0 : \begin{cases} p_0(u_1^{(1)}, v_1^{(1)}, u_2^{(1)}, v_2^{(1)}) = \left(\frac{1}{2\pi\sigma^2}\right)^2 e^{-[(u_1^{(1)}-x_1)^2+(v_1^{(1)}-y_1)^2]/2\sigma^2} e^{-[(u_2^{(1)}-x_2)^2+(v_2^{(1)}-y_2)^2]/2\sigma^2} \\ p_0(u_1^{(2)}, v_1^{(2)}, u_2^{(2)}, v_2^{(2)}) = \left(\frac{1}{2\pi\sigma^2}\right)^2 e^{-[(u_1^{(2)}-x_1)^2+(v_1^{(2)}-y_1)^2]/2\sigma^2} e^{-[(u_2^{(2)}-x_2)^2+(v_2^{(2)}-y_2)^2]/2\sigma^2} \end{cases}$$

$$H_1 : \begin{cases} p_1(u_1^{(1)}, v_1^{(1)}, u_2^{(1)}, v_2^{(1)}) = \left(\frac{1}{2\pi\sigma^2}\right)^2 e^{-[(u_1^{(1)}-x_1)^2+(v_1^{(1)}-y_1)^2]/2\sigma^2} e^{-[(u_2^{(1)}-x_2)^2+(v_2^{(1)}-y_2)^2]/2\sigma^2} \\ p_1(u_1^{(2)}, v_1^{(2)}, u_2^{(2)}, v_2^{(2)}) = \left(\frac{1}{2\pi\sigma^2}\right)^2 e^{-[(u_1^{(2)}-x_2)^2+(v_1^{(2)}-y_2)^2]/2\sigma^2} e^{-[(u_2^{(2)}-x_1)^2+(v_2^{(2)}-y_1)^2]/2\sigma^2}. \end{cases}$$

Thus, the negatives of the loglikelihood functions are sums of squared distances, as follows

$$\begin{aligned} -\Lambda(x_1, y_1, x_2, y_2|H_0) &= [(u_1^{(1)} - x_1)^2 + (v_1^{(1)} - y_1)^2] + [(u_2^{(1)} - x_2)^2 + (v_2^{(1)} - y_2)^2] \\ &\quad + [(u_1^{(2)} - x_1)^2 + (v_1^{(2)} - y_1)^2] + [(u_2^{(2)} - x_2)^2 + (v_2^{(2)} - y_2)^2] \\ -\Lambda(x_1, y_1, x_2, y_2|H_1) &= [(u_1^{(1)} - x_1)^2 + (v_1^{(1)} - y_1)^2] + [(u_2^{(1)} - x_2)^2 + (v_2^{(1)} - y_2)^2] \\ &\quad + [(u_1^{(2)} - x_2)^2 + (v_1^{(2)} - y_2)^2] + [(u_2^{(2)} - x_1)^2 + (v_2^{(2)} - y_1)^2]. \end{aligned}$$

Now we are ready to maximize the likelihood by first minimizing the right side of each equation over (x_1, y_1, x_2, y_2) and then choosing the hypothesis for which the resulting term is minimum. Although this minimization can be found analytically, it is more informative to perform this minimization by geometric reasoning. In the first equation, only the first and third terms depend on (x_1, y_1), and the sum of these two terms is the sum of two squared distances of the form

$$d(P^{(1)}, Q^{(1)})^2 + d(P^{(2)}, Q^{(1)})^2$$

where the measured points $P^{(i)}$ are equal to $(u_1^{(i)}, v_1^{(i)})$, and the actual point $Q^{(1)}$ is equal to (x_1, y_1). Clearly, the minimum occurs if $Q^{(1)}$ is midway between $P^{(1)}$ and $P^{(2)}$, and so each distance is equal to $d(P^{(1)}, P^{(2)})/2$. Therefore

$$\min_{Q^{(1)}} \left[d(P^{(1)}, Q^{(1)})^2 + d(P^{(2)}, Q^{(1)})^2\right] = \tfrac{1}{2} d(P^{(1)}, P^{(2)})^2.$$

After this minimization, applied as well to (x_2, y_2), the negative loglikelihoods are

$$\begin{aligned} -\Lambda_0 &= \tfrac{1}{2} d(P_1^{(1)}, P_1^{(2)})^2 + \tfrac{1}{2} d(P_2^{(1)}, P_2^{(2)})^2 \\ -\Lambda_1 &= \tfrac{1}{2} d(P_1^{(1)}, P_2^{(2)})^2 + \tfrac{1}{2} d(P_2^{(1)}, P_1^{(2)})^2. \end{aligned}$$

The decision between the two hypotheses is now made based on which of these two quantities is smaller. This question can be expressed conveniently by defining the following matrix

$$M = \begin{bmatrix} d_{11}^2 & d_{12}^2 \\ d_{21}^2 & d_{22}^2 \end{bmatrix}$$

where the elements d_{ij}^2 of M are the four squared distances $d(P_i^{(1)}, P_j^{(2)})^2$. The decision is then based on a comparison between the trace of M and the trace of the matrix with the two columns of M interchanged.

When there are n targets and two scans, there are $n!$ ways of assigning the targets in the second scan to targets in the first scan. One can again form the matrix of pairwise squared distances, given by

$$M = \begin{bmatrix} d_{11}^2 & \cdots & d_{1n}^2 \\ \vdots & & \\ d_{n1}^2 & \cdots & d_{nn}^2 \end{bmatrix}$$

where $d_{ij} = d(P_i^{(1)}, P_j^{(2)})$. The natural generalization of the decision rule to n targets is to minimize the trace of the matrix M by permutation of its columns. There are $n!$ permutations of the columns and the permutation that minimizes the trace corresponds to maximum-likelihood association of targets in the first and second scans. This is a satisfactory and widely used decision rule. To see that it is the maximum-likelihood decision rule, first observe that there are $n!$ hypotheses corresponding to the $n!$ permutations of the points in the second scan. For a hypothesis that assigns $P_i^{(1)}$ and $P_j^{(2)}$ to the same target, the maximum-likelihood rule requires that the position of that target be estimated at the midpoint of $P_i^{(1)}$ and $P_j^{(2)}$. The distance $d(P_i^{(1)}, P_j^{(2)})$ becomes the appropriate statistic.

In the case of multiple scans, the appropriate statistic is a little more complicated. For example, with three scans, the statistic is

$$d_{ijk} = \min_{Q} \left[d(P_i^{(1)}, Q)^2 + d(P_j^{(2)}, Q)^2 + d(P_k^{(3)}, Q)^2 \right].$$

This minimum occurs at the point $Q(x, y)$ where

$$x = \frac{1}{3}(x_1 + x_2 + x_3), \qquad y = \frac{1}{3}(y_1 + y_2 + y_3).$$

The three-dimensional matrix M with elements d_{ijk} is then defined. The permutation of the j and k indices that minimizes the trace of M is the maximum-likelihood solution.

This decision procedure can be extended to moving targets if the data set is rich enough and if the motion model is appropriately constrained. For example, with three scans, one can model the velocity as constant. For a hypothesis that assigns $P_i^{(1)}$, $P_j^{(2)}$, and $P_k^{(3)}$ to the same target, a least squares straight line is fit to the three points. Then d_{ijk}^2 is the cumulative squared error from the straight-line fit. Finally, minimize the trace of the three-dimensional matrix with elements d_{ijk}^2 by choice of permutation. The minimizing permutation defines the assignment.

14.4 The assignment problem

To start with the simplest possible problem, we shall deal with only two scans, as discussed in Section 14.3 and illustrated in Figure 14.4. Let i index the targets in the first scan, and let j index the targets in the second scan. The task is to pair each target in the first scan with a target in the second scan so that the sum of the pairwise squared euclidean distances is minimized. The value of the minimum sum squared euclidean distance itself is not normally of interest. It is the pairing that yields this minimum which is of interest.

The problem can be restated as a standard matrix problem, called the *assignment problem*. Let d_{ij} be the euclidean distance from the ith point of the first scan to the jth point of the second scan, and let $\boldsymbol{M}$ be the matrix whose ij entry is d_{ij}^2. The *trace* of a square matrix is defined as the sum of its diagonal elements. A new matrix can be obtained by permuting the rows of $\boldsymbol{M}$, and that matrix also has a trace. Thus, one can permute the rows of $\boldsymbol{M}$ to minimize the trace. The solution of the assignment problem is given by the row permutation that minimizes the trace. An equivalent statement of this problem is to find the column permutation that minimizes the trace.

For example, let

$$\boldsymbol{M} = \begin{bmatrix} 1 & 4 & 8 \\ 12 & 5 & 3 \\ 5 & 2 & 1 \end{bmatrix}.$$

The trace of the matrix is 7. There are 3! ways to permute the rows of $\boldsymbol{M}$. By interchanging the second column and third column of $\boldsymbol{M}$, the trace becomes 6. This is the minimum value of the trace over all permutations of columns, as can be seen by examining each of the six possible column permutations. (Equivalently, the trace can be minimized by interchanging the second and third rows.)

In a larger problem, with an n by n matrix, there are $n!$ possible permutations of columns. For large n ($n = 1000$ is not unreasonable in some applications), it is not practical to exhaustively try all possibilities. Instead, fast computational algorithms are needed to search for the minimizing permutation.

More generally, for the L-dimensional assignment problem, described by an L-dimensional n by n n array,

$$M = C_{i_1 i_2 \ldots i_L},$$

each of the indices, except i_1, should be permuted to minimize the sum of the diagonal elements. The minimizing permutation is the solution to the n-dimensional assignment problem. Thus there are $(n!)^{L-1}$ possibilities to be minimized over. If n is large, it is not possible to try all possibilities. An efficient computational procedure is needed to find the minimum.

An equivalent statement of the problem is given as follows. Let $M = [C_{i_1 i_2 \ldots i_n}]$ be an L-dimensional n by ... by n matrix of nonnegative real numbers called *costs*. Let $Z = [Z_{i_1 i_2 \ldots i_n}]$ be a zero-one matrix, that is, every element of Z is either a zero or a one. The problem is to find the zero-one matrix Z that minimizes the expression

$$\sum_{i_1=0}^{n-1} \sum_{i_2=0}^{n-1} \cdots \sum_{i_\ell=0}^{n-1} C_{i_1 i_2 \ldots i_\ell} Z_{i_1 i_2 \ldots i_\ell},$$

and subject to

$$\sum_{i_2=0}^{n-1} \sum_{i_3=0}^{n-1} \cdots \sum_{i_\ell=0}^{n-1} Z_{i_1 i_2 \ldots i_\ell} = 1 \qquad i_1 = 0, \ldots, n-1$$

$$\sum_{i_1=0}^{n-1} \sum_{i_3=0}^{n-1} \cdots \sum_{i_\ell=0}^{n-1} Z_{i_1 i_2 \ldots i_\ell} = 1 \qquad i_2 = 0, \ldots, n-1$$

$$\vdots$$

$$\sum_{i_1=0}^{n-1} \sum_{i_2=0}^{n-1} \cdots \sum_{i_\ell=0}^{n-1} Z_{i_1 i_2 \ldots i_\ell} = 1 \qquad i_\ell = 0, \ldots, n-1.$$

The constraints say that each "row" must contain exactly a single one. The matrix Z is a way of specifying a permutation.

The two-dimensional version of the assignment problem in this form is to find the zero-one matrix $Z = [Z_{ij}]$ that minimizes

$$\sum_{i=0}^{I-1} \sum_{j=0}^{J-1} C_{ij} Z_{ij},$$

subject to the constraints that all rows of Z sum to one and all columns of Z sum to one. Equivalently, every row has a single one, and every column has a single one.

This alternative formulation of the assignment problem replaces the notion of a column permutation by the notion of a zero-one matrix. The locations of the ones in the array define the permutation. These elements are the elements that would be permuted into the diagonal to obtain the permutation with the minimum trace. This formulation facilitates the introduction of optimization methods based on lagrange multipliers. In such methods an L-dimensional assignment problem is reduced to an $(L-1)$-dimensional assignment problem by appending a constraint with the aid of a lagrange multiplier. This is called *lagrangian relaxation.* The $(L-1)$-dimensional problem is solved for many values of the lagrange multiplier in order to find the particular value that gives a solution satisfying the constraint. In this way the L-dimensional assignment problem is replaced by a search over a large number of $(L-1)$-dimensional assignment problems. In turn, by recursion, the $(L-1)$-dimensional problem is solved the same way. In this way, an L-dimensional assignment problem is reduced to a large stack of two-dimensional assignment problems. Then we only need an algorithm for the two-dimensional assignment problem.

One widely used algorithm for the two-dimensional assignment problem is the *Munkres algorithm.* The key to this algorithm is the fact that any constant can be added to all elements of that row or that column of $\boldsymbol{M}$ without affecting which permutation minimizes the trace. Consequently, by subtracting the appropriate constant from all elements of any row (or any column), the minimum element of that row (or that column) can be changed to a zero with no essential change in the problem. The Munkres algorithm combines two elementary operations: adding and subtracting constant rows or constant columns, and permutations.

The Munkres algorithm starts with the matrix $\boldsymbol{M}$ and iteratively makes changes in various elements to eventually obtain the desired zero-one matrix $\boldsymbol{Z}$. In the description of the Munkres algorithm, each row or column of the array may be flagged for further reference; this is referred to as a *covered row* or *covered column.* Each element in a covered row or covered column is referred to as a *covered element.* Individual elements of the array may also be marked by being starred $(*)$ or primed $(')$. The locations of the stars designate the locations in which the matrix $\boldsymbol{Z}$ will contain a one. The elements marked by a prime are candidates for a star, and eventually each prime will either be replaced by a star or will be removed.

The Munkres algorithm consists of a number of iterations. The result of each iteration consists of one or more elements of $\boldsymbol{M}$ being marked by asterisks, or one or more elements of $\boldsymbol{M}$ being marked by primes. The algorithm iterates until there is exactly one star in every row and one star in every column. The set of locations of these stars constitutes the solution to the problem.

Proposition 14.4.1 *Let $\boldsymbol{c} = (c_i)$ and $\boldsymbol{r} = (r_j)$ be any column and row of the nonnegative square matrix $\boldsymbol{A}$ whose elements are denoted a_{ij}. Then the matrix $\boldsymbol{B}$ with* *elements $b_{ij} = a_{ij} - c_i - r_j$ has its trace minimized by the same row permutation as does **A**.*

Proof: Let π be a permutation of indices. Then

$$\sum_{i=1}^{n} b_{i\pi(i)} = \sum_{i=1}^{n} a_{i\pi(i)} - \sum_{i=1}^{n} c_i - \sum_{i=1}^{n} r_{\pi(i)}.$$

Because the second two terms do not depend on π, the proposition is proved. □

It is clear that if each row and each column of the nonnegative matrix $\boldsymbol{C}$ has only one zero, then the location of these zeros mark the location of the ones in the permutation matrix $\mathbf{Z}$. The computations defined in Proposition 14.4.1 lead to the following examples:

$$\begin{bmatrix} 1 & 4 & 8 \\ 12 & 5 & 3 \\ 5 & 2 & 1 \end{bmatrix} \rightarrow \begin{bmatrix} 0 & 3 & 7 \\ 9 & 2 & 0 \\ 4 & 1 & 0 \end{bmatrix} \rightarrow \begin{bmatrix} 0 & 2 & 7 \\ 9 & 1 & 0 \\ 4 & 0 & 0 \end{bmatrix}.$$

Because the right side does not have a single zero in each row and column, it does not yet reveal the permutation. A different example is

$$\begin{bmatrix} 1 & 4 & 8 \\ 12 & 5 & 3 \\ 5 & 1 & 2 \end{bmatrix} \rightarrow \begin{bmatrix} 0 & 3 & 7 \\ 9 & 2 & 0 \\ 4 & 0 & 1 \end{bmatrix} \rightarrow \begin{bmatrix} 0 & 3 & 7 \\ 9 & 2 & 0 \\ 4 & 0 & 1 \end{bmatrix}.$$

Because the right side has a single zero in each row and column, the permutation is revealed.

Two elements of a matrix will be called unaligned if they do not lie in the same row or column. The Munkres algorithm will produce a matrix of starred elements in which all the starred elements are unaligned. These stars mark the ones in the permutation matrix $\mathbf{Z}$.

Theorem 14.4.2 *If the largest set of unaligned zero elements in a matrix **A** has size n, then all zero elements of **A** lie in a set of m rows or columns.*

Proof: By permutation of rows, the m rows containing these zeros can be made to be the first m rows. By column permutation these zeros can be made to lie in the first m columns. Clearly all zeros must lie in the first m rows or the first m columns because any zero that did not could be used to enlarge the set of m unaligned zeros, contrary to assumption. □

The Munkres algorithm is as follows:

Step 0: To prepare the matrix, subtract the smallest element of each row from all elements of that row, then subtract the smallest element of each column from all elements of that column.

Step 1: Star every zero of $\boldsymbol{M}$ that is unaligned with any other zero. If n zeros are starred, halt. Otherwise flag every column with a starred zero and go to Step 2.

Step 2: Choose a zero in an unflagged column and mark it. If there is none, go to Step 4. If there is no starred zero in the row of this marked zero, go to Step 3. Otherwise, flag this row and unflag the column of this marked zero. Repeat Step 2.

Step 3: Find the smallest element of the matrix not in a flagged row or flagged column. Add this element to each flagged row, then subtract it from each flagged column. Return to Step 1.

Step 4: Starting with the uncovered, primed zero, find a starred zero (if any) in that column. Then, in the row of that starred zero, find a primed zero. Continue to alternate between primed zeros and starred zeros in this way by alternating moves along columns and rows until a primed zero is reached with no starred zero in its column. Unstar each starred zero of this sequence and star each primed zero of the sequence. Remove all primes and covers and return to Step 1.

For an example of the Munkres algorithm, consider the matrix

$$\boldsymbol{M} = \begin{bmatrix} 7 & 5 & 13 \\ 5 & 4 & 1 \\ 9 & 3 & 2 \end{bmatrix}.$$

The algorithm first makes the following changes

$$\begin{bmatrix} 7 & 5 & 13 \\ 5 & 4 & 1 \\ 9 & 3 & 2 \end{bmatrix} \rightarrow \begin{bmatrix} 2 & 0 & 8 \\ 4 & 3 & 0 \\ 7 & 1 & 0 \end{bmatrix} \rightarrow \begin{bmatrix} 0 & 0 & 8 \\ 2 & 3 & 0 \\ 5 & 1 & 0 \end{bmatrix}.$$

Upon reaching Step 4, the algorithm computes

$$\begin{bmatrix} \overline{0^*} & \overline{0'} & \overline{6} \\ 2 & 3 & \overline{0^*} \\ 5 & 1 & 0 \end{bmatrix} \rightarrow \begin{bmatrix} \overline{0^*} & \overline{0'} & \overline{8} \\ 1 & 2 & \overline{0^*} \\ 4 & 0 & \overline{0} \end{bmatrix}.$$

Upon returning to Step 1, the algorithm finds

$$\begin{bmatrix} \overline{0^*} & \overline{0} & \overline{8} \\ \overline{1} & \overline{2} & \overline{0^*} \\ \overline{4} & \overline{0^*} & \overline{0} \end{bmatrix}.$$

Because all elements are covered, the algorithm terminates. The three starred zeros determine the locations of the ones in $\mathbf{Z}$.

14.5 Multilateration

Multilateration is a form of triangulation in which multiple curves of position are intersected to determine the location of a point. We will allow these curves of position to be of various kinds, such as circles of constant range, shown in Figure 14.5; lines of bearing, shown in Figure 14.6; or hyperbolas or ellipses. Other situations may use

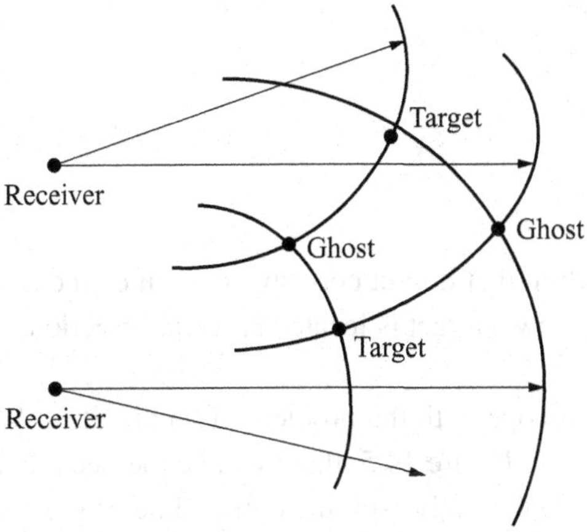

Figure 14.5 Bistatic range ambiguities

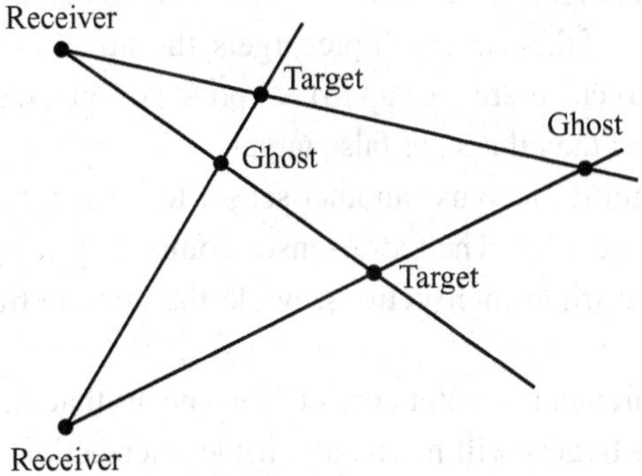

Figure 14.6 Bistatic angle ambiguities

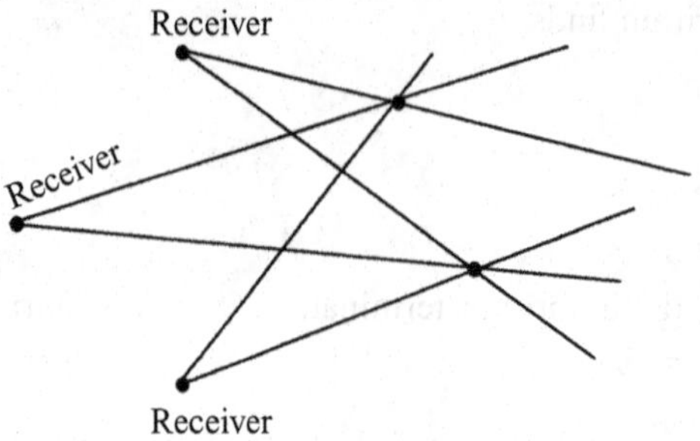

Figure 14.7 Resolution of ambiguities

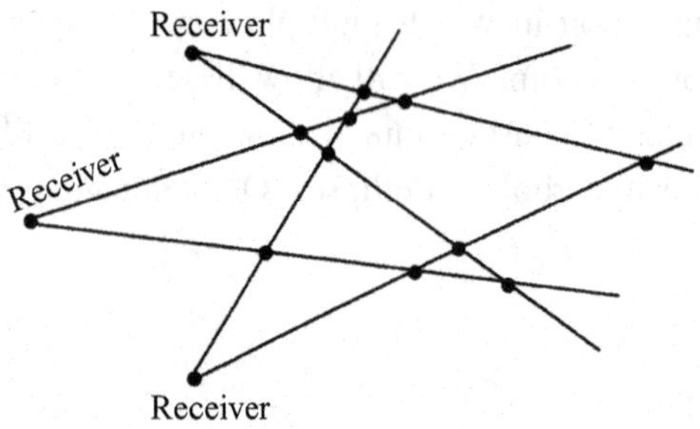

Figure 14.8 Perturbed lines of position

the intersections of curves of position that do not correspond to an elementary curve. Figure 14.5 and Figure 14.6 show how a target is located at the intersection of several curves.

All multilateration systems must cope with the problem of ambiguities. Whenever there are multiple targets, as shown in Figure 14.5, there will be the need to associate a particular curve of position from one family with the correct line of position of the other family. In the absence of any side information telling how to accomplish this, there will be ambiguities, as indicated in the figures. From the data itself, there are two pairs of solutions in Figure 14.5 and Figure 14.6, and there is no way to distinguish the two true targets from the two ghosts. If there are multiple targets, the situation becomes even more complicated. With N targets, there are (up to) N^2 points of intersection. Of these N are true target positions and the others are false targets.

One way of resolving the ambiguities is to use another sensor to form a third set of lines of position, as shown in Figure 14.7. The extra sensor completely resolves the problem; the true targets are at the triple intersections, while the ambiguities are at double intersections.

In a realistic problem, the measurements are not perfect. This means that the lines of position will be perturbed, and the targets will not be at a triple intersection. Instead, targets will be near a small cluster of three double intersections, as shown in Figure 14.8. This, then, leads to an instance of the assignment problem.

Problems

14.1 Suppose a radar detects an object at a point, $\boldsymbol{x} = (x, y, z)$, in three-dimensional space. The error in the three-dimensional location is a vector gaussian random variable with the three by three covariance matrix $\boldsymbol{\Sigma}_1$. If the true position of the object is $\boldsymbol{x}_1$, then the joint probability density function is

$$p(\boldsymbol{x}|\boldsymbol{x}_1) = \frac{1}{\sqrt{\det(2\pi\boldsymbol{\Sigma}_1)}} e^{-(\boldsymbol{x}-\boldsymbol{x}_1)^\dagger \boldsymbol{\Sigma}_1^{-1}(\boldsymbol{x}-\boldsymbol{x}_1)}.$$

A second radar with independent errors also locates an object with gaussian-distributed errors with the covariance matrix $\boldsymbol{\Sigma}_2$. A decision is to be made as to whether these two objects are the same object. Set up the appropriate hypothesis-testing problem and give expressions for the probabilities of error as a function of the threshold. Does the problem require an assumption about the vector difference $\boldsymbol{x}_1 - \boldsymbol{x}_2$? Can the problem be made parametric in this difference?

14.2 A two-dimensional gaussian random variable with the probability density function $p(x, y)$ has zero mean and covariance matrix $\boldsymbol{\Sigma}$. The *circular error probability* is defined as the value of R satisfying

$$\frac{1}{2} = \int_{-\infty}^{\infty}\int_{-\infty}^{\infty} \operatorname{circ}\left(\frac{x}{2R}, \frac{y}{2R}\right) p(x, y)\, dx\, dy.$$

Calculate the circular error probability as a function of λ_1 and λ_1/λ_2 where λ_1 and λ_2 are the largest and the smallest eigenvalues of $\boldsymbol{\Sigma}$, respectively.

14.3 Verify that the maximum-likelihood solution to the target-sorting problem in the presence of gaussian position error for n targets and L scans reduces to the L-dimensional, n by n assignment problem.

14.4 Given a three-target, two-scan, target-sorting problem in the presence of gaussian noise, suppose that one data point from the third scan is missing. Find a maximum-likelihood solution to this sorting problem.

14.5 Given three points $P^{(1)} = (x_1, y_1)$, $P^{(2)} = (x_2, y_2)$, $P^{(3)} = (x_3, y_3)$ in the plane, let

$$d(P^{(i)}, Q) = \sqrt{(x_i - x)^2 + (y_i - y)^2}$$

where Q denotes the point (x, y). Find the point $\widehat{Q} = (\widehat{x}, \widehat{y})$ defined by

$$(\widehat{x}, \widehat{y}) = \operatorname{argmin}_{(x,y)} \left[\sum_{i=1}^{3} d(P^{(i)}, Q)^2\right].$$

14.6 Given the three points $P^{(1)}$, $P^{(2)}$, $P^{(3)}$, and $\widehat{Q}$, as in Problem 14.5, determine whether $d(P^{(1)}, P^{(2)})^2 + d(P^{(2)}, P^{(3)})^2 + d(P^{(3)}, P^{(1)})^2$ is an adequate approximation to $\sum_{i=1}^{3} d(P^{(i)}, \widehat{Q})^2$.

Notes

The need to recognize tracks in fragmentary data is at least as old as radar, and this need can be found even in prehistoric human activities. As a formal mathematical discipline, however, track assignment took shape only recently; the papers of Wax (1955) and Sittler (1964) can be considered to start the development of a formal theory. Nahi (1969) studied the problem of false measurements in a recursive estimator. Singer, Sea, and Housewright (1974) presented an optimal, maximum-likelihood tracking algorithm that requires data memory to grow with time as data is accumulated. The maximum-likelihood approach was further studied by Stein and Blackman (1975). Morefield (1977) posed the tracking problem as an integer programming problem that clarified the connection with the general assignment problem. The tracking problem has been extensively studied in a series of papers by Bar-Shalom (1978). Iterative methods of maintaining an established track are closely related to the filtering methods of Kalman and Bucy (1960, 1961) and do often employ the Kalman filter.

The assignment problem has its own literature as a topic in optimization theory. When treated by the methods of complexity theory, it is found to be nonpolynomial hard in the number of data points. Radar applications, however, do not normally require that the worst-case assignment problem be solved – only that typical or nearly typical problems be solved and, for these typical problems, occasional missed assignments can be tolerated.

The best-known algorithm for the two-dimensional assignment problem is the Munkres (1957) algorithm, which was based on earlier work by the Hungarian mathematician Egervary, as reported by Kuhn (1955), and later augmented to nonsquare matrices by Bourgeois and La Salle (1971). An alternative algorithm has been developed by Bertsekas (1988).

15 Phase noise and phase distortion

A coherent image-formation system is degraded if the coherence of the received waveform is imperfect. This reduction in coherence is due to an anomalous phase angle in the received signal, which is referred to as *phase error*. Phase errors can be in either the time domain or the frequency domain, and they may be described by either a deterministic model or a random model. Random phase errors in the time domain arise because the phase varies randomly with time. Random phase errors in the frequency domain arise because the phase of the Fourier transform varies randomly with frequency. We will consider both unknown deterministic phase errors and random phase errors in both the time domain and the frequency domain.

Random phase errors in the time domain appear as complex time-varying exponentials multiplying the received complex baseband signal and are called *phase noise*. Phase noise may set a limit on the maximum waveform duration that can be processed coherently. Random phase errors in the frequency domain appear as complex exponentials multiplying the Fourier transform of the received complex baseband signal and are called *phase distortion*. Phase distortion may set a limit on the maximum bandwidth that a single signal can occupy. We shall primarily study phase noise in this chapter. Some of the lessons learned from studying phase noise can be used to understand phase distortion.

The source of phase noise is any unintentional phase modulation that is introduced into the received signal, either in the transmitter, the propagation medium, or the receiver. There are four major sources of phase noise: motion of the transmitter or the receiver, such as antenna vibration or scanning; phase errors arising in the local oscillators and mixing circuits; phase errors arising in the carrier recovery circuitry; and phase errors due to inhomogeneities and aberrations in the propagation medium, such as those caused by atmospheric turbulence.

We shall study the effect of phase errors in two ways. First, in Section 15.1, we will model the phase error $\theta(t)$ as slowly varying and described by several unknown parameters. We will compute the performance loss as a function of the unknown parameters. Then, in the following sections, we will model the phase error $\theta(t)$ as a random process, possibly one that varies rapidly. In Section 15.2, we will compute the expected performance loss based on a probabilistic model of $\theta(t)$.

15.1 Quadratic-phase errors

Let $s(t)$ be a waveform, possibly complex, and of finite energy. Let $\theta(t)$ be an unknown real function of time. The signal $s(t)$ contaminated by the phase error $\theta(t)$ is given by

$$v(t) = s(t)e^{j\theta(t)}.$$

The next section will model $\theta(t)$ as a random process. This section models $\theta(t)$ as a slowly varying deterministic – though unknown – function that can be approximated by the first terms of a Taylor series expansion.

Let

$$\theta(t) = \theta_0 + \dot{\theta}_0 t + \tfrac{1}{2}\ddot{\theta}_0 t^2 + \ldots$$

The coefficients are arbitrary, unknown deterministic parameters. Suppose that $\theta(t)$ is adequately approximated by the first three terms of this series. We will study the performance as a function of the value of the parameters θ_0, $\dot{\theta}_0$, and $\ddot{\theta}_0$ by studying the effect of these parameters in turn on the Fourier transform, the matched filter, and the sample cross-ambiguity function.

If the parameter θ_0 is an arbitrary constant in $[0, 2\pi]$, then the term $e^{j\theta_0}$ can be brought outside the integral defining the Fourier transform, the matched filter, or the sample cross-ambiguity function. In each of these instances, this phase term drops out when taking the magnitude of that function. In an estimation problem, for example, the effect of θ_0 is to require that a noncoherent detector and estimator be used. We have studied coherence in some detail, so we need not consider the parameter θ_0 further.

The analysis of the parameter $\dot{\theta}_0$ is trivial. The slope parameter $\dot{\theta}_0$ will appear as a false frequency shift. It will simply cause the Fourier transform to be offset by $\dot{\theta}_0$ in the frequency axis, and the cross-ambiguity function to be translated by $\dot{\theta}_0$ in the doppler direction.

The parameter $\ddot{\theta}_0$ is a more interesting parameter. The term $\frac{1}{2}\ddot{\theta}_0 t^2$ is called the *quadratic-phase error*, and will also be written as $\frac{1}{2}\alpha t^2$. The quadratic-phase error is the first term in the Taylor series expansion of the phase error that affects the shape of the cross-ambiguity function. The quadratic term is also the dominant term among the terms neglected in the Fresnel and Fraunhofer approximations. An analysis of the effect of quadratic-phase error is needed to justify those approximations.

We shall first examine the effect of quadratic-phase error on the Fourier transform. Suppose the desired signal is a rectangle function,

$$s(t) = \text{rect}(t),$$

but the signal is actually contaminated by an unknown quadratic-phase error. Let

$$\begin{aligned} v(t) &= s(t)e^{j\pi\alpha t^2} \\ &= \text{rect}(t)e^{j\pi\alpha t^2} \end{aligned}$$

be the actual signal. The Fourier transform $V(f)$ is

$$V(f) = \int_{-1/2}^{1/2} e^{j\pi\alpha t^2} e^{-j2\pi f t}\, dt.$$

Define the loss at $f = 0$ as

$$L = \frac{V(0)}{S(0)}.$$

Then

$$\begin{aligned} L &= \left| \int_{-1/2}^{1/2} e^{j\pi\alpha t^2}\, dt \right| \\ &= 2\int_0^{1/2} \cos(\pi\alpha t^2)\, dt. \end{aligned}$$

The total phase change over time T due to the quadratic-phase term is $\Delta\theta = \alpha T^2/2$ cycles. Over half of the pulse duration $T = 1/2$, and the half-interval phase change is $\Delta\theta = \alpha/8$. The loss can be written

$$L = \frac{1}{\sqrt{2\Delta\theta}} \int_0^{\sqrt{2\Delta\theta}} \cos(\pi x^2/2)\, dx$$

with $\Delta\theta$ in cycles. The integral cannot be integrated in closed form but can be expressed in terms of standard functions. Define the *Fresnel cosine integral* as

$$C(\eta) = \int_0^{\eta} \cos\frac{\pi t^2}{2}\, dt,$$

and the *Fresnel sine integral* as

$$S(\eta) = \int_0^{\eta} \sin\frac{\pi t^2}{2}\, dt.$$

Numerically evaluated tables of $C(\eta)$ and $S(\eta)$ are widely available. Figure 15.1 shows a graph of $S(\eta)$ versus $C(\eta)$ with η as a parameter. This graph is known as the *Cornu spiral.*

The loss due to quadratic-phase error is plotted in Figure 15.2. In many applications, the loss due to the quadratic-phase term is less than one decibel, if the half-interval phase change is less than 30°, which is acceptable in many applications.

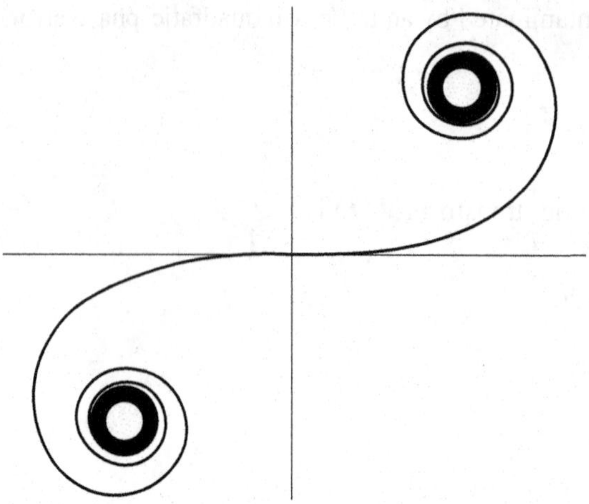

Figure 15.1 The Cornu spiral

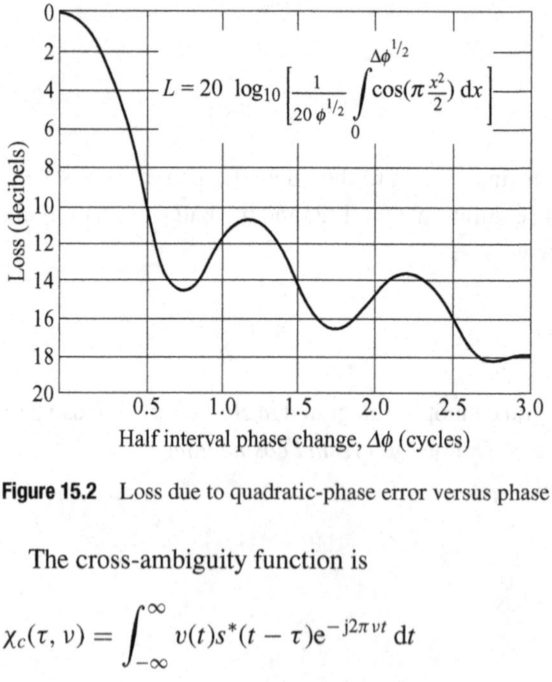

Figure 15.2 Loss due to quadratic-phase error versus phase change

The cross-ambiguity function is

$$\chi_c(\tau, \nu) = \int_{-\infty}^{\infty} v(t)s^*(t-\tau)e^{-j2\pi\nu t}\,dt$$
$$= \int_{-\infty}^{\infty} s(t)s^*(t-\tau)e^{j\pi\alpha t^2}e^{-j2\pi\nu t}\,dt.$$

The presence of the quadratic-phase term means that $\chi_c(\tau, \nu)$ is no longer a simple translation of $\chi(\tau, \nu)$. For small α, $\chi_c(\tau, \nu)$ will be nearly equal to a translation of $\chi(\tau, \nu)$ and will have a peak similar to the peak of $\chi(\tau, \nu)$. However, because of α, the peak of $\chi_c(\tau, \nu)$ will have less amplitude, and its curvature will not be as sharp. For larger α, the peak can collapse so much that it loses its identity.

First, we will evaluate the loss in peak amplitude. Define the loss as a ratio:

$$\begin{aligned} L &= \frac{|\chi_c(0,0)|}{\chi(0,0)} \\ &= \frac{1}{E_p}\left|\int_{-\infty}^{\infty} |s(t)|^2 e^{j\pi\alpha t^2}\, dt\right|. \end{aligned}$$

To proceed, we must specify $|s(t)|^2$. We will choose $|s(t)|^2 = 1$ for $t \in [-T/2, T/2]$ and otherwise equal to zero. This choice is not as restrictive as it might seem. For any $|s(t)|^2$, as long as the fluctuations in $|s(t)|^2$ are rapid compared to the rate at which the phase changes, the fluctuations will have little effect on the value of L. Even if $s(t)$ is a pulse train of low duty cycle, for the purposes of computing L, it is adequately approximated by setting $|s(t)|^2 = 1$, provided the number of pulses is not too small.

In addition to attenuating the peak, phase noise may also change the location and shape of the peak. The primary change affects the curvature in the ν direction. The curvature in the τ direction is essentially unaffected.

The location of the peak in the ν direction is given by

$$\bar{t} = \frac{1}{E_p}\int_{-T/2}^{T/2} t|s(t)|^2 e^{j\pi\alpha t^2}\, dt,$$

and the curvature is described by

$$\overline{t^2} - \bar{t}^2 = \frac{1}{E_p}\int_{-T/2}^{T/2} t^2|s(t)|^2 e^{j\pi\alpha t^2}\, dt - \bar{t}^2.$$

Again, one can choose

$$s(t) = \begin{cases} 1 & |t| \le T/2 \\ 0 & |t| > T/2. \end{cases}$$

For purposes of the present analysis, this is a good approximation.

15.2 Phase noise and coherence

A pulse contaminated by both phase noise and additive noise is given by

$$v(t) = s(t)e^{j\theta(t)} + n(t).$$

In this section, we will study the effect of phase noise on the signal by modeling phase noise $\theta(t)$ as a stationary noise process. The pulse contaminated by only phase noise is

$$v(t) = s(t)e^{j\theta(t)}.$$

The phase-noise process $\theta(t)$ is taken to be a real covariance stationary random process with mean $\bar{\theta}$ and the known correlation function

$$\begin{aligned}\phi(\tau) &= \mathrm{E}[\theta(t)\theta(t+\tau)] \\ &= \sigma_\theta^2 \rho(\tau).\end{aligned}$$

This second-order description is only a partial description of the phase-noise process because higher-order moments are not specified. It is a complete description if the phase noise is gaussian because then higher-order properties are implied by the second-order properties. If the phase noise is not gaussian, the second-order description may still be an adequate description if a small-angle approximation for the phase noise can be made within the analysis.

Suppose that $\theta(t)$ is a gaussian random process with probability density function

$$p(\theta) = \frac{1}{\sqrt{2\pi}\sigma} \mathrm{e}^{-\theta^2/2\sigma_\theta^2}.$$

Recall that, for a gaussian random variable of mean $\overline{\theta}$ and variance σ_θ^2, the characteristic function $\mathrm{E}[\mathrm{e}^{\mathrm{j}\theta(t)}]$ has the form of a Fourier transform. The following theorem is the key to the analysis of gaussian phase noise.

Theorem 15.2.1 *Let θ be a gaussian random variable with mean $\overline{\theta}$ and variance σ_θ^2. Then*

$$\mathrm{E}[\mathrm{e}^{\mathrm{j}a\theta}] = \mathrm{e}^{\mathrm{j}a\overline{\theta}} \mathrm{e}^{-a^2\sigma_\theta^2/2}.$$

Proof: The expectation is defined as

$$\mathrm{E}[\mathrm{e}^{\mathrm{j}a\theta}] = \int_{-\infty}^{\infty} p(\theta)\mathrm{e}^{\mathrm{j}a\theta}\, d\theta,$$

which has the form of a Fourier transform of the gaussian function $p(\theta)$ in the variable θ. The tabulated Fourier transform of a gaussian pulse provides the conclusion of the theorem. □

Corollary 15.2.2 *Let $\theta(t)$ be a stationary, gaussian random process. Then*

$$\mathrm{E}[\mathrm{e}^{\mathrm{j}\theta(t)}] = \mathrm{e}^{\mathrm{j}\overline{\theta}} \mathrm{e}^{-\sigma_\theta^2/2}.$$

Proof: Set a equal to one in the theorem. □

Given the pair of complex random processes of the form

$$\begin{aligned}v_1(t) &= \mathrm{e}^{\mathrm{j}[\theta_1(t)]} \\ v_2(t) &= \mathrm{e}^{\mathrm{j}[\theta_2(t)]},\end{aligned}$$

the *coherence* is defined as the correlation

$$\begin{aligned}\Gamma &= \mathrm{E}[v_1(t)v_2^*(t)]\\ &= \mathrm{E}\left[\mathrm{e}^{\mathrm{j}[\theta_1(t)-\theta_2(t)]}\right].\end{aligned}$$

Thus coherence is another name for correlation when applied to complex exponentials. Obviously, if $\theta_2(t) = \theta_1(t)$, then $\Gamma = 1$, while if $\theta_2(t) = \theta_1(t) + \pi$, then $\Gamma = -1$.

Let $\theta_1(t)$ and $\theta_2(t)$ be gaussian random processes of zero mean, equal variance σ^2, and correlation $\rho = \mathrm{E}[\theta_1(t)\theta_2(t)]/\sigma^2$. Because $\theta_1(t) - \theta_2(t)$ is itself a gaussian random process with the variance $\mathrm{E}[\theta_1(t) - \theta_2(t)]^2 = 2\sigma^2(1-\rho)$, Corollary 15.2.2 gives

$$\Gamma = \mathrm{e}^{-2\sigma^2(1-\rho)}$$

for the coherence.

15.3 Phase noise and the Fourier transform

Let $s(t)$ be any pulse with energy $E_p = 1$ and mean $\bar{t} = 0$. Let

$$v(t) = s(t)\mathrm{e}^{\mathrm{j}\theta(t)}$$

where $\theta(t)$ is a stationary gaussian process. The Fourier transform of $v(t)$ is

$$\begin{aligned}V(f) &= \int_{-\infty}^{\infty} v(t)\mathrm{e}^{-\mathrm{j}2\pi ft}\,\mathrm{d}t\\ &= \int_{-\infty}^{\infty} s(t)\mathrm{e}^{\mathrm{j}\theta(t)}\mathrm{e}^{-\mathrm{j}2\pi ft}\,\mathrm{d}t.\end{aligned}$$

We are interested in the distortion in $V(f)$ caused by the phase noise $\theta(t)$. If $\theta(t) = 0$, then $V(f) = S(f)$ and there is no phase noise.

The effect of phase noise on the Fourier transform of the pulse $s(t)$ can be characterized by several scalar parameters: the loss of amplitude at the origin, the change in the location of the maximum or the mean of the transform, and the width of the transform. Because $\theta(t)$ is random, $V(f)$ is also random. The expectation is

$$\begin{aligned}\mathrm{E}[V(f)] &= \mathrm{E}\left[\int_{-\infty}^{\infty} s(t)\mathrm{e}^{\mathrm{j}\theta(t)}\mathrm{e}^{-\mathrm{j}2\pi ft}\,\mathrm{d}t\right]\\ &= \mathrm{e}^{-\sigma_\theta^2/2}S(f).\end{aligned}$$

Thus, in expectation, the phase noise attenuates the Fourier transform.

Because the energy in $v(t)$ is the same as the energy in $s(t)$, the energy theorem says that the energy in $V(f)$ is the same as the energy in $S(f)$. Because $V(f)$ has, on average, less amplitude than $S(f)$, we must expect that it is wider, on average. This is quantified in the next proposition.

Proposition 15.3.1 *Let $s(t)$ be a real-valued pulse with Gabor bandwidth $B_G^2(s)$. The expected Gabor bandwidth of the pulse contaminated by phase noise of power density spectrum $\Phi(f)$ is*

$$E[B_G^2(v)] = \int_{-\infty}^{\infty} f^2\Phi(f)\,df + B_G^2(s).$$

Proof: Because the energy of pulse $v(t)$ equals one, its Gabor bandwidth is

$$\begin{aligned}
B_G^2(v) &= \int_{-\infty}^{\infty} |\dot{v}(t)|^2\,dt \\
&= \int_{-\infty}^{\infty} |j\dot{\theta}(t)s(t)e^{j\theta(t)} + \dot{s}(t)e^{j\theta(t)}|^2\,dt \\
&= \int_{-\infty}^{\infty} |j\dot{\theta}(t)s(t) + \dot{s}(t)|^2\,dt.
\end{aligned}$$

The expectation is

$$\begin{aligned}
E[B_G^2(v)] &= \int_{-\infty}^{\infty} E|j\dot{\theta}(t)s(t) + \dot{s}(t)|^2\,dt \\
&= \int_{-\infty}^{\infty} |s(t)|^2 E[\dot{\theta}(t)]^2\,dt + \int_{-\infty}^{\infty} |\dot{s}(t)|^2\,dt.
\end{aligned}$$

The second term is the Gabor bandwidth of $s(t)$. The first term is evaluated by using

$$E[\dot{\theta}(t)]^2 = \phi''(0) = \int_{-\infty}^{\infty} f^2\Phi(f)\,df.$$

This completes the proof of the proposition. □

15.4 Phase noise and the matched filter

We can easily calculate the expected loss in the output of a filter, $g(t)$, due to phase noise. Let $u_s(t) = g(t) * v(t)$ be the output of the filter $g(t)$ in the presence of phase noise. We will define the loss as

$$L = \frac{E[u_s(t)]}{u(t)}$$

where $u(t) = g(t) * s(t)$ is the filter output in the absence of phase noise. Although it appears to be a function of t, L is actually independent of t.

Theorem 15.4.1 *Let $v(t) = s(t)e^{j\theta(t)}$ where $s(t)$ is a known pulse and $\theta(t)$ is a zero-mean, stationary, gaussian random noise process with the variance σ_θ^2. Then*

$$L = e^{-\sigma_\theta^2/2},$$

which is independent of t.

Proof:

$$E[u_s(t)] = E\left[\int_{-\infty}^{\infty} g(t-\xi)s(\xi)e^{j\theta(\xi)}\,d\xi\right]$$
$$= \int_{-\infty}^{\infty} g(t-\xi)s(\xi)E[e^{j\theta(\xi)}]\,d\xi.$$

Because

$$E[e^{j\theta(\xi)}] = e^{-\sigma_\theta^2/2},$$

the proof is complete. □

We can conclude from Theorem 15.4.1 that the expected output of the filter is an attenuated version of the filter output in the absence of phase noise. We can regard the effect of the phase noise as an effective reduction in signal strength, which strength degrades very quickly with σ_θ^2. This means that situations with large phase noise are rarely of practical interest which will help justify the assumption of small σ_θ^2 in problems that we cannot solve otherwise.

Proposition 15.4.2 *For small values of σ_θ^2, the variance in the output of a matched filter at $t = 0$ due to the phase noise of power density spectrum $N_\theta(f)$ is approximated by*

$$\sigma^2 \approx e^{-\sigma_\theta^2} \int_{-\infty}^{\infty} N_\theta(f)|W(f)|^2\,df$$

where $W(f)$ is the Fourier transform of $|s(t)|^2$.

Proof: In the presence of phase noise, the output of the matched filter at the peak is given by

$$u_s(0) = \int_{-\infty}^{\infty} |s(\xi)|^2 e^{j\theta(\xi)}\,d\xi,$$

and

$$E[|u(0)|^2] = E\left[\int_{-\infty}^{\infty}\int_{-\infty}^{\infty} |s(\xi)|^2|s(\xi')|^2 e^{j(\theta(\xi)-\theta(\xi'))}\,d\xi\,d\xi'\right]$$
$$= \int_{-\infty}^{\infty}\int_{-\infty}^{\infty} |s(\xi)|^2|s(\xi')|^2 e^{-(\phi(0)-\phi(\xi-\xi'))}\,d\xi\,d\xi'$$

where $\phi(\tau)$ is the correlation function of $\theta(t)$. The variance is

$$\sigma^2 = E[|u(0)|^2] - |E|u(0)||^2$$
$$\sigma^2 = \int_{-\infty}^{\infty}\int_{-\infty}^{\infty} |s(\xi)|^2|s(\xi')|^2 e^{-\phi(0)}(e^{\phi(\xi-\xi')} - 1)\,d\xi\,d\xi'.$$

Let $\eta = \xi - \xi'$. Then

$$\sigma^2 = e^{-\phi(0)} \int_{-\infty}^{\infty} (e^{\phi(\eta)} - 1) \int_{-\infty}^{\infty} |s(\xi)|^2 |s(\xi - \eta)|^2 \, d\xi \, d\eta.$$

Under the assumption that σ_θ^2 is small, $\phi(\eta)$ is small as well. Then $e^{\phi(\eta)} - 1$ can be approximated by $\phi(\eta)$. Finally, use Parseval's theorem to complete the proof. □

We can approximate the exponential in the formula for σ^2 by one. This gives

$$\sigma^2 \approx \int_{-\infty}^{\infty} N_\theta(f) |W(f)|^2 \, df.$$

This formula is the same as would be obtained if additive thermal noise were passed through the filter $W(f)$. We can obtain the same formula another way by starting with the approximation

$$\begin{aligned} v(t) &= s(t) e^{j\theta(t)} \\ &\approx s(t) + j\theta(t) s(t). \end{aligned}$$

The matched-filter output is

$$u(t) = \int_{-\infty}^{\infty} |s(t)|^2 \, dt + j \int \theta(t) |s(t)|^2 \, dt,$$

and the variance is

$$\begin{aligned} \sigma^2 &= E \int_{-\infty}^{\infty} \int_{-\infty}^{\infty} \theta(\xi) |s(\xi)|^2 \theta(\xi') |s(\xi')|^2 \, d\xi \, d\xi' \\ &= \int_{-\infty}^{\infty} \phi(\eta) \int_{-\infty}^{\infty} |s(\xi)|^2 |s(\xi - \eta)|^2 \, d\xi \, d\eta, \end{aligned}$$

which leads to the same formula as before.

For example, if $|s(t)|^2$ is a rectangular pulse of duration T, then

$$\sigma^2 = \int_{-\infty}^{\infty} N_\theta(f) T^2 \mathrm{sinc}^2(fT) \, df.$$

If, further, the phase noise is white with $N_\theta(f) = N_\theta/2$, then

$$\sigma^2 = \frac{N_\theta}{2} T = \frac{N_\theta}{2} E_p.$$

The signal at the output of the filter is equal to E_p, so the signal-to-phase-noise power ratio is

$$\frac{S}{N} = \left(\frac{N_\theta}{2}\right)^{-1} E_p.$$

Now we may compare the effect of phase noise to additive thermal noise. Although phase noise is different from additive noise, insofar as its effect on the peak output of the matched filter, phase noise behaves as additive white noise of the power density

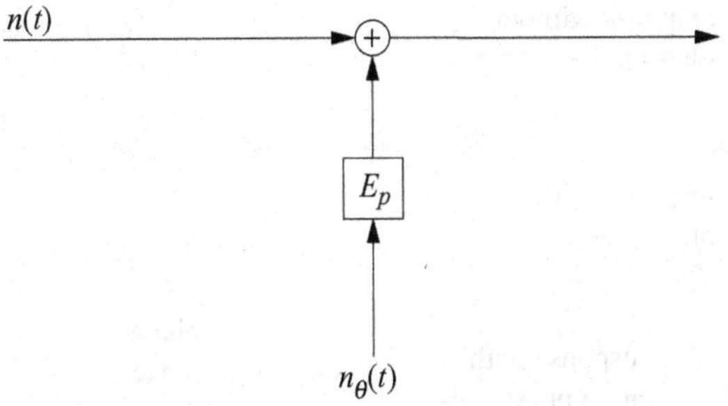

Figure 15.3 Modeling phase noise as an additive noise

spectrum $E_p N_\theta$. When a system has both additive thermal noise and phase noise, we may combine them by introducing an additive noise that is equivalent to the phase noise, as shown in Figure 15.3:

$$\frac{S}{N} = \frac{2E_p}{N_0 + E_p N_\theta}.$$

This is a useful summary formula, but it must be used with care because the two kinds of noise are not really interchangeable. The summary formula makes it obvious that increasing E_p will not increase performance indefinitely. Eventually, the phase noise will dominate the additive noise.

15.5 Phase noise and the ambiguity function

We shall study the effect of phase noise on the sample cross-ambiguity function $\chi_c(\tau, \nu)$ given by

$$\begin{aligned}\chi_c(\tau, \nu) &= \int_{-\infty}^{\infty} s(t)v^*(t+\tau)\mathrm{e}^{-\mathrm{j}2\pi\nu t}\,\mathrm{d}t \\ &= \int_{-\infty}^{\infty} s(t)s^*(t+\tau)\mathrm{e}^{-\mathrm{j}\theta(t+\tau)}\mathrm{e}^{-\mathrm{j}2\pi\nu t}\,\mathrm{d}t\end{aligned}$$

or on the sample cross-ambiguity surface $|\chi_c(\tau, \nu)|$. Because $\theta(t)$ is a random process, the sample cross-ambiguity function $\chi_c(\tau, \nu)$ becomes a random process through its dependence on $\theta(t)$. As illustrated in one dimension in Figure 15.4, the main lobe becomes degraded by phase noise in several ways: as the variance of the phase noise is increased, the maximum of the peak of the main lobe may be moved, its amplitude is reduced, and the width may be broadened. Any reduction in amplitude is referred to as a *coherence loss*; any increase in width is referred to as a *resolution loss*; and the change in shape is referred to in terms of an equivalent noise power. The phase

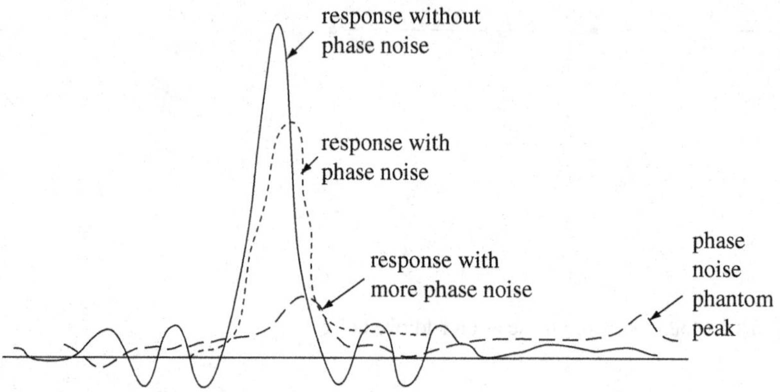

Figure 15.4 Illustrating the effects of phase noise

noise may also produce additional random sidelobes about the main lobe that can be detrimental in some applications.

If we consider a radar image formed from the received pulse $s(t)$ as a function of the noise variance σ_θ^2, we can imagine σ_θ^2 as increasing from zero to understand how the image is degraded. As σ_θ^2 increases from zero, the image intensity begins to fade slowly because the main lobe is attenuated, and the position of the maximum may move from the correct coordinates (0, 0) to the error coordinates (τ_e, ν_e). The peak, however, does not initially lose its sharpness. Eventually, resolution loss begins because the main lobe widens. The image not only continues to fade as phase noise is further increased, but the sharpness is dissipated. While the desired image is fading, undesired background clutter is increasing because of spurious sidelobes caused by the phase noise.

The effect of phase noise can be measured in many ways. Depending on the application, we may wish to know various expectations such as $E[\chi_c(0, 0)]$, $E[|\chi_c(0, 0)|]$, $E[|\chi_c(0, 0)|^2]$, $E[\tau_e]$, $\text{var}[\tau_e]$, $E[\nu_e]$, or $\text{var}[\nu_e]$. Another significant effect is the effective loss in signal-to-noise ratio, defined as

$$L = \frac{E[\chi_c(0, 0)]}{\chi_c(0, 0)}.$$

Because $\chi_c(\tau, \nu)$ is related nonlinearly to $\theta(t)$, the first and second moments of $\chi_c(\tau, \nu)$ depend on all moments of $\theta(t)$. A general statistical analysis is not possible if only the first and second moments of $\theta(t)$ are known and may be impractical even when higher moments are known. There are two approaches that can be taken: either assume that θ is small so that a small-angle approximation to the complex exponential can be used to linearize the relationship, or assume a specific probability density function for θ, usually gaussian.

15.6 Effect of phase distortion

Let $s(t)$ be a pulse, possibly complex. Suppose a received pulse is ideally $s(t)$ but is actually contaminated by phase distortion. The Fourier transform of the received pulse is given by

$$V(f) = S(f)e^{j\theta(f)}.$$

The phase distortion $\theta(f)$ is a frequency-dependent phase shift that may be introduced by phase errors in the transfer function of a filter that the pulse has passed through. Phase distortion can be studied in a way that is parallel to the study of phase noise.

15.7 Array errors

A uniform pulse train consisting of N equispaced, identical pulses centered at the origin has the Fourier transform

$$\begin{aligned} P(f) &= S(f)\frac{\sin N\pi fT}{\sin \pi fT} \\ &= S(f)\mathrm{dirc}_N(fT). \end{aligned}$$

The Dirichlet function $\mathrm{dirc}_N(t)$ was studied in Section 2.3. In this section we will study the effect of errors in spacing on the dirichlet function. Let

$$p(t) = \sum_{\ell=0}^{N-1} s(t - \ell T + T_{\ell e} + \tfrac{1}{2}(N-1)T)$$

where $T_{\ell e}$ is a small error in the time of the ℓth pulse. Then

$$P(f) = S(f)e^{j\pi f(N-1)T}\sum_{\ell=0}^{N-1} e^{-j2\pi f\ell T}e^{-j2\pi fT_{\ell e}}.$$

Now take the error $T_{\ell e}$ to be zero-mean gaussian random variables with variance σ_T^2. The expectation of $P(f)$ is

$$E[P(f)] = S(f)e^{j2\pi f(N-1)T}\sum_{\ell=0}^{N-1} e^{-j2\pi f\ell T}E[e^{-j2T\pi fT_{\ell e}}].$$

To evaluate the expectation, we use Theorem 15.2.1 to write

$$\begin{aligned} E[P(f)] &= S(f)e^{j\pi f(N-1)T}e^{-\sigma_\theta^2/2}\sum_{\ell=0}^{N-1} e^{-j2\pi f\ell T} \\ &= [S(f)\mathrm{dirc}_N(fT)]e^{-\sigma_\theta^2/2}. \end{aligned}$$

The grating lobes are attenuated in expectation by $e^{-\sigma_\theta^2/2}$.

Problems

15.1 The variance in the value of the peak of the ambiguity function $\chi(\tau, \nu)$ due to phase noise is given approximately by

$$\sigma^2 = \int_{-\infty}^{\infty} N_\theta(f)|W(f)|^2\,\mathrm{d}f$$

where $W(f)$ is the Fourier transform of $|s(t)|^2$.

a. Let

$$s(t) = \left(\frac{T_r}{T}\right)^{\frac{1}{2}} \sum_{n=0}^{N-1} p(t - nT_r)$$

where $p(t)$ is a rectangular pulse of width $T \ll T_r$. Find an approximate expression for σ^2.

b. Suppose that the spectrum of the phase noise is confined to low frequencies, that is, $N_\theta(f)$ is approximately zero for large frequencies. Show that σ^2 could be calculated just as easily by treating $s(t)$ as a single rectangular pulse of width approximately NT_r and height 1. In terms of T_r and T, what does it mean to say that $N_\theta(f)$ is confined to low frequencies?

15.2 The Gabor bandwidth of the pulse $s(t) = \text{rect}\,(t)$ is infinite because the sidelobes of $S(f)$ decay as f^{-1}. What is $S(f)$ if $s(t) = \text{rect}\,(t) * \text{sinc}(2t)$? What is the Gabor bandwidth of $s(t)$?

15.3 Let

$$\widetilde{s}(t) = \pm\text{rect}(t)\cos 2\pi f_0 t$$

and

$$\widetilde{v}(t) = \widetilde{s}(t)e^{j2\pi\theta(t)}$$

where $\theta(t)$ is a gaussian process with the power density spectrum $N(f)$.

a. Give a test for determining the sign of $\widetilde{s}(t)$ from the received signal $\widetilde{v}(t)$.

b. Give an approximate formula for the probability of error as a function $N(f)$.

Notes

The study of the effect of phase errors has a long history, starting primarily in the field of optics. The Cornu spiral is named after French physicist Marie Alfred Cornu (1841–1902). Cornu introduced this construction to study the effect of the phase approximation in Fresnel diffraction.

References

Abbe, E. (1873). Beiträge zur Theorie des Mikroskops und der Mikroskopischen Wahrnehmung. *Archiv. Mikroskopische Anat.*, **9**, 413–68.

Ables, J. G. (1968). Fourier transform photography: A new method for X-ray astronomy. *Proceedings of the Astronomical Society of Australia*, **4**, 172–3.

Ali, S. M. and S. D. Silvey. (1966). A general class of coefficients of divergence of one distribution from another. *Journal of the Royal Statistical Society*, **28**, 131–42.

Arimoto, S. (1972). An algorithm for computing the capacity of an arbitrary discrete memoryless channel. *IEEE Transactions on Information Theory*, **IT-18**, 14–20.

Armitage, J. D. and A. W. Lohmann. (1965). Character recognition by incoherent spatial filtering. *Applied Optics*, **4**, 461–7.

Armstrong, E. H. (1936). A method of reducing disturbances in radio signaling by a system of frequency modulation. *Proceedings of the IRE*, **24**, 689–740.

Auslander, L. and R. Tolimieri. (1985). Radar ambiguity functions and group theory. *Siam Journal of Mathematical Analysis*, **16**, 577–601.

Ayers, G. R. and J. C. Dainty. (1982). Iterative blind deconvolution method and its applications. *Applied Optics*, **21**, 2758–69.

Bader, T. R. (1960). Wideband signal processing for emitter location. *Proceedings of SPIE Symposium, Advances in Optical Information Processing*, **1296**.

Baggeroer, A. B., W. A. Kuperman, and P. N. Mikhalevsky. (1993). An overview of matched field methods in ocean acoustics. *IEEE Journal of Oceanic Engineering*, **OE-18**, 401–24.

Banerjee, P. P. (1985). A simple derivation of the Fresnel diffraction formula. *Proceedings of the IEEE*, **73**, 1859–60.

Barabell, A. J., *et al.*, (1984). *Performance Comparison of Superresolution Array Processing Algorithms*, MIT Lincoln Laboratory Project Report, **TST-72**.

Barbarosa, S. and A. Farina. (1990). A novel procedure for detecting and focusing moving objects with SAR based on the Wigner–Ville Distribution. *Proceedings of the 1990 IEEE International Radar Conference*, 44–50.

Bar-Shalom, Y. (1978). Tracking methods in a multitarget environment. *IEEE Transactions on Automatic Control*, **AC-23**, 618–26.

Bates, R. H. T. (1984). Uniqueness of solutions to two-dimensional Fourier phase problems of localized and positive images. *Computer Vision, Graphics, and Image Processing*, **24**, 205–7.

Bates, R. H. T. and M. J. McDonnell. (1986). *Image Restoration and Reconstruction*. Oxford: Oxford University Press.

Bello, P. (1960). Joint estimation of delay, doppler and doppler rate. *IRE Transactions on Information Theory*, **IT-6**, 330–41.

Bernfeld, M. (1984). Chirp doppler radar. *Proceedings of the IEEE*, **72**, 540–1.

Bertsekas, D. P. (1988). The auction algorithm: A distributed relaxation method for the assignment problem. *Annals of Operations Research*, **14**, 105–23.

Blackman, S. S. and R. Poopoli. (1999). *Design and Analysis of Modern Tracking Systems*. Dedham, MA: Artech House.

Blahut, R. E. (1972). Computation of channel capacity and rate distortion functions. *IEEE Transactions on Information Theory*, **IT-18**, 460–73.

Bloch, F. (1946). Nuclear induction. *Physics Review*, **70**, 460–74.

Boerner, W. M., C.-M. Ho, and B.-Y. Foo. (1981). Use of Radon's projection theory in electromagnetic inverse scattering. *IEEE Transactions on Antennas and Propagation*, **AP-29**, 336–41.

Bojarski, N. M. (1982). A survey of physical optics inverse scattering identity. *IEEE Transactions on Antennas and Propagation*, **AP-30**, 980–9.

Bourgeois, F. and J. C. La Salle. (1971). An extension of the Munkres algorithm for the assignment problem to rectangular matrices. *Communications of the ACM*, **14**, 802–6.

Bracewell, R. N. (1956). Strip integration in radio astronomy. *Australian Journal of Physics*, **9**, 198–217.

(1958). Radio interferometry of discrete sources. *Proceedings of the Institute of Radio Engineers*, **46**, 97–105.

(1958). Restoration in the presence of errors. *Proceedings of the IRE*, **46**, 106–11.

(1995). *Two-Dimensional Imaging*. Upper Saddle River, NJ: Prentice-Hall.

Bracewell, R. N. and A. C. Riddle. (1967). Inversion of fan-beam scans in radio astronomy. *The Astrophysics Journal*, **150**, 427–34.

Bragg, W. L. (1929). The determination of parameters in crystal structures by means of Fourier analysis. *Proceedings of the Royal Society, A*, **123**, 537–59.

(1942). The X-ray microscope. *Nature*, **149**, 470–1.

Bresler, Y. and A. Macouski. (1987). Three-dimensional reconstruction from projections with incomplete and noisy data by object estimation. *IEEE Transactions on Acoustics, Speech, and Signal Processing*, **ASSP-35**, 1139–52.

Bricogne, G. (1984). Maximum entropy and the foundations of direct methods. *Acta Crystallographia*, **A40**, 410–45.

Brown, W. M. and R. Fredericks. (1969). Range-doppler imaging with motion through resolution cells. *IEEE Transactions on Aerospace and Electronic Systems*, **AES-5**, 98–102.

Brown, W. M. and C. J. Palermo. (1963). Effects of phase errors on the ambiguity function. *IEEE International Convention Record*, **Pt. 4**, 118–23.

Brown, W. M. and L. J. Porcello. (1969). An introduction to synthetic aperture radar. *IEEE Spectrum*, **6**, 52–62.

Bruck, Y. M. and L. G. Sodin. (1979). On the ambiguity of the image reconstruction problem. *Optical Communications*, **30**, 304–8.

Budinger, T. F. (1980). Physical attributes of single-photon tomography. *Journal of Nuclear Medicine*, **21**, 579–92.

Burckhardt, C. B. (1978). Speckle in ultrasound B-mode scans. *IEEE Transactions on Sonics and Ultrasonics*, **SU-25**, 1–6.

Burg, J., D. Luenberger and D. Weaver. (1982). Estimation of structured covariance matrices. *Proceedings of the IEEE*, **70**, 963–74.

Butler, J. and R. Lowe. (1961). Beam-forming matrix simplifies design of electronically scanned antennas. *Electronic Design*, **9**, 170–3.

Byrne, C. L. (1993). Iterative image reconstruction algorithms based on cross-entropy minimization. *IEEE Transactions on Image Processing*, **IP-2**, 96–103.

Campbell, D. B. (1971). Radar interferometric observations of Venus. Ph.D. dissertation, Cornell University.

Capon, J. (1964). Optimum weighting functions for the detection of sampled signals in noise. *IRE Transactions on Information Theory*, **IT-10**, 152–9.

(1969). High-resolution frequency-wavenumber spectrum analysis. *Proceedings of the IEEE*, **57**, 1408–18.

Carney, P. S. and J. C. Schotland. (2003). Near-field tomography. In G. Uhlmann, ed., *Inside Out*. Cambridge: Cambridge University Press, pp. 131–66.

Chen, C. C. and H. C. Andrews. (1980). Multi-frequency imaging of radar turntable data. *IEEE Transactions on Aerospace and Electronic Systems*, **AES-16**, 15–22.

Chen, V. C. (1994). Radar ambiguity function, time-varying matched filter, and optimum wavelet correlator. *Optical Engineering*, **33**, 2212–17.

Chen, V. C. and S. Qian. (1998). Joint time–frequency transform for radar range-doppler imaging. *IEEE Transactions on Aerospace and Electronic Systems*, **AES-34**, 486–99.

Chen, Z., Y. Zhao, S. M. Srinivas, *et al.* (1999). Optical doppler tomography. *IEEE Journal of Selected Topics in Quantum Electronics*, **QE-5**, 1134–41.

Chestnut, P. C. (1982). Emitter location accuracy using TDOA and differential doppler. *IEEE Transactions on Aerospace and Electronic Systems*, **AES-18**, 214–18.

Cho, Z. H. (1974). General view on 3–4 image reconstruction and computerized transverse axial tomography. *IEEE Transactions on Nuclear Science*, **NS-21**, 44–71.

Choi, H. and D. C. Munson, Jr. (1998). Direct Fourier reconstruction in tomography and synthetic aperture radar. *International Journal on Imaging Systems Technology*, **9**, 1–13.

Christiansen, W. N. and J. A. Warburton. (1955). The distribution of brightness over the solar disk at a wavelength of 21 cm III – The quiet sun. Two-dimensional observations. *Australian Journal of Physics*, **8**, 474–86.

Classen, T. A. C. M. and W. F. Mecklenbrauker. (1980). The Wigner distribution: A tool for time–frequency signal analysis. *Phillips Journal of Research*, **35**, Parts 1, 2, and 3, 217–50, 276–300, and 372–89.

Clem, T. R. (1995). Superconducting magnetic sensors operating from a moving platform. *IEEE Transactions on Applied Supercomputing*, **AS-5**, 2124–8.

Coble, M. R. (1992). High resolution radar imaging of a rotating sphere. M.S. dissertation, University of Illinois, Urbana, IL.

Cochran, W., F. H. C. Crick, and V. Vand. (1952). Structure of synthetic polypeptides. I: The transform of atoms on a helix. *Acta Cystallographia*, **5**, 581–6.

Cook, C. and M. Bernfeld. (1967). *Radar Signals: An Introduction to Theory and Applications*. New York: Academic Press.

Cooley, J. W. and J. W. Tukey. (1965). An algorithm for the machine computation of complex Fourier series. *Mathematics of Computation*, **19**, 297–301.

Cormack, A. M. (1963). Representation of a function by its line integrals, with some radiological applications. *Journal of Applied Physics*, **34**, 2722–7.

Costas, J. P. (1975). Medium constraints on sonar design and performance. *EASCON Convention Record*, pp. 68A–68L.

(1984). A study of a class of detection waveforms having nearly ideal range-doppler ambiguity properties. *Proceedings of the IEEE*, **72**, 996–1009.

Cover, T. M. (1984). An algorithm for maximizing expected log investment return. *IEEE Transactions on Information Theory*, **IT-30**, 369–73.

Crowther, R. A., D. J. DeRosier, and A. Klug. (1970). The reconstruction of a three-dimensional structure from projections and its application to electron microscopy. *Proceedings of the Royal Society of London*, **A317**, 319–40.

Csiszár, I. (1991). Why least squares and maximum entropy – An axiomatic approach to inverse problems. *Annals of Statistics*, **19**, 2033–66.

Csiszár, I. and G. Tusnady. (1984). Information geometry and alternating decisions. *Statistical Decisions*, Supplementary issue #1, 205–207.

Cutrona, L. J., E. N. Leith, C. J. Palermo, and L. J. Porcello. (1960). Optical data processing and filtering systems. *IRE Transactions on Information Theory*, **IT-6**, 386–400.

Cutrona, L. J., W. E. Vivian, E. N. Leith, and G. O. Hall. (1961). A high-resolution radar combat–surveillance system. *IRE Transactions Military Electronics*, **MIL-5**, 127–31.

Dainty, J. C. and J. R. Fienup. (1987). Phase retrieval and image reconstruction for astronomy. In H. Stark, ed., *Image Recovery: Theory and Applications*. New York: Academic Press, pp. 231–75.

DeBuda, R. (1970). Signals that can be calculated from their ambiguity function. *IEEE Transactions on Information Theory*, **IT-16**, 195–202.

DeGraaf, S. R. (1998). SAR imaging via modern 2D spectral estimation methods. *IEEE Transactions on Image Processing*, **IP-7**, 729–61.

Delong, D. F. (1983). *Multiple Signal Direction Finding with Thinned Linear Arrays*. MIT Lincoln Laboratory Project Report, **TST-68**.

Dempster, A. D., N. M. Laird, and D. B. Rubin. (1977). Maximum likelihood from incomplete data via the EM algorithm. *Journal of the Royal Statistical Society*, **B39**, 1–38.

DeRosier, D. J. and A. Klug. (1968). Reconstruction of three-dimensional structures from electron micrographs. *Nature*, **217**, 130–4.

Dicke, R. H. (1968). Scatter-hole cameras for X-Rays and gamma rays. *Journal of Astrophysics*, **153**, L101–L106.

Dines, K. A. and R. J. Lytle. (1979). Computerized geophysical tomography. *Proceedings of the IEEE*, **67**, 1065–73.

Dolph, C. L. (1946). A current distribution for broadband arrays which optimizes the relationship between beam width and sidelobe level. *Proceedings of the IRE*, **34**, 335–48.

Donoho, D. L., I. M. Johnstone, J. C. Hoch, and A. S. Stern. (1992). Maximum entropy and the nearly black object. *Journal of the Royal Statistical Society B*, 41–81.

Dowski, E. and W. Cathey. (1995). Extended depth of field through wavefront coding. *Applied Optics*, **34**, 1859–66.

Dragone, C. (1987). Use of imaging with spatial filtering in reflector antennas. *IEEE Transactions on Antennas and Propagation*, **AP-35**, 258–67.

Duffieux, P. M. (1946). L'Integrale de Fourier et ses applications a l'optique. In *Faculte des Sciences*. Paris: Besancon.

Dugundji, J. (1958). Envelopes and pre-envelopes of real waveforms. *IRE Transactions on Information Theory*, **IT-4**, 53–7.

Durnin, J. (1987). Exact solutions for nondiffracting beams. I: The scalar theory. *Journal of the Optical Society of America*, **4**, 651–4.

Dziewonski, A. and J. Woodhouse. (1987). Global images of the Earth's interior. *Science*, **236**, 37–48.

Elias, P. (1953). Optics and communication theory. *Journal of the Optical Society of America*, **43**, 229–32.

Elias, P., D. S. Grey, and D. Z. Robinson. (1952). Fourier treatment of optical processes. *Journal of the Optical Society of America*, **42**, 127–34.

Emerson, R. C. (1954). *Some Pulsed Doppler, MTI, and AMTI Techniques*. Rand Corp. report **R-274**.

Emslie, A. G. (1946). *Moving Target Indication on MEW*. MIT Radiation Laboratory report 1080.

Ermert, H. and R. Karg. (1979). Multi-frequency acoustical holography. *IEEE Transactions on Sonics and Ultrasonics*, **SU-26**, 279–86.

Fan, H. and J. L. C. Sanz. (1985). Comments on 'Direct Fourier reconstruction in computer tomography. *IEEE Transactions on Acoustics, Speech, and Signal Processing*, **ASSP-33**, 446–9.

Feig, E. and F. A. Grünbaum. (1986). Tomographic methods in range-doppler radar. *Inverse Problems*, **2**, 185–95.

Fenimore, E. E. and T. M. Cannon. (1978). Coded aperture imaging with uniformly redundant arrays. *Applied Optics*, **17**, 337–47.

Fienup, J. R. (1978). Reconstruction of an object from the modulus of its Fourier transform. *Optics Letters*, **3**, 27–9.

(1981). Reconstruction and synthesis applications of an iterative algorithm. *SPICE Transformations in Optical Signal Processing*, **373**, 147–60.

(1982). Phase retrieval algorithms: A comparison. *Journal of Applied Optics*, **21**, 2758–69.

Fisher, M. L. (1981). The Lagrangian relaxation method for solving integer programming problems. *Management Science*, **27**, 1–18.

Fomalont, E. B. (1973). Earth-rotation aperture synthesis. *Proceedings of the IEEE*, **61**, 1211–18.

Foucault, L. (1858). *Ann. de l'Observ. Imp. de Paris*, **5**, 203.

Frahm, C. P. (1972). *Inversion of the Magnetic Field Gradient Equations for a Magnetic Dipole Field*, Naval Coastal Systems Laboratory informal report, 135–72.

Franklin, R. E. and A. Klug. (1955). The splitting of layer lines in X-ray fibre diagrams of helical structures: Application to tobacco mosaic virus. *Acta Crystallographia*, **8**, 777–80.

Frieze, A. M. and J. Yadegar. (1981). An algorithm for solving three-dimensional assignment problems with application to scheduling a teaching practice. *Journal of the Operational Research Society*, **32**, 989–95.

Friis, H. T. (1946). A note on a simple transmission formula. *Proceedings of the IRE*, **34**, 254–6.

Gaarder, N. T. (1968). Scattering function estimation. *IEEE Transactions on Information Theory*, **IT-14**, 684–93.

Gabor, D. (1946). Theory of communication. *Journal of the Institute of Electrical Engineers*, pt. III, **93**, 429–41.

(1948). A new microscope principle. *Nature*, **161**, 777–8.

(1949). Microscopy by reconstructed wavefronts, part I. *Proceedings of the Royal Society*, **A197**, 454–87.

(1951). Microscopy by reconstructed wavefronts, part II. *Proceedings of Physical Society*, **B64**, 449–69.

Geoffrion, A. M. (1974). Lagrangian relaxation for integer programming. In M. L. Balinski, ed., *Mathematical Programming Study 2: Approaches to Integer Programming*. Amsterdam: North Holland Publishing Company.

Gerchberg, R. W. and W. O. Saxton. (1972). A practical algorithm for the determination of phase from image and diffraction plane pictures. *Optics*, **35**, 237–46.

Ghiglia, D. C., L. A. Romero, and G. A. Mastin. (1996). Systematic approach to two-dimensional blind deconvolution by zero-sheet separation. *Journal of the Optical Society of America*, **10**, 1024–36.

Golay, M. J. E. (1961). Complementary series. *IRE Transactions on Information Theory*, **IT-7**, 82–7.

Goldstein, G. B. (1973). False alarm regulation in log-normal and weibull clutter. *IEEE Transactions on Aerospace and Electronic Systems*, **AES-9**, 84–92.

Golomb, S. W. and H. Taylor. (1982). Two-dimensional synchronization patterns for minimum ambiguity. *IEEE Transactions on Information Theory*, **IT-28**, 600–4.

(1984). Constructions and properties of Costas arrays. *Proceedings of the IEEE*, **72**, 1143–63.

Goodman, J. W. (1976). Some fundamental properties of speckle. *Journal of the Optical Society of America*, **66**, 1145–50.

Goodman, N. R. (1963). Statistical analysis based on a certain multivariate complex Gaussian distribution. *Annals of Mathematical Statistics*, **34**, 152–77.

Gordon, R. (1974). A tutorial on ART (Algebraic Reconstruction Techniques). *IEEE Transactions on Nuclear Science*, **NS-21**, 78–93.

Graham, L. C. (1974). Synthetic interferometer radar for topographic mapping. *Proceedings of the IEEE*, **62**, 763–8.

Green, Jr., P. E. (1962). *Radar Astronomy Measurement Techniques*. Technical report No. 282, Lincoln Laboratory, MIT, Cambridge, MA.

Green, T. J. and J. H. Shaprio. (1994). Maximum-likelihood laser radar range profiling with the expectation-maximization algorithm. *Optical Engineering*, **33**, 865–73.

Grenander, U. (1981). *Abstract Inference*. New York: Wiley.

Grünbaum, F. A. (1984). A remark on radar ambiguity functions. *IEEE Transactions on Information Theory*, **IT-30**, 126–7.

Gull, S. F. and G. J. Daniell. (1979). The maximum entropy method. C. van Schooneveld, ed. In *Image Formation from Coherence Functions in Astronomy*, Dordrecht: D. Reidel, pp. 219–25.

Hahn, E. L. (1960). Detection of sea-water motion by nuclear precession. *Journal of Geophysical Research*, **65**, 776–7.

Hansen, V. G. (1970). Detection performance of some non-parametric rank tests and an application to radar. *IEEE Transactions on Information Theory*, **IT-16**, 609–18.

Harker, D. and J. S. Kasher. (1948). Phases of Fourier coefficients directly from the crystal diffraction data. *Acta Cystallographia*, **1**, 70–5.

Harris, F. J. (1978). On the use of windows for harmonic analysis with the discrete Fourier transform. *Proceedings of the IEEE*, **66**, 51–83.

Hauptman, H. (1986). The direct methods of X-ray crystallography. *Science*, **233**, 178–83.

Hauptman, H. and J. Karle. (1953). *Solution of the Phase Problem I: The Antisymmetrical Crystal.* Western Springs, IL: American Crystallographic Association, Monograph 3, Polycrystal Book Service.

Hayes, M. H. (1982). The reconstruction of a multidimensional sequence from the phase or magnitude of its Fourier transform. *IEEE Transactions on Acoustics, Speech, and Signal Processing*, **ASSP-30**, 140–54.

Hayes, M. H. and J. H. McClellan. (1982). Reducible polynomials in more than one variable. *Proceedings of the IEEE*, **70** (2), 197–8.

Helstrom, C. W. (1960). *Statistical Theory of Signal Detection*. Oxford: Pergamon Press.

Hero, A. O. and J. A. Fessler. (1995). Convergence in norm for alternating expectation-maximization (EM) type algorithms. *Statistica Sinica*, **5**, 41–54.

Hirasawa, H. and N. Kobayashi. (1986). Terrain height measurement by synthetic aperture radar with an interferometer. *International Journal on Remote Sensing*, **7**, 339–48.

Högbom, J. A. (1974). Aperture synthesis with a nonregular distribution of interferometer baselines. *Astronomy and Astrophysics Supplement Series*, **15**, 417–26.

Hounsfield, G. N. (1972). *A Method of and Apparatus for Examination of a Body by Radiation such as X-ray or Gamma Radiation*, British Patent No. 1283915, London.

(1973). *Method and Apparatus for Measuring X or Radiation Absorption or Transmission at Plural Angles and Analyzing the Data*, U.S. Patent 3 778 614.

Ishii, M., J. Leigh, and J. Schotland. (1995). Optical diffusion imaging using a direct inversion method. *Physics Review E*, **52**, 4361.

Izraelevitz, D. and J. S. Lim. (1987). A new direct algorithm for image reconstruction from Fourier transform magnitude. *IEEE Transactions on Acoustics, Speech, and Signal Processing*, **ASSP-35**, 511–19.

Jakowatz, Jr., C. V. and P. A. Thompson. (1992). *A Three-Dimensional Tomographic Formulation for Spotlight Mode – Synthetic Aperture Radar*. Sandia National Laboratories, Albuquerque, NM.

Jakowatz, C. V. and P. A. Thompson. (1995). A new look at spotlight mode synthetic aperture radar as tomography. *IEEE Transactions on Image Processing*, **IP-4**, 699–703.

Jackson, P. L. (1965). Diffractive processing of geophysical data. *Applied Optics*, **4**, 419–27.

Jain, A. K. and S. Ranganath. (1981). Applications of two-dimensional spectral estimation in image restoration. *Proceedings of the IEEE International Conference on Acoustics, Speech, and Signal Processing*, 1113–16.

Jaynes, E. T. (1957). Information theory and statistical mechanics. *Physics Review*, **106**, 620–30.

Jennison, R. C. (1958). A phase sensitive interferometer technique for the measurement of the Fourier transforms of spatial brightness distributions of small angular extent. *Monthly Notices of the Royal Astronomical Society*, **118**, 276–84.

Jensen, H., L. C. Graham, L. J. Porcello, and E. N. Leith. (1977). Side-looking airborne radar. *Scientific American*, **237**, 84–95.

Johnson, D. H. (1982). The application of spectral estimation methods to bearing estimation problems. *Proceedings of the IEEE*, **70**, 1018–28.

Kak, A. C. (1979). Computerized tomography with X-ray, emission, and ultra-sound sources. *Proceedings of the IEEE*, **67**, 1245–71.

Kak, A. C. and M. Slaney. (1988). *Principles of Computerized Tomographic Imaging*. New York: IEEE Press.

Kalman, R. E. (1960). A new approach to linear filtering and prediction problems. *Transactions of the American Society of Mechanical Engineers*, Series D, Journal of Basic Engineering, **82D**, 35–46.

Kalman, R. E. and R. S. Bucy. (1961). New results in linear filtering and prediction theory. *Transactions of the American Society of Mechanical Engineers*, Series D, Journal of Basic Engineering, **83**, 95–108.

Karle, J. and H. Hauptman. (1950). The phases and magnitudes of the structure factors. *Acta Crystallographia*, **3**, 181–7.

Kashyap, R. L. and A. R. Rao. (1976). *Dynamic Stochastic Models from Empirical Data*. New York: Academic.

Kay, I. and R. A. Silverman. (1957). On the uncertainty relation for real signals. *Information and Control*, **1**, 64–75.

Kell, R. R. (1965). On the derivation of bistatic RCS from monostatic measurements. *Proceedings of the IEEE*, **53**, 983–8.

Klauder, J. R. (1960). The design of radar signals having both high range resolution and high velocity resolution. *Bell System Technical Journal*, **39**, 809–20.

Klauder, J. R., A. C. Price, S. Darlington, and W. J. Albersheim. (1960). The theory and design of chirp radars. *Bell System Technical Journal*, **39**, 745–808.

Klug, A. and J. E. Berger. (1964). An optical method for the analysis of periodicities in electron micrographs, and some observations on the mechanism of negative staining. *Journal of Molecular Biology*, **10**, 565–9.

Klug, A., F. H. C. Crick, and H. W. Wyckoff. (1958). Diffraction by helical structures. *Acta Crystallographia*, **11**, 199–213.

Knox, K. T. and B. J. Thompson (1974). Recovery of images from atmospherically degraded short-exposure photographs. *Astrophysics Journal Letters*, **193**, L45–L48.

Kozma, A. (1966). Photographic recording of spatially modulated coherent light. *Journal of the Optical Society of America*, **56**, 428–32.

Kozma, A. and D. L. Kelly. (1965). Spatial filtering for detection of signals submerged in noise. *Applied Optics*, **4**, p. 387.

Kuhl, D. E. and R. Q. Edwards. (1963). Image separation radioisotope scanning. *Radiology*, **80**, 653–61.

Kuhn, H. W. (1955). The Hungarian method for the assignment problem. *Naval Research Logistics Quarterly*, **2**, 83–97.

Kulkarni, M. D., C. W. Thomas, and J. A. Izatt. (1997). Image enhancement in optical coherence tomography using deconvolution. *Electronics Letters*, **33**, 1365–1467.

Kullback, S. (1959). *Information Theory and Statistics*. New York: Wiley, and New York: Dover, 1968.

Kumar, A., D. Welti, and R. Ernst. (1975). NMR Fourier zeugmatography. *Journal of Magnetic Resonance*, **18**, 69–83.

Labeyrie, A. (1970). Attainment of diffraction-limited resolution in large telescopes by Fourier analyzing speckle patterns in star images. *Astronomy and Astrophysics*, **6**, 85–7.

Lane, R. G. (1992). Blind deconvolution of speckle images. *Journal of the Optical Society of America*, **9**, 1508–14.

Lane, R. G. and R. H. T. Bates. (1987). Automatic multidimensional deconvolution. *Journal of the Optical Society of America*, **A4**, 180–8.

Lane, R. G., W. R. Fright, and R. H. T. Bates. (1987). Direct phase retrieval. *IEEE Transactions on Acoustics, Speech, and Signal Processing*, **ASSP-35**, 520–25.

Lange, K. and R. Carson. (1984). EM reconstruction algorithms for emission and transmission tomography. *Journal of Computer Assisted Tomography*, **8**, 306–16.

Lanterman, A. D. (2000). Statistical radar imaging of diffuse and specular targets using an expectation-maximization algorithm. *Algorithms for Synthetic Aperture Radar Imagery VII*, SPIE Proceedings 4053, Orlando, FL.

Largened, R. L., J. Beyond, and D. E. Boeck. (1990). Identification and restoration of noisy blurred images using the expectation maximization algorithm. *IEEE Transactions on Acoustics, Speech, and Signal Processing*, **ASSP-38**, 1180–91.

Lauterbur, P. C. (1973). Image formation by induced local interactions: Examples employing nuclear magnetic resonance. *Nature*, **242**, 190–1.

Leith, E. N. and J. Upatnieks. (1962). Reconstructed wavefronts and communication theory. *Journal of the Optical Society of America*, **52**, 1123–30.

(1964). Wavefront reconstruction with diffused illumination and three-dimensional objects. *Journal of the Optical Society of America*, **54**, 1295–1301.

Lerner, R. M. (1958). Signals with uniform ambiguity functions. *IRE National Convention Record*, Part 4, pp. 27–33.

Lewis, R. M. (1969). Physical optics inverse diffraction. *IEEE Transactions on Antennas and Propagation*, **AP-17**, 308–14.

Liang, Z.-P. and P. C. Lauterbur. (2000). *Principles of Magnetic Resonance Imaging – A Signal Processing Perspective*. Piscataway, NJ: IEEE Press.

Lighthill, M. J. (1958). *Introduction to Fourier Analysis and Generalized Functions*. Cambridge: Cambridge University Press.

Lockwood, G., J. Talman, and S. Brunke. (1998). Real-time 3-D ultrasound imaging using sparse synthetic aperture beamforming. *IEEE Transactions on Ultrasonics, Ferroelectronics, and Frequency Control*, **UFFC-45**, 1077–87.

Lohmann, A. (1977). Incoherent optical processing of complex data. *Applied Optics*, **16**, 261–3.

Lu, J.-Y. and J. F. Greenleaf. (1990). Ultrasonic nondiffracting transducer for medical imaging. *IEEE Transactions on Ultrasonics, Ferroelectronics, and Frequency Control*, **UFFC-37**, 438–47.

Lucy, L. (1974). An iterative technique for the rectification of observed distributions. *The Astronomical Journal*, **79**, 745–54.

Manasse, R. (1959). *The Use of Radar Interferometric Measurements to Study Planets*. Lincoln Laboratory, MIT, Cambridge, MA, Group Report No. 312–24.

(1961). *The Use of Pulse Coding to Discriminate Against Clutter*. Lincoln Laboratory MIT, Cambridge, MA, Technical report 312–12.

Marcum, J. R. (1960). A statistical theory of target detection of pulsed radar. *IRE Transactions on Information Theory*, **IT-6**, 145–267.

McEwan, N. J. and P. F. Goldsmith. (1989). Gaussian beam techniques for illuminating reflector antennas. *IEEE Transactions on Antennas and Propagation*, **AP-37**, 297–303.

McPherson, A. (1989). Macromolecular crystals. *Scientific American*, pp. 62–9, March.

Mersereau, R. M. (1973). Recovering multidimensional signals from their projections. *Computer Graphics and Image Processing*, **1**, 179–95.

(1979). The processing of hexagonally-sampled two-dimensional signals. *Proceedings of the IEEE*, **67**, 930–49.

Mersereau, R. M. and A. V. Oppenheim. (1974). Digital reconstruction of multidimensional signals from their projections. *Proceedings of the IEEE*, **62**, 1319–38.

Michelson, A. A. (1890). On the application of interference methods to astronomical measurements. *Philosophical Magazine*, **30**, 1–21.

(1921). On the application of interference methods to astronomical measurements. *Astrophysics Journal*, **53**, 249–59.

Middleton, D. and D. Van Meter. (1955). On optimum multiple-alternative detection of signals in noise. *IRE Transactions on Information Theory*, **IT-1**, 1–9.

Millane, R. P. (1990). Phase retrieval in crystallography and optics. *Journal of the Optical Society of America*, **7** (3), 394–411.

Miller, M. I. and D. L. Snyder. (1987). The role of likelihood and entropy in incomplete-data problems: Applications to estimating point-process intensities and Toeplitz constrained covariances. *Proceedings of the IEEE*, **75**, 892–907.

Miller, M. I., D. L. Snyder, and T. R. Miller. (1985). Maximum likelihood reconstruction for single photon emission computed tomography. *IEEE Transactions on Nuclear Science*, **NS-32**, 769–78.

Moran, P. R. (1982). A flow velocity zeugmatographic interlace for NMR imaging in humans. *Magnetic Resonance Imaging*, **2**, 555–66.

Morefield, C. L. (1977). Application of 0–1 integer programming to multitarget tracking problems. *IEEE Transactions on Automatic Control*, **AC-22**, 302–312.

Moulin, P. (1990). A method of sieves for radar imaging and spectral estimation. D.Sc. Dissertation, Washington University, St. Louis.

Moulin, P., J. A. O'Sullivan, and D. L. Snyder. (1992). A method of sieves for multiresolution spectrum estimation and radar imaging. *IEEE Transactions on Information Theory*, **IT-38**, 801–813.

Mouyan, Z. and R. Unbehauen. (1997). Methods for reconstruction of 2-D sequences from Fourier transform magnitude. *IEEE Transactions on Image Processing*, **IP-6**, 222–33.

Mueller, R. K., M. Kaveh, and G. Wade (1979). Reconstructive tomography and applications to ultrasonics. *Proceedings of the IEEE*, **67**, 567–87.

Munkres, J. (1957). Algorithm for the assignment and transportation problems. *Siam Journal*, **5**, 32–8.

Munson, Jr., D. C. (1998). Computational imaging. In A. Vardy, ed., *Codes, Curves, and Signals: Common Threads in Communications*. Dordrect: Kluwer Academic.

Munson, Jr., D. C., J. D. O'Brien, and W. K. Jenkins. (1983). A tomographic formulation of spotlight-mode synthetic aperture radar. *Proceedings of the IEEE*, **71**, 917–25.

Munson, Jr., D. C. and J. L. C. Sanz. (1984). Image reconstruction from frequency-offset Fourier data. *Proceedings of the IEEE*, **72**, 661–99.

Munson, D. C. and R. L. Visentin. (1989). A signal processing view of strip-mapping synthetic aperture radar. *IEEE Transactions on Acoustics, Speech, and Signal Processing*, **ASSP-37**, 2131–47.

Murino, V. and A. Trucco. (2000). Three-dimensional image generation and processing in underwater acoustic vision. *Proceedings of the IEEE*, **88**, 1903–48.

Nahi, N. E. (1969). Optimal recursive estimation with uncertain observations. *IEEE Transactions on Information Theory*, **IT-15**, 457–62.

Natterer, F. (1986). *The Mathematics of Computerized Tomography*. New York: Wiley.

Nayar, S. K. and Y. Nakagawa. (1994). Shape from focus. *IEEE Transactions on Pattern Analysis and Machine Intelligence*, **PAMI-16**, 824–31.

Newell, A. C. (1988). Error analysis technique for planar near field measurements. *IEEE Transactions on Antennas and Propagation*, **AP-36**, 754–68.

Neyman, J. and E. Pearson. (1933). On the problem of the most efficient tests of statistical hypotheses. *Philosophical Transactions of the Royal Society*, series A, **231**, p. 289.

Nilsson, N. J. (1961). On the optimum range resolution of radar signals in noise. *IRE Transactions on Information Theory*, **IT-7**, 245–53.

North, D. O. (1943). *An Analysis of the Factors which Determine Signal/Noise Discrimination in Pulsed-Carrier Systems*. RCA technical report PTR-6C.

Nyquist, H. (1928). Certain topics in telegraph transmission theory. *Transactions of the AIEE*, **47**, 617–44.

Ollinger, J. M. and J. A. Fessler. (1997). Positron-emission tomography. *IEEE Signal Processing Magazine*, **14**, 43–55.

O'Neill, E. L. (1956). Spatial filtering in optics. *IRE Transactions on Information Theory*, **IT-2**, 56–65.

O'Neill, E. L. (ed). (1962). *Communication and Information Theory Aspects of Modern Optics*. Syracuse, NY: General Electric Co., Electronics Laboratory.

O'Neill, E. L. and A. Walter. (1963). The question of phase in image formation. *Optica Acta*, **10**, 33–40.

O'Sullivan, J. A. (1998). Alternating minimization algorithms: From Blahut–Arimoto to expectation-maximization. In A. Vardy, ed., *Codes, Curves, and Signals*. Dordrect: Kluwer.

(2002). Iterative algorithms for maximum likelihood sequence detection. In R. E. Blahut and R. Koetter, eds., *Codes, Graphs, and Systems*. Dordrect: Kluwer.

O'Sullivan, J. A., R. E. Blahut, and D. L. Synder. (1998). Information-theoretic image formation. *IEEE Transactions on Information Theory*, **IT-44**, 2094–2123.

Papoulis, A. (1974). Ambiguity functions in Fourier optics. *Journal of the Optical Society of America*, **64**, 779–88.

Pasedach, K. and E. Haase. (1981). Random and guided generation of coherent two-dimensional codes. *Optics Communications*, **36**, 423–28.

Patterson, A. L. (1935). A direct method for the components of the interatomic distances in crystals. *Zeitschrift Furkristallographie, Kristallgeometrie, Kristalphysik, Kristallchemie*, **A90**, 517–42.

Petersen, D. P. and D. Middleton. (1962). Sampling and reconstruction of wave-number-limited functions in N-dimensional Euclidean Space. *Information and Control*, **5**, 279–323.

Peterson, W. W. and T. G. Birdsall. (1953). *The Theory of Signal Detectability, Parts I and II*. University of Michigan technical report 13. Also with W. Fox, *IRE Transactions on Information Theory*, **IT-4**, 171–212, 1954.

Politte, A. G. and D. L. Snyder. (1991). Corrections for accidental coincidences in maximum-likelihood image reconstruction for positron-emission tomography. *IEEE Transaction on Medical Imaging*, **MI-10**, 82–9.

Poor, H. V. (1994). *An Introduction to Signal Detection and Estimation*, 2nd edn. New York: Springer-Verlag.

Porcello, L. J. (1970). Turbulence-induced phase errors in synthetic-aperture radars. *IEEE Transactions on Aerospace and Electronic Systems*, **AE6-6**, 636–44.

Porter, A. B. (1906). On the diffraction theory of microscopic vision. *London, Edinburgh, Dublin Philosphical Magazine*, **11**, 154–66.

Price, R. and E. M. Hofstetter. (1965). Bounds on the volume and height distribution of the ambiguity function. *IEEE Transactions on Information Theory*, **IT-11**, 207–14.

Radon, J. (1917). Über die Bestimmung von Funktionen durch ihre Integralwerte langs gewisser Mannigfaltigkeiten. *Berichte Sachsische Akademie der Wissenschaften. Leipzig, Math.–Phys. KI*, **69**, 262–267. English translation: On the determination of functions from their integrals along certain manifolds. Appendix A of *The Radon Transform and Some of Its Applications*, S. R. Deans, Wiley, NY, 1983.

Rattey, R. A. and A. G. Lindgren. (1981). Sampling the 2-D radon transform. *IEEE Transactions on Acoustics, Speech, and Signal Processing*, **ASSP-29**, 994–1002.

Rayleigh, Lord. (1879). *Philosophical Magazine*, **8**, p. 261.

Readhead, A. C. S. and P. N. Wilkinson. (1978). The mapping of compact radio sources from *VLBI* data. *Astrophysics Journal*, **223**, 25–36.

Reimers, P. and J. Goebbels. (1983). New possibilities of nondestructive evaluation by X-ray computed tomography. *Materials Evaluation*, **41**, 732–7.

Rhodes, D. R. (1974). *Synthesis of Planar Antenna Sources*. Oxford: Clarendon Press.

Richardson, W. H. (1972). Bayesian-based iterative method of image restoration. *Journal of the Optical Society of America*, **62**, 55–9.

Rihaczek, A. W. (1965). Radar signal design for target resolution. *Proceedings of the IEEE*, **53**, 116–28.

(1967). Radar resolution of moving targets. *IEEE Transactions on Information Theory*, **IT-13**, 51–6.

Röntgen, W. K. (1896). On a new kind of rays. (English Translation) *Nature*, **53**, 274–6.

Ron, M. Y. and R. Unbehauen. (1995). New algorithms of two-dimensional blind convolution. *Optical Engineering*, **34**, 2945–56.

Root, W. L. (1987). Ill-posedness and precision in object-field reconstruction problems. *Journal of the Optical Society of America*, **4**, 171–9.

Ryle, M. (1952). A new radio interferometer and its application for the observation of weak radio stars. *Proceedings of the Royal Society of London*, **A211**, 351–75.

Ryle, M. and A. Hewish (1960). The synthesis of large radio telescopes. *Monthly Notices of the Royal Astronomical Society*, **120**, 220–30.

Sayre, D. (1952). Some implications of a theorem due to Shannon. *Acta Crystallographia*, **5**, p. 843.

(1952). The squaring method: A new method for phase determination. *Acta Crystallographia*, **5**, 60–5.

Schmitt, J. M. (1999). Optical coherence tomography (OCT): A review. *IEEE Journal of Selected Topics in Quantum Electronics*, **QE-5**, 1205–15.

Schotland, J. (1997). Continuous-wave diffusion imaging. *Journal of the Optical Society of America*, **14**, 275.

Schulz, T. J. (1993). Multiframe blind convolution of astronomical images. *Journal of the Optical Society of America A*, **10**, 1064–73.

Schulz, T. J. and D. L. Snyder. (1991). Imaging a randomly moving object from quantum-limited data: Applications to image recoveryfrom second- and third-order autocorrelations. *Journal of the Optical Society of America A*, **8**, 801–7.

(1992). Image recovery from correlations. *Journal of the Optical Society of America A*, **9**, 1266–72.

Schwarz, U. J. (1978). Mathematical-statistical description of the iterative beam removing technique (method clean). *Astronomy and Astrophysics*, **65**, 345–56.

(1979). The Method 'CLEAN' – Use, misuse, and variations. In C. van Schoonveld, ed., *Image Formation from Coherence Functions in Astronomy*. Dordrecht: Reidel, pp. 261–75.

Schweppe, F. C. (1968). Sensor-array data processing for multiple signal sources. *IEEE Transactions on Information Theory*, **IT-14**, 294–305.

Scudder, H. J. (1978). Introduction to computer aided tomography. *Proceedings of the IEEE*, **66** (6), 628–37.

Sekine, M., S. Ohtani, and T. Muska. (1981). Weibull-distributed ground clutter. *IEEE Transactions on Aerospace and Electronic Systems*, **AES-17**, 596–8.

Selin, I. (1965). Detection of coherent radar returns of unknown Doppler shift. *IEEE Transactions on Information Theory*, **IT-11**, 396–400.

Shannon, C. E. (1948). A mathematical theory of communication. *Bell System Technical Journal*, **27**, 379–423, 623–56.

Shapiro, J. (1982). Target reflectivity theory for coherent laser radars. *Applied Optics*, **21**, 3398–407.

Shapiro, J., B. A. Capron, and R. C. Harney. (1981). Imaging and target detection with a heterodyne-reception optical radar. *Applied Optics*, **20**, 3292–313.

Shepp, L. A. (1980). Computerized tomography and nuclear magnetic resonance. *Journal of Computer Assisted Tomography*, **4** (1), 94–107.

Shepp, L. A. and B. F. Logan. (1974). The Fourier reconstruction of a head section. *IEEE Transactions on Nuclear Science*, **NS-21**, 21–43.

Shepp, L. A. and Y. Vardi. (1982). Maximum likelihood reconstruction for emission tomography. *IEEE Transactions on Medical Imaging*, **MI-1**, 113–122.

Sherwin, C. W., J. P. Ruina, and R. D. Rawcliffe. (1962). Some early developments in synthetic-aperture radar systems. *IRE Transactions on Military Electronics*, **MIL-6**, 111–115.

Shore, J. E. and R. W. Johnson. (1980). Axiomatic derivation of the principle of maximum entropy and the principle of minimum cross entropy. *IEEE Transactions on Information Theory*, **IT-26**, 26–37.

Siebert, W. M. (1956). A radar detection philosophy. *IRE Transactions on Information Theory*, **IT-2**, 204–221.

(1958). Studies of Woodward's uncertainty function. *Massachusetts Institute of Technology, Research Laboratory Electronics Quarterly Progress Report*, April.

Silva, M. T. and E. A. Robinson. (1979). *Deconvolution of Geophysical Time Series in the Exploration of Oil and Natural Gas*. Amsterdam: Elsevier.

Simpson, R. G., H. H. Barrett, J. A. Suback, and H. D. Fisher. (1975). Digital processing of annual coded-aperture imagery. *Optical Engineering*, **14**, 490–4.

Singer, Jr., G. T. (1971). NMR spin-echo flow measurements. *Journal of Applied Physiology*, **42**, p. 938.

Singer, J. R., F. A. Grünbaum, P. Kohn, and J. P. Zubelli. (1990). Image reconstruction of the interior of bodies that diffuse radiation. *Science*, **244**, 990–3.

Singer, R. A., R. G. Sea, and K. B. Housewright. (1974). Derivation and evaluation of improved tracking filters for use in dense multitarget environments. *IEEE Transactions on Information Theory*, **IT-20**, 423–32.

Sittler, R. W. (1964). An optimal data association problem in surveillance theory. *IEEE Transactions on Military Electronics*, **MIL-8**, 125–39.

Skinner, G. K. (1988). X-ray imaging with coded masks. *Scientific American*, 66–71.

Slepian, D. (1967). *Restoration of Photographs Blurred by Image Motion. Bell System Technical Journal*, **46**, 2353–3362.

Slepian, D. and H. O. Pollak. (1961). *Prolate Spheroidal Wave Functions, Fourier Analysis and Uncertainty, part I. Bell System Technical Journal*, **40**, 43–64.

Smith, B. D. (1985). Image reconstruction from cone-beam projections: Necessary and sufficient conditions and reconstruction methods. *IEEE Transactions on Medical Imaging*, **MI-4**, 14–25.

Snyder, D. L. and J. R. Cox, Jr. (1977). An overview of reconstructive tomography and limitations imposed by a finite number of projections. In: M. Ter-Pogossian *et al.*, eds., *Reconstruction Tomography in Diagnostic Radiology and Nuclear Medicine*. Maryland: University Park Press.

Snyder, D. L. and P. M. Fishman. (1975). How to track a swarm of fireflies by observing their flashes. *IEEE Transactions on Information Theory*, **IT-22**, 692–5.

Snyder, D. L. and M. I. Miller. (1985). The use of sieves to stabilize images produced with the EM algorithm for emission tomography. *IEEE Transactions on Nuclear Science*, **NS-32**, 3864–71.

(1991). *Random Point Processes in Time and Space*. Dordrect: Springer-Verlag.

Snyder, D. L., J. A. O'Sullivan, and M. I. Miller. (1989). The use of maximum-likelihood estimation for forming images of diffuse radar targets from delay-doppler data. *IEEE Transactions on Information Theory*, **IT-35**, 536–48.

Snyder, D. L. and D. G. Politte. (1983). Image reconstruction from list-mode data in an emission tomography system having time-of-flight measurements. *IEEE Transactions on Nuclear Science*, **NS-30**, 1843–9.

Snyder, D. L. and T. J. Schulz. (1990). High-resolution imaging at low-light levels through weak turbulence. *Journal of the Optical Society of America A*, **7**, 1251–65.

Snyder, D. L., T. J. Schulz, and J. A. O'Sullivan. (1992). Deblurring subject to nonnegativity constraints. *IEEE Transactions on Signal Processing*, **SP-40**, 1143–50.

Solomon, D. C. (1976). The X-ray transform. *Journal of Mathematical Analysis and Applications*, **56**, 61–83.

Sorenson, H. W. (1980). *Parameter Estimation: Principles and Problems*. New York: Marcel-Dekker.

Soumekh, M. (1997). Moving target detection in foliage using along track monopulse synthetic aperture radar imaging. *IEEE Transactions on Image Processing*, **IP-6**, 1148–63.

Stark, H. (1979). Sampling theorems in polar coordinates. *Journal of the Optical Society of America*, **69**, 1519–25.

Stark, L. (1974). Microwave theory of phased-array antennas – A review. *Proceedings of the IEEE*, **62**, 1661–1701.

Staudaher, F. M. (1970). Airborne MTI. In M. I. Skolnik, ed., *Radar Handbook*. New York: McGraw-Hill, chapter 18.

Stein, J. J. and S. S. Blackman. (1975). Generalized correlation of multi-target track data. *IEEE Transactions on Aerospace and Electronic Systems*, **AES-11**, 1207–17.

Stein, S. (1981). Algorithms for ambiguity function processing. *IEEE Transactions on Acoustics, Speech, and Signal Processing*, **ASSP-29**, 588–99.

Stout, G. H. and L. H. Jensen. (1989). *X-Ray Structure Determination*. New York: Wiley.

Stutt, C. A. (1964). Some results on real-part/imaginary-part and magnitude-phase relations in ambiguity functions. *IEEE Transactions on Information Theory*, **IT-10**, 321–7.

Sullivan, III, W. T. (ed). (1984). *The Early Years of Radio Astronomy*. Cambridge: Cambridge University Press.

Sussman, S. M. (1962). Least-square synthesis of radar ambiguity functions. *IRE Transactions on Information Theory*, **IT-8**, 246–54.

Swenson, Jr., G. W. and N. C. Mathur. (1968). The interferometer in radio astronomy. *Proceedings of the IEEE*, **56**, 2114–30.

Swerling, P. (1957). Detection of fluctuating pulsed signals in the presence of noise. *IRE Transactions on Information Theory*, **IT-3**, 175–8.

(1960). Probability of detection for fluctuating models. *IRE Transactions on Information Theory*, **IT-6**, 269–308.

(1964). Parameter estimation accuracy formulas. *IRE Transactions on Information Theory*, **IT-10**, 302–14.

Synge, E. (1928). A suggested method for extending microscopic resolution into the ultramicroscopic region. *Philosophical Magazine*, **6**, 356–62.

Tagfors, T. and D. Campbell. (1973). Mapping of planetary surfaces by radar. *Proceedings of the IEEE*, **61**, 1219–25.

Taxt, T. (1995). Restoration of medical ultrasound images using two-dimensional homomorphic deconvolution. *IEEE Transactions on Ultrasonics, Ferroelectronics, and Frequency Control*, **UFFC-42**, 543–54.

Ter-Pogossian, M. M., M. E. Raichle, and B. E. Soble. (1980). Positron-emission tomography. *Scientific American*, **243**, 140.

Thomas, J. B. and J. K. Wolf. (1962). On the statistical detection problem for multiple signals. *IRE Transactions on Information Theory*, **IT-8**, 274–80.

Thompson, A. R., J. M. Moran, and G. W. Swenson, Jr. (1986). *Interferometry and Synthesis in Radio Astronomy*. New York: Wiley, 2nd edn, 2001.

Tikhonov, A. and V. Arsenin. (1977). *Solutions of Ill-Posed Problems*. Washington: Winston.

Titchmarsch, E. C. (1937). *Introduction to the Theory of Fourier Integrals*. Oxford: Clarendon Press.

Tolimieri, R. and S. Winograd. (1985). Computing the ambiguity surface. *IEEE Transactions on Acoustics, Speech, and Signal Processing*, **ASSP-33**, 1239–45.

Tur, M., K. C. Chin, and J. W. Goodman. (1982). When is speckle noise multiplicative? *Applied Optics*, **21**, 1157–9.

Tuy, H. K. (1983). An inversion formula for cone-beam reconstruction. *Siam Journal of Applied Mathematics*, **43**, 546–51.

Urkowitz, H. (1953). Filters for detection of small targets in clutter. *Journal of Applied Physics*, **24**, 1024–31.

van Cittert, P. H. (1934). Die wahrscheinliche Schwingungsverteilung von einer . . . oder mittels einer Linse beleuchteten Ebene. *Physica*, **1**, 201–10.

Van Trees, H. L. (1971). *Detection, Estimation and Modulation Theory, Part III: Radar-Sonar Signal Processing and Gaussian Signals in Noise*. New York: Wiley.

(1971). *Detection, Estimation, and Modulation Theory*, vol. 3. New York: Wiley.

Vander Lugt, A. B. (1964). Signal detection by complex spatial filtering. *IEEE Transactions on Information Theory*, **IT-10**, 139–45.

Vardi, Y. and D. Lee, (1993). From image deblurring to optimal investments: Maximum likelihood solutions for positive linear inverse problems. *Journal of the Royal Statistics Society, B*, **55**, 569–612.

Vardi, Y., L. A. Shepp, and L. Kaufman. (1985). A statistical model for positron emission tomography. *Journal of the American Statistical Association*, **80**, 8–35.

Ville, J. (1948). Theory and application of the notion of the complex signal. *Cables et Transmission*, **2**, 67–74.

Walker, J. L. (1980). Range-Doppler imaging of rotating objects. *IEEE Transactions on Aerospace and Electronic Systems*, **AES-16**, 23–52.

Watson, J. D. and F. H. C. Crick. (1953). A structure for deoxyribose nucleic acid. *Nature*, No. 4356, 737–8.

Wax, N. (1955). Signal-to-noise improvement and the statistics of tracking populations. *Journal of Applied Physics*, **26**, 586–95.

Webb, J. L. H. and D. C. Munson, Jr. (1995). Radar imaging of three-dimensional surfaces using limited data. *Proceedings of the IEEE International Conference on Image Processing*, Washington, D.C., pp. 136–9.

Webb, J. L. H., D. C. Munson, Jr., and N. J. Stacy. (1998). High-resolution planetary imaging via spotlight-mode synthetic aperture radar. *IEEE Transactions on Image Processing*, **IP-7**, 1571–82.

Whittaker, E. T. (1915). On the functions which are represented by the expansions of the interpolation theory. *Proceedings of the Royal Society*, Edinburgh, **35**, 181–4.

Wiener, N. (1949). *Extrapolation, Interpolation, and Smoothing of Stationary Time Series with Engineering Applications*. New York: Wiley.

Wilcox, C. H. (1960). *The Synthesis Problem for Radar Ambiguity Functions*. University of Wisconsin, Mathematical Research Center, Technical summary report 157.

Wolf, E. (1969). Three-dimensional structure determination of semi-transparent objects from holographic data. *Optics Communications*, **1**, 153–6.

(1996). Principles and development of diffraction tomography. In A. Consortini, ed., *Trends in Optics*, pp. 83–110. San Diego: Academic Press.

Wolfe, P. (1975). A method of conjugate subgradients for minimizing nondifferentiable functions. *Mathematical Programming*, Study 3, 147–73.

Woodward, P. M. (1953). *Probability and Information Theory, with Applications to Radar*. New York: McGraw-Hill.

Wu, C. F. J. (1983). On the convergence properties of the EM algorithm. *Annals of Statistics*, **11**, 95–103.

Wynn, W. M. (1972). *Dipole Tracking with a Gradiometer*. NSRDL/PC Informal report 3493.

Wynn, W. M., C. P. Frahm, P. J. Caroll, *et al.* (1975). Advanced superconducting gradiometer/magnetometer arrays and a novel signal processing technique. *IEEE Transactions on Magnetics*, **MAG-11**, 701–7.

Zadeh, L. A. and J. R. Ragazzini. (1952). Optimum filters for the detection of signals in noise. *Proceedings of the IRE*, **40**, 1123–31.

Zebker, H. A. and R. M. Goldstein. (1986). Topographic mapping from interferometric synthetic aperture radar observations. *Journal of Geophysical Research*, **91**, 4993–9.

Zernike, F. (1935). Das Phasekontrastverfahren bei der Mikroskopischen Beobachtung. *Z. Tech. Physik*, **16**, 454–7.

(1938). The concept of degree of coherence and its application to optical problems. *Physica*, **5**, 785–95.

(1942). Phase contrast, a new method for the microscopic observation of transparent objects. *Physica*, **9**, 686–98, 974–86.

Index

Bold indicates a reference to a section.

Printed in the United States
By Bookmasters